PURINE AND PYRIMIDINE METABOLISM IN MAN V

Part B: Basic Science Aspects

ADVANCES IN EXPERIMENTAL MEDICINE AND BIOLOGY

Recent Volumes in this Series

A Continuation Order Plan is available for this series. A continuation order will bring delivery of each new volume immediately upon publication. Volumes are billed only upon actual shipment. For further information please contact the publisher.

PURINE AND PYRIMIDINE METABOLISM IN MAN V
Part B: Basic Science Aspects

FESTSCHRIFT FOR J. E. SEEGMILLER

Edited by

W. L. Nyhan
University of California, San Diego
La Jolla, California

L. F. Thompson
Scripps Clinic and Research Foundation
La Jolla, California

and

R. W. E. Watts
M. R. C. Clinical Research Center
Harrow, England

PLENUM PRESS • NEW YORK AND LONDON

Library of Congress Cataloging in Publication Data

International Symposium on Human Purine and Pyrimidine Metabolism (5th: 1985: San Diego, Calif.)
   Purine and pyrimidine metabolism in man V.

   "Proceedings of the Fifth International Symposium on Human Purine and Pyrimidine Metabolism, held July 28–August 1, 1985, in San Diego, California"—T.p. verso.
   Publication dedicated as a festschrift to J. E. Seegmiller.
   Includes bibliographies and index.
   1. Purines—Metabolism—Disorders—Congresses. 2. Pyrimidines—Metabolism—Disorders—Congresses. 3. Seegmiller, J. E.—Congresses. I. Nyhan, William L., 1926–   . II. Thompson, L. F. (Linda Frances), 1947-   . III. Watts, R. W. E. IV. Seegmiller, J. E. V. Title. [DNLM: 1. Purine—Pyrimidine Metabolism, Inborn Errors —congresses. 2. Purines—metabolism—congresses. 3. Pyrimidines—metabolism—congresses. W3 IN918RP 5th 1985p/WD 205.5.P8 1985p]
   RC632.P87I58   1985   616.3'9   85-32557
   ISBN 0-306-42230-1 (pt. A)
   ISBN 0-306-42231-X (pt. B)

Proceedings of the Fifth International Symposium on Human Purine and Pyrimidine Metabolism, held July 28–August 1, 1985, in San Diego, California

© 1986 Plenum Press, New York
A Division of Plenum Publishing Corporation
233 Spring Street, New York, N.Y. 10013

Printed in the United States of America

J. E. Seegmiller

PREFACE

The Fifth International Symposium on Human Purine and Pyrimidine Metabolism was held in San Diego, California (U.S.A.) in July and August of 1985. Previous meetings in this series were held in Tel Aviv (Israel), Baden (Austria), Madrid (Spain) and Maastricht (The Netherlands). The proceedings of each of these meetings were published by Plenum. The next meeting will be in Japan.

This Symposium differed from those that went before in that it permitted us to honor Dr. J. E. Seegmiller, Professor of Medicine at the University of California San Diego, for his many contributions to our understanding of purine metabolism in man. This publication is dedicated as a Festschrift to Jay. Dr. Richard W. E. Watts delivered the keynote address outlining in scholarly fashion the history of Dr. Seegmiller's accomplishments in research on purine metabolism and the great number of currently active scientists in this field who have worked with him. This address is published as the first contribution to Volume I. Dr. Dewitt Stetten, Jr., was scheduled to be the speaker at our banquet. Unfortunately, he could not be with us. Dr. Seegmiller has written an appreciation of Dr. Stetten and his contributions to our field, and this has been published following Dr. Watts' paper.

The growth of knowledge in purine and pyrimidine metabolism continues to be exponential. The variety of subjects included in these volumes is impressive. New or previously unrecognized disorders of purine metabolism continue to be uncovered. An entire section on disorders of purine and pyrimidine metabolism other than deficiency of HPRT is led off by two papers on adenylosuccinase deficiency. Among the disorders of pyrimidine metabolism there are papers on orotic aciduria and dihydrothymine dehydrogenase deficiency. Clinical and biochemical studies of gout and urolithiasis continue to be actively pursued. At the same time the study of purine metabolism has become an integral feature of immunology. The importance of purines in clinical oncology was first demonstrated with the synthesis by George Hitchings of 6-mercaptopurine. Its continuing impact on hematology and oncology is seen throughout these volumes, particularly in the effects of inhibition of adenosine deaminase on T cells and on T cell leukemia. This publication has implications for internal medicine, pediatrics, urology, biochemistry, immunology, genetics, hematology, and oncology.

Modern molecular biology and techniques involving recombinant DNA were evident in papers on HPRT and on adenosine deaminase, as well as in studies on APRT and UMP synthase. The genes for HPRT, adenosine deaminase and purine nucleoside phosphorylase have been cloned. The background for ultimate approaches to gene therapy in man was provided in papers from Dr. Seegmiller's laboratory on the insertion of HPRT cDNA into human bone marrow cells and on metabolic cooperation.

Purine receptors have been discovered in the central nervous system, in lymphocytes and in a variety of other tissues. There are also adenosine receptors in Leishmania and a number of purine riboside analogs are under study as potential therapeutic agents in leishmaniasis.

We acknowledge the support of the distinguished members of the Executive Committee and the Scientific Committee. The Executive Committee was comprised of Drs. J.E. Seegmiller (La Jolla), B.T. Emmerson (Brisbane), H.A. Simmonds (London), W.N. Kelley (Ann Arbor), and G.H. Hitchings (Durham). The members of the Scientific Committee were Drs. M.A. Becker (Chicago), M.S. Coleman (Lexington), G.J. Delespesse (Winnipeg), G.B. Elion (Durham), R.M. Fox (Sydney), M.W. Grever (Columbus), W. Gutensohn (Münich), E.H. Harley (Cape Town), J.F. Henderson (Edmonton), E.W. Holmes (Durham), J.M. Lowenstein (Waltham), D.W. Martin (San Francisco), P. Merra Khan (Leiden), B.S. Mitchell (Ann Arbor), M.M. Müller (Vienna), N. Ogasawara (Aichi), D. Patterson (Denver), W.E. Plunkett (Houston), K.O. Raivo (Helsinki), A. Rapado (Madrid), J.T. Scott (London), P.A. Simkin (Seattle), L.B. Sorenson (Chicago), O. Sperling (Tel Aviv), E.R. Tully (Cork), B. Ullman (Lexington), J.P.R.M. van Laarhoven (Nijmegen), R.W.E. Watts (Harrow), and N. Zöllner (Münich).

We are particularly indebted to the members of the Local Planning Committee who did so much of the work that made this meeting a success. They included Drs. G.R. Bartlett, G.R. Boss, D.A. Carson, T. Friedmann, H.E. Gruber, S.S. Matsumoto, W.L. Nyhan, T.M. Page, L.F. Thompson, and R.C. Willis. Dr. Randy Willis particularly deserves credit for spearheading this committee and doing a major amount of the work that turned this meeting into a reality.

We would like to express our special appreciation to Maureen Kearns who handled all of the administrative details from the beginning of planning to the end of the meeting. We especially thank Mrs. Dorothy MacElhose for her fine editorial assistance. She assembled and typed the Table of Contents and the Index, as well as this Preface, and corrected all of the typographical errors we found in the manuscripts.

We are pleased to acknowledge the support to the Symposium of the following sponsors: Burroughs Wellcome Co., The Wellcome Foundation, Merck, Sharp and Dohme, The Joan Kroc Foundation, Upjohn Company, and Calbiochem/Behring.

The Symposium was supported by USPHS Grant No. AM34557 from the National Institute of Arthritis, Diabetes, and Digestive and Kidney Diseases, National Institutes of Health, Bethesda, Maryland, U.S.A.

W.L. Nyhan

L.F. Thompson

R.W.E. Watts

CONTENTS

V.   DEOXYNUCLEOTIDE AND NUCLEOSIDE TOXICITY AND METABOLISM

VI. ENZYMES

VII. PURINE AND PYRIMIDINE METABOLISM DURING LYMPHOCYTE DIFFERENTIATION

XIII. S-ADENOSYLMETHIONINE METABOLISM

XIV. INTERFERON

ADENOSINE RECEPTORS ON HUMAN T LYMPHOCYTES AND HUMAN THYMOCYTES

Winand N.M. Dinjens,[1] Rineke van Doorn,[1] Jan P.R.M. van Laarhoven,[2] Dirk Roos,[1] Wim P. Zeijlemaker[1] and Chris H.M.M. de
Bruijn[3]

[1]Central Laboratory of the Netherlands Red Cross Blood Trans-
fusion Service, incorporating the Laboratory for Experimental
and Clinical Immunology of the University of Amsterdam
Amsterdam, The Netherlands; [2]Laboratory of Experimental and
Chemical Endocrinology, University Hospital St. Radboud, Nij-
megen, The Netherlands; [3]Project Group Biomedical and Health
Care Technology, University of Technology, Eindhoven, The
Netherlands

INTRODUCTION

Adenosine affects the immune system in several ways by virtue of its
effects on lymphoid cells. The effects of adenosine on lymphoid cells are
mediated by intracellular metabolism and/or the binding to a cell-surface
receptor. In lymphoid cells, binding of adenosine to a putative cell membrane receptor leads to an increase in the intracellular level of cyclic
AMP (cAMP)[1-3] and therefore this receptor is classified as an A_2 (Ra) adenosine receptor.[4]

Our studies deal with the role of adenosine in the differentiation of
human T lymphocytes. Therefore, we study the metabolic pathways of adenosine and the binding of adenosine to a cell-surface receptor.

The aim of the present study is the demonstration of the adenosine A_2
receptor on human T-lymphoid cells. As an indication of adenosine-receptor
binding, we measured the intracellular level of cAMP before and after incubation with adenosine. To prevent degradation of adenosine by adenosine
deaminase (ADA) during the incubation, the ADA inhibitor coformycin was
used. We also used the non-metabolizable adenosine analogue 2-chloroadeno-
sine. To prevent breakdown of cAMP after the incubation, an inhibitor of
phosphodiesterase (Ro 20-1724 from Hoffmann-La Roche) was added.

MATERIALS AND METHODS

Chemicals

Adenosine and 2-chloroadenosine were from Sigma Chemical Company, St.
Louis, MO, USA. Coformycin was from Meiji Seiko-Kaisho Ltd., Tokyo, Japan
and Percoll was from Pharmacia, Uppsala, Sweden. Ro 20-1724 was a gift from
Hoffmann-La Roche.

Preparation of Human Peripheral Blood T Lymphocytes

T lymphocytes from human peripheral blood were prepared by nylon wool filtration of mononuclear leukocytes from blood of normal healthy donors.[5] In short, acid-citrate-dextrose blood was obtained by collection of 50 parts of venous blood into 7 parts of a solution containing 2.7 g of anydrous disodium citrate and 2.3 g of anhydrous glucose per 100 ml (U.S.P. Formula A). The blood was diluted 1:1 with phosphate-buffered saline (PBS; 140 mM NaCl, 9.2 mM Na_2HPO_4, 1.3 mM KH_2PO_4, pH 7.4) containing 5 g of human serum albumin per liter and 13 mM trisodium citrate. This suspension was centrifuged (20 min, 1000 g, 20°C) over Percoll (d = 1.076 g/cm^3) at room temperature, and the leukocytes were collected from the top layer. After two low-speed (10 min, 600 g, 20°C) washing steps with PBS to remove platelets, the contaminating erythrocytes were lysed (5 min, 0°C) with ice-cold ammonium-chloride solution (155 mM NH_4Cl, 10 mM $KHCO_3$, 0.1 mM disodium EDTA, pH 7.4). The leukocytes were washed, centrifuged (10 min, 600 g, 4°C) and resuspended in Earle's Balanced Salt Solution (EBSS) with 5% (v/v) fetal calf serum and 9.25 mM magnesium chloride. After filtration (at 37°C) through a nylon wool column, the final cell suspension contained more than 90% E-rosette-forming cells.

Preparation of Human Thymocytes

Human thymocytes were prepared according to a method described by Astaldi et al.[6] Thymus fragments were obtained from children (1.5 to 6 years of age) undergoing cardiac surgery. After removal of adipose tissue, connective tissue and blood vessels, the thymus fragments were cut into small pieces, which subsequently were gently pressed through a metal sieve. The thymocytes were washed (10 min, 600 g, 20°C) in EBSS.

Experimental Procedure

T lymphocytes ($2x10^6$) or thymocytes ($10x10^6$) were incubated in 100 µl of EBSS in the presence of various combinations of stimulators and inhibitors, for various periods of time. Incubations were terminated by the addition of 25 µl of ice-cold 250 mM Tris-HCl, 20 mM EDTA, pH 7.5. Thereafter, the assay tubes were immediately frozen in liquid nitrogen. Next, the tubes were heated for 3 min in boiling water. The contens were mixed, and denatured protein was spun down. The amount of cAMP was measured in the supernatant (50 µl) with a cAMP assay kit (TRK 432, Amersham International Ltd., Amersham, UK).

RESULTS

In Figures 1 and 2 the effect of adenosine and other compounds on the cAMP levels in human T lymphocytes and thymocytes are shown. Although there were considerable inter-experiment variations in the absolute levels of cAMP, the same overall effects were seen in all experiments.

Figure 1 shows the results of a representative experiment with human T lymphocytes. Incubation of T lymphocytes with 10 µM adenosine, 10 µM adenosine + 50 µM coformycin (an ADA inhibitor) or with 10 µM 2-chloroadenosine (a non-metabolizable adenosine analogue) resulted in a prolonged cAMP increase (continuing for more than 15 min) as compared with the control cAMP level. Coformycin itself had no effect on the cAMP level in T lymphocytes (data not shown). The strongest cAMP increase was observed when T lymphocytes were incubated with 10 µM 2-chloroadenosine + 0.5 mM Ro 20-1742 (a phosphodiesterase inhibitor). Although the addition of Ro 20-1724 by itself gave some increase in cAMP, the magnitude of the effect was much lower.

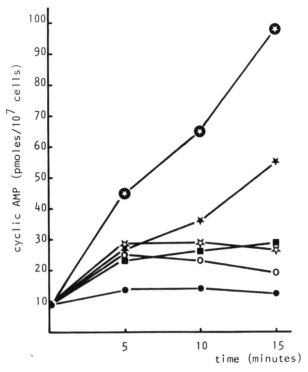

T lymphocytes were incubated for various times under the following conditions:

(●—●) without addition
(○—○) 10 μM adenosine
(■—■) 10 μM adenosine + 50 μM coformycin
(✿—✿) 10 μM 2-chloroadenosine
(✦—✦) 0.5 mM Ro 20-1724
(✪-✪) 10 μM 2-chloroadenosine + 0.5 mM Ro 20-1724

Fig. 1: cAMP Levels in Human Lymphocytes.

In Figure 2 the results of a representative experiment with human thymocytes are shown. Incubation of thymocytes with 10 μM adenosine resulted in a rapid increase in the intracellular level of cAMP, which then rapidly returned to control levels. When the thymocytes were incubated with 10 μM adenosine + 50 μM coformycin or with 10 μM 2-chloroadenosine we observed a prolonged cAMP increase. Coformycin alone had no effect on the control cAMP level (data not shown). The strongest prolonged cAMP increase was observed when thymocytes were incubated with 10 μM 2-chloroadenosine + 0.5 mM Ro 20-1724. Under these conditions, the rise of the cAMP level continued for more than 10 min. Addition of 0.5 mM Ro 20-1724 by itself had only a marginal effect on the cAMP level in human thymocytes.

DISCUSSION

In this paper evidence is presented that adenosine A_2 receptors are

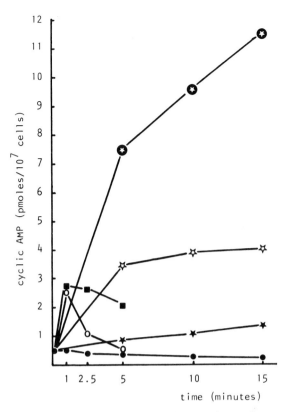

Thymocytes were incubated for various times
under the following conditions:

(●—●) without addition
(○—○) 10 μM adenosine
(■—■) 10 μM adenosine + 50 μM coformycin
(☆—☆) 10 μM 2-chloroadenosine
(✶—✶) 0.5 mM Ro 20-1724
(✪-✪) 10 μM 2-chloroadenosine + 0.5 mM Ro 20-1724

Fig. 2:   cAMP Levels in Human Thymocytes.

present on human T lymphocytes and human thymocytes. This is deduced from
the adenosine-induced cAMP increases in both cell types.

The observed inter-donor variation in cAMP levels of human T lymphocytes has also been observed in human blood mononuclear cells by other investigators.[7] The cAMP levels are also affected by the in vitro experimental
conditions. A variety of cell separation methods, incubation media, incubation times and different termination and quantitation methods have been
described which may be responsible for the considerable range of the
reported basal cAMP levels.[8-10]

The basal intracellular level of cAMP in human T lymphocytes is much
higher than that in human thymocytes. This difference cannot be fully explained by the diffference in rise between the two cell types. In both T
lymphocytes and thymocytes incubation with adenosine resulted in an increase
in the cAMP level. Thus, both cell types apparently possess A_2 adenosine
receptors.

In human thymocytes, incubation with 10 µM adenosine induces a rapid increase in the cAMP level, followed by a decrease. Because this decrease is much less pronounced in the presence of the ADA inhibitor coformycin, it is probably due to breakdown of adenosine by the thymocytes. Indeed, thymocytes possess a high ADA activity.[11] In contrast, this effect is only marginal in T lymphocytes, which have a much lower ADA activity.[11]

In T lymphocytes a strong increase in the amount of cAMP was observed when the breakdown of cAMP was prevented by the addition of Ro 20-1724, an inhibitor of phosphodiesterase. This compound has no affinity for the adenosine receptor. Addition of Ro 20-1724 to thymocytes had only a marginal effect on the cAMP level. These data indicate that the basal adenylate cyclase activity in human T lymphocytes is higher than that in human thymocytes.

In conclusion, in the present paper we present evidence for the occurrence of A_2 adenosine receptors on human T lymphocytes and human thymocytes. The binding of adenosine to this receptor and the subsequent increase in the cAMP level may play an important role with regard to the effects of adenosine on T-lymphoid cells.[12-15]

ACKNOWLEDGEMENTS

This study was supported by grant no. 13-40-61 from the Foundation for Medical Research (FUNGO), which is subsidized by the Netherlands Organization for the Advancement of Pure Research (ZWO). The participation in the ISHPPM was made possible by a grant from the foundation Simonsfund.

REFERENCES

1. G. Marone, M. Plaut, and L. M. Lichtenstein, Characterization of a specific adenosine receptor on human lymphocytes, J. Immunol. 121:2153 (1978).
2. A. L. Schwartz, R. C. Stern, and S. H. Polmar, Demonstration of an adenosine receptor on human lymphocytes in vitro and its possible role in the adenosine deaminase-deficient form of severe combined immunodeficiency, Clin. Immunol. Immunopathol. 9:499 (1978).
3. R. J. van de Griend, A. Astaldi, P. Wijermans, R. van Doorn, and D. Roos, Low β-adrenergic receptor concentration in human thymocytes, Clin. Exp. Immunol. 53:53 (1983).
4. C. Londos, D. M. F. Cooper, and J. Wolff, Subclasses of external adenosine receptors, Proc. Natl. Acad. Sci. USA 77:2551 (1980).
5. D. Roos and J. A. Loos, Changes in the carbohydrate metabolism of mitogenically stimulated human peripheral lymphocytes. I. Stimulation by phytohaemagglutinin, Biochim. Biophys. Acta 222:565 (1970).
6. G. C. B. Astaldi, A. Astaldi, M. Groenewoud, P. Wijermans, P. T. A. Schellekens, and V. P. Eijsvoogel, Effect of a human serum thymic factor on hydrocortisone-treated thymocytes, Eur. J. Immunol. 7:836 (1977).
7. M. Svenson, H. Permin, and A. Wirk, Basal cyclic AMP levels in human blood mononuclear cells, Eur. J. Clin. Invest. 14:268 (1984).
8. H. G. Morris, S. A. Rusnak, J. C. Selner, K. Barzens, and J. Barnes, Adrenergic desensitation in leukocytes of normal and asthmatic subjects, J. Cycl. Nucl. Res. 3:439 (1977).

9. C. Rochette-Egly and J. Kempf, Cyclic nucleotides and calcium in human lymphocytes induced to divide, J. Physiol. 77:721 (1981).
10. B. J. Goffstein, L. K. Gordon, H. J. Wedner, and J. P. Atkinson, Cyclic AMP concentrations in human peripheral blood lymphocytes, J. Lab. Clin. Med. 96:1002 (1980).
11. H. J. Schuurman, J. P. R. M. van Laarhoven, R. Broekhuizen, G. T. Spierenburg, P. Brekelmans, C. G. Figdor, C. H. M. M. de Bruijn, and L. Kater, Lymphocyte maturation in the human thymus, Scand. J. Immunol. 18:539 (1983).
12. G. Wolberg, T. P. Zimmerman, K. Hiemstra, M. Winston, and L. C. Chu, Adenosine inhibition of lymphocyte-mediated cytolysis: possible role of cyclic adenosine monophosphate, Science 187:957 (1975).
13. F. F. Snyder, J. Mendelsohn, and J. E. Seegmiller, Phytohemagglutinin-stimulated human lymphocytes, J. Clin. Invest. 58:654 (1976).
14. R. E. Birch and S. H. Polmar, Pharmacological modification of immunoregulatory T lymphocytes. I. Effect of adenosine, H_1 and H_2 histamine agonists upon T lymphocyte regulation of B lymphocyte differentiation in vitro, Clin. Exp. Immunol. 48:218 (1982).
15. G. Sandberg, Regulation of thymocyte proliferation by endogenous adenosine and adenosine deaminase, Int. J. Immunopharmac. 5:259 (1983).

ADENOSINE RECEPTORS ON HUMAN LYMPHOCYTES

Gianni Marone, Sergio Vigorita, Massimo Triggiani and Mario Condorelli

Department of Medicine, University of Naples, II School of Medicine, Via S. Pansini 5, 80131 Naples, Italy

INTRODUCTION

Evidence is accumulating that adenosine functions as an important immunoregulatory autacoid (1,2). Adenosine inhibits the mitotic response of human lymphocytes (3,4), lymphocyte-mediated cytolysis (5), super-oxide anion generation by neutrophils (6), platelet aggregation (7) and mediator release from human basophils (8-10). The nucleoside has been shown to modulate a number of T lymphocyte responses (11,12). Adenosine has been implicated in the pathophysiology of patients with severe combined immune deficiency associated with a deficit of adenosine deaminase (13), in systemic lupus erythematosus (14), in bronchial asthma (15,16) and in a variety of cardiovascular diseases (17).

The mechanism underlying adenosine-induced modulation of immune responses has been the subject of considerable controversy. Among the various mechanisms postulated is an adenosine-induced alteration of cAMP metabolism in human inflammatory cells (1,4,12,18). Over the past few years adenosine receptors that modulate adenylate cyclase have been found in a variety of tissues (18,19). There are at least two subclasses of adenosine receptors that modulate either inhibition or stimulation of adenylate cyclase. For instance, the inhibitor receptor (called R_i or A_1), which is more sensitive to the agonists (20-100 nM) and prefers (-)-N^6-(R-phenylisopropyl)-adenosine {(-)-R-PIA} over 5'-N-ethylcarbo-xamideadenosine (NECA) (20-22), and the stimulatory receptor (R_a or A_2), which is effective over a range of 0.1-100 μM and prefers NECA over (-)-R-PIA (20-22). Both subclasses are antagonized by methylxanthines and are located on the outer cell surface (10,18,20-22). We have recently reported the presence of both adenosine A_1/R_i and A_2/R_a receptors on different subpopulations of human leukocytes (1,2).

In addition to these two subclasses of adenosine receptors, an inhibitory adenosine-related P-site has also been identified (20). 2',5'-dideoxyadenosine (DDA), high concentrations of adenosine and certain modified ribose analogs of adenosine interact with the P-site (20-23). Interactions at the P-site inhibit adenylate cyclase activity and the inhibition is enhanced in the presence of Mn^{2+} (20-22). The P-site is probably located on the cytoplasmic surface of the plasma membrane. In contrast to A_1/R_i and A_2/R_a receptors, the P-site is not inhibited by methylxanthines (20-22).

We have recently investigated different subpopulations of human leukocytes for the presence of the P-site. The results support the hypothesis that an adenosine P-site is present on subpopulations of human lymphocytes that possess adenosine A_2/R_a receptor and receptors for histamine and β-adrenergic agonists. The adenosine P-site also appears to be present on human polymorphonuclear leukocytes (PMNs).

EFFECT OF P-SITE AGONISTS ON THE LEVEL OF cAMP IN HUMAN LYMPHOCYTES

2',5'-dideoxyadenosine (DDA), 9-β-D-xylofuranosyladenosine (XFA), and 9'-β-D-arabinofuranosyladenosine (ARA) are adenosine analogs that interact with the P-site (20-22). Concentrations above 10^{-5} M of DDA caused a dose-dependent decrease in the cAMP content of human lymphocytes. XFA and ARA, two P-site effectors, were less potent than DDA, as previously observed in other tissues (20-22).

EFFECT OF P-SITE AGONISTS ON THE cAMP INCREASES INDUCED BY PGE_1, ISO-PROTERENOL, HISTAMINE, AND ADENOSINE

PGE_1, isoproterenol, histamine and adenosine all increase lymphocyte cAMP levels presumably by interacting with a specific membrane receptor (18,23-25). Therefore, we examined the effects on lymphocyte cAMP levels of these adenylate cyclase agonists in the presence of various concentrations of DDA. DDA (10^{-6} - 2×10^{-4} M) decreased the cAMP content of human lymphocytes and blocked the stimulatory effect of PGE_1. The inhibitory effect of DDA was not confined to interaction with PGE_1: DDA also blocked the stimulatory effect of isoproterenol on cAMP metabolism, which requires the activation of the β-adrenergic receptor. Histamine and adenosine increased lymphocyte cAMP levels, presumably by interacting with specific H_2 and A_2/R_a receptors, respectively (2,18,25). DDA almost completely blocked both histamine- and adenosine-induced increases of lymphocyte cAMP levels. All experiments were preceded by a 5-min preincubation period with DDA, kinetic studies having indicated that the inhibitory effect of DDA is extremely rapid. The kinetics of inhibition of activated adenylate cyclase by DDA in lymphocytes is similar to the characteristics of the P-site in different systems (26).

The ability of P-site effectors to inhibit the effect of adenylate cyclase agonists was not confined to DDA. Interestingly enough we found that high concentrations of adenosine itself dose-dependently inhibited the effect of many adenylate cyclase agonists. Figure 1 shows that adenosine (10^{-5} - 10^{-3} M) also blocked the stimulatory effect of NECA on cAMP metabolism. NECA is a selective agonist of adenosine A_2/R_a receptor on human lymphocytes (1,2). Therefore, the results of the experiment shown in Figure 1 indicate that high concentrations of adenosine might activate the P-site and modulate the stimulation of adenosine A_2/R_a receptor. The biological significance of this remarkable observation is unknown and deserves further investigation. We also found that high concentrations of adenosine blocked the stimulatory effect of isoproterenol, histamine and PGE_1 on cAMP metabolism (2).

EFFECT OF THEOPHYLLINE

Theophylline and other methylxanthines block the effect of adenosine and its analogs on both the stimulatory (A_2/R_a) and the inhibitory (A_1/R_i) receptors (8,10,20). In contrast, the P-site, which is probably located on the cytoplasmic surface of the plasma membrane (20), is insensitive to methylxanthines (20-22). We found that low concentrations of theophylline, which do not inhibit cAMP phosphodiesterase (PDE)

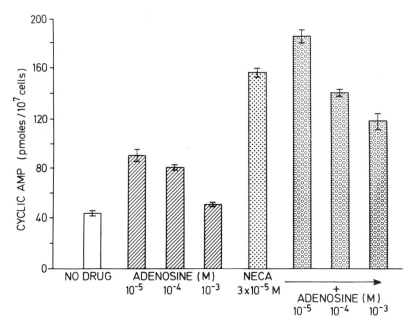

Fig. 1. The effect of adenosine, alone and in combination with NECA, on the level of cyclic AMP in human lymphocytes. The cells were preincubated 5 min with adenosine. NECA was then added and cells were incubated for an additional 8 min. Each bar represents the mean $\pm$ S.E.M. of triplicate determinations.

activity, do not block the inhibition induced by DDA or by high concentrations of adenosine of human lymphocyte cAMP changes caused by adenylate cyclase agonists. Therefore the inhibition of the effect of adenylate cyclase agonists by DDA is apparently unrelated to interaction with the A_1/R_i receptor, which is antagonized by methylxanthines (19).

EFFECT OF Mn^{2+}

A characteristic feature of P-site-mediated inhibition is that the sensitivity to P-site effectors is increased by agents that increase adenylate cyclase activity, such as Mn ions (20,22). We found that the inhibition caused by DDA on the stimulatory effect of adenylate cyclase agonists was potentiated in the presence of Mn^{2+}. Thus human lymphocytes appear to contain a P-site that exhibits the characteristic cation dependence seen in other systems (20,22).

EFFECT OF DDA ON THE ACTIVATION OF ADENYLATE CYCLASE OF HUMAN PMNs

It is well established that only PGE_1 and forskolin are able to increase the cAMP content of human PMNs in the absence of PDE inhibitors (23,24,27). We have investigated the presence of an inhibitory adenosine P-site in human PMNs by evaluating the interaction between DDA and high

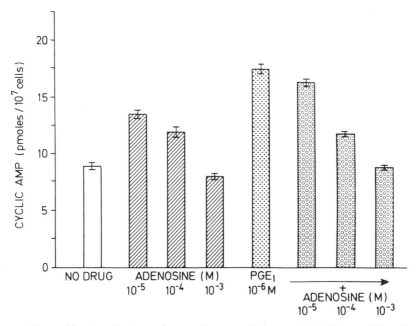

Fig. 2.   The effect of adenosine, alone and in combination with PGE_1 on the level of cyclic AMP in human polymorphonuclear leukocytes. The cells were preincubated 5 min with adenosine. PGE_1 was then added and the cells were incubated for an additional 8 min. Each bar represents the mean $\pm$ S.E.M. of triplicate determinations.

concentrations of adenosine and both PGE_1 and forskolin. The highest concentrations (2×10^{-3} M) of adenosine produced a significant decrease in the intracellular level of cAMP in PMNs. In contrast, lower concentrations of adenosine ($10^{-6} - 2 \times 10^{-5}$ M) produced a small but significant increase in the basal level of cAMP in PMNs. Figure 2 shows that adenosine ($10^{-5} - 2 \times 10^{-3}$ M) dose-dependently suppressed the cAMP accumulation produced by PGE_1 in human PMNs. Similar results were obtained with lower concentrations ($10^{-5} - 2 \times 10^{-4}$ M) of DDA, which is a selective agonist of the P-site. DDA and high concentrations of adenosine also blocked the increase of cAMP in PMNs produced by forskolin. The effects described were not blocked by low concentrations of theophylline. These results suggest that human PMNs possess an adenosine P-site.

CONCLUDING REMARKS

The findings reported here suggest that high concentrations of adenosine, DDA and other modified ribose analogs of adenosine, which interact with a P-site, modulate cAMP metabolism in human lymphocytes and PMNs.

We have previously shown that adenosine and its analogs, in micromolar concentrations, increase intracellular cAMP in human lymphocytes and PMNs by activating a membrane receptor that possesses properties similar to those of an adenosine A_2/R_a receptor (1,2,18). In addition, human lymphocytes and PMNs also possess an adenosine A_1/R_i receptor whose activation inhibits the effects of many adenylate cyclase agonists (23,27). The present results indicate that high concentrations (i.e., millimolar) of adenosine and its analogs modified in the ribose moiety inhibit the effects of agonists of adenylate cyclase in human lymphocytes and PMNs by activating an inhibitory adenosine P-site. This conclusion derives from a number of observations. The inhibitory effect of DDA on lymphocyte cAMP metabolism is also exerted by ARA and XFA, which are known agonists of the P-site in a variety of tissues (20,22). Furthermore, the inhibitory effect of DDA was observed when adenylate cyclase activity was stimulated by such agonists as isoproterenol, NECA, PGE_1, histamine and forskolin. We have confirmed the finding of Londos and Wolff (20) that the inhibitory effect of DDA on the activation of adenylate cyclase is enhanced by the presence of Mn^{2+}.

The inhibitory effects of high concentrations of adenosine and its analogs on cAMP accumulation induced by adenylate cyclase agonists are distinct from the activation of the adenosine A_1/R_i receptor. We have shown that (-)-R-PIA, a potent agonist of adenosine A_1/R_i receptors, partially blocks the effects of adenylate cyclase agonists in human lymphocytes (1,2). The inhibitory effect of DDA is clearly distinct from the activation of adenosine A_1/R_i receptor. Its effect is not blocked by theophylline or by other methylxanthines, which are, in low concentrations, specific antagonists of adenosine A_1/R_i receptors in human lymphocytes (1,2,18). Furthermore, the shape of the dose-response curve of (-)-R-PIA on adenylate cyclase agonists is clearly different from that of DDA (19). Finally, adenosine and DDA are active at the P-site in submillimolar concentrations, whereas the activation of adenosine A_1/R_i receptor requires nanomolar concentrations of the agonists (1,2).

The adenosine P-site is probably present on different subpopulations of human peripheral blood leukocytes. In fact, we have shown that high concentrations of adenosine and DDA inhibit the effects of many adenylate cyclase agonists including isoproterenol, histamine, NECA, and PGE_1. Each of these agonists acts on different hormone receptors present on the membrane of human lymphocytes. It is not clear if subpopulations of human lymphocytes bearing different membrane receptors for these agonists exist. However, high concentrations of adenosine and DDA inhibit the accumulation of PMN cAMP induced by PGE_1 and forskolin. Therefore, the adenosine P-site appears to be present also on human PMNs.

Previous investigations have characterized the properties of the adenosine P-site in a variety of tissues (19,20,22). The present report describes for the first time the presence of an adenosine P-site on human inflammatory cells. The properties of this site are similar to those previously described by Londos and Wolff in other systems (19,20,22). Although the biological significance of the P-site is unknown, its presence in human inflammatory cells suggests that its activation might play a role in the modulation of inflammatory reactions.

In an earlier investigation we showed that micromolar concentrations of adenosine and NECA activate the A_2/R_a receptor and increase the cAMP levels in different subpopulations of human lymphocytes (1,2,18). Lower concentrations, i.e., nanomolar, of adenosine and (-)-R-PIA interact with high affinity with the A_1/R_i receptor and inhibit the stimulating effect of many adenylate cyclase agonists. The present results indicate that high concentrations, i.e., millimolar, of adenosine and DDA activate the P-site in human leukocytes and inhibit the effect of many adenylate cyclase agonists. These observations suggest that different concentrations of adenosine might exert distinct and opposite effects on the adenylate cyclase-cAMP system. It is well established that immunological and non-immunological stimuli modify the local and serum concentrations of adenosine (28-30). It has been suggested that intracellular cAMP levels are probably of physiological importance in modulating several aspects of the immune response (2,10,14,18,31-33). The results of this group of investigations, therefore, might indicate that adenosine, a natural nucleoside, plays a fundamental role in the control of the immune response in man by modifying cAMP levels in inflammatory cells.

ACKNOWLEDGMENTS

This work was supported in part by grants from the C.N.R. (83.00430.04 and 84.01756.04), the Ministero Sanità and the M.P.I. (Rome, Italy).

REFERENCES

1. G. Marone, R. Petracca, and M. Condorelli, Adenosine receptors on human inflammatory cells, in: Purine Metabolism in Man - IV - Part A: Clinical and Therapeutic Aspects; Regulatory Mechanisms, C.H.M.M. De Bruijn, H.A. Simmonds, and M.M. Müller, eds., Plenum Press, New York, p. 501 (1984).
2. G. Marone, R. Petracca, and S. Vigorita, Adenosine receptors on human inflammatory cells, Int. Archs. Allergy appl. Immunol. 77: 259 (1985).
3. R. Hirschhorn, J. Grossman, and G. Weissmann, Effect of cyclic 3',5'-adenosine monophosphate and theophylline on lymphocyte transformation, Proc. Soc. Exp. Biol. Med. 133: 1361 (1970).
4. D.A. Carson, and J.E. Seegmiller, Effect of adenosine deaminase inhibition upon human lymphocyte blastogenesis, J. Clin. Invest. 57: 274 (1976).
5. G. Wolberg, T.P. Zimmerman, K. Hiemstra, M. Winston, and L.C. Chu, Adenosine inhibition of lymphocyte-mediated cytolysis: possible role of cyclic adenosine monophosphate, Science 187: 957 (1975).
6. B.N. Cronstein, S.B. Kramer, G. Weissmann, and R. Hirschhorn, Adenosine: a physiological modulator of superoxide anion generation by human neutrophils, J. Exp. Med. 158: 1160 (1983).
7. G. Marone, S. Quattrin, M. Masturzo, O. Finizio, A. Genovese, and M. Condorelli, Adenosin-rezeptoren von menschlichen plaettchen: charakterisierung des A_2/R_a rezeptors und des P-lokus, in: Thrombose- und Hämostaseforschung 1984, E.A. Beck, ed., F.K.

Schattauer Verlag, Stuttgart, p. 221 (1984).

8. G. Marone, S.R. Findlay, and L.M. Lichtenstein, Adenosine receptor on human basophils: modulation of histamine release, J. Immunol. 123: 1473 (1979).

9. G. Marone, A. Kagey-Sobotka, and L.M. Lichtenstein, IgE-mediated histamine release from human basophils: differences between antigen E- and anti-IgE-induced secretion, Int. Archs. Allergy appl. Immunol. 65: 339 (1981).

10. G. Marone, S. Vigorita, C. Antonelli, G. Torella, A. Genovese, and M. Condorelli, Evidence for an adenosine A_2/R_a receptor on human basophils, Life Sci. 36: 339 (1985).

11. R.E. Birch, and S.H. Polmar, Induction of Fc_γ receptors on a subpopulation of human T lymphocytes by adenosine and impromidine, an H_2-histamine agonist, Cell. Immunol. 57: 455 (1981).

12. G.M. Kammer, and S.A. Rudolph, Regulation of human T lymphocyte surface antigen mobility by purinergic receptors, J. Immunol. 133: 3298 (1984).

13. R. Hirschhorn, Adenosine deaminase deficiency and immunodeficiencies, Fed. Proc. 36: 2166 (1977).

14. R. Mandler, R.E. Birch, S.H. Polmar, G.M. Kammer, and S.A. Rudolph, Abnormal adenosine-induced immunosuppression and cAMP metabolism in T lymphocytes of patients with systemic lupus erythematosus, Proc. Natl. Acad. Sci. USA 79: 7542 (1982).

15. R. Ronchetti, N. Lucarini, P. Lucarelli, F. Martinez, F. Macrì, E. Carapella, and E. Bottini, A genetic basis for heterogeneity of asthma syndrome in pediatric ages: adenosine deaminase phenotypes, J. Allergy Clin. Immunol. 74: 81 (1984).

16. S.T. Holgate, J.S. Mann, and M.J. Cushley, Adenosine as a bronchoconstrictor mediator in asthma and its antagonism by methylxanthines, J. All. Clin. Immunol. 74: 302 (1984).

17. R.M. Berne, The role of adenosine in the regulation of coronary blood flow, Circ. Res. 47: 807 (1980).

18. G. Marone, M. Plaut, and L.M. Lichtenstein, Characterization of a specific adenosine receptor on human lymphocytes, J. Immunol. 121: 2153 (1978).

19. J. Wolff, C. Londos, and D.M.F. Cooper, Adenosine receptors and the regulation of adenylate cyclase, in : Advances in Cyclic Nucleotide Research, vol. 14, J.E. Dumont, P. Greengard, and G.A. Robison, eds., Raven Press, New York, p. 199 (1981).

20. C. Londos, and J. Wolff, Two distinct adenosine-sensitive sites on adenylate cyclase, Proc. Natl. Acad. Sci. USA 74: 5482 (1977).

21. D. Van Calker, M. Müller, and B. Hamprecht, Adenosine inhibits the accumulation of cyclic AMP in cultured brain cells, Nature 276: 839 (1978).

22. C. Londos, D.M.F. Cooper, and J. Wolff, Subclasses of external adenosine receptors, Proc. Natl. Acad. Sci. USA 77: 2551 (1980).

23. G. Marone, L.L. Thomas, and L.M. Lichtenstein, The role of agonists that activate adenylate cyclase in the control of cAMP metabolism and enzyme release by human polymorphonuclear leukocytes, J. Immunol. 125: 2277 (1980).

24. G. Marone, L.M. Lichtenstein, and M. Plaut, Hydrocortisone and human lymphocytes: increases in cyclic adenosine 3':5'-monophosphate and potentiation of adenylate cyclase-activating agents, J. Pharmacol. Exp. Ther. 215: 469 (1980).

25. M. Plaut, G. Marone, and E. Gillespie, The role of cyclic AMP in modulating cytotoxic T lymphocytes. II. Sequential changes during culture in responsiveness of cytotoxic lymphocytes to cyclic AMP-active agents, J. Immunol. 131: 2945 (1983).

26. P.M. Lad, T.B. Nielsen, C. Londos, M.S. Preston, and M. Rodbell, Independent mechanisms of adenosine activation and inhibition of

the turkey erythrocyte adenylate cyclase system, J. Biol. Chem. 255: 10841 (1980).

27. G. Marone, A. Kagey-Sobotka, and L.M. Lichtenstein, Effects of arachidonic acid and its metabolites on antigen-induced histamine release from human basophils in vitro, J. Immunol. 123: 1669 (1979).

28. B.B. Fredholm, Release of adenosine from rat lung by antigen and compound 48/80, Acta Physiol. Scand. 111: 507 (1981).

29. D.L. Marquardt, H.E. Gruber, and S.L. Wasserman, Adenosine release from stimulated mast cells, Proc. Natl. Acad. Sci. USA 81: 6192 (1984).

30. A. Sollevi, J. Ostergren, P. Hjemdahl, B.B. Fredholm, and B. Fagrell, The effect of dipyridamole on plasma adenosine levels and skin micro-circulation in man, in: Purine Metabolism in Man -IV-Part A: Clinical and Therapeutics Aspects; Regulatory Mechanisms, C.H.M.M. De Bruijn, H.A. Simmonds, and M.M. Müller, eds., Plenum Press, New York, p. 547 (1984).

31. G. Marone, The role of basophils and mast cells in the pathogenesis of pulmonary diseases, Int. Archs. Allergy appl. Immunol. 76 (suppl. 1): 70 (1985).

32. G. Marone, G. Ambrosio, D. Bonaduce, A. Genovese, M. Triggiani, and M. Condorelli, Inhibition of IgE-mediated histamine release from human basophils and mast cells by fenoterol, Int. Archs. Allergy appl. Immunol. 74: 356 (1984).

33. G. Marone, M. Columbo, L. Soppelsa, and M. Condorelli, The mechanism of basophil histamine release induced by pepstatin A, J. Immunol. 133: 1542 (1984).

INHIBITION OF MAST CELL MEDIATOR RELEASE BY

5-AMINO-4-IMIDAZOLECARBOXAMIDE RIBOSIDE

Diana L. Marquardt, and Harry E. Gruber

University of California San Diego School of Medicine
San Diego, CA 92103

INTRODUCTION

Stimulated mast cells break down $ATP_1$and release adenosine within 60 seconds after antigen or A23187 challenge[1], and adenosine itself potentiates mast cell pre-formed mediator release[2]. Because 5-amino-4-imidazolecarboxamide riboside (AICA riboside) has been shown to alter metabolism of adenosine and accelerate the repletion of ATP pools in other tissues[3], its effect on mast cell function was examined. We chose to study the effects of acute and chronic AICA riboside exposure in mast cells to determine whether changes in cell purine content or metabolism would alter cell function.

Mouse bone marrow-derived mast cells have been useful in studies requiring days of exposure to pharmacologic agents[4,5] in that they may be maintained for weeks in tissue culture. They also resemble human lung mast cells in their ability to generate leukotriene and prostaglandin products of arachidonic acid metabolism. Therefore, this "mucosal-type" mast cell was chosen in order to undertake the studies described.

RESULTS

Mouse bone marrow-derived mast cells were challenged with the calcium ionophore, A23187 in the presence of 100µM AICA riboside either added simultaneously with the secretagogue or present for an one hour incubation prior to challenge. Under these conditions of short-term exposure, AICA riboside did not alter mast cell β-hexosaminidase release. However, when the cells were grown in the presence of 100µM AICA riboside for 6 days, the cells maintained their viability and intracellular stores of β-hexosaminidase, yet released 50% less β-hexosaminidase than control cells grown in media alone (Figure 1). This inhibition persisted in the additional presence of adenosine, but was not evident if the cells were cultured in AICA.

When various concentrations of AICA riboside were present in the mast cell culture media for 6 days, a dose-dependent inhibition of A23187-stimulated β-hexosaminidase release was observed. This inhibition was initially evident at 10µM AICA riboside and was maximal at 100µM (Figure 2). At concentrations greater than 100µM AICA riboside, cell growth was

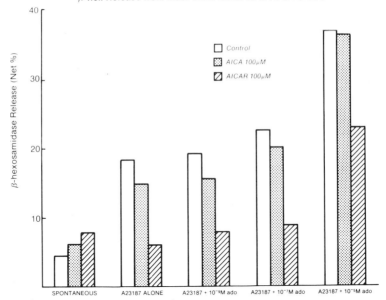

(Figure 1. Effect of chronic AICA - or AICA riboside - exposure on mast cell β-hexosaminidase release. Mast cells cultured in the presence of AICA (100µM) 🏁 or AICA riboside
(100µM) 🏁 for 4-8 days were washed and challenged with A23187
for 10 minutes. Shown is net % β-hexosaminidase released from
each population of cells.)

severely depressed. In studying the time course of the AICA riboside
effect, little change in mediator release was noted 24 hours after administration, but by 48 hours a significant inhibition was present which persisted for at least 8 days with continued AICA riboside exposure.

In a manner similar to that seen with attenuation of β-hexosaminidase
release, mast cells were cultured in AICA riboside for 6 days generated
50% less leukotriene C_4 20 minutes after A23187 challenge than cells cultured in media alone. The release of adenosine and the intracellular ATP
levels before and after stimulation were not affected to a significant
degree by chronic AICA riboside administration.

However, cells cultured in AICA riboside displayed an interesting
alteration in their intracellular nucleotide content compared to control
cells. An unusual ribonucleotide triphosphate, termed ZTP was present in
AICA riboside-cultured cells at an average concentration of 0.065
nmoles/10^6 cells (Figure 3). This substance has been previously identified as a regulatory molecule in formyl-depleted cells[6].

CONCLUSIONS

Alterations of mast cell purine content and/or metabolism appear to
be able to alter mast cell secretory responses as well. The mechanism of
this action is not well understood, but it is possible that a change in
the mast cell nucleotide content profile may alter its ability to release
pre-formed, granule-associated mediators and to generate arachidonic acid
metabolites. This global inhibition of mast cell mediator release by the
relatively non-toxic agent, AICA riboside, may prove to be important in
the treatment of allergic diseases.

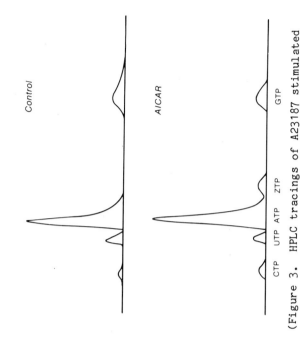

(Figure 3. HPLC tracings of A23187 stimulated control and AICA riboside-treated mast cell pellets. Mast cells cultured in medial alone or AICA riboside (100μM) for 6 days were challenged with A23187 for 60 seconds, centrifuged, extracted, and analyzed for nucleotide content by HPLC. A ZTP peak was identified in AICA riboside-treated cells.)

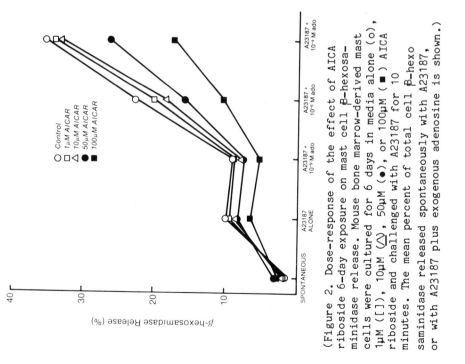

(Figure 2. Dose-response of the effect of AICA riboside 6-day exposure on mast cell β-hexosaminidase release. Mouse bone marrow-derived mast cells were cultured for 6 days in media alone (o), 1μM ([]), 10μM (△), 50μM (●), or 100μM (■) AICA riboside and challenged with A23187 for 10 minutes. The mean percent of total cell β-hexosaminidase released spontaneously with A23187, or with A23187 plus exogenous adenosine is shown.)

17

REFERENCES

1. D. L. Marquardt, H. E. Gruber, and S. I. Wasserman, Adenosine release
    from stimulated mast cells, Proc. Natl. Acad. Sci. USA. 81:6192 (1984).

2. D. L. Marquardt, C. W. Parker, and T. J. Sullivan, Potentiation of mast
    cell mediator release by adenosine, J. Immunol. 120:871 (1978).

3. R. L. Sabina, K. H. Kernstine, R. L. Boyd, E. W. Holmes, and J. L.
    Swain, Metabolism of 5-amino-4imidazolecarboxamide riboside
    in cardiac and skeletal muscle, J. Biol. Chem. 257:10178 (1982).

4. D. L. Marquardt, L. L. Walker, and S. I. Wasserman, Adenosine receptors
    on mouse bone marrow-derived mast cells: functional significance and
    regulation by aminophylline, J. Immunol. 133:932 (1984).

5. D. L. Marquardt, L. L. Walker, and S. I. Wasserman, Cromolyn
    inhibition of mouse bone marrow-derived mast cell mediator release,
    J. Allergy Clin. Immunol. 75:192 (1985).

6. R. L. Sabina, E. W. Holmes, and M. A. Becker, The enzymatic synthesis of
    5-amino-4-imidazolecarboxamide riboside triphosphate (ZTP), Science.
    223:1193 (1984).

PURINE METABOLISM BY GUINEA-PIG ILEUM

D.R. Webster, G. Boston, and D.M. Paton

Department of Pharmacology and Clinical Pharmacology
University of Auckland School of Medicine
Private Bag, Auckland, New Zealand

INTRODUCTION

Adenosine and the 5'-adenine nucleotides inhibit cholinergic neurotransmission in guinea-pig ileum through actions at presynaptic P_1-purinoceptors
(Paton et al., 1985). However, it was not possible to show whether the
adenine nucleotides acted directly on the receptor or whether their actions
were indirect, subsequent to hydrolysis of the nucleotides to adenosine.
Activation of adenosine A_2 receptors in VA13 fibroblasts by adenine nucleotides required conversion of the nucleotides to adenosine (Bruns, 1980).
Moody et al. (1984) demonstrated breakdown of 5'-ATP by guinea-pig ileum and
suggested that P_1-purinoceptor stimulation by 5'-ATP is partly direct and
partly by the degradation products 5'-AMP and adenosine.

If the action of a nucleotide is dependent on hydrolysis to adenosine, it
should be modified in the following ways (Paton, 1985): the addition of
exogenous adenosine deaminase should reduce its action, (b) inhibition of
endogenous adenosine deaminase may potentiate its action, and (c) inhibition
of nucleoside transport would be expected to potentiate its action. Addition
of exogenous adenosine deaminase reduced the presynaptic actions of only
adenosine and 5'-AMP in guinea-pig ileum, inhibition of this enzyme potentiated the actions of 5'-ADP and 5'-ATP while inhibitors of adenosine transport potentiated the actions of all the nucleotides studied (Paton et al.,
1985).

The aims of the present study were to examine using HPLC the metabolic
fate of exogenous adenosine and 2'-AMP, 5'-AMP and NAD in guinea-pig ileum as
reflected in the purines present in the incubation media, to correlate the
concentrations of adenosine and the adenine nucleotides with the observed
inhibition of cholinergic transmission, and to examine the effects on both
these variables of inhibition of adenosine deamination. 2'-AMP was used as
it has been reported to be a stable analogue of 5'-AMP in some tissues (e.g.,
rat heart, Baer and Drummond, 1968). NAD^+ was used as an example of a
dinucleotide. Separate specific receptors have been reported to exist for
NAD^+ in guinea-pig brain (Richards et al., 1983).

METHODS AND MATERIALS

Tissue Preparation

Guinea-pigs were killed and 100mg segments of terminal ileum prepared as described (Webster et al., this symposium).

Recording of Longitudinal Contractions

Ileal segments were suspended and stimulated as described (Webster et al., this symposium). Isometric contractions (twitch responses) were recorded with a force-displacement transducer (Grass FTO3C) and displayed on a polygraph (Grass model 7D).

Determination of Purine Metabolism

At 20 min intervals for 80 min after the tissue was suspended in the organ bath, the bath was drained and the medium replaced with fresh Krebs solution. The tissue was then stimulated at 0.2Hz for 5 min to determine the basal response, then the medium was changed to fresh Krebs medium (3.0ml) or Krebs medium containing the purine with or without luM deoxycoformycin, and stimulation at 0.2Hz was continued for 15 min. Samples of the medium were taken with a Hamilton microlitre syringe at 30 sec, 1,2,4,6,8,10 and 15 min. TCA (10μl, 50%) was added to the medium samples while mixing, then extracted with water-saturated diethylether until the pH was in the range 4-5.

Medium

The composition of the modified Krebs solution was as described (Webster et al., this symposium). 1μM Propranolol and 1μM phentolamine were added to prevent actions at adrenoceptors.

Drugs

Adenosine and nucleotides used in the study were obtained from Sigma Chemical Co. Deoxycoformycin was a gift from Dr. J.F. Henderson, while Ciba Pharmaceuticals and ICI New Zealand Ltd. provided phentolamine and propranolol respectively.

Analysis of Purines

Purine analysis was by reversed-phase HPLC as described (Webster et al., this symposium).

Statistical Analysis

Results were analysed by Students t-test (unpaired). Results were regarded as significant when $P < 0.05$.

Abbreviations

The following abbreviations have been used: ADA, adenosine deaminase 2'-, or 5'-AMP, adenosine 2', or 5'-monophosphate; AR, adenosine; HR, inosine; H, hypoxanthine; x, xanthine; UA, uric acid; dCF, deoxycoformycin; NAD^+, beta-nicotinamide adenine dinucleotide.

RESULTS

Spontaneous efflux of purines occurred from the isolated ileum preparations for several hours after suspension in the organ baths, but had fallen to less than 10pmol/mg tissue/min after 80 min (Webster et al., 1985). Purines were therefore not added to the bath until after this time. Since the uricase activity of homogenised ileum was extremely small, the total purine degradation product consisted of 5'-AMP+AR+HR+H+X+UA. During equilibration and stimulation at 0.2Hz, ileal pieces lost an average of 45.4% ± 1.8% in weight. This was associated with release of purine degrading enzymes into the medium. Nucleotidases, ADA, purine nucleoside phosphorylase and xanthine oxidase were all released into the medium, since 5'-AMP added to medium after the ileum sample had been removed was degraded to uric acid at room temperature. Medium samples were therefore stabilised by precipitating proteins with TCA.

Changes in the purine content of the incubation media are shown in Figures 1-5. Figure A shows the metabolism of added purine during continuous stimulation at 0.2Hz during 15 min and Figure B the metabolism of added purine in the presence of 1μM dCF.

Spontaneous purine efflux occurred during stimulation at 0.2Hz (Fig. 1A), the major component being uric acid. A small amount of xanthine was also found. This has been summed with uric acid for ease of comparison. The amount of AR and HR in the medium remained < 0.1μM. Addition of dCF produced a small but statistically significant increase in the amount of adenosine observed in the medium with a concomitant decrease in the concentration of hypoxanthine and inosine.

Purines were added to the incubation medium at 2.0μM since this had been found to be about the IC_{50} value for adenosine and the 5'-adenine nucleotides in this tissue (Paton and Webster, 1984). Added adenosine rapidly disappeared from the medium (Figure 2A), less than 20% remained after 15 min. Large amounts of UA+X and H+HR were found in the medium at all time points. When dCF was present (Figure 2B), the concentration of adenosine was constant and significantly higher at all time-points than in the absence of dCF, and the concentration of H+HR was significantly lower.

5'-AMP was hydrolysed to adenosine which was then deaminated (Fig. 3A). The adenosine concentration was about 0.4μmol/l throughout the incubation while the 5'-AMP concentration fell from 1.59 ± 0.06μM from 30 sec to 15 min. Addition of dCF (Figure 3B) did not alter the rate of hydrolysis of 5'-AMP. However, the adenosine concentration was significantly higher throughout the incubation and the concentration of deaminated products significantly lower.

2'-AMP was metabolised in a similar way to 5'-AMP (Figure 4A). The initial disappearance from the medium was very fast (the added concentration of 2'-AMP to the medium was 2.07 ± 0.12μM), and more deaminated products and less adenosine were found in the medium than when 5'-AMP was the substrate (P<0.05), all time-points). The concentration of adenosine was about 0.2μM throughout the experiment. When adenosine deaminase was inhibited, (Figure 4B), the concentration of adenosine was significantly higher and that of H+HR significantly lower, exactly analogous to the situation found with 5'-AMP.

The dinucleotide NAD^+ was also metabolised during stimulation at 0.2Hz (Figure 5A), the products being 5'-AMP, AR and further degradation products. In the presence of dCF, adenosine was significantly higher only after 4 min (Figure 5B).

Inhibition of twitch responses is shown with the concentration of active purine metabolites in Figures 1C-5C. In the control situation (Figure 1C), the twitch response and the adenosine concentration were constant throughout stimulation. In the presence of dCF, both the inhibition of twitch response and the adenosine concentration were increased slightly.

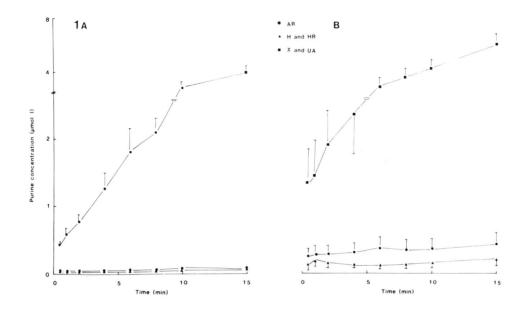

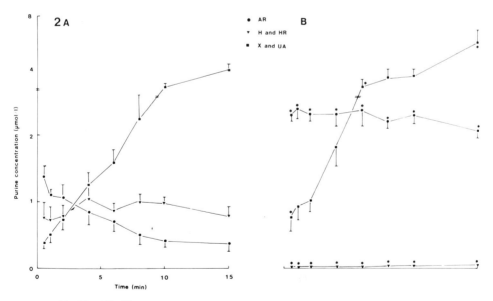

Figure 1A-5A, 1B-5B.
    Purine metabolism by guinea-pig ileal segments stimulated at 0.2Hz.
Purines were added at 2µM. Figure 1A has no added purine, 2A, AR; 3A,
5'AMP; 4A, 2'AMP and 5A, NAD. B represents the metabolism under the
same conditions with the addition of 1µM dCF.

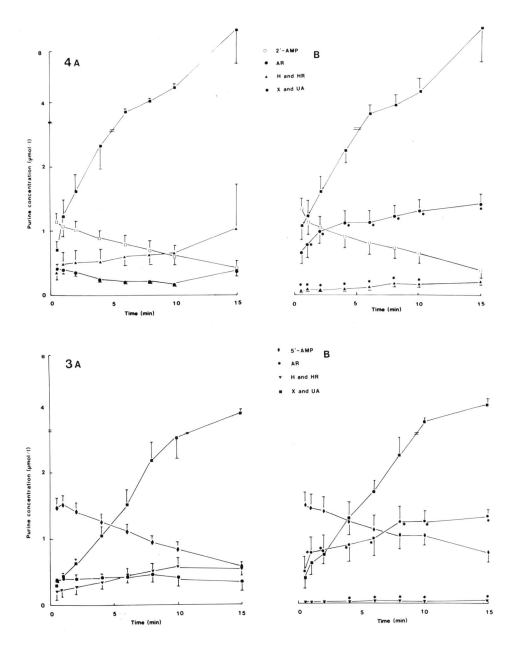

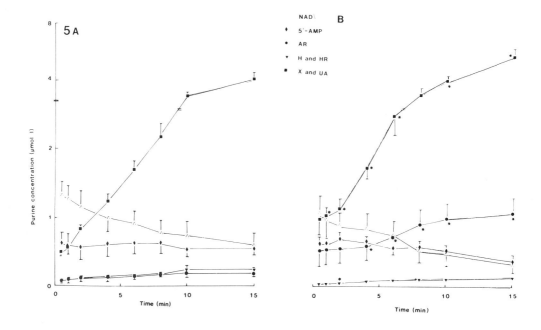

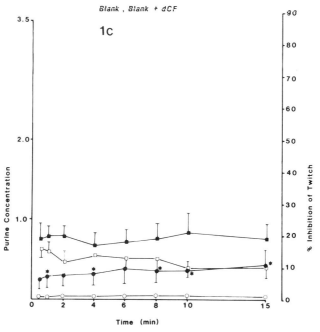

Figure 1C-5C show the relationship of purine concentration to inhibition of twitch response during 15 min stimulation at 0.2Hz. □ , ■ represents inhibition of twitch response and inhibition in the presence of dCF. o in Fig. 1C, 2C represents AR; 3C, AR + 5'AMP; 4C, AR + 2'AMP and 5C, AR + 5'AMP + NAD. ● Represents the same purines in the presence of dCF. * indicates P < 0.01, and the error bars show the S.E.M.

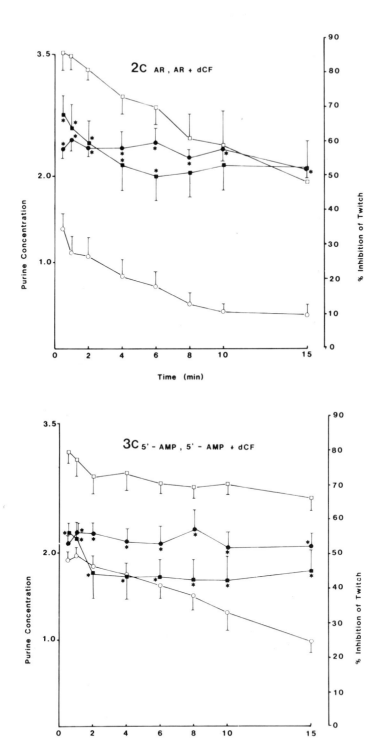

25

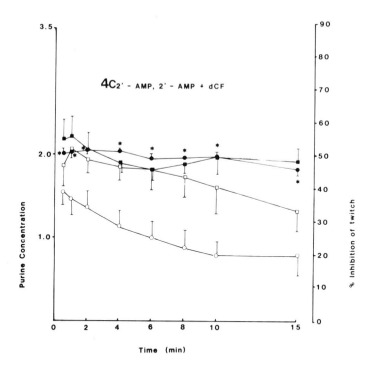

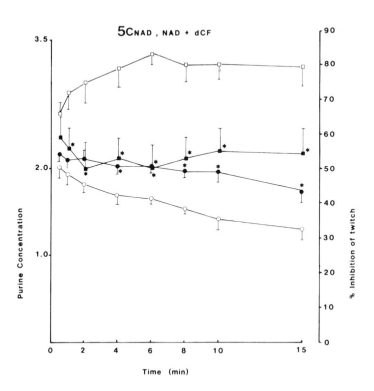

When adenosine was added (Figure 2C), the inhibition of twitch response declined over 15 min from 86 ± 5% at 30 sec to 48 ± 9% at 15 min. The adenosine concentration declined in parallel with the fall in inhibition of the twitch response. When ADA was inhibited, the inhibition of twitch declined initially then was constant from 4 min as was the adenosine concentration.

The situation is more complex when the inhibition of twitch by 5'- and 2'-AMP is considered, since the nucleotide and/or adenosine may be contributing to the observed inhibition of the twitch. Plotting the variables alone and in combination showed that a simple sum of nucleotide + adenosine correlated best with the inhibition of twitch response. This is shown in Figure 3C for 5'-AMP. The inhibition of the twitch decreased in parallel with a decrease in (5'-AMP + AR) concentration and both were constant when ADA was inhibited. Figure 4C shows that the relationship of purine concentration to inhibition of twitch response was the same for 2'-AMP as it was for 5'-AMP.

Since NAD^+ was hydrolysed to both 5'-AMP and adenosine, three purines thought to inhibit the twitch response in guinea-pig ileum are present in the medium during stimulation. None of the possible combinations of concentrations produced parallel inhibition-concentration graphs. In Figure 5C, the sum $(NAD^+ + 5'\text{-AMP} + AR)$ concentration is shown. The inhibition of twitch response increased through the time of stimulation while the combined purine concentration decreased. In the presence of dCF, the situation was similar to that for adenosine and the other nucleotides studied with little change in either inhibition of twitch response or purine concentration throughout.

DISCUSSION

Since the purine end-product in this tissue was uric acid, and there was negligible IMP in medium or tissue samples (Webster et al., 1985), each molecule of HR, H, X and UA was formed via a molecule of adenosine. There is considerable contribution to the measured purine from endogenous sources since uric acid was observed with stimulation in medium alone (Figure 1). Considerable endogenous adenosine production has also been noted in this tissue stimulated at 3Hz (Gustafsson et al., 1981).

Inhibition of ADA with dCF increased adenosine concentration significantly in the control and added-purine experiments, with a concomitant decrease in the concentration of the deaminated products, hypoxanthine and inosine. The presence of dCF in the medium altered inhibition of twitch response with added adenosine up to 6 min, and 5'-AMP and NAD^+. In all cases inhibition by purine was reduced by dCF. The concentration of adenosine in the medium was increased in all cases, and the adenosine-nucleotide concentration/time profiles were very similar for 5'-AMP and 2'-AMP (Figures 3A,B and 4A,B).

The observation that dCF reduces the inhibition of twitch response in this experimental situation is unexplained. When adenosine alone is added, the inhibition of response decreases in parallel with the decrease in adenosine concentration (Figure 2C). In most of the experiments presented here the summed concentration of active purines parallels the changes in inhibition of twitch response. Inhibition of ADA by dCF has been previously found to only potentiate responses to 5'-AMP and 5'-ADP. The difference may possibly relate to the addition of inhibitor with the purine in this study, without allowing previous time for equilibration. However 20 min preincubation with dCF made only 10% difference in incorporation of adenosine into nucleotides in Erlich ascites cells (Henderson et al., 1977), and deamination appeared to be inhibited in this preparation as judged by rate of loss of adenosine from the medium.

This study has therefore shown conclusively that endogenous adenosine and other purines are released in considerable quantity from this preparation and must be taken into account in interpretation of the effects of exogenous purines. While the inhibition of presynaptic cholinergic neurotransmission appears to be related to the concentration of adenosine when this purine is added, this is not so for the nucleotides. It has been shown that the nucleotides studied (5'-AMP; 2'-AMP and NAD^+) are all extensively hydrolysed to adenosine and the adenosine concentrations are increased when adenosine deaminase is inhibited. However, conclusive evidence for the mode of action of the adenine nucleotides in P_1-purinoceptors in guinea-pig ileum is still not available.

ACKNOWLEDGEMENTS

This work was supported by a grant from the Medical Research Council of New Zealand.

REFERENCES

Baer, H.P., and Drummond, G.I., 1968, Catabolism of adenine nucleotides by the isolated perfused rat heart, Proc. Soc. exp. Biol. Med., 127:33.
Bruns, R.F., 1980, Adenosine receptor activation by adenine nucleotides requires conversion of the nucleotides to adenosine, Naunyn-Schmeideberg's Arch. Pharmacol., 315:5.
Gustafsson, L., Fredholm, B.B., and Hedqvist, P., 1981, Theophylline interferes with the modulatory role of endogenous adenosine on cholinergic neurotransmission in guinea-pig ileum, Acta Physiol. Scand., 111:269.
Henderson, J.F., Brox, L., Zombor, G., Hunting, D., and Lomax, C.A., 1977, Specificity of adenosine deaminase inhibitors, Biochem. Pharmacol., 26:1967.
Moody, C.J., Meghji, P., and Burnstock, G., 1984, Stimulation of P_1-purinoceptors by ATP depends partly on its conversion to AMP and adenosine and partly on direct action. Eur. J. Pharmacol., 97:47.
Paton, D.M., 1985, Classification of adenosine receptors in peripheral tissues. In: "Methods in Pharmacology, vol VI, Adenosine." Paton, D.M. ed., Plenum. New York. In press.
Paton, D.M., 1981, Structure-activity relations for presynaptic inhibition of noradrenergic and cholinergic transmission by adenosine: evidence for action on A_1 receptors. J. Auton. Pharmacol. 1:287.
Paton, D.M., Broome, S.G., and Webster, D.R., 1985, Mechanism of presynaptic inhibition of cholinergic transmission in guinea-pig ileum by adenine nucleotides. Meth. Find. Clin. Exp. Pharmacol., 7:65.
Paton, D.M., and Webster, D.R., 1984, On the classification of adenosine and purinergic receptors in rat atria and in peripheral adrenergic and cholinergic nerves. In: "Neuronal and Extraneuronal Events in Autonomic Pharmacology." Fleming, W.W. et al., ed., Raven Press, New York.
Richards, C.D., Snell, C.R., and Snell, P.H., 1983, Nicotinamide adenine dinucleotide depresses synaptic transmission in the hippocampus and has specific binding sites on the synaptic membranes. Br. J. Pharmacol., 79:553.

SUPPRESSION BY 5'-METHYLTHIOADENOSINE OF HISTAMINE OR LEUKOTRIENE-INDUCED

CONTRACTION IN ISOLATED GUINEA PIG TRACHEAL RINGS

Y. Nishida, S. Suzuki and T. Miyamoto

Department of Medicine and Physical Therapy
Faculty of Medicine, University of Tokyo
7-3-1 Hongo, Bunkyo-ku, Tokyo, Japan

INTRODUCTION

5'-Methylthioadenosine (MTA) is a naturally occurring purine nucleoside, which is produced from S-adenosylmethionine by several biosynthetic routes (1)(2) and metabolized by MTA phosphorylase to S-methylthioribose-1-phosphate and adenine.(3) MTA has been shown to have numerous effects on cell metabolism and function(2) and recent interest has focused extensively on its anti-proliferative action.(4)(5) Snyder et al(6) suggested that MTA stimulates the adenosine receptor. There are many reports that, in the human subjects, many cells including tracheal smooth muscle cells, have the adenosine receptor.(7) The action of adenosine on tracheal muscle cells in vitro has been intensively investigated(8), but there are no reports concerning the action of MTA on tracheal smooth muscle preparations.

In this experiment, we studied the effect of MTA on isolated guinea pig tracheal strips and the mechanisms that may mediate this are discussed.

MATERIALS AND METHOD

MTA, indomethacin, tetrodotoxin, theophylline, propranolol were purchased from the Sigma Chemical Corp. Leukotriene D_4 was purchased from Paesel GMBH Corp. Male guinea pigs weighing 500-600 g, were killed by a blow on the head. The trachea was rapidly isolated and cut into 16 rings. The tracheal chains were prepared using eight rings in a series, which were suspended in a 10 ml organ bath containing atropinized Krebs-Henseleit solution of the following composition: 1.17×10^{-1} M NaCl, 5.36×10^{-3} M KCl, 2.52×10^{-3} M $CaCl_2$, 1.16×10^{-3} M $MgSO_4$, 1.16×10^{-3} M NaH_2PO_4, 2.52×10^{-2} M $NaHCO_3$ and 1.11×10^{-2} M glucose. The medium was kept at 37°C and continuously gassed with 5% carbon dioxide and 95% oxygen. The chains were prepared under a resting tension of 0.3 g and allowed to stabilize for 1 h before the start of the experiment. The changes in muscle tension were recorded through a strain gauge (Nihon-Koden Cor.) and displayed on a polygraph. A standard contraction was achieved by adding 1 µg/ml histamine.

The tracheal relaxations were studied in resting tissue by adding varying concentrations of MTA, after which the MTA action on the tracheal rings, contracted by histamine or leukotriene D_4, were investigated. The effects of MTA on histamine or leukotriene D_4-induced tracheal contraction were also studied. Firstly the tracheal contractions were measured by adding varying concentrations of histamine or leukotriene D_4 with or without 100 µM MTA.

Secondly the tracheal contraction was achieved with 1 μg/ml histamine (ED 50) or 1 ng leukotriene D_4 and the addition of varying concentrations of MTA. These experiments were done in triplicate using different tracheal rings.

We also carried out a series of experiments to investicate the mechanism of action of MTA. Propranolol (0.1×10^{-5} M), tetrodotoxin (1.5×10^{-6} M), indomethacin (1 μg/ml), dipyridamole (2×10^{-4} M) and theophylline (1×10^{-4} M) were added to the bath 10 min before the addition of MTA and their effect on the MTA response to histamine induced-contraction was studied.

RESULTS

Fig I shows that relaxation of resting tracheal rings occurred after a latent period following the addition of MTA to the medium. Relaxation caused by MTA was dose-dependent over a concentration range of 50 to 1000 μM i.e. 0.07 g at 50 μM, 0.20 g at 200 μM and 0.30 g at 500 μM. Another characteristic feature of the MTA mediated dilatation effects was its latent period. When small dose of MTA was added to the medium, the latent period was elongated in a dose dependent manner, i.e. 3.0 min. at 50 μM and 1.0 min. at 500 μM. Addition of MTA to the contracted tracheal rings induced by histamine or leukotriene D_4 caused significant relaxation, which was more than twice that of resting tracheal rings.

Figure 2 shows the dose-response curve of contraction of histamine or leukotriene D_4 with or without the addition of 100 μM MTA. Pretreatment of guinea pig tracheal rings with 100 μM MTA inhibited the contraction produced by varying doses of histamine or leukotriene D_4. The inhibition rate caused by 100 μM MTA was dependent on histamine or leukotriene D_4 concentrations, i.e. 51% at 0.25 μg/ml histamine, 22% at 2.0 μg/ml histamine, and 57% at 0.5 ng leukotriene D_4, 46% at 2 ng leukotriene D_4.

The inhibitory effects of varying doses of MTA on the contraction by fixed standard dose (1 μg/ml) of histamine or (1 ng) of leukotriene D_4 are also shown in Figure 2. The tracheal contractions induced by histamine or leukotriene D_4 were inhibited by MTA in a dose-dependent manner, i.e. 22% at 10 μM, 55% at 50 μM for histamine and 50% at 100 μM, 70% at 200 μM for leukotriene D_4. 500 μM or MTA completely blocked both histamine and leukotriene D_4 contractions.

Figure III shows that the effect of MTA was not blocked by pretreatment of the tracheal rings with propranolol and tetrodotoxin and indomethacin and dipyridamole also had no effect. In contrast, theophylline, at a concentration of 100 μM, suppressed the actions of MTA to about 10% of control.

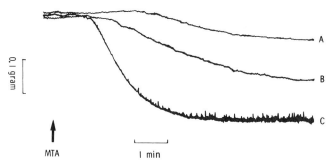

Fig. I Relaxation of the tracheal rings after treatment with methylthioadenosine. (A: 50 μM, B: 200 μM, C: 500 μM)

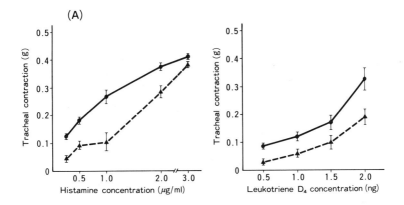

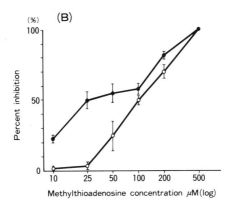

Fig. II Inhibitory effect of methylthioadenosine on contractile response
to histamine in isolated guinea pig tracheal rings. (A) After
preincubation with 100 μM methyladenosine, varying doses of
histamine and or leukotriene D₄ were added. (●————● =
control ▲---------▲ = methylthioadenosine) (B) Varying
concentrations of methylthioadenosine were preincubated, then
1 μg/ml of histamine (●————●) and or 1 ng leukotriene
D₄ (o————o) were added.
Results are expressed as the mean ± SD of three times experiments.

DISCUSSION

In these experiments, treatment of isolated guinea pig tracheal rings with relatively high concentrations of MTA produced relaxation. Pretreatment with MTA in low concentrations suppressed histamine and leukotriene D_4 induced contraction. Previously, MTA has been shown to affect a variety of biological systems. In vitro, MTA inhibited many enzymes such as s-adenosylhomocysteine hydrolase, spermine synthetase and phosphodiesterase.[9][10][11]

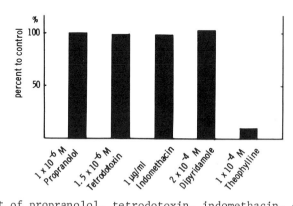

Fig.III Effect of propranolol, tetrodotoxin, indomethacin, dipyridamole and theophylline on the action of methylthioadenosine in tracheal contraction induced by histamine.

The pharmacologic effects of MTA also reported include depression of blood pressure in rabbits, relaxation of intestinal strips in rabbits and contraction of isolated guinea pig uterus. However, the mechanism of action is uncertain. Dipyridamole, a blocker of purine nucleoside uptake, had no effect, suggesting that MTA may act on cell surface receptors. Pretreatment of the tracheal rings with propranolol, a β-adrenergic receptor antagonist, did not block MTA and it seems unlikely that it stimulates the β-adrenergic receptor, although adenylate cyclase is an important regulator of bronchial muscle tension. Tetrodotoxin, an inhibitor of neurotransmitter release, had no effect on the actions of MTA suggesting that MTA does not produce its relaxant effect by causing the release of neurotransmitters. MTA was also little affected by indomethacin, so it seems that prostaglandin synthesis is not involved. However, theophylline, an adenosine receptor antagonist, blocked the action of MTA. These findings strongly suggested that MTA may stimulate the adenosine receptor of tracheal muscle cells.

Concentrations of MTA in the cells and/or extracellular fluid are unknown. Therefore, it is not clear as to what extent MTA may have a physiological role in the tracheal relaxation. In addition, the pharmacotherapeutic implications of MTA in vivo must be further investicated.

REFERENCES

1. A.E.Pegg, and H.G.Williams-Ashman,
   On the role of S-adenosyl-L-methionine in the biosynthesis of spermidine
   by rat prostate.
   J. Biol. Chem., 244, 682-293 (1969).
2. H.G.Williams-Ashman, J.Seidenfeld, and P.Galletti,
   Trends in the biochemical pharmacology of 5'-deoxy-5'-methylthioadenosine
   Biochem. Pharm., 31, 277-288 (1982).
3. N.Kamatani, and D.A.Carson,
   Dependence of adenine production upon polyamine synthesis in cultured
   human lymphoblasts.
   Biochim. Biophys. Acta., 675, 344-350 (1981).
4. L.Christa, L.Thuillier, and J-I,Perignon,
   5'-Deoxy-5'-methylthioadenosine inhibition of rat T lymphocyte phosphodiesterase: Correlation with inhibition of ConA induced proliferation.
   Biochem. Biophys. Res. Commun., 113, 425-432 (1983).
5. R.Dante, M.Arnaud, and A.Niveleau,
   Effects of 5'-deoxy-5'methylthioadenosine on the metabolism of
   S-adenosylmethionine.
   Biochem. Biophys. Res. Commun., 114, 214-221 (1983).
6. S.J.Snyder, R.F.Bruns, J.W.Daly, and R.B.Innis,
   Multiple neurotransmitter receptors in the brain; amines, adenosine and
   cholecystokinin.
   Fed. Proc., 40, 142-146 (1981).
7. B.B.Fredholm, K.Brodin, and K.Strandberg,
   On the mechanism of relaxation of tracheal muscle by theophylline and
   other cyclic nucleotide phosphodiesterase inhibitors.
   Acta Pharmac. Tox., 45, 336-344 (1979).
8. R.B.Clark, and M.N.Seney,
   Regulation of adenylate cyclase from cultured human cell lines by
   adenosine.
   J. Biol. Chem., 251, 4239-4246 (1976).
9. A.J.Ferro, A.A.Vandebark, and M.R.MacDonald,
   Inactivation of S-adenosyl homocysteine hydrolase by 5'-deoxy-5'-
   methylthioadenosine.
   Biochem. Biophys. Res. Commun., 100, 523-531 (1981).
10. I.H.Fox, T.D.Palella, D.Thompson, and C.Herring,
    Adenosine metabolism: Modification by S-adenosylhomocysteine and 5'-
    methylthioadenosine.
    Arch. Biochem. Biophys., 215, 302-308 (1982).
11. A.E.Pegg, R.T.Borchardt, and J.K.Coward,
    Effects of inhibitors of spermidine and spermine synthesis on polyamine
    concentrations and growth of transformed mouse fibroblasts.
    Biochem. J., 194, 79-89 (1981).
12. P.L.Ewing, and F.Schlenk,
    Some pharmacological actions of adeninethiomethylpentose.
    J. Pharmac. Exp. Therap., 79, 164-168 (1943).

ADENOSINE RECEPTORS ON HUMAN BASOPHILS AND LUNG MAST CELLS

Gianni Marone*, Massimo Triggiani*, Anne Kagey-Sobotka**,
Lawrence M. Lichtenstein** and Mario Condorelli*

*Department of Medicine, University of Naples, II School
of Medicine, Via S. Pansini 5, 80131 Naples, Italy, and
**Division of Clinical Immunology, Department of
Medicine, The Johns Hopkins University School of
Medicine, 5601 Loch Raven Boulevard, Baltimore, MD
21239, U.S.A.

Adenosine is a natural nucleoside that plays an important physiological role in the regulation of adenosine 3',5'-monophosphate (cAMP) metabolism in various mammalian tissues (1-3). Adenosine has recently been implicated in the stimulus-response coupling of many inflammatory cells. Adenosine receptors and/or hormone-like responses to adenosine have been reported in lymphocytes (1,3,4,5), polymorphonuclear leukocytes (6), mast cells (7-9), platelets (10), and macrophages (11). In human basophils, adenosine inhibits IgE-mediated histamine release, presumably by increasing intracellular cAMP (12).

A variety of adenosine analogs have been used to classify adenosine receptors into A_2/R_a and A_1/R_i subtypes (13). The former is a low affinity membrane receptor that activates adenylate cyclase (14), while the latter is a high affinity receptor that exerts an inhibitory effect on adenylate cyclase (14,15). Both types of receptors are inhibited by methylxanthines (1,4,12,13) and show an absolute dependence on the presence of GTP (16,17) as would be expected for receptor-mediated modulations of adenylate cyclase (18,19). High affinity specific binding of adenosine analogs and antagonists has been observed in membranes from several cells (20,21). These studies led to the characterization of a number of adenosine analogs that bind with typical characteristics to A_2/R_a or A_1/R_i receptors (22). The inhibitory receptor, A_1/R_i, is more sensitive to the agonists (20 - 100 nM) and prefers $(-)-N^6-(R-phenyl-isopropyl)$-adenosine { (-)-R-PIA } to 5'-N-ethylcarboxamideadenosine (NECA), whereas the stimulatory receptor, A_2/R_a, is effective over a range of 0.1 - 30 microM and prefers NECA to (-)-R-PIA (15,19,22).

The immunological activation of human basophil leukocytes and mast cells initiates a complex sequence of biochemical events resulting in the secretion and/or de novo synthesis of chemical mediators (23-26). There is now compelling evidence that chemical mediators immunologically released from human basophils and mast cells play a fundamental role in the pathogenesis of allergic and/or inflammatory disorders (26-29). Human basophils and mast cells isolated from lung tissue release histamine and generate eicosanoids upon IgE-dependent activation (26-30). These cells synthetize de novo a series of metabolically-related products of arachidonic acid metabolism including peptide leukotriene C$_4$ (LTC$_4$). LTC$_4$ appears to be the main product of human basophils and lung mast cells generated through 5-lipoxygenase activity (31). It possesses a variety of biologically relevant properties (32) suggesting that, together with preformed mediators, i.e. histamine, it participates in many aspects of inflammatory disorders.

We have recently explored the effect of adenosine and its analogs on mediator release from human basophils and lung mast cells. The results support the hypothesis that this natural nucleoside and its analogs exert opposite effects on mediator release from the two different types of cells.

In a first series of experiments, various adenosine derivatives were compared for their effects on mediator release from human basophils challenged with antigen. Unless otherwise specified, all experiments included a 15-min preincubation of cells with adenosine derivatives before the addition of antigen. Figure 1 shows the effect of various concentrations of NECA, 2-chloroadenosine, adenosine, (-)-R-PIA, and (+)-S-PIA on antigen-induced mediator release from human basophils. A constant finding of these experiments was the potent inhibitory effect of NECA, which is consistent with its high affinity for the A$_2$/R$_a$ receptor demonstrated in other systems (2,22). Furthermore, the inhibitory effect of NECA > 2-chloroadenosine > adenosine > (-)-R-PIA > (+)-S-PIA, in order of potency, was also consistent with their marked differential affinity for A$_2$/R$_a$ receptor (15,22) and their ability to increase intracellular cAMP in human leukocytes (1,4,33).

It is likely that adenosine and its analogs produced their effects by interacting with a receptor on the basophil surface. Theophylline and other methylxanthines have been shown to antagonize both A$_2$/R$_a$ and A$_1$/R$_i$ adenosine receptors in several cell types (1,4,12,13). In fact, theophylline (10^{-5} M) by itself had no effect on mediator release, but it did antagonize the inhibitory action of adenosine: e.g., 10^{-5} M theophylline almost completely suppressed the effect of 3 x 10^{-7} M adenosine (34). At higher concentrations adenosine progressively overcame the block by theophylline. This finding confirms the previous observation made by our group that methylxanthines are competitive antagonists of the adenosine A$_2$/R$_a$ receptor in human basophils as well as in human lymphocytes (1,12,34).

We have recently extended these observations by showing that the activation of adenosine A$_2$/R$_a$ receptor on human basophils leads to the inhibition of the synthesis of immunoreactive LTC$_4$. The inhibition caused by adenosine was due to a real blockade of LTC$_4$ synthesis since the concentrations of LTC$_4$ metabolites (i.e., LTD$_4$ and LTE$_4$) were not increased by pretreatment with the nucleoside. Again, in the basophil system, theophylline acts as a competitive antagonist of the inhibition of LTC$_4$ release.

Recently it became possible to isolate and purify mast cells from human lung tissue (30). This led to the discovery of several morphological, biochemical and functional differences between human basophils and lung mast cells (26,34). We therefore evaluated the effect of adenosine and its analogs on the release of chemical mediators from purified human lung mast cells challenged with anti-IgE. Mast cells purified from human lung release histamine and de novo synthesize LTC_4 (31). We were surprised to find that low concentrations of adenosine dose-dependently enhanced the IgE-mediated release of histamine from human lung mast cells. Furthermore, we found that NECA, a specific A_2/R_a agonist, and (-)-R-PIA, a selective A_1/R_i agonist possessed the same

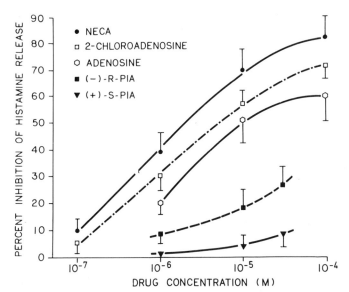

Fig.1.   The effect of various concentrations of adenosine and its analogs on antigen-induced secretion during the "first stage" of histamine release from human basophils (28). Each point represents the mean $\pm$ S.E.M. obtained from twelve experiments.

enhancing effect as adenosine. Figure 2 shows a typical experiment of the effect of (-)-R-PIA, adenosine and NECA on IgE-mediated release of histamine from purified human lung mast cells. This observation raises the possibility that the enhancing effect of adenosine and its analogs is not mediated by the interaction with the classical A_1/R_i or A_2/R_a receptor. Whatever the underlying mechanism, these findings might be clinically relevant since the effect of adenosine occurs at submicromolar concentrations comparable to those found in the serum of normal controls and asthmatic subjects (C.H.M.M. de Bruijn and G. Marone, unpublished results).

Fredholm was the first to demonstrate that adenosine is released from lung tissue exposed to hypoxic conditions or antigen challenge (36). A significant release of adenosine from mast cells stimulated with Ca^{2+} ionophore A23187 has also been documented (37). These observations raise the possibility that this endogenous source of adenosine may act locally through a positive feedback mechanism to potentiate immediate hypersensitivity reactions. This hypothesis is still largely speculative. Mast cells purified from human lung and challenged with an immunological stimulus might yield rewarding information regarding this feedback mechanism.

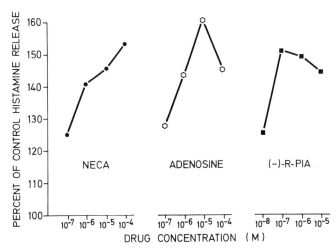

Fig. 2. The effect of various concentrations of adenosine, NECA and (-)-R-PIA on anti-IgE-induced histamine secretion from human lung mast cells. Each point represents the mean of duplicate determinations obtained in a typical experiment.

Another discrepancy between the basophil and the mast cell system was the effect of theophylline. In fact, in the mast cell system, preincubation with low concentrations of theophylline did not completely suppress the adenosine-induced enhancement of IgE-mediated histamine release.

A final observation made in the human lung mast cell system was obtained with dipyridamole. We have previously shown that this drug effectively inhibits the uptake of exogenous adenosine in human inflammatory cells (1) and some of its therapeutic actions are ascribed to this property. We found that low concentrations of dipyridamole does not block the enhancing effect of adenosine on IgE-mediated release of chemical mediators from human lung mast cells. This suggests that the potentiating effect of adenosine is exerted at the level of the cytoplasmic membrane of these cells. More pragmatically these results suggest that the use of dipyridamole in clinical conditions characterized by enhanced mediator release from lung mast cells, such as allergic disorders, might be negatively affected by the coincident use of dipyridamole.

Several observations can be drawn from these studies. We have previously shown that adenosine inhibits IgE-mediated histamine release from human basophils (12) probably by increasing intracellular cAMP (1,4,38). The present results indicate that adenosine and its analogs inhibit the release of histamine and LTC_4 by acting on a cell-surface receptor that possesses properties similar to those of an adenosine A_2/R_a receptor. The inhibitory effects of NECA $>$ 2-chloroadenosine $>$ adenosine $>$ (-)-R-PIA $>$ (+)-S-PIA are consistent with their different affinity for A_2/R_a receptor (15,22) linked to adenylate cyclase (19). In contrast, adenosine, NECA and (-)-R-PIA enhance to a similar extent anti-IgE-induced mediator release from human lung mast cells. The present results suggest that the nucleosides do not exert their enhancing effect by activating a surface receptor whose properties are typical of A_2/R_a or A_1/R_i. This raises the possibility of the existence of subtypes of A_1/R_i and A_2/R_a receptors. In this context it is interesting to note that heterogeneity of adenosine A_1/R_i receptors has been reported in brain tissue (39).

It has been shown that low concentrations of theophylline and of methylxanthines, in general, are competitive antagonists of cell-surface adenosine receptors (1,12,34). In human basophils, theophylline and methylxanthines act as competitive antagonists, which indicates the presence of a specific cell surface receptor. In contrast, in the lung mast cells methylxanthines do not appear to act as competitive antagonists of the enhancing effect of adenosine. Further investigations are required to characterize better the mechanism of action of adenosine on the metabolism of human lung mast cells.

In conclusion, it is evident that the natural nucleoside adenosine exerts opposite effects on the release of chemical mediator from two cell types that, despite a different origin, have some similarities. Differences between the two cell types have begun to emerge from recent studies (26,35). The present work provides yet another example of these differences. Adenosine is a natural nucleoside whose concentrations increase following tissue hypoxia (40) or by blocking its uptake with dipyridamole (41). This is the first example of an endogenous autacoid that exerts opposite effects on mediator release from human basophils and lung mast cells.

Adenosine is a natural nucleoside already implicated in the pathogenesis of immunodeficiency (42), autoimmune diseases (43) and bronchial asthma (44). It is possible that by interacting with different subtypes of adenosine receptors it plays a complex role in the fine tuning of inflammatory cells.

ACKNOWLEDGMENTS

The technical assistance of Mr. Oreste Marino and Mrs. Yaffa Nif is gratefully acknowledged. This work was supported in part by grants from the C.N.R. (83.00430.04 and 84.01756.04), the Ministero Sanità and the M.P.I. (Rome, Italy).

REFERENCES

1. G. Marone, M. Plaut, and L.M. Lichtenstein, Characterization of a specific adenosine receptor on human lymphocytes, J. Immunol. 121: 2153 (1978).
2. J.W. Daly, R.F. Bruns, and S.H. Snyder, Adenosine receptors in the central nervous system: relationship to the central actions of methylxanthines, Life Sci. 28: 2083 (1981).
3. A.L. Schwartz, R.C. Stern, and S.H. Polmar, Demonstration of an adenosine receptor on human lymphocytes in vitro and its possible role in the adenosine deaminase-deficient form of severe combined immunodeficiency, Clin. Immunol. Immunopathol. 9: 499 (1978).
4. G. Marone, R. Petracca, and M. Condorelli, Adenosine receptors on human inflammatory cells, in: Purine Metabolism in Man -IV- Part A: Clinical and Therapeutic Aspects; Regulatory Mechanisms, C.H.M.M. De Bruijn, H.A. Simmonds, and M.M. Müller, eds., Plenum Press, New York, p. 501 (1984).
5. G. Wolberg, T.P. Zimmerman, K. Hiemstra, M. Winston, and L.C. Chu, Adenosine inhibition of lymphocyte-mediated cytolysis: possible role of cyclic adenosine monophosphate, Science 187: 957 (1975).
6. B.N. Cronstein, S.B. Kramer, G. Weissmann, and R. Hirschhorn, Adenosine: a physiological modulator of superoxide anion generation by human neutrophils, J. Exp. Med. 158: 1160 (1983).
7. D.L. Marquardt, C.W. Parker, and T.J. Sullivan, Potentiation of mast cell mediator release by adenosine, J. Immunol. 120: 871 (1978).
8. S.P. Peters, E.S. Schulman, R.P. Schleimer, D.W. MacGlashan, Jr., H.H. Newball, and L.M. Lichtenstein, Dispersed human lung mast cells. Pharmacologic aspects and comparison with human lung tissue fragments, Am. Rev. Resp. Dis. 126: 1034 (1982).
9. D.L. Marquardt, L.L. Walker, and S.I. Wasserman, Adenosine receptors on mouse bone marrow-derived mast cells: functional significance and regulation by aminophylline, J. Immunol. 133: 932 (1984).
10. G. Marone, S. Quattrin, M. Masturzo, O. Finizio, A. Genovese, and M. Condorelli, Adenosin-rezeptoren von menschlichen plaettchen: charakterisierung des A_2/R_a rezeptors und des P-lokus, in: Thrombose- und Hämostaseforschung 1984, E.A. Beck, ed., F.K. Schattauer Verlag, Stuttgart, p. 221 (1984).
11. M.C. Pike, and R. Snyderman, Transmethylation reactions regulate affinity and functional activity of chemotactic factor receptors on macrophages, Cell 28: 107 (1982).
12. G. Marone, S.R. Findlay, and L.M. Lichtenstein, Adenosine receptor on human basophils: modulation of histamine release, J. Immunol. 123: 1473 (1979).
13. C. Londos, and J. Wolff, Two distinct adenosine-sensitive sites on adenylate cyclase, Proc. Natl. Acad. Sci. USA. 74: 5482 (1977).
14. D. Van Calker, M. Müller, and B. Hamprecht, Adenosine regulates via two different types of receptors, the accumulation of cyclic AMP in cultured brain cells, J. Neurochem. 33: 999 (1979).
15. C. Londos, D.M.F. Cooper, and J. Wolff, Subclasses of external adenosine receptors, Proc. Natl. Acad. Sci. USA 77: 2551 (1980).
16. D.M.F. Cooper, and C. Londos, Evaluation of the effects of adenosine on hepatic and adipocyte adenylate cyclase under conditions where adenosine is not generated endogenously, J. Cycl. Nucl. Res. 5: 289 (1979).
17. D.M.F. Cooper, C. Londos, and M. Rodbell, Adenosine receptor-mediated inhibition of rat cerebral cortical adenylate cyclase by a GTP-dependent process, Mol. Pharmacol. 18: 598 (1980).
18. M. Rodbell, The role of hormone receptors and GTP-regulatory

proteins in membrane transduction, Nature 284: 17 (1980).

19. J. Wolff, C. Londos, and D.M.F. Cooper, Adenosine receptors and the regulation of adenylate cyclase, in: Advances in Cyclic Nucleotide Research, vol. 14, J.E. Dumont, P. Greengard, and G.A. Robison, eds., Raven Press, New York, p. 199 (1981).

20. R.R. Goodman, M.J. Cooper, M. Gavish, and S.H. Snyder, Guanine nucleotide and cation regulation of the binding of (^{3}H)cyclohexyladenosine and (^{3}H)diethylphenylxanthine to adenosine A_1 receptors in brain membranes, Mol. Pharmacol. 21: 329 (1982).

21. R.F. Bruns, J.W. Daly, and S.H. Snyder, Adenosine receptor binding: structure-activity analysis generates extremely potent xanthine antagonists, Proc. Natl. Acad. Sci. USA 80: 2077 (1983).

22. J.W. Daly, Adenosine receptors: characterization with radioactive ligands, in: Physiology and Pharmacology of Adenosine Derivatives, J.W. Daly, J.W. Phillis, Y. Kuroda, H. Shimizu and M. Ui, eds., Raven Press, New York, p. 59 (1983).

23. G. Marone, S. Poto, R. Petracca, M. Triggiani, E. de Lutio di Castelguidone, and M. Condorelli, Activation of human basophils by staphylococcal protein A. I. The role of cyclic AMP, arachidonic acid metabolites, microtubules and microfilaments, Clin. Exp. Immunol. 50: 661 (1982).

24. G. Marone, S. Hammarström, and L.M. Lichtenstein, An inhibitor of lipoxygenase inhibits histamine release from human basophils, Clin. Immunol. Immunopathol. 17: 117 (1980).

25. G. Marone, S.R. Findlay, and L.M. Lichtenstein, Modulation of histamine release from human basophils in vitro by physiological concentrations of zinc, J. Pharmacol. Exp. Ther. 217: 292 (1981).

26. G. Marone, The role of basophils and mast cells in the pathogenesis of pulmonary diseases, Int. Archs. Allergy appl. Immunol. 76 (Suppl. 1): 70 (1985).

27. G. Marone, A. Kagey-Sobotka, and L.M. Lichtenstein, IgE-mediated histamine release from human basophils: differences between antigen E- and anti-IgE-induced secretion, Int. Archs. Allergy appl. Immunol. 65: 339 (1981).

28. G. Marone, G. Ambrosio, D. Bonaduce, A. Genovese, M. Triggiani, and M. Condorelli, Inhibition of IgE-mediated histamine release from human basophils and mast cells by fenoterol, Int. Archs. Allergy appl. Immunol. 74: 356 (1984).

29. G. Marone, M. Columbo, L. Soppelsa, and M. Condorelli, The mechanism of basophil histamine release induced by pepstatin A, J. Immunol. 133: 1542 (1984).

30. E.S. Schulman, D.W. MacGlashan, Jr., S.P. Peters, R.P. Schleimer, H.H. Newball, and L.M. Lichtenstein, Human lung mast cells: purification and characterization, J. Immunol. 129: 2662 (1982).

31. D.W. MacGlashan, Jr., R.P. Schleimer, S.P. Peters, E.S. Schulman, G.K. Adams III, H.H. Newball, and L.M. Lichtenstein, Generation of leukotrienes by purified human lung mast cells, J. Clin. Invest. 70: 747 (1982).

32. S. Hammarström, Leukotrienes, in: Advances in Clinical Immunology. The Role of Chemical Mediators in Pulmonary and Cardiac Diseases, M. Condorelli, G. Marone, and L.M. Lichtenstein, eds., O.I.C. Medical Press, Florence, p. 77 (1984).

33. G. Marone, R. Petracca, and S. Vigorita, Adenosine receptors on human inflammatory cells, Int. Archs. Allergy appl. Immunol. 77: 259 (1985).

34. G. Marone, S. Vigorita, C. Antonelli, G. Torella, A. Genovese, and M. Condorelli, Evidence for an adenosine A_2/R_a receptor on human basophils, Life Sci. 36: 339 (1985).

35. R.P. Schleimer, D.W. MacGlashan, Jr., S.P. Peters, K.A. Knauer, E.S. Schulman, G.K. Adams III, A. Kagey-Sobotka, and L.M.

Lichtenstein, In vivo and in vitro studies of human anaphylaxis, in: Advances in Clinical Immunology. The Role of Chemical Mediators in Pulmonary and Cardiac Diseases, M. Condorelli, G. Marone, and L.M. Lichtenstein, eds., O.I.C. Medical Press, Florence, p. 65 (1984).

36. B.B Fredholm, Release of adenosine from rat lung by antigen and compound 48/80, Acta Physiol. Scand. 111: 507 (1981).

37. D.L. Marquardt, H.E. Gruber, and S.I. Wasserman, Adenosine release from stimulated mast cells, Proc. Natl. Acad. Sci. USA 81: 6192 (1984).

38. G. Marone, M. Plaut, and L.M. Lichtenstein, The role of adenosine in the control of immune function, Ric. Clin. Lab. 10: 303 (1980).

39. K.M.M. Murphy, and S.H. Snyder, Heterogeneity of adenosine A_1 receptor binding in brain tissue, Mol. Pharmacol. 22: 250 (1982).

40. R.M. Berne, The role of adenosine in the regulation of coronary blood flow, Circ. Res. 47: 807 (1980).

41. A. Sollevi, J. Ostergren, P. Hjemdahl, B.B. Fredholm, and B. Fagrell, The effect of dipyridamole on plasma adenosine levels and skin micro-circulation in man, in: Purine Metabolism in Man -IV- Part A: Clinical and Therapeutic Aspects; Regulatory Mechanisms, C.H.M.M. De Bruijn, H.A. Simmonds, and M.M. Müller, eds., Plenum Press, New York, p. 547 (1984).

42. B.S. Mitchell, and W.N. Kelley, Purinogenic immunodeficiency diseases: clinical features and molecular mechanisms, Ann. Int. Med. 92: 826 (1980).

43. R. Mandler, R.E. Birch, S.H. Polmar, G.M. Kammer, and S.A. Rudolph, Abnormal adenosine-induced immunosuppression and cAMP metabolism in T lymphocytes of patients with systemic lupus erythematosus, Proc. Natl. Acad. Sci. USA 79: 7542 (1982).

44. M.J. Cushley, A.E. Tattersfield, and S.T. Holgate, Inhaled adenosine and guanosine on airway resistance in normal and asthmatic subjects, Br. J. Clin. Pharmacol. 15: 161 (1983).

HYPOXANTHINE UPTAKE BY ISOLATED BRAIN MICROVESSELS

P.Cardelli-Cangiano, A.Fiori, A.Giacomello, R.Strom,
and C.Salerno

Department of Human Biopathology and Institute of Biological
Chemistry, University of Rome, and C.N.R. Centre for
Molecular Biology, Rome, Italy

INTRODUCTION

Brain is among the most active tissues in nucleotide and nucleic acid synthesis (1) but the role of the blood brain barrier in regulating the transport of purine and pyrimidine has not, so far, well defined. The existence of a presynaptic purinergic modulation of trasmitter release in the central nervous system (1) emphasize a possibly important role of purine transport at the blood-brain barrier. Previous reports (2) have demonstrated the existence, on the luminal side of the blood-brain barrier, of two saturable purine transport systems, one is specific for adenine and hypoxanthine transport and the other one accounts for nucleoside transport. Recently Betz (3) has shown in isolated brain microvessels the existence of xanthine oxidase activity, as well as the capability of the isolated vessels to take up hypoxanthine which is subsequently degraded to xanthine and uric acid or incorporated in other compounds by energy dependent processes. The aim of this report was to establish whether, as in other tissues, in isolated brain microvessels, which are an in vitro model of the blood brain barrier, hypoxanthine transport and metabolism is regulated by the levels of inorganic phosphate and/or phosphate esters.

MATERIALS AND METHODS

Chemicals: ^{14}C- and ^{3}H-hypoxanthine, ^{14}C-sucrose, and Aquasol-2^R were obtained from New England Nuclear Corp. (Boston, MA, USA). N-Hydroxyethyl-piperazine-N-2-ethanesulfonic acid (HEPES) was from Sigma Chemical Corp. (St.Louis, MO, USA); carboxyfluorescein diacetate was obtained from Molecular Probes (Junction City, OR, USA). All other products were from Merck (Darmstadt, FRG) or from Fluka (Buchs, Switzerland).

Preparation of isolated brain microvessels: microvessels were isolated from the gray matter of fresh bovine brain essentially as previously described (4,5). Briefly, the gray matter was homogenized by hand in a buffer consisting of 122 mM NaCl, 15 mM NaHCO$_3$, 10 mM glucose, 3 mM KCl, 1.4 mM CaCl$_2$, 1.2 mM MgSO$_4$, 0.4 mM K$_2$HPO$_4$, pH 7.4 and equilibrated with 95% O$_2$ plus 5% CO$_2$. The homogenate was poured on a nylon sieve (86 um pore nylon sieve) and washed with a spray of cold buffer. The material retained on the sieve was re-homogenized and washed again with cold buffer. The isolated microvessels were then resuspended in the appropriate buffer and kept until use at 0°C. After the isolation step the microvessels were

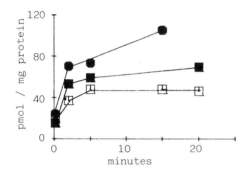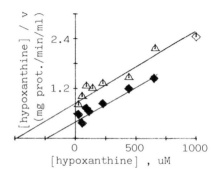

Fig. 1. (left) Time course of ^{14}C hypoxanthine uptake by microvessels suspended in media containing different phosphate concentrations. Immediately after the isolation step, the microvessels were washed and resuspended in buffers containing phosphate (⊡, none; ■, 1 mM; ●, 10 mM). After 10 min incubation at 37°C, 16 uM ^{14}C hypoxanthine was added to the capillary suspensions and the uptake of radioactive material by microvessels was followed within the first 20 min. Each point represents the mean of three different determinations.

Fig. 2. (right) Hanes plots of initial velocity of hypoxanthine uptake by isolated brain microvessels suspended in either phosphate-free medium (△) or 10 mM phosphate (◆). Other experimental conditions were as reported in fig. 1.

resuspended in HEPES-free buffers containing different concentrations of inorganic phosphate ranging from 0 to 10 mM. The osmolarity of the buffers were maintained by modifying the concentration of $NaHCO_3$. Phase contrast light microscopy and scanning electron microscopy showed that the isolated brain microvessels were free from nerve or glial cell contamination. The preparations were found not only to be impermeable to trypan blue and to ^{14}C sucrose but, after preloading with carboxyfluorescein diacetate, the efflux of carboxyfluorescein did not exceed 20% within the first 30 min at 37°C. The ATP content under standard conditions was around 500 pmoles/mg protein. Our preparations were found to be enriched with respect to the gray matter of some enzymatic activity such as γ-glutamyltranspeptidase and alkaline phosphatase (4,5).

Enzyme assay: for the assay of enzymatic activity the microvessels were homogenized in appropriate buffers. Hypoxanthine-guanine phosphoribosyltransferase, alkaline phosphatase, and γ-glutamyltranspetidase were assayed as previously described (4-6).

Uptake experiment: the uptake of radiolabeled hypoxanthine was performed essentially as described for radiolabeled amino acids in previous reports (4,5).

Chromatographic analysis: the supernatant from trichloroacetic acid-precipitated microvessels (after suitable preincubation with radiolabeled hypoxanthine) was subjected to paper chromatography (6).

RESULTS

Enzymatic activities of the microvessel preparations: brain microvessels, as compared to gray matter, are notably enriched in some enzyme activities such as γ-glutamyltranspeptidase, alkaline phosphatase and xanthine oxidase. In our preparations we also found a detectable activity of hypoxanthine-guanine phosphoribosyltransferase.

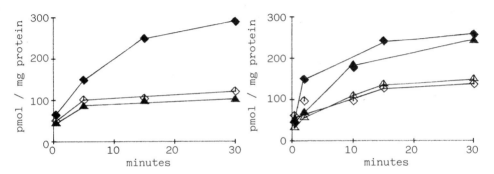

Fig. 3. (left) Effect of phloretine on hypoxanthine uptake by isolated microvessels. Phloretine concentration was: ◇◆ , none; ▲ , 1 mM. Phosphate concentration was: ◇ , none; ◆▲ ; 10 mM. ^{14}C Hypoxanthine concentration was 32 uM. Other conditions were as in fig. 1.

Fig. 4. (right) Effect of 2,4-dinitrophenol on hypoxanthine uptake by isolated microvessels. 2,4-Dinitrophenol concentration was: ◇◆ ,
none; △▲ , 100 uM. Phosphate concentration was: ◇△ , none; ◆▲ ,
10 mM. All other conditions were as in fig. 3.

Regulation of hypoxanthine uptake: upon incubation at 37°C, radiolabeled
hypoxanthine was rapidly taken up by the isolated microvessels, this uptake
being enhanced by increasing phosphate ion concentration in the incubation
medium (fig. 1). Preincubation of the microvessels in the absence of phosphate ions reduced but did not abolished the amount of radioactivity within
the cells. Analysis of the kinetic parameters showed that the initial rate
of uptake was a hyperbolic function of external hypoxanthine concentration.
The presence of inorganic phosphate decreased the apparent Km without affecting the apparent Vmax of hypoxanthine uptake (fig. 2, table I). The
phosphate dependent component of hypoxanthine uptake was inhibited by addition of phloretine in the incubation medium (fig. 3) while addition of the
uncoupler agent 2,4-dinitrophenol was without any effect on the final
levels (may be with just a slight slowering) of both the phosphate-dependent
and the phosphate-independent components (fig. 4). The absence of sodium
ions, as well as the substitution of glucose with sucrose in the phosphate-
containing buffer, reduced the rate of hypoxanthine uptake.

Intracellular metabolism of hypoxanthine: as previously described for
other cells, the transport of hypoxanthine was strictly linked to its intracellular metabolism. The use of chromatography allowed us to distinguish
between the different metabolic compounds formed. In phosphate-containing
medium after 30 min incubation in the presence of labeled hypoxanthine the
radioactivity was mainly recovered as IMP. When microvessels were exposed

Table I. Kinetic parameters of hypoxanthine transport
by isolated microvessels in the presence or
in the absence of phosphate ions.

[phosphate ions] (mM)	Vmax (pmol/mg protein/min)	Km (uM)
0	660 ± 10	475 ± 11
10	633 ± 15	195 ± 16

to phloretine in the same experimental conditions, IMP synthesis was prevented with the formation of other unidentified compounds, presumably (3) xanthine and uric acid. The same was true in the absence of glucose. Upon deprival of phosphate ions, instead, the IMP synthesis at 37°C dropped from approximately 100 pmols/30 min/mg protein (out of 200 pmols of hypoxanthine taken up by the microvessels) to around 20 pmols/30 min/mg protein (out of 105). Addition of 2,4-dinitrophenol decreased the conversion of hypoxanthine into IMP.

DISCUSSION

This study supports the idea that brain capillaries contain hypoxanthine guanine phosphoribosyltransferase in addition to xanthine oxidase (3). Extracellular hypoxanthine can be transported into the cells by a system which is modulated by the concentration of phosphate ions. The phosphate-dependent component of transport appears to be both sodium-dependent and glucose-requiring. Intracellular hypoxanthine is rapidly converted, most likely via the phosphoribosyltransferase pathway, to the corresponding nucleotide. In the absence of glucose, or upon addition of phloretine which prevents glucose entry, as well as in the absence of sodium ions, less hypoxanthine is taken up by the microvessels, and it is no more transformed into IMP, but converted into final catabolites. No effect is instead caused, either on the final intracellular level of hypoxanthine or on its transformation into IMP, by the addition of the uncoupler agent 2,4-dinitrophenol. The absence of phosphate ions, which considerably reduces the uptake of hypoxanthine, does not prevent its partial transformation into the corresponding nucleotide, thus suggesting that even under such conditions the phosphoribosyltransferase pathway could be actively working.

REFERENCES

1. D.M. Paton, Presynaptic neuromodulation mediated by purinergic receptors, in: "Purinergic Receptors", G. Burnstock, ed., Chapman and Hall, London (1981).
2. E.M. Cornford and W.H. Pardridge, Independent blood-brain barrier transport system for nuclei acid precursors, Biochim. Biophys. Acta 394:211 (1975).
3. A.C. Betz, Identification of hypoxanthine transport and xanthine oxidase activity in brain capillaries, J. Neurochem. 44:574 (1985).
4. C. Cangiano, P. Cardelli-Cangiano, J.H. James, F. Rossi Fanelli, M.A. Patrizi, K.A. Brackett, R. Strom, and J.E. Fischer, Brain microvessels take up neutral amino acids in exchange for glutamine, J. Biol. Chem. 258:8949 (1983).
5. P. Cardelli-Cangiano, C. Cangiano, J.H. James, F. Ceci, J.E. Fischer, and R. Strom, Effect of ammonia on amino acid uptake by brain microvessels, J. Biol. Chem. 259:5295 (1984).
6. A. Giacomello and C. Salerno, Human hypoxanthine guanine phosphoribosyltransferase: steady-state kinetics of the forward and reverse reactions, J. Biol. Chem. 253:6038 (1978).

MORPHINE: SITES OF ACTION IN GUANOSINE NUCLEOSIDE PATHWAY

Major L. Cohn, Faye A. Eggerding, Antonio F. Machado,
Stephan J. Cohn, Bart D. Waxman and Timothy Delaney

Drew/UCLA School of Medicine
Department of Anesthesiology Research
Los Angeles, CA USA

INTRODUCTION

In 1973, based on evidence yielded by morphine binding studies, Pert
and Snyder first proposed that the pharmacologic actions of morphine are
effected through a reversible binding of the opiate to specific receptors
in the central nervous system (CNS). Subsequently, investigators reported
several opiate receptor types (Martin et al, 1976; Lord et al, 1977); more
recently, evidence for opiate receptor subtypes has also been presented
(Pasternak, 1984). Despite these substantial advancements, however, surprisingly little is known concerning the cascade of biochemical events
triggered by the binding of opiate alkaloids to receptor sites in the CNS
(Miller, 1984). Most studies intended to elucidate the biochemical actions
of opiates are primarily concerned with the effects of these compounds on
the activities of the various enzymes involved in adenylate nucleoside
metabolism (Palmer, 1983). The numerous contradictory reports in this line
of research have, however, clearly demonstrated the futility of attempts
to correlate opiate activity to brain levels of the adenylate cyclic nucleotide (Wollemann, 1981; Palmer, 1983). Similarly, there is little con-
sensus regarding the effects of opiates on guanylate cyclic nucleotide
concentrations, a subject on which comparatively few investigations have
thus far been conducted (Palmer, 1983).

Our interest in the guanylate compounds derives from our behavioral
finding that, administered intracerebroventricularly into rats, the dibutyryl analog of guanosine 3':5'-cyclic monophosphate (dibutyryl cGMP)
is a potent analgetic agent which does not elicit the adverse effects
associated with morphine and related compounds (Cohn et al, 1978). The
analgetic properties of dibutyryl cGMP strongly implicate endogenous cGMP
and/or one or more of its metabolic products as key factor(s) in the
central regulation of analgesia. In the present investigation, incubation
studies of human brain slices were performed in an attempt to determine
the effects of morphine on cGMP metabolism.

METHODS AND MATERIALS

Large sections of specific brain regions (frontal, temporal and
occipital cortex) were sliced on a McIlwain Tissue Chopper set at 260 um.
Wet weights of the slices were recorded. Slices were then placed in an

IL237 Tonometer and preincubated for 20 minutes either without (control) or with morphine (1.5×10^{-2} mM). Control and treated incubations were conducted simultaneously in three separate tonometers. Treated incubations were performed with the addition of K^+ (50 mM) as this level of K^+ has previously been demonstrated to enhance the effects of morphine on cyclic GMP metabolism in brain tissue (Cohn et al, 1984). Throughout the procedure, the tonometers permitted the maintenance of constant environmental conditions at $37^\circ C$ with a flow of 20% O_2, 5% CO_2 and balance N_2 at a rate of 500 ml/min. The incubations were carried out in a total of six ml of Krebs-Ringer bicarbonate/glucose (KRBG) buffer at pH 7.35 (Partington et al, 1980). One control sample was withdrawn immediately prior to adding the substrate, cGMP (1.5×10^{-1} mM). Samples of 0.5 ml of incubation mixture were withdrawn 1, 2, 5, 10, 15, 30, and 60 min after addition of substrate; straightway after each sampling, an aliquot of 0.5 ml of KRBG buffer without (control) or with morphine and K^+ was added as replacement. After proteinaceous material was removed by ultrafiltration (0.45 u filter), the samples were quick-frozen until assayed.

Our high performance liquid chromatographic (HPLC) instrumentation consisted of: a liquid chromatograph with Auto-Sampler, temperature control oven, diode array detector, and microcomputer; a DPU Multichannel Integrator directly interfaced with a second microcomputer which is used for storage and analysis of data; and an Ultrasphere ODS Column (4.6 mm x 15 cm) with reversed-phase, 5 u C-18 packing. Standard compounds, diluted with ultrapure water to concentrations suitable for absorbance measurements (3.1-5.8×10^{-2} mM), were subsequently filtered with a Millipore Sample Clarification Kit. All elution solvents were filtered with a Millipore Solvent Clarification Kit and allowed to equilibrate through the column for a minimum of one hour prior to sample injection. The mobile phase was comprised of 93.5% KH_2PO_4 buffer (0.02 M) at pH 3.7 and 6.5% MeOH delivered through the column at a flow rate of 1.5 ml/min. Chromatographic analyses were carried out at $41^\circ C$ with external standards consisting of the appropriate purine nucleotides, nucleosides and bases whose retention times (Fig. 1) served for the initial identification of unknown peaks representing compounds eluted from incubated brain slice samples. Peak identification was confirmed by data obtained with the built-in photodiode array detector which has the capability of measuring spectral absorbance of each compound over the entire range of wavelengths from 190 to 600 nm, simultaneously. Peak purity was determined by comparing spectral analyses from the upslope, apex and downslope of each peak.

RESULTS

Analysis of brain slices incubated with cGMP as substrate resulted in peaks representing 5'-GMP, guanine, xanthine, cGMP, inosine, and guanosine appearing successively at retention times consistent with those of external standards (Fig. 1). In addition, our analytical evidence confirmed the sequence and demonstrated the rate of cGMP metabolism in the control incubation (Fig. 2a). The conversion of substrate cGMP to its mandatory metabolic product, 5'-GMP, was completed within 15 min of incubation; next, 5'-GMP was degraded to guanosine which gradually accumulated throughout the 60 min incubation. Guanosine was subsequently converted to guanine which was detected only in minute amounts; this observation suggests a rapid deamination of the latter compound to xanthine.

In the presence of the depolarizing agent, K^+ (50 mM), morphine, added to the incubation mixture, significantly decreased the rate of cGMP degradation and increased the latent accumulation of guanosine by 50-100% above control values (Fig. 2b). Additionally, the consistent appearance of chromatographic peaks representing inosine was not altered by the

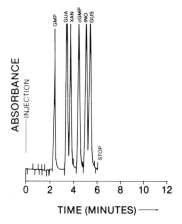

Fig. 1   External Standards for HPLC analysis of cGMP metabolism. Shown are the retention times of guanosine monophosphate (5'-GMP), guanine (GUA), xanthine (XAN), cyclic GMP (cGMP), inosine (INO), and guanosine (GUS) all separated in less than six min with an ODS reversed-phase column.

presence of morphine; uric acid was not detected in samples from either control or treated brain slice incubations.

DISCUSSION

High performance liquid chromatography is currently recognized as a reliable technique for separating, identifying and quantifying purine compounds extracted from biological tissues. In the present study, spectral analysis allowed the ascertainment of purity and confirmed the initial identification of peaks. Quantification of identified eluants was accomplished via automatic integration of peaks representing unknowns with peaks representing standards.

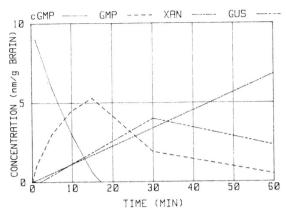

Fig. 2a Graph representing cGMP metabolism in control
incubations of human brain slice. Substrate cGMP
is converted to 5'-GMP in 15 min. Note the marked
accumulation of guanosine. Similar data were obtained from each brain region tested. The appearance of inosine and guanine is not depicted here.

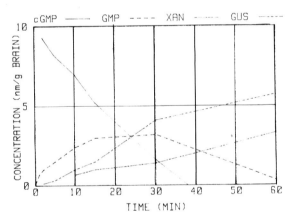

Fig. 2b Graph representing cGMP metabolism in human brain
slice incubated with morphine (1.5×10^{-2} mM) and K^+
(50 mM). Here, substrate cGMP hydrolysis is
markedly delayed. Guanosine accumulates significantly above control values throughout the incubation. This accumulation is most significant
after 30 min of incubation ($p < 0.05$).

    In treated brain slice incubations, brain tissue was pre-incubated with
morphine for 20 min in order to allow for the opiate's slow onset of action
in biological systems. The delayed conversion of cGMP to 5'-GMP in the

presence of morphine strongly suggests that the opiate inhibits cGMP phosphodiesterase activity. In light of the fact that this enzyme is calcium-calmodulin dependent, our finding is clearly supported by reports that acute injections of morphine result in depleted brain calcium levels (Cardenas and Ross, 1975). It is worthy of note that our failure to detect uric acid in samples of incubated brain slices is in accordance with the fact that the enzyme, xanthine oxidase, has never been reported in mammalian brain.

In control incubations, the gradual rise of guanosine was probably due to the fact that nucleoside phosphorylase, which catalyzes the conversion of guanosine to guanine, is rate limiting; however, in the presence of morphine, guanosine accumulation was greatly delayed but significantly enhanced over control values. Therefore, the evidence strongly indicates that, in our system, activities of both cGMP phosphodiesterase and 5'-nucleotidase were inhibited by morphine. Though we have not yet clarified the mechanisms involved, this dual inhibition would account for both the latency and increase in guanosine accumulation.

Throughout the course of control incubations performed with both human and rat brain slices, a consistent appearance of inosine was noted. This observation prompted rat brain slice incubation studies using the various metabolites of cGMP as substrates. Samples from incubations performed with guanosine, guanine or xanthine as substrate yielded no chromatographic evidence of inosine; however, like that of its precursor, cGMP, the metabolism of 5'-GMP resulted in the appearance of inosine peaks on all chromatograms. This phenomenon was unaltered by the addition of morphine to the incubation mixture (Cohn et al, 1983). The analytical data which resulted from these studies support our prior report of the presence, in rat brain, of 5'-GMP reductase (Moreno et al, 1982), an enzyme which had previously been identified in human erythrocytes and calf thymus (Spector et al, 1979) as well as, more recently, in human thymocytes (Barankiewicz and Cohen, 1984).

The strong evidence of a direct pathway through which 5'-GMP is converted to 5'-IMP may provide a valuable clue into some aspects of the biochemical actions of morphine. It may be postulated that the relationship between the purine metabolic pathways is the mechanism by which morphine elicits dependence and tolerance. This hypothesis is based on the fact that inosine can be readily salvaged to cAMP which is known to intensify and accelerate opiate-induced dependence and tolerance in vivo (Ho et al, 1972; Collier and Roy, 1974).

This study was supported by NIH (MBS Program) Grant RR-08140.

REFERENCES

Barankiewicz, J., and Cohen, A., 1984, Nucleotide catabolism and nucleoside cycles in human thymocytes. Role of orthophosphate, Biochem. J., 219:197-203.
Cardenas, H. L., and Ross, D. H., 1975, Morphine induced calcium depletion in discrete regions of rat brain, J. Neurochem., 24:487-493.
Cohn, M. L., Cohn, M., and Taylor, F. H., 1978, Cyclic GMP: a central nervous system regulator of analgesia, Science, 199:319-322.
Cohn, M. L., Cohn, M., Larrinaga, J., Wooten, D. J., Samora, J., and Fernandez, G., 1983, Morphine: a new site of action in purine nucleoside pathways, Society for Neuroscience, 9:89.
Cohn, M. L., Yakel, J., Samora, J., Larrinaga, J., Fernandez, G., and Wooten, D. J., 1984, Morphine and calcium channel blockers: effects on cGMP metabolism in rat brain, Society for Neuroscience, 10:592.
Collier, H. O. J., and Roy, A. C., 1974, Hypothesis: inhibition of E prostaglandin-sensitive adenyl cyclase as the mechanism of

morphine analgesia, Prostaglandins, 7:361-373.

Ho, I. K., Loh, H. H., and Way, E. L., 1972, Effects of cyclic AMP on morphine analgesia, tolerance and physical dependence, Nature, 238:397-398.

Lord, J. A., Waterfield, A. A., Hughes, J., and Kosterlitz, H. W., 1977, Endogenous opioid peptides: multiple agonists and receptors, Nature, 267:495-499.

Martin, W. R., Eades, C. G., Thompson, J. A., Huppler, R. E., and Gilbert, P. E., 1976, The effects of morphine- and nalorphine- like drugs in the nondependent and morphine-dependent chronic spinal dog, J. Pharmacol. Exp. Ther., 197:517-532.

Miller, R., 1984, How do opiates act?, Trends NeuroSci., 7:184-185.

Moreno, J., Cohn, M., Cohn, M. L., 1982, GMP reductase activity in rat brain slices, Minority Bioscience Symposium, 10:161.

Palmer, G. C., 1983, Effects of psychoactive drugs on cyclic nucleotides in the central nervous system, in: "Progress in Neurobiology," G. A. Kerkut and J. W. Phillis, ed., 21:1-133.

Partington, C. R., Edwards, M. W., and Daly, J. W., 1980, Regulation of cyclic AMP formation in brain tissue by alpha-adrenergic receptors: requisite intermediacy of prostaglandins of the E series, Proc. Natl. Acad. Sci. USA, 77:3024-3028.

Pasternak, G., 1984, The contribution of multiple-opiate receptors to spinal and supraspinal pain control, IVth World Congress on Pain, Seattle, Washington, August 31-September 5.

Pert, C. B., and Snyder, S. H., 1973, Opiate receptor: demonstration in nervous tissue, Science, 179:1011-1014.

Spector, T., Jones, T. E., and Miller, R. L., 1979, Reaction mechanism and specificity of human GMP reductase. Substrates, inhibitors, activators, and inactivators, J. Biol. Chem., 254:2308-2315.

Wollemann, M., 1981, Endogenous opioids and cyclic AMP, Prog. Neurobiol., 16:145-154.

PURINE AND MONOAMINE METABOLITES IN CEREBROSPINAL FLUID

N. Lawrence Edwards [†], Faye S. Silverstein*, and
Michael V. Johnston*

[†]Department of Medicine, University of Florida
and the VAMC, Gainesville, Florida, USA

*Department of Pediatrics and Neurology, University
of Michigan, Ann Arbor, Michigan, USA

INTRODUCTION

The study of human cerebral metabolism is hampered by a paucity of techniques that allow serial observations. Determination of cerebrospinal
fluid (CSF) metabolites has been a technique used to diagnose neurologic
disease states, to assess severity and progression of disease and to judge
the efficacy of therapeutic interventions. The measurement of CSF metabolites of the serotonin and dopamine pathways is felt to accurately reflect
aberrations in those CNS processes. We have recently demonstrated reduced
CSF HVA levels in Lesch-Nyhan syndrome (1). This observation parallels the
post-mortem studies by Lloyd et al (2) where reduced dopamine and HVA levels
were found in Lesch-Nyhan striatum.

Alterations in CSF oxypurine levels have been reported in childhood
hypoxia, acute and chronic ischemic brain disease of adults, memory loss,
depression, and the Lesch-Nyhan syndrome. In adult ischemic disease the
levels of CSF oxypurines correlated well with the extent of injury to
brain tissues (3). Finally, a highly significant correlation between
CSF purines (either hypoxanthine or xanthine) and CSF monoamines (either
HVA or HIAA) in depressed patients suggested a direct modulation of monoamine neurotransmitter release by purines (4).

We sought to determine if measurement of CSF purines or monoamine
neurotransmitters would be useful in assessing neurologic dysfunction in
the Lesch-Nyhan syndrome or aid our understanding of the mechanisms of
neural involvement.

METHODS AND RESULTS

Four boys with the Lesch-Nyhan syndrome were observed intermittently
in the Clinical Research Center at the University of Michigan over a 5
year period. During that time a total of 26 CSF samples were obtained from
these patients. Control CSF samples were obtained from neonates and age-
matched subjects (ages 0.3-12.0 years) who were being routinely evaluated

for CNS abnormalities but found to be normal. HVA, HIAA and oxypurines were measured by HPLC. The method of monoamine determination has been previously published (1). Hypoxanthine and xanthine were measured by reverse-phase chromatography using 0.02 M potassium phosphate buffer, pH 6.0.

Relationships between CSF purines (hypoxanthine and xanthine) and neurotransmitter (HVA and HIAA) were looked for in the neonates, age-matched controls, and the Lesch-Nyhan patients using Pearson's coefficient of correlation (linear). The data were also examined for hyperbolic and other non-linear relations but none was found.

Table 1 shows the statistical results (r values) of the data analyzed. Positive "r" values indicate a direct relationship while negative values indicate an inverse relation.

TABLE 1. RELATIONSHIPS BETWEEN CSF PURINES AND MONOAMINE NEUROTRANSMITTERS (PEARSON'S CORRELATION COEFFICIENT)

			Lesch-Nyhan		
Relation	Neonate	Age-Matched Control	Off Allop.	On Allop.	Combined
	n = 14	n = 12	n = 7	n = 19	n = 26
Purine-Purine					
Hypoxanth vs Xanth	0.42	0.62	0.08	0.65	0.57
Monoamine-Monoamine					
HIAA vs HVA	0.94	0.80	0.99	0.95	0.91
Monoamine-Purine					
HIAA vs Hypoxanth	0.58	0.03	-0.41	-0.08	-0.38
HVA vs Hypoxanth	0.58	0.06	-0.39	-0.09	-0.34
HIAA vs Xanth	0.30	0.39	-0.86	0.12	-0.26
HVA vs Xanth	0.30	0.27	-0.81	0.26	-0.17

Good correlations ($r \geq 0.80$) are observed between CSF HIAA and HVA in the neonates, age-matched controls and Lesch-Nyhan patient. The control and Lesch-Nyhan data are also shown in Figure 1. Good correlations (inverse) are also observed in Lesch-Nyhan patients (off allopurinol) between xanthine and both HVA and HIAA. This correlation is not seen in the Lesch-Nyhan patients taking allopurinol or in the controls (Table 1).

Fair correlation (0.50-0.80) is found between CSF hypoxanthine and xanthine in the controls (r = 0.62) and the Lesch-Nyhan subjects taking allopurinol (r = 0.65). When CSF purine-purine, neurotransmitter-neurotransmitter, and neurotransmitter-purine relationships were examined, the only other fair correlation observed was in the neonates between hypoxanthine and either HVA or HIAA.

We have recently published our observations of an inverse relationship between age and both HVA and HIAA in controls and Lesch-Nyhan subjects (1). When comparing age and CSF hypoxanthine, no correlation is observed for either controls or Lesch-Nyhan patients (Table 2). CSF xanthine on the other hand correlates inversely with age in controls (r = -0.56) and directly (r = 0.78) in Lesch-Nyhan not taking allopurinol.

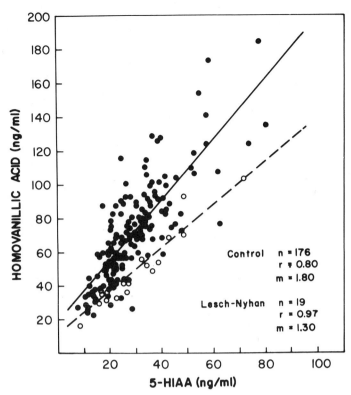

Figure 1. A close linear correlation exists between spinal fluid HVA and
5-HIAA levels in both the control subjects (n = 176) and the Lesch-Nyhan
patients (19 determinations in 4 subjects). There is a marked difference
in the slope of the correlation generated by the control data when compared
to the Lesch-Nyhan slope suggesting a relative dysfunction of the dopaminergic (HVA) pathway. The control data had a correlation coefficient
(Pearson's) of 0.80 and slope of 1.8. The Lesch-Nyhan data produced a
correlation coefficient of 0.91 and a slope of 1.3.

DISCUSSION

Serial determinations of spinal fluid hypoxanthine, xanthine, homo-
vamillic acid (HVA) and 5-hydroxyindoleacetic acid (HIAA) were made in four
patients with the Lesch-Nyhan syndrome over a 5-year period. Control
spinal fluids for age matched comparison were obtained from neurologic
and non-neurologic pediatric patients. A direct linear correlation is
present between CSF concentrations of the serotonin metabolite, HIAA and
the dopamine metabolite, HVA, in both the control and Lesch-Nyhan subjects.
Linear correlation is also present between hypoxanthine and xanthine in the

controls and also in the Lesch-Nyhan patients (on allopurinol).

TABLE 2. EFFECT OF AGE ON CSF PURINES (PEARSON'S CORRELATION COEFFICIENT)

	Age-Matched Controls n = 12	Lesch-Nyhan Off Allop.	Lesch-Nyhan On Allop.
Age vs Hx	0.01	-0.09	0.34
Age vs Xanth	-0.56	0.78	-0.03
Age vs Hx:Xanth	0.83	-0.53	0.35

Previous reports in adults have demonstrated good correlation between CSF purines and CSF monoamine neurotransmitters (3). In this pediatric age group no such correlation was noted except for a significant inverse relationship between xanthine and either HIAA or HVA. This is less likely caused by modulation between the purines and neurotransmitter metabolites than it is a function of age. In the pediatric age group CSF xanthine levels decline with age whereas in the Lesch-Nyhan xanthine increases.

These findings illustrate the importance of serial determinations of CSF metabolites since indeed the normal and Lesch-Nyhan range for CSF xanthine, HVA and HIAA overlap greatly. Other unknown peaks on the HPLC profile of Lesch-Nyhan spinal fluid will need to be identified before the actual importance of this type of monitoring can be ascertained.

REFERENCES

1. F. S. Silverstein, M. V. Johnston, R. J. Hutchinson, and N. L. Edwards. Lesch-Nyhan syndrome: CSF neurotransmitter abnormalities. Neurology 35:907 (1985).
2. K. G. Lloyd, O. Hornykiewicz, L. Davidson, K. Shammak, I. Fasley, M. Shibuya, W. N. Kelley, and I. H. Fox. Biochemical evidence of dysfunction of brain neurotransmitters in the Lesch-Nyhan syndrome. N. Engl. J. Med. 305:1106 (1981).
3. R. Hallgren, F. Niklasson, A. Terent, A. Akerblom, and E. Widerlov. Oxypurines in cerebrospinal fluid as indices of distured brain metabolism. Stroke 14:382 (1983).
4. F. Niklasson, H. Agren, and R. Hallgren. Purine and monoamine metabolites in cerebrospinal fluid parallel purinergic and monoaminergic activation in depressive illness? J. Neurol., Neurosurg., Psych. 46:255 (1983).

CHARACTERISTICS OF HIGH AFFINITY AND LOW AFFINITY

ADENOSINE BINDING SITES IN HUMAN CEREBRAL CORTEX

David John and Irving H. Fox

From the Human Purine Research Center
Department of Internal Medicine and
Biological Chemistry, Clinical Research Center
University Hospital, The University of Michigan
Ann Arbor, Michigan 48109

SUMMARY

The binding characteristics of human brain cortical membrane fractions were evaluated to test the hypothesis that there are A_1 and A_2
adenosine binding sites. The ligands used were 2-chloro[8-^{3}H]adenosine
and N^6-[adenine-2,8-^{3}H]cyclohexyladenosine.

Binding of chloroadenosine to human brain cortical membranes was time
dependent, reversible and concentration dependent. The K_d calculated for
chloroadenosine by Scatchard analysis of equilibrium data was 280 nM, with
a Bmax of 1.6 pmoles/mg protein, suggesting a single class of binding
sites. The specificity of chloroadenosine binding was assessed by the
ability of adenosine analogs to compete for binding sites. Using this
approach, the apparent K_d was estimated to be 0.74 µM for 5'-N-ethyl-
carboxamideadenosine, 1 µM cyclohexyladenosine, and 13 µM for N^6-(L-2-
phenylisopropyl)adenosine. Isobutylmethylxanthine and theophylline,
receptor antagonists, had apparent K_d values of 84 µM and 105 µM, respectively. Hill slope factors ranged from 0.3 to 0.6. Chloroadenosine
binding to human brain cortical membranes approached equilibrium at 90
minutes, with a $T_{1/2}$ of 10 minutes. The k_{ob} was 0.080 min^{-1} and the k_1 was
7.5×10^4 min^{-1} M^{-1}. Reversibility of chloroadenosine binding at
equilibrium was completed at approximately 10 minutes with a k_2 value of
0.074 min^{-1}. The K_d calculated from the rate constants was 990 nM.

Cyclohexyladenosine binding was concentration dependent. The K_d calculated for cyclohexyladenosine via Scatchard analysis of equilibrium data
was 5 nM with a B_{max} of 0.35 pmoles/mg protein. Cyclohexyladenosine

binding was displaced by 3 known receptor agonists: N^6-(L-2-phenyliso propyl)adenosine (K_d 4 nM), 2-chloroadenosine (K_d 10 nM) and 5'-N-ethyl-carboxamideadenosine (K_d 6 nM). The apparent K_d values for the agonists were 1 to 3 orders of magnitude lower with this ligand as compared to radioactive chloroadenosine. Binding was also displaced by 2 known antagonists, isobutylmethylxanthine and theophylline, with apparent K_d values of 4 μM and 8 μM, respectively. Hill slope factors ranged from 0.5 to 0.8.

Our data support the existence of two adenosine binding sites in human cortex compatable with the low affinity (A_2) and high affinity (A_1) adenosine receptors.

INTRODUCTION

Adenosine analogues and xanthine derivatives have definite behavioral properties in mice (1). Parenteral adenosine analogs cause profound depression of locomotor activity in mice, while xanthine derivatives are behavioral stimulants.

The behavioral properties of adenosine analogs and xanthine derivatives have been correlated with the modification of adenylate cyclase activity and the binding properties of these compounds. Adenosine and its analogs cause a marked accumulation of cyclic AMP in brain slices from all regions (2). In contrast, no stimulation of adenylate cyclase is evident in brain membranes, with the exception of striatal membrane (3,4). These opposite properties of adenosine and its analogues on brain adenylate cyclase in slices and membrane preparations have been interpreted as compatible with two types of adenosine receptors, the high affinity A_1 receptor inhibitory for adenylate cyclase and the low affinity A_2 receptor stimulatory for adenylate cyclase (5-7). The binding properties of the A_1 receptor have been related to the behavioral properties of adenosine analogues and methylxanthine antagonists (1,2,8).

What is the relevance of these observations in mouse and rat brain to the human brain? To elucidate the adenosine binding properties of the human brain, we have performed studies to test the hypothesis that there are A_1 and A_2 adenosine binding sites in human cerebral cortex. To accomplish this goal we have used two radioactive ligands, N^6-[adenine-2,8-^{3}H]cyclohexyladenosine, which binds to A_1 sites, and 2-chloro[8-^{3}H]-adenosine, which binds to A_1 and A_2 sites (3).

METHODS

Fresh human brain cortex was obtained at necropsy from patients who had expired within the previous six hours from non-neurologic, acute disorders. The cortex was rinsed in cold physiologic saline and then was immediately homogenized and and prepared as previously described (9-11).

The binding assay was performed in triplicate in a total volume of 850 µl containing 400 µg protein, 80 nM [^{3}H]-chloroadenosine or 20 nM [^{3}H]-cyclohexyl-adenosine and incubation buffer as previously described (9-11). For experiments using radioactive chloroadenosine, incubations were performed for 40 minutes at 4°C; for those experiments utilizing radioactive cyclohexyladenosine, incubation time was 90 minutes at 37°C. Duplicate sets of incubation mixtures had 10 µM nonisotopic ligand added. For experiments utilizing radioactive chloroadenosine, this ligand was chloroadenosine and for assays with radioactive cyclohexyladenosine the ligand was N^6-(L-2-phenylisopropyl)adenosine. Specific binding was 53 to 70% of the total bound counts with radioactive chloroadenosine and was 65 to 87% of the total bound counts with radioactive cyclohexyl-adenosine. Calculations were performed as previously described (10,11). Data analysis was performed using Clinfo on Vax 11/730. Hill plots were fit to a linear regression and the p values for the correlation coefficients were estimated using the F statistic.

RESULTS

Affinity of [^{3}H]Chloroadenosine Binding Sites

There was a concentration dependent increase in chloroadenosine binding (Fig. 1). Analysis of this data by Scatchard plot was compatible with a single class of binding sites with a K_d of 284 nM (Fig. 1). This analysis gives an estimate of the concentration of binding sites of 646 fmoles per 400 µg protein in the incubation medium or 1.6 pmoles ligand bound/mg protein.

Specificity of [^{3}H]Chloroadenosine Binding

The ability of other adenosine analogs to compete for binding sites on human brain cortical membranes was studied (Table 1, Fig. 2). Agonist apparent K_d values were determined as follows: 5'-N-ethylcarboxamideadenosine, 0.74 µM; N^6-(L-2-phenylisopropyl)adenosine, 13 µM; and N^6-cyclohexyladenosine, 1 µM. The apparent K_d values for theophylline and isobutylmethylxanthine, receptor antagonists, were 84 and 107 µM, respectively.

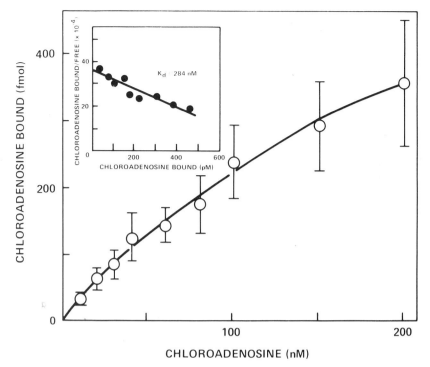

Fig. 1. Concentration Dependence of Chloroadenosine Binding To
Human Brain Membranes. There is a concentration
dependent increase in chloroadenosine binding.
Analysis of this data by Scatchard plot indicates a
single class of binding sites with a K_d of 284 nM.
This analysis gives an estimate of the concentration
of binding sites of 646 fmol for the 400 μg protein in
the incubation medium or 1.6 pmol ligand bound/mg
protein.

Hill plots had correlation coefficients ranging from 0.95 - 0.97 (p <
0.001). The Hill slope factors ranged from 0.3 to 0.6. These observations suggest either heterogeneity of receptor sites or negatively
cooperative interactions.

Binding Kinetics

    Chloroadenosine binding to human brain cortical membranes approached
equilibrium at approximately 90 minutes with a $T_{1/2}$ of 10 minutes (Fig.
3). The observed forward rate constant was 0.080 min^{-1}, giving a second
order rate constant (k_1) of 7.5 x 10^4 $min^{-1}M^{-1}$. At equilibrium, the
addition of 10 μM chloroadenosine resulted in a rapid displacement of
[^{3}H]chloroadenosine which was complete at approximately 10 minutes. There
was a $T_{1/2}$ of reversal of approximately 2 minutes (Fig. 4). The first
order rate constant (k_2) for the reversal of [^{3}H]chloroadenosine binding
was 0.074 min^{-1}. The K_d calculated from both rate constants was 987 nM.

Table 1. KINETIC CONSTANT DETERMINATIONS FROM BINDING
STUDIES OF THE ADENOSINE RECEPTOR

Compound:	Slope Factor[a]	Observed[b] IC_{50} (nM)	Apparent[c] K_d (nM)
Ligand - [^{3}H]chloroadenosine			
5 - N - ethylcarboxamideadenosine	0.3	950	740
N^6 - (L - 2 - phenylisopropyl)adenosine	0.5	16,200	12,700
N^6 - cyclohexyladenosine	0.6	1,450	1,130
Theophylline	0.4	134,900	105,400
Isobutylmethylxanthine	0.6	107,000	83,600
Ligand - [^{3}H]cyclohexyladenosine			
5 - N - ethylcarboxamideadenosine	0.6	32	6
N^6 - (L - 2 - phenylisopropyl)adenosine	0.7	20	4
2 - chloroadenosine	0.8	50	10
Theophylline	0.5	38,900	7,800
Isobutylmethylxanthine	0.5	19,500	3,900

[a]Slope factor from the Hill plot.
[b]Observed IC_{50} values were calculated from the Hill equation derived from each set of data as described in Methods.
[c]The apparent K_d values were estimated by the following method:

$$K_d = \frac{IC_{50}}{\left[1 + \frac{[ligand]}{K_d \ ligand}\right]}$$

where the ligand is chloroadenosine (80 nM) or cyclohexyladenosine (20 nM) with K_d's of 284 nM and 5 nM, respectively.

Saturability and Affinity of [^{3}H]Cyclohexyladenosine Binding Sites

Saturation of binding sites occurred with 104 fmoles ligand bound per 400 μg protein or 0.26 pmoles ligand bound/mg protein (Fig. 5). At a ligand concentration of approximately 4 nM, half the binding sites were occupied. This value represents an estimate of the dissociation constant of [^{3}H]cyclohexyadenosine for the binding sites. Subsequent Scatchard analysis of this data indicates a single class of binding sites with a K_d of 5 nM (Fig. 5). This analysis gives an estimate of the concentration of binding sites of 139 fmoles per 400 μg protein in the incubation medium or 0.35 pmoles ligand bound/mg protein.

Specificity of [3H]Cyclohexyladenosine Binding

The inhibition of [3H]cyclohexyladenosine binding to human brain cortical membranes by adenosine receptor agonists and antagonists was studied (Table 1, Fig. 6). N^6-(L-2-phenylisopropyl)adenosine was the most potent agonist with an apparent K_d of 4 nM; chloroadenosine had an

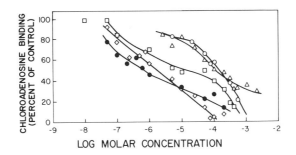

LOG MOLAR CONCENTRATION

Fig. 2. Inhibition of Chloroadenosine Binding By Adenosine
Receptor Agonists and Antagonists In Human Brain
Membranes   The inhibition of [3H]chloroadenosine
binding by adenosine receptor agonists and antagonists
was studied (● , NECA; □ PIA; ◇ , CHA; △ ,
theophylline; ○ , IBMX). N-ethylcarboxamide-
adenosine (NECA) was a more potent agonist than L-
phenylisopropyladenosine (PIA) with an IC50 o~ [1]µM.
Isobutylmethyl-xanthine (IBMX) and theophylline had
similar IC_{50} values of 107 µM and 135 µM,
respectively.

apparent K_d of 10 nM and 5'-N-ethylcarboxamideadenosine an apparent K_d of 6 nM. The two antagonists studied, isobutylmethylxanthine and theophylline, had apparent K_d values of 4 and 8 µM, respectively.

Hill plots had correlation coefficients ranging from 0.90 to 0.99 ($p < 0.002$). The Hill slope factors were below 1.0 and ranged from 0.5 to 0.8. These observations suggest either heterogeneity of receptor sites or negatively cooperative interactions.

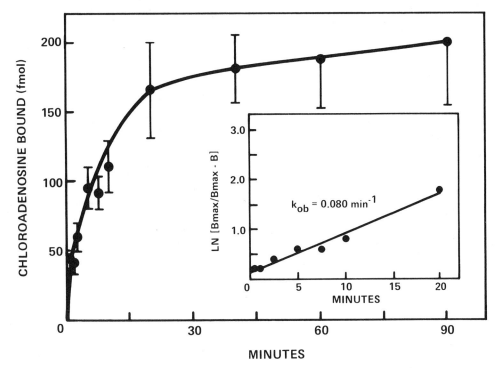

Fig. 3. Time Dependence of Chloroadenosine Binding to Human
Brain Membranes. [^{3}H]chloroadenosine binding to human
brain cortical membranes reached equilibrium at
approximately 90 minutes with a T $1/2$ of 10 minutes.
The observed forward rate constant was 0.080 min.$^{-1}$,
giving a second order rate constant (k_1) of 7.5 x 10^4
min^{-1} M^{-1}.

DISCUSSION

Two subclasses of external adenosine receptors have been defined, Ra
and Ri or A_1 and A_2 receptors (3,5-7). Both receptors are GTP-dependent
and are inhibited by methylxanthines. The Ri or A_1 receptor is a high
affinity receptor which inhibits adenylate cyclase. In contrast, the Ra
or A_2 receptor is a low affinity extracellular adenosine receptor which
activates adenylate cyclase. High affinity specific binding of adenosine
analogs has been observed in membranes from brain, fat cells, and testes
(9,11-18). Low affinity specific binding has been observed in human
placenta, platelets and rat liver (10,19,20).

Our observations suggest that [^{3}H]chloroadenosine, an adenosine
receptor agonist, identifies binding sites in human brain cortical
membranes with characteristics of an adenosine receptor. The binding is
time dependent, reversible and concentration dependent. Using chloroadenosine as ligand we identified sites which bind 5'-N-ethylcarboxamide-

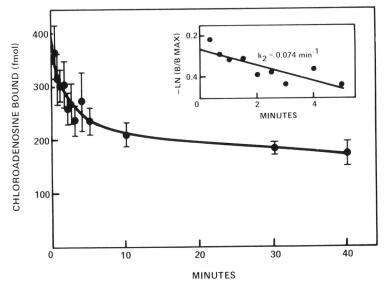

Fig. 4. Reversibility of Chloroadenosine Binding to Human
Brain Membranes. At equilibrium, the addition of 10
µM chloroadenosine resulted in a rapid displacement of
[^{3}H]chloroadenosine which was completed at
approximately 10 minutes. The vertical axis indicates
total binding. The half-life of reversibility was
approximately 2 minutes. The first order rate
constant (k_2) for the reversal of [^{3}H]chloroadenosine
binding was 0.074 min^{-1} (Inset).

adenosine with a higher affinity than N^6-L-(phenylisopropyl)adenosine.
This hierarchy of potency and the K_d values of agonists conform to the low
affinity receptor (Ra or A_2) as defined by Londos et al. (5-7). These
properties are similar to binding studies recently reported for both human
placenta and human platelet (10,19). Our estimate of the K_d of 987 nM
calculated from the rate constants was in agreement with the K_d of 284 nM
measured by Scatchard analysis. The density of binding was 1.6 pmol/mg
protein which was similar to the value for human placenta (10). Our
studies contrast with experiments in rat brain cortical microsomes
(9,12,21) and cultured mouse fetal brain cells (22). Using chloroadenosine as ligand, K_d values have ranged from 1.4 to 24 nM. Differences
in experimental conditions make absolute comparison of affinities
difficult. Densities of receptors in these latter studies ranged from 0.2
to 0.5 pmol/mg protein (3,19). The density of receptors in rat brain
using 1,3-diethyl-8-phenylxanthine as ligand was 1.3 pmol/mg protein (23).
The latter ligand should bind theoretically to both high and low affinity
adenosine binding sites. While these observations are qualitatively
useful, the quantitative estimates are only rough approximations as a
result of the binding to both receptors by these ligands.

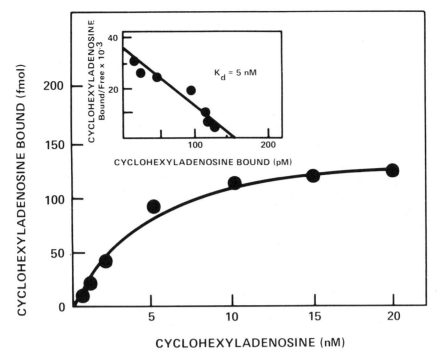

Fig. 5. Concentration Dependence of Cyclohexyladenosine
Binding To Human Brain Membranes. There is a
concentration dependent increase in chloroadenosine
binding. Saturation of binding sites with 104 fmol
ligand bound/400 µg protein or 0.26 pmol ligand
bound/mg protein. Analysis of this data by Scatchard
plot indicates a single class of binding sites with a
K_d of 5 nM. This analysis gives an estimate of the
concentration of binding sites of 139 fmol for the 400
µg protein in the incubation medium or 0.35 pmol
ligand bound/mg protein.

Utilizing [^{3}H]cyclohexyladenosine as ligand we observed evidence for
a high affinity binding site in human brain cortical membranes with a Kd
of 5 nM. Radioactive cyclohexyladenosine is known to bind primarily to
the high affinity (R_i or A_1) receptor. The hierarchy of agonist potencies
was L-phenylisopropyl-adenosine > 5'-N-ethylcarboxamideadenosine > 2-
chloroadenosine. These relative potencies and the K_d values of agonists
are consistent with the high affinity receptor (5-7). These properties
are similar to observations in brain membranes from mouse (1), rat (23-
27), guinea pig (16), rabbit (26), cattle (13,17,18,23,24,27) and human
(24) sources. The K_d of 5.0 nM for human cerebral cortex with

[³H]cyclohexyladenosine as ligand is virtually identical to a previous study (24). Density of binding sites in our study, 0.35 pmol/mg protein, is similar to the ratios observed in previous studies with this ligand (3,19), but is lower than 0.74 pmol/mg protein measured with N^6-(L-2-phenylisopropyl)-adenosine as ligand (23). The high affinity receptor has

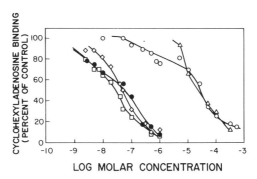

LOG MOLAR CONCENTRATION

Fig. 6.   Inhibition of Cyclohexyladenosine Binding by Adenosine Receptor Agonists and Antagonists in Human Brain Membranes. The inhibition of [³H]cyclohexyladenosine binding by adenosine receptor agonists and antagonists was studied (● , NECA; □ , PIA; ◇ , CHLOR; △ , theophylline; ○ , IBMX). L-phenylisopropyl-adenosine (L-PIA) was the most potent agonist with an IC_{50} of 20 nM, chloroadenosine (CHLOR) had an IC_{50} of 50 nM and N-ethylcarboxamide-adenosine (NECA) an IC_{50} of 32 nM. Isobutylmethylxanthine and theophylline had IC_{50} values of 4 and 8 μM, respectively (Table 1).

been localized using autoradiographic techniques and [³H]cyclohexyl-adenosine as ligand. Increased density of the high affinity receptor occurs in molecular layers of the cerebellum, molecular and polymorphic layers of the hippocampus and dentate gyrus, the medial geniculate body, certain thalamic nuclei and the lateral septum (28).

Our competitive binding experiments yielded Hill slope factors which were substantially lower than 1. Hill plots of this type cannot distinguish between heterogeneity of receptor sites or negatively cooperative interactions. The different apparent affinity constants and relative potency of binding seen with the two radioactive ligands suggest that there are two types of adenosine receptors being studied. In addition, there is evidence for two affinity states of Ri receptor in rat brain membranes related to the presence of guanine nucleotide (23). Therefore, it is possible that the Hill slope factors less than 1 represent two affinity states of both the high and low affinity adenosine binding sites in human brain.

The relationships between adenosine receptor agonist and antagonist binding, inhibition and stimulation of brain adenylate cyclase, and behavioral properties of these compounds remain to be defined. These studies of mammalian brain adenosine receptors are clearly relevant to human cerebral physiology, since our experiments provide evidence for the existence of high and low affinity adenosine binding sites in human cerebral cortex. Whether alteration of adenosine receptors occurs in human disease is not clear. Future studies of the adenosine receptor in the Lesch-Nyhan Syndrome will be important to distinguish whether an abnormality of the adenosine receptor exists (29), as suggested by studies in the rat (30).

REFERENCES

1. J. N. Crawley, J. Patel, and P. J. Marangos, Behavioral characterization of two long-lasting adenosine analogs: sedative properties and interaction with diazepam, Life Science 29:2623-2630 (1981).
2. J. W. Daly, R. F. Bruns, and S. H. Snyder, Adenosine receptors in the central nervous system: relationship to the central actions of methylxanthines, Life Science 28:2083-2097 (1981).
3. J. W. Daly, Binding of radioactive ligands to adenosine receptors in the central nervous system, in: "Regulatory Function of Adenosine," R. M. Berne, T. W. Rall, and R. Rubio, ed., Martinus Nijhoff Publishers, Boston (1983).
4 D. M. F. Cooper, C. Londos, and M. Rodbell, Adenosine receptor-mediated inhibition of rat cerebral cortical adenylate cyclase by a GTP-dependent process, Mol Pharmacol 18:598-601 (1980).
5. C. Londos and J. Wolff, Two distinct adenosine-sensitive sites on adenylate cyclase, Proc Natl Acad Sci USA 74:5482-5486 (1977).
6. C. Londos, D. M. F. Cooper, W. Schlegel, and M. Rodbell, Adenosine analogs inhibit adipocyte adenylate cyclase by a GTP-dependent process: basis for actions of adenosine and methylxanthines on cyclic AMP production and lipolysis, Proc Natl Acad Sci USA 75:5362-5366 (1978).
7. C. Londos, D. M. F. Cooper, and J. Wolff, Subclasses of external adenosine receptors, Proc Natl Acad Sci USA 77:2551-2554 (1980).
8. S. H. Snyder, J. J. Katims, Z. Annau, R. F. Bruns, and J. N. Daly, Adenosine receptors and behavioral actions of methylxanthines, Proc Natl Acad Sci USA 78:3260-3264 (1981).
9. M. Williams and E. A. Risley, Chemical characterization of putative central purinergic receptors by using 2-chloro[^{3}H]adenosine, a stable analog of adenosine, Proc Natl Acad Sci USA 77:6892-6896 (1980).
10. I. H. Fox and L. Kurpis, Binding characteristics of an adenosine receptor in human placenta, J Biol Chem 258:6952-6955 (1983).

11. D. John and I. H. Fox, Characteristics of high affinity and low affinity adenosine binding sites in human cerebral cortex, J Lab Clin Med (In Press).

12. P. H. Wu, J. W. Phillis, K. Balls, and B. Rinaldi, Specific binding of 2-[^{3}H]chloroadenosine to rat brain cortical membranes, Can J Physiol Pharmacol 58:576-579 (1980).

13. R. F. Bruns, J. W. Daly, and S. H. Snyder, Adenosine receptors in brain membranes: Binding of N^6-cyclchexyl[^{3}H]adenosine and 1,3-diethyl-8-[^{3}H]phenylxanthine, Proc Natl Acad Sci USA 77:5547-5551 (1980).

14. U. Schwabe and T. Trost, Characterization of adenosine receptors in rat brain by (-)[^{3}H]N^6-phenylisopropyladenosine, Naunyn-Schmiedeberg's Arch Pharmacol 313:179-187 (1980).

15. T. Trost and U. Schwabe, Adenosine receptors in fat cells: identification by (-)-N^6-[^{3}H]phenylisopropyladenosine binding, Mol Pharmacol 19:228-235 (1981).

16. K. M. M. Murphy and S. H. Snyder, Adenosine receptors in rat testes: labeling with ^{3}H-cyclohexyladenosine, Life Sci 28:917-920 (1981).

17. R. R. Goodman, M. J. Cooper, M. Gavish, and S. H. Snyder, Guanine nucleotide and cation regulation of the binding of [^{3}H]cyclohexyl-adenosine and [^{3}H]diethylphenylxanthine to adenosine A$_1$ receptors in brain membranes, Mol Pharmacol 21:329-335 (1982).

18. J. Patel, P. J. Marangos, J. Stivers, and F. K. Goodwin, Characterization of adenosine receptors in rat brain using N^6-cyclohexyl[^{3}H]-adenosine, Brain Res 237:203-214 (1982).

19. U. Schwabe, General aspects of binding of ligands to adenosine receptors, in: "Regulatory Function of Adenosine," R. M. Berne, T. W. Rall, and R. Rubio, eds., Martinus Nijhoff Publishers, Boston (1983).

20. Y. Lenschow, E. Hüttemann, D. Ukema, and U. Schwabe, Study of R$_a$ adenosine receptors in human platelets by radioligand binding, Naunyn - Schmiedeberg's Arch Pharmacol 321 (Suppl.):R31 (1982).

21. J. W. Phillis and P. H. Wu, Specific binding of ^{3}H-2-chloroadenosine to rat brain cortical membranes: adenosine receptors, Can J Physiol Pharmacol 58:576-579 (1980).

22. D. Van Calker, M. Muller, and B. Hamprecht, Adenosine regulates via two different types of receptors, the accumulation of cyclic AMP in cultured brain cells, J Neurochem 33:999-1005 (1979).

23. M. J. Lohse, V. Lenschow, and U. Schwabe, Two affinity states of Ri adenosine receptors in brain membranes: analysis of guanine nucleotide and temperature effect on radioligand binding, Mol Pharmacol 26:1-9 (1984).

24. K. M. M. Murphy and S. H. Snyder, Heterogeneity of adenosine A$_1$ receptor binding in brain tissue, Mol Pharmacol 22:250-257 (1982).

25. J. Prémont, M. Preez, G. Blanc, J. P. Tassin, A. Thierry, D. Hervé, and J. Bockaert, Adenosine-sensitive adenylate cyclase in rat brain homogenates: kinetic characteristics, specificity, topographical, subcellular and cellular division, Mol Pharmacol 16:790-804 (1979).

26. M. B. Anand-Srivastava and R. A. Johnson, Regulation of adenosine-sensitive adenylate cyclase from rat brain striatum, J Neurochem 35:905-914 (1980).

27. D. R. Haubrich, M. Williams, and G. G. Yarbrough, 2-chloroadenosine inhibits brain acetylcholine turnover in vivo, Can J Physiol Pharmacol 59:1196-1198 (1981).

28. R. R. Goodman and S. H. Snyder, Autoradiographic localization of adenosine receptors in rat brain using [^{3}H]cyclohexyladenosine, J Neuroscience 2:1230-1241 (1982).

29. I. J. Kopin, Neurotransmitters and the Lesch-Nyhan syndrome, New Engl J Med 305:1148 (1981).

30. R. D. Green, H. K. Poundfit, and S. M. H. Yeung, Modulation of striatal dopaminergic function by local injection of 5'-N-ethylcarboxamide-adenosine, Science 218:58-60 (1982).

ACKNOWLEDGMENTS

We wish to thank Bonnie Taylor for technical assistance, Drs. R.F. Bruns and J. Bristol of Warner Lambert/Parke Davis, Ann Arbor, Michigan, for 5'-N-ethylcarboxamideadenosine, Linda Allen for typing the manuscript, and Steve Schmaltz for assistance with CLINFO. This research was supported by a grant from the American Heart Association and the United States Public Health service grants 2-R01-AM-19674 and 5-M01-RR-00042.

HYPOXANTHINE TRANSPORT THROUGH

HUMAN ERYTHROCYTE MEMBRANES

E.Capuozzo, M.C.Gigante, C.Salerno, and C.Crifò

Institutes of Biological Chemistry and Rheumatology
University of Rome, and C.N.R. Centre for Molecular
Biology, Rome, Italy

INTRODUCTION

It is well known that purine bases can be taken up and utilized by the cells in nucleotide synthesis (1). In bacterial cells, purine bases from external medium are phosphoribosylated and transported through the plasma membrane in one step catalyzed by phosphoribosyltransferases which are located in the pericytoplasmic compartment (2). A specific phosphoribosyltransferase seems to be also involved in hypoxanthine uptake by microsome vesicles purified from transformed eucaryotic cell lines (3). In intact eucaryotic cell, however, the conversion of extracellular purines to intracellular nucleotides is thought to involve most likely the tandem operation of carrier-mediated facilitated transport and appropriate phosphoribosyltransferases rather than a single step group translocation catalyzed by these transferases. In line with this hypothesis is the observation that the lack of hypoxanthine-guanine phosphoribosyltransferase does not affect the capacity of eucaryotic cells to transport hypoxanthine or guanine (4,5).

The data available still do not allow a final decision concerning the cellular localization and the role of the specific phosphoribosyltransferase on hypoxanthine uptake by human erythrocytes (6,7). The kinetics of radioactivity release from 14-C hypoxanthine-loaded human erythrocytes, suspended in a phosphate-containing medium, suggested that hypoxanthine transport could be mediated by a two-component system, only one being saturable (8). Since it is well known (9-11) that in the presence of inorganic phosphate hypoxanthine taken up by the cells is quite completely converted to IMP, one or both the kinetic components characterizing hypoxanthine release from human erythrocytes could be related to the IMP-to-purine base conversion. We report here the kinetics of 14-C hypoxanthine uptake by intact human erythrocytes suspended in a phosphate-free medium, i.e. in conditions which make negligible 14-C hypoxanthine phosphoribosylation.

EXPERIMENTAL PROCEDURES

Human erythrocytes were prepared from blood freshly drawn in heparin and washed three times with isotonic glucose-NaCl solution (5 mM glucose and 155 mM NaCl). Plasma and buffy coat were removed by aspiration. Protease-treated erythrocytes were prepared by incubating the packed cells at 37°C for 60 min in the presence of 0.1 mg/ml Proteinase K. At the end of

71

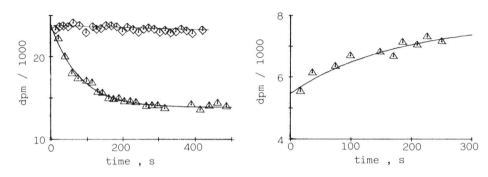

Fig. 1. (left) Time course of hypoxanthine uptake by human erythrocytes.
The cells were suspended in ice-cold isotonic glucose-NaCl solution
containing 9.2 uM 14-C hypoxanthine. At regular intervals, samples
of the incubation mixture were removed and the separated medium was
employed for liquid scintillation counting. ⚠ : native erythrocytes;
◈: erythrocytes preincubated at 37°C for 60 min in the presence of
0.1 mg/ml Proteinase K.

Fig. 2. (right) Time course of hypoxanthine release from hypoxanthine-loaded
erythrocytes. The cells were preincubated for 10 min in a ice-cold
isotonic glucose-NaCl solution containing 9.2 uM 14-C hypoxanthine,
separated from the extracellular radioactive material by centrifugation through a n-dibutyl phthalate layer, and transferred in an
equal volume of ice-cold isotonic glucose-NaCl solution. At regular
intervals, samples of the incubation mixture were removed and the
separated medium was employed for liquid scintillation counting.

the incubation period, the packed cells were cooled at 0-5°C and immediately
used for incorporation experiments.

The packed cells were suspended in an equal volume of isotonic glucose-
NaCl solution containing 0.001-1 mM 14-C hypoxanthine and incubated in a
thermostated bath. At regular intervals, samples of the incubation mixture
were removed. Erythrocytes were separated from extracellular substances by
centrifugation through a layer of n-dibutyl phthalate (12). Erythrocyte
sedimentation required centrifugation times of 5-10 s. The medium, on the
top of the n-dibutyl phthalate layer, and the erythrocyte pellet, on the
bottom of the centrifuge tube, were employed for liquid scintillation counting and paper chromatography (9). The residual suspending medium present
in red cell pellet (usually around 2% of the cell volume) was determined
using extracellular space markers (12). Hemoglobin in the suspending medium,
determined by the cyanmethemoglobin method (13), was less than 2% of the
total hemoglobin content of the incubation mixture.

RESULTS

Human erythrocytes, suspended in isotonic glucose-NaCl solution containing 14-C hypoxanthine, took up the labeled purine base. At temperature
around 0°C, the time course of hypoxanthine uptake behaved as a monoexponential process (fig. 1). The relaxation time of the uptake was independent of
initial hypoxanthine concentration and an inverse function of the temperature
of the incubation medium. At temperatures higher than 15°C, the uptake was
too fast to allow an accurate study of its kinetics. If the erythrocytes
(separated from the extracellular radioactive material by centrifugation
through a n-dibutyl phthalate layer) were transferred in an isotonic glucose-
NaCl medium not containing hypoxanthine, then the label taken up by the
cells was released in the suspending medium. At temperature around 0°C, the

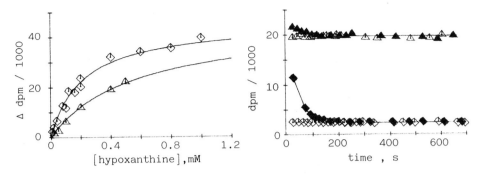

Fig. 3. (left) Total amount of radioactive material taken up by erythrocytes as a function of initial 14-C hypoxanthine concentration in the suspending medium. Cold adenine concentration was: ◈ , none; ▵ 0.1 mM. Other conditions were as described in fig. 1.

Fig. 4. (right) Time course of hypoxanthine uptake by human erythrocytes incubated at 20°C. ▵ , native erythrocytes; ▲ , erythrocytes preincubated at 37°C for 60 min in the presence of 0.1 mg/ml Proteinase K; ◈ , erythrocytes preincubated at 37°C for 60 min in the presence of 20 mM phosphate; ◆ , erythrocytes preincubated at 37°C for 60 min in the presence of 0.1 mg/ml Proteinase K and 20 mM phosphate. 14-C hypoxanthine concentration was 18 uM. Other conditions were as described in fig. 1.

time course of leakage of radioactive material from 14-C hypoxanthine-loaded erythrocytes behaved as a monoexponential process (fig. 2).

Experiments performed at equilibrium showed that the total amount of radioactive material taken up by erythrocytes was a hyperbolic function of 14-C hypoxanthine concentration in the suspending medium. In the presence of cold hypoxanthine or adenine, erythrocytes took up a less quantity of 14-C hypoxanthine. The inhibition effect exerted by the cold purine bases could be overcome by increasing the concentration of labeled hypoxanthine in the suspending medium (fig. 3).

Hypoxanthine accounted for more than 95% of the total radioactive material taken up by erythrocytes from an isotonic medium containing glucose, NaCl, and the labeled purine base. If inorganic phosphate was present in the suspending medium, then about all the extracellular hypoxanthine was incorporated in erythrocytes and IMP became the chief radioactive intracellular compound. Hypoxanthine incorporation and IMP synthesis in the presence of inorganic phosphate were relatively slow processes even at 37°C (11). Radioactive IMP was rapidly sythesized by erythrocytes, if the cells were preincubated in a medium containing inorganic phosphate but not hypoxanthine and then transferred into a phosphate-free medium containing 14-C hypoxanthine.

If erythrocytes were treated with Proteinase K, then the uptake of 14-C hypoxanthine from a phosphate-free medium slowed markedly down (figs. 1,4). A decrease in the rate of 14-C hypoxanthine uptake was also observed if protease-treated erythrocytes were preincubated in a phosphate-containing medium before addition of the labeled purine base to the cell suspension. When inorganic phosphate and 14-C hypoxanthine were added together to the cell suspension, the rate of hypoxanthine incorporation and IMP synthesis did not appreciably change using native or proteinase-treated erythrocytes.

DISCUSSION

Our results can be interpreted by assuming that extracellular hypoxanthine reversibly binds receptors present in human erythrocytes probably on the outer surface of the cell membrane. The hypoxanthine-receptor complex most likely undergoes rearrangement(s) which is (are) the rate limiting step of the overall binding process and could be related to the transport of the purine base through the cell membrane.

In the absence of inorganic phosphate in the suspending medium, hypoxanthine receptor appears to be saturated by relatively low purine base concentration. When the cells are suspended in a medium containing inorganic phosphate and, thus, phosphoribosylpyrophosphate becomes available for nucleotide synthesis (1,9), hypoxanthine is phosphoribosylated to IMP. It can be suggested that under these conditions the receptor gets rid of hypoxanthine, crosses the cell membrane, and takes up new exogenous purine base.

In line with this hypothesis are the observations that in the presence of inorganic phosphate about all the extracellular hypoxanthine is taken up by erythrocytes and that the enzymatic hydrolysis of proteins on the outer surface of erythrocyte membrane slows down both the uptake of the purine base by the hypoxanthine receptor in native cells and the IMP synthesis in phosphoribosylpyrophosphate-enriched cells.

REFERENCES

1. A.W. Murray, The biological significance of purine salvage, Ann. Rev. Biochem. 40:773 (1971).
2. J.Hochstadt-Ozer and E.R. Stadtman, The regulation of purine utilization in bacteria, J. Biol. Chem. 246:5294 (1971).
3. D.C. Quinlan and J. Hochstadt, Uptake of hypoxanthine and inosine by purified membrane vesicles from Balb/c 3T3 and Balb/c SV-3T3 cells, Fed. Proc. 33:1359 (1974).
4. B.L. Alford and E.M. Barnes, Hypoxanthine transport by cultured Chinese hamster lung fibroblasts, J. Biol. Chem. 251:4823 (1976).
5. J.M. Zylka and P.G.W PLagemann, Purine and pyrimidine transport by cultured Novikoff cells, J. Biol. Chem. 250:5756 (1975).
6. C.H.M.M. DeBruyn and T.L. Oei, Purine metabolism in intact erythrocytes from controls and HG-PRT deficient individuals, Advn. Exp. Med. Biol. 41A:223 (1974).
7. W. Gutensohn, Hypoxanthine phosphoribosyltransferase and hypoxanthine uptake in human erythrocytes, Hoppe-Seyler's Z. Physiol. Chem. 356: 1105 (1975).
8. U.V. Lassen, Hypoxanthine transport in human erythrocytes, Biochim. Biophys. Acta 135:146 (1967).
9. A. Giacomello and C. Salerno, Hypoxanthine uptake by human erythrocytes, FEBS letters 107:203 (1979).
10. A. Hershko, A. Razin, and J. Mager, Regulation of the synthesis of phosphoribosylpyrophosphate in intact red blood cells and in cell-free preparations, Biochim. Biophys. Acta 184:64 (1969).
11. C. Salerno, G. Gerber, and A. Giacomello, Regulatory aspects of hypoxanthine uptake by human erythrocytes, these proceedings.
12. J.D. Young and J.C. Ellory, Flux measurements, in: "Red Cell Membranes - Methodological Approach", J.D. Young and J.C. Ellory, eds., Academic Press, New York (1982).
13. M.M. Wintrobe, "Clinical Hematology", Lea and Febiger, Philadelphia (1961).

REGULATORY ASPECTS OF HYPOXANTHINE UPTAKE

BY HUMAN ERYTHROCYTES

C.Salerno, G.Gerber,and A.Giacomello

Institutes of Rheumatology and Biological Chemistry
University of Rome, and C.N.R. Centre for Molecular Biology
Rome, Italy; Institute of Physiological and Biological
Chemistry, Humboldt University, Berlin, DDR

INTRODUCTION

Mature human erythrocytes are incapable of de novo synthesis of purine rings, thus the turnover results from continual entry and release of purines into the plasma. The synthesis of IMP from exogenous hypoxanthine is thought to be primarily governed by the catalytic capacity of P-Rib-PP synthetase and the intracellular avaibility of P-Rib-PP (1-3).

The activity of P-Rib-PP synthetase is dependent not only on concentrations of substrates but also on concentrations of a number of cellular metabolites. Experiments performed with cell-free systems (4) reveal that the enzyme has an absolute requirement for inorganic phosphate and magnesium ions. The purified enzyme is inhibited by P-Rib-PP and 2,3-DPG competitively with respect to ribose 5-phosphate. P-Rib-PP formation in intact erythrocytes is stimulated by a sufficiently high inorganic phosphate level in the suspending medium (5). Methylene blue can enhance the rate of P-Rib-PP synthesis in the human red cells presumably by accelerating the regeneration of NADP in the oxidative pathway of glucose metabolism and thereby increasing the intracellular avaibility of ribose 5-phosphate (6). However, the rate-limiting role of the intracellular supply of ribose 5-phosphate manifests itself only when the optimum requirement of the system for inorganic phosphate has been satisfied.

It seems unlikely that P-Rib-PP is an important inhibitor of its own synthesis under normal conditions since the Ki of P-Rib-PP is approximately 10 times higher than its concentration in human cells (4). In contrast to P-Rib-PP, 2,3-DPG may be an important factor in the control of P-Rib-PP synthesis since its concentration in human erythrocytes is approximately equal to the Ki for the synthetase. It was suggested (4,6) that a shift in the oxigenation level of hemoglobin in the erythrocytes could substantially alter P-Rib-PP synthesis as a result of concomitant changes in free intracellular 2,3-DPG concentration. It was reported (7) that, at low phosphate concentration, the formation of nucleotides in intact erythrocytes was much more efficient under an atmosphere of nitrogen than in air. A surprising result was the inhibition of nucleotide synthesis under nitrogen atmosphere

Abbreviations: P-Rib-PP, 5-phosphoribosyl-1-pyrophosphate; 2,3-DPG, 2,3-diphosphoglycerate; Tris, tris(hydroxymethyl)aminomethane.

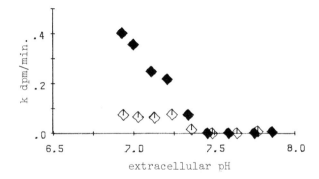

Fig. 1. Initial rate of 14-C hypoxanthine incorporation by intact human
erythrocytes suspended in isotonic saline solution containing 3 mM
phosphate as a function of pH of incubation medium. The cells were
equilibrated in nitrogen atmosphere (◆) or in oxygen atmosphere
(◇). All other conditions were as described under "Experimental
Procedures".

at high phosphate concentration. Since this finding could be attributed,
at least in part, to a shift of the intracellular pH due to the binding of
inorganic phosphate to hemoglobin (8), we decided to study the influence
of oxygen partial tension on IMP synthesis at different phosphate concentrations.

EXPERIMENTAL PROCEDURES

    Human erythrocytes were prepared from blood freshly drawn in heparin
and washed three times with isotonic Tris-NaCl-glucose buffer of a desidered
pH. Glucose concentration was 5 mM. The first washing solution contained
100 mM Tris. In all other cases a 50 mM-Tris solution was used. The packed
cells were equilibrated with the appropriate gas atmosphere and suspended in
an equal volume of Tris-NaCl-glucose isotonic medium containing 3-40 mM
sodium phosphate and 5-20 uM 14-C labeled hypoxanthine (60 Ci/mol). The
cell suspension was incubated for 15-90 min at 37°C under a layer of viscous
paraffin. At regular intervals, samples of the incubation mixture were
removed and the medium, separated from the cells by centrifugation, was employed for liquid scintillation counting and pH determination. The radioactive compounds present within the erythrocytes were identified by paper
chromatography as previously described (5). Intracellular pH was determined
in cell-free extracts obtained by freezing and thawing packed erythrocytes.

RESULTS AND DISCUSSION

    Human erythrocytes suspended in isotonic buffer containing Tris, phos-
phate, NaCl, glucose, and 14-C hypoxanthine took up the radioactive purine
base. Hypoxanthine in the suspending medium and IMP in the erythrocytes
were the only radioactive compounds present in detectable amount during the
incubation period. The amount of incorporated radioactive IMP was always
equal, within the experimental error, to the decrease in radioactive hypoxanthine in the suspending medium. The oxygenation level of hemoglobin,
the pH of the intracellular compartment, and the pH of the suspending medium
remained roughly constant with time. Control experiments performed at
physiological pH in air showed that paraffin layer on the top of the incubation mixture and Tris in the suspending medium did not appreciably affect
the time course of hypoxanthine incorporation by erythrocytes.

    Under the experimental conditions employed, the rate of hypoxanthine incorporation was constant with time, a direct function of phosphate concentration, and independent of hypoxanthine concentration. The incorporation rate markedly increased by decreasing the pH and/or the oxygen partial tension in the incubation mixture (fig. 1). The effect exerted by oxygen was more evident in the acidic pH range and could be overcome by removing the gas. At pH higher than 7.4, the rate of hypoxanthine incorporation was very low even in anaerobic conditions. Carbon monoxide inhibited the incorporation of the labeled purine base as well as pure oxygen atmosphere.

These results confirm that oxygen can lower the rate of IMP synthesis from exogenous hypoxanthine in human erythrocytes presumably by increasing the free intracellular 2,3-DPG concentration. Differences in the intracellular level of this organic polyanion can account, at least in part, also for the strong dependence on pH of the rate of IMP synthesis. It is well known (8) that the salt bridges formed between 2,3-DPG and deoxyhemoglobin are stabilized by an increase in pK of positively charged groups of the protein and a decrease in pK of the phosphate groups of 2,3-DPG. At neutral pH, 2,3-DPG binding is accompained by uptake of protons; thus, an increase in pH causes dissociation of 2,3-DPG from deoxyhemoglobin. At physiological pH, the intracellular concentration of free 2,3-DPG is most likely sufficiently high to completely inhibit P-Rib-PP synthetase at least in aerobic conditions. The problem, then, is how human erythrocytes make sufficient quantities of P-Rib-PP for nucleotide synthesis in vivo. It was suggested (7) that P-Rib-PP is mainly synthesized in the venous circulation. Another possibility worthy of further study is that the salvage of the phosphoribosyl moiety of nucleotides is of major importance in erythrocytes. The phosphoribosyl moiety of catabolized nucleotides could be directly transferred to purine bases taken up by the cells (9); thus, the synthesis of new nucleotides could not require P-Rib-PP consumption.

REFERENCES

1. A.W. Murray, The biological significance of purine salvage, Ann. Rev. Biochem. 40:773 (1971).
2. J.B. Wyngaarden and W.N. Kelley, "Gout and Hyperuricemia", Grune and Stratton, New York (1976).
3. A. Hershko, A. Razin, and J. Mager, Regulation of the synthesis of 5-phosphoribosyl-1-pyrophosphate in intact red blood cells and in cell-free preparations, Biochim. Biophys. Acta 184:64 (1969).
4. I.H. Fox and W.N. Kelley, Human phosphoribosylpyrophosphate synthetase: distribution, purification and properties, J. Biol. Chem. 247:5739 (1972).
5. A. Giacomello and C. Salerno, Hypoxanthine uptake by human erythrocytes, FEBS lett. 107:203 (1979).
6. J.B. Wyngaarden and W.N. Kelley, Gout, in: "The Metabolic Basis of Inherited Disease", J.B. Stanbury, J.B. Wyngaarden, D.S. Fredrickson, J.L. Goldstein, and M.S. Brown, eds., McGraw-Hill Book, New York (1983).
7. N. Sciaky, A. Razin, B. Gazit, and J. Mager, Regulatory aspects of the synthesis of 5-phosphoribosyl-1-pyrophosphate in human red blood cells, Adv. Exp. Med. Biol. 41A:87 (1974).
8. K. Imai, "Allosteric Effects in Haemoglobin", University Press, Cambridge (1982).
9. C. Salerno and A. Giacomello, Hypoxanthine-guanine exchange by intact human erythrocytes, Biochemistry 24:1306 (1985).

MUTANT MOUSE CELLS WITH NITROBENZYLTHIOINOSINE-INSENSITIVE NUCLEOSIDE TRANSPORT FUNCTIONS

Bruce Aronow and Buddy Ullman

Department of Biochemistry
University of Kentucky Medical Center
Lexington, KY 40536-0084

SUMMARY

From a mutagenized population of wildtype S49 T lymphoma cells, clones were generated that were resistant to the physiological effects of the potent inhibitor of nucleoside transport, 4-nitrobenzyl-6-thioinosine (NBMPR). NBMPR protected wildtype cells from the cytotoxic effects of a spectrum of nucleosides, whereas two mutant clones, KAB1 and KAB5, were still sensitive to nucleoside-mediated cytotoxicity in the presence of NBMPR. In addition, NBMPR prevented wildtype cells from surviving in hypoxanthine-amethopterin-thymidine containing medium, whereas KAB1 and KAB5 cells grew normally. Rapid sampling transport studies indicated that mutant cells, unlike wildtype parental cells, had acquired a substantial NBMPR-insensitive nucleoside transport component. Binding studies with [^{3}H]NBMPR indicated that KAB5 cells were 70-75% deficient in the number of NBMPR binding sites, whereas KAB1 cells possessed a wildtype complement of NBMPR binding sites. The characterization of the KAB1 and KAB5 cell lines suggested that the NBMPR binding site in wildtype S49 cells is genetically distinguishable from the nucleoside carrier site.

INTRODUCTION

In order to exert many of their physiological and cytotoxic effects in mammalian cells, nucleosides must first permeate across their plasma membranes. The study of nucleoside transport in mammalian cells has been greatly enhanced by the existence of specific high affinity inhibitors of the nucleoside transporter in mammalian cells, including the 6-substituted thiopurine ribonucleosides (1) and dipyridamole (2). 4-Nitrobenzyl-6-thioinosine (NBMPR) is a highly specific competitive inhibitor of nucleoside transport which binds to animal cell surfaces with an apparent Kd around 0.1-1.0nM (3,4). Not all mammalian cells, however, possess high affinity NBMPR binding sites or nucleoside transporters which are completely sensitive to NBMPR (5-10). Certain mammalian cell lines possess NBMPR-insensitive components ranging from 2% to 100% of the total cellular nucleoside transport capacity (6-10). Furthermore, a nucleoside transport-deficient cell line derived from S49 cells lacks high affinity NBMPR binding sites (11). Whether the NBMPR-sensitive and NBMPR-insensitive nucleoside transporters are genetically identical or distinct is not clear.

To attempt to resolve whether or not two structurally distinct transporters exist we have taken advantage of this high affinity interaction of NBMPR with the nucleoside transport function in S49 murine lymphoma cells to isolate mutants which were resistant to the physiological effects of NBMPR. Growth rate determinations, rapid transport measurements, and binding site quantitations indicated that mutant cells were not completely responsive to NBMPR and had gained an NBMPR-insensitive nucleoside transporter function. The results of these studies may ultimately have important chemotherapeutic implications in the clinical utilization of transport inhibitors to modulate nucleoside cytotoxicity.

EXPERIMENTAL PROCEDURES

Cell culture   The growth characteristics and lymphocytic properties of wildtype S49 cells have been described previously in great detail (12). The KAB1 and KAB5 clones were isolated from approximately 10^6 mutagen-treated wildtype cells in soft agarose containing 0.5mM hypoxanthine, 0.4μM methotrexate, 30μM thymidine, 30μM deoxycytidine, and 30μM NBMPR. To determine the growth sensitivities of wildtype and mutant cells to various nucleosides and nucleoside analogs in the presence and absence of inhibitors of nucleoside transport, cells were incubated in Costar multiwell (24 well) tissue culture plates and enumerated on a Model ZB1 Coulter Counter as described previously (13).

Nucleoside transport measurements   Nucleoside transport was measured by the rapid sampling technique described by Aronow et al. through a layer of inert oil (14).

Measurements of 4-nitrobenzylthioinosine binding.   The binding of [^{3}H]NBMPR to wildtype and mutant cells was measured by the procedure of Aronow et al. (14). The differences between the amount of radiolabeled NBMPR associated with the cells in the presence and absence of 20μM nonradiolabelled NBMPR were considered to be a measure of the specifically bound NBMPR.

RESULTS

Selection and isolation of mutant clones.   In order to dissect genetically the high affinity interactions of inhibitors of nucleoside transport with the nucleoside transporter, we attempted to devise selective procedures to isolate variants resistant to the physiological effects of these inhibitors. Since the cytotoxic effects of 0.4μM methotrexate in the presence of 0.5mM hypoxanthine can be reversed by thymidine, a pyrimidine nucleoside, the ability of NBMPR to prevent thymidine rescue of methotrexate toxicity was determined. Wildtype cells in the presence of 10μM NBMPR did not survive in hypoxanthine-methotrexate regardless of the thymidine concentration (HAT medium). Two clones, KAB1 and KAB5, were isolated from semi-solid agarose as described (14) and grew normally in HAT medium containing 10μM NBMPR (Figure 1). To determine whether the altered apparent sensitivities of KAB1 and KAB5 cells to NBMPR were specific for thymidine or general for a spectrum of nucleosides, the ability of NBMPR to protect wildtype and mutant cells from the cytotoxic effects of deoxyguanosine, 5-fluorouridine, adenosine-EHNA, 6-thioguanosine, and deoxyadenosine-EHNA were examined. Wildtype and mutant cells were approximately equally sensitive to any of the five nucleosides in the absence of NBMPR. For all five nucleosides, NBMPR protected wildtype cells from the nucleoside-mediated cytotoxicity, whereas the KAB1 and KAB5 cells were unprotected or only slightly protected from nucleoside toxicity by NBMPR (data not shown). Conversely, dipyridamole protected wildtype and mutant

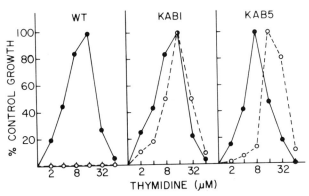

Figure 1. Ability of wildtype and mutant cells to survive
HAT medium in the absence or presence of NBMPR. The ability
of wildtype, KAB1, and KAB5 cells to grow in culture medium
containing 0.5mM hypoxanthine, 0.4µM methotrexate, and
increasing thymidine concentrations were determined in the
absence (●) or presence (O) of 10µM NBMPR. The cells were
counted after 72 hr and the results processed as described
in Materials and Methods.

cells from nucleoside-mediated growth inhibition equally. Thus, the
altered responsiveness of KAB1 and KAB5 cells to NBMPR appeared to be a
general phenomenon for all nucleosides yet specific to NBMPR and not to
other inhibitors of nucleoside transport.

Transport of nucleosides. Since the nucleosides to which KAB1 and
KAB5 cells responded abnormally are all metabolized by biochemically and
genetically distinct pathways, we hypothesized that the altered
responsiveness of mutant cells to NBMPR might be attributed to a
genetically altered nucleoside transport function. Thus, the rates of
transport of [³H]adenosine (Figure 2A) and [³H]cytidine (Figure 2B)
as a function of NBMPR concentration were measured in wildtype and mutant
cells. Transport of either nucleoside into wildtype cells could be
completely blocked by 0.1-1.0µM NBMPR and was inhibited 50% by 10nM
NBMPR. In KAB1 and KAB5 cells, however, transport was refractory to
complete inhibition by NBMPR. Approximately 5-20% of the total
nucleoside transport capacity of KAB1 and KAB5 cells was insensitive to
inhibition by 10µM NBMPR, Figure 2. The responses of the mutant cells to
inhibition of nucleoside transport by NBMPR suggested that the nucleoside
transporter in mutant cells had gained an NBMPR-insensitive component not
observed in wildtype parental cells. This NBMPR-insensitive nucleoside
transport component of KAB1 and KAB5 cells had similar apparent K_m
values for adenosine and cytidine (data not shown) as the NBMPR-sensitive
component.

Nitrobenzylthioinosine binding. To further characterize the nature
of the mutations in KAB1 and KAB5 cells at the molecular level, the
interaction of [³H]NBMPR with specific cell surface binding components
was examined in wildtype and mutant cells. The results indicated
differences between the KAB1 and KAB5 cells lines. Wildtype and KAB1
cells had the same number of NBMPR binding sites, while KAB5 cells

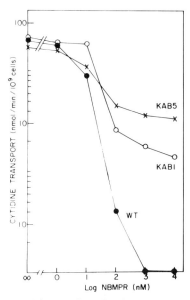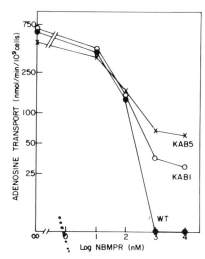

Figure 2. Cytidine and adenosine transport in wildtype and mutant cells. The effects of increasing NBMPR concentrations on the abilities of wildtype (●) KAB1 (O), and KAB5 (X) cells to transport either 50μM [3H]cytidine (left panel) or 50μM [3H]adenosine (right panel) were determined.

possessed about one third of the wildtype complement of NBMPR binding sites. A fourth cell line, AE_1 (15), which is genetically deficient in nucleoside transport, did not contain any measurable high affinity [3H]NBMPR binding sites (10). The dissociation constants for NBMPR with the receptors on the mutant cells did not differ by more than 2-fold from that obtained with parental cells. Similarly, nonradiolabelled nucleosides and dipyridamole displaced [3H]NBMPR from the NBMPR receptor with equal efficacy in all 3 cell lines (data not shown).

DISCUSSION

Whether NBMPR binds to the nucleoside carrier site of the nucleoside transport function albeit with much higher affinity or to a second site has not been definitively demonstrated. In S49 cells (11), NBMPR binding abolishes nucleoside transport. In other cell lines, however, the extent to which NBMPR inhibits the transport of nucleosides varies considerably. Some cell lines lack cell surface binding sites for NBMPR (16), while others bind but are unaffected by NBMPR (7). Moreover, some cells appear to possess both NBMPR-sensitive and NBMPR-insensitive nucleoside transport components (5,7,9,10). An important question to resolve therefore is whether NBMPR-sensitive and NBMPR-insensitive transporters are genetically identical except for their ability to bind and interact with NBMPR and related compounds. To address this question, the nucleoside transporter function in S49 cells has been manipulated genetically to isolate variants which had gained an NBMPR-insensitive nucleoside transporter function. Two types of mutants were isolated. The first type of mutant appeared to have lost ~50% of the NBMPR binding sites, while the second and predominant phenotype possessed the normal complement of ligand binding sites on the cell surface. By measuring nucleoside influx during short time intervals, we determined that KAB1 and KAB5 cells possessed both NBMPR-sensitive and NBMPR-insensitive components. The NBMPR-sensitive and

NBMPR-insensitive nucleoside transport systems in mutant cells had similar substrate affinities and substrate specificities and similar sensitivities to inhibition by 10μM dipyridamole and 1.5mM N-ethylmaleimide. However, the NBMPR-insensitive transporters were more sensitive to 0.5mM p-chloromercuribenzenesulfonate (data not shown). Despite clear and consistent differences in the number of NBMPR binding sites between KAB1 and KAB5 cells, the NBMPR-insensitive component in both mutants comprised about 5-20% of the total nucleoside transport activity. Thus, KAB1 and KAB5 cells have been manipulated genetically to behave like CHO, P388, L1210, and L929 cells (16) which contain both NBMPR-resistant and NBMPR-sensitive components.

It appears that there are two distinct mechanisms for generating an NBMPR-insensitive transporter. The first type of mutation resulted in decreased NBMPR binding with unaltered transport capacity. The second type of mutation, i.e., that in the KAB1 cells, appeared to be in some genetically determined transducing function required for linking NBMPR binding to its inhibitory effects on nucleoside transport. These data support the existence of a single nucleoside transporter which can probably exist in two conformations, either coupled or uncoupled to the NBMPR binding site in a genetically alterable fashion. Moreover, the data strongly suggest that the determinants on the nucleoside transporter which bind nucleosides and those which bind NBMPR can be distinguished genetically.

ACKNOWLEDGEMENTS

This work was supported by Grant RO1 CA32580 from the National Institutes of Health. B.U. is a recipient of a Research Career Development Award from the National Institutes of Health.

REFERENCES

1. Paterson, A. R. P. and Oliver, J. M. (1971) Can. J. Biochem. 49, 271-274.
2. Scholtissek, C. (1968) Biochim. Biophys. Acta 158, 435-447.
3. Pickard, M. A., Brown, R. R., Paul, B. and Paterson, A. R. P. (1973) Can. J. Biochem. 51, 666-672.
4. Cass, C. E., Gaudette, L. A. and Paterson, A. R. P. (1971) Biochim. Biophys. Acta 345, 1-10.
5. Wohlheuter, R. M., Marz, R. and Plagemann, P. G. W. (1978) J. Membrane Biol. 42, 247-264.
6. Dahlig-Harley, E., Eilam, Y., Paterson, A. R. P. and Cass, C. E. (1981) Biochem. J. 200, 295-305.
7. Plagemann, P. G. W. and Wohlheuter, R. M. (1984) Biochim. Biophys. Acta 773, 39-52.
8. Eilam, Y. and Cabantchik, Z. I. (1977) J. Cell Physiol. 89, 831-838.
9. Belt, J. A. (1983) Biochem. Biophys. Res. Commun. 110, 417-412.
10. Belt, J. A. (1983) Molec. Pharmacol. 24, 479-484.
11. Cass, C. E., Kolassa, N., Uehara, Y., Dahlig-Harley, E., Harley, E. R. and Paterson, A. R. P. (1981) Biochim. Biophys. Acta 649, 769-777.
12. Ullman, B., Cohen, A. and Martin, D. W., Jr. (1976) Cell 9, 205-211.
13. Ullman, B. (1983) J. Biol. Chem. 258, 523-528.
14 Aronow, B, Allen, K., Patrick, J., and Ullman, B. (1985) J. Biol. Chem. 260, 6226-6233.
15. Cohen, A., Ullman, B. and Martin, D. W., Jr. (1979) J. Biol. Chem. 254, 112-116.
16. Paterson, A. R. P., Jakobs, E. S., Harley, E. R., Cass, C. E. and Robins, M. J. (1983) in Development of Target Oriented Anticancer Drugs, (Chang, Y. C. and Goz, B., eds.) pp. 41-56, Raven Press, New York.

CARRIER MEDIATED UPTAKE OF DEOXYGUANOSINE IN RAT LIVER MITOCHONDRIA

Linda Watkins and Roger A. Lewis

Department of Biochemistry
University of Nevada, Reno
Reno, NV 89557

INTRODUCTION

It has been known for a number of years that mitochondria contain enzymes important for the synthesis of DNA. A distinct mitochondrial deoxyguanosine kinase was first documented by Gower et al. (1) in calf thymus. Mitochondrial deoxycytidine/deoxythymidine kinase (2) was demonstrated in blast cells of leukemia patients and HeLa mitochondria contain a deoxythymidine kinase (3). Mouse, human and chick mitochondria possess a deoxypyrimidine kinase (3). A recent report by Fabianowska-Majewska and Greger (4) shows that rat liver mitochondria contain deoxythymidine, deoxcytidine, deoxyadenosine and deoxyguanosine kinase activities, however few details regarding these activities were reported.

That thymidine can be converted in mitochondria to dTMP, dTDP and dTTP followed by subsequent incorporation into mitochondrial DNA has been demonstrated (5,6). A preliminary study shows that deoxyguanosine is metabolized by mitochondria into its corresponding deoxynucleotides, dGMP, dGDP and dGTP (7).

These reports indicate that deoxynucleotide metabolism is active in mitochondria and that deoxynucleosides may be important precursors for mitochondrial DNA. This paper describes the uptake of deoxyguanosine by rat liver mitochondria and indicates that this deoxynucleoside is taken up by a carrier mediated mechanism.

METHODS

Intact rat liver mitochondria were isolated by the method of Chappell and Hansford (8) and incubated at 37° C for 30 sec in a buffer containing 100 mM KCl, 75 mM sucrose, 5 mM KH_2PO_4, 0.3-0.5 mg protein, 5 mM HEPES, pH 7.0 and 10 μM [^{3}H]dG (2 μCi) in a final volume of 125 μl. The reaction was terminated by the addition of 4 ml of a rinse soution which contained 250 mM sucrose, 1 mM $MgCl_2$ and 3 mM deoxyguanosine. This mixture was rinsed by filtration on GF/B filters with an additional 15 ml of the rinse buffer. The filters were air dried and quantified by liquid scintillation techniques.

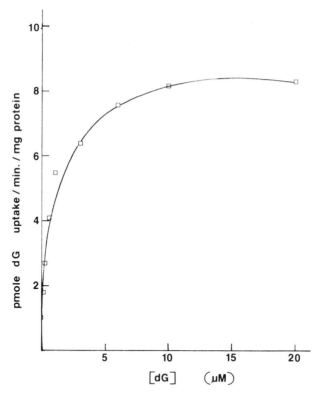

Figure 1. Substrate Saturation of Deoxyguanosine Uptake

RESULTS AND DISCUSSION

Figure 1 shows that deoxyguanosine was taken up by intact rat liver mitochondria by a process which was substrate dependent and saturable. A Km of 0.44 μM deoxyguanosine and a V_{max} of 13.7 pmole/min/mg protein were calculated by a Hanes plot of the kinetic data. Saturability with substrate of deoxyguanosine uptake is an indication that this is a carrier mediated process. The data shown below add further support to this conclusion.

Table I. Specificity of the Deoxyguanosine Uptake Process

Added Nucleoside (100 μM)	Specific Activity pmol uptake/min/mg protein	% Activity
none	18.78 + 1.97	100
dAdo	15.40 + 1.94	82
dCyt	18.40 + 0.80	98
dThy	16.24 + 0.46	86
dIno	10.43 + 1.13	56
Guo	19.27 + 1.13	103

Table II. The Effect of Known Nucleoside Transport Inhibitors

Addition	Specific Activity (pmol uptake/min/mg protein)	% Activity
none	13.66 + 1.16	
cytochalasin B (33 μM)	11.67 + 1.06	85
none	19.42 + 0.33	
NBMPR (33 μM)	19.19 + 1.90	99
none	1.84 + 0.05	
NBTGR (10 μM)	2.00 + 0.24	108

To test the specificity of deoxyguanosine uptake, competition by other deoxynucleosides and guanosine was tested. Table I shows that the transport system is relatively specific for deoxyguanosine. Only deoxyinosine, at a 10:1 excess, caused a significant lowering of deoxyguanosine uptake. Even guanosine was an ineffective inhibitor, suggesting the transporter can distinguish between the 2'-deoxy and hydroxy forms of the nucleoside.

Cytochalasin B, 6-(2-hydroxy-5-nitrobenzyl)-thioinosine (NBMPR) and S-(p-nitrobenzyl)-6-thioguanosine (NBTGR) are commonly used inhibitors of nucleoside transport. When these compounds were assayed as inhibitors of deoxyguanosine uptake, none caused a significant change from the control values (Table II). These data indicate that the mitochondrial transporter for deoxyguanosine is different from that located in the plasma membrane. Likewise, these results further emphasize the strict substrate specificity shown in Table I.

To measure the importance of ATP hydrolysis on the uptake process, reaction mixtures were supplemented with the nonhydrolyzable analogue AMP-PCP. Table III reveals that in the presence of this compound, which competes with ATP and effectively blocks ATP hydrolysis, the uptake of deoxyguanosine was strongly inhibited. Almost complete inhibition was measured at 10 mM AMP-PCP.

Since the hydrolysis of ATP is important to deoxyguanosine uptake, then the products of hydrolysis may function as product-inhibitors of the process. When ADP was tested for its influence on deoxyguanosine uptake, it was shown to be inhibitory in a concentration dependent manner (Table IV).

Table III. Inhibition of Deoxyguanosine Uptake by AMP-PCP

Addition	Specific Activity (pmol uptake/min/mg protein)	% Activity
none	9.46 + 0.70	100
AMP-PCP (1 mM)	4.76 + 0.56	50
AMP-PCP (10 mM)	0.96 + 0.22	10

Table IV. The Effect of ADP on Deoxyguanosine Uptake

[ADP] (mM)	Specific Activity (pmol uptake/min/mg protein)	% Activity
0	21.43 + 1.60	100
0.001	16.64 + 0.40	78
0.010	16.52 + 1.58	77
0.100	17.40 + 2.35	81
1.000	11.03 + 0.98	51

CONCLUSIONS

The observations that deoxyguanosine uptake is a saturable event and that it is inhibited by deoxyinosine, AMP-PCP or ADP lead to the conclusion that the uptake of deoxyguanosine is a carrier mediated process. Furthermore, uptake requires the hydroysis of ATP, however the specific role of that hydrolysis has not been identified. The uptake is specific for deoxyguanosine with deoxyinosine being the only nucleoside which caused any competing inhibition. Likewise, cytochalasin B, NBMPR and NBTGR are not inhibitory to the process. Collectively, these data show that the uptake of deoxyguanosine by the mitochondria is different from that measured in the plasma membrane.

ACKNOWLEDGEMENT

This is a contribution of the Nevada Adgricultural Experiment Station.

REFERENCES

1. Gower, W.R., Jr., Carr, M.C. and Ives, D.H. (1979) J. Biol. Chem. 254:2180.
2. Cheng, Y., Domin, B and Lee, L (1977) Biochim. Biophys. Acta 481:481.
3. Leung, W., Dubbs, D.R., Trkula, D. and Kit, S. (1975) J. Virol. 16:486.
4. Fabianowska-Majewska, K. and Greger, J. (1982) Enzyme 27:124.
5. Mitra, R.S. and Bernstein, I.A. (1970) J. Biol. Chem. 245:1255.
6. Tsiftsoglou, A.S. and Georgatsos, J.G. (1971) Biochim. Biophys. Acta 262:239.
7. Lewis, R.A. and Watkins, L.F. (1984) in Purine and Pyrimidine Metabolism in Man-IV, Part B. Plenum Press, New York, p. 79.
8. Chappell, J.B. and Hansford, R.G. in Subcellular Components. Preparation and Fractionation, 2nd Ed. Butterworths, London.

MECHANISM OF THYMINELESS DEATH

Mehran Goulian, Bruce M. Bleile, Lisa M. Dickey, Robert H. Grafstrom, Holly A. Ingraham, Scott A. Neynaber, Mark S. Peterson, and Ben Y. Tseng

University of California, San Diego
Department of Medicine (M-013-G)
La Jolla, California 92093

Approximately 30 years ago it was observed by Seymour Cohen that an E. coli mutant that required thymine became non-viable within one generation time after transfer to a medium lacking thymine.[1] This was in contrast to the prolonged tolerance for depletion of precursors for protein or RNA synthesis, or requirements for energy metabolism and carbon sources.

There are a number of examples of phenomena which can be explained, at least in part, by the toxicity of thymine deprivation. These include the antibacterial effects of sulfa drugs, which inhibit synthesis of folate, a requirement for synthesis of thymidylate; dihydrofolate reductase inhibitors, which limit the supply of the active, i.e. reduced, form of folate, and are used to enhance the antibacterial (trimethoprim) or antiprotozool (pyrimethamine) activity of sulfa drugs, or as antiproliferactive agents (methotrexate, MTX) in clinical treatment of tumors or immunosuppression; direct inhibition of thymidylate synthase, an effect of the antitumor agent, 5-FUra, resulting from one of its intracellular products, 5-FdUMP; and clinical deficiencies of folate and vitamin B12 (the latter causing a secondary deficiency of folate) resulting in the thymineless death of blood cell precursors in the bone marrow.

There have been many studies on "thymineless death", initially mostly in bacteria but subsequently in eukaryotes, as well (see review[2]). Thymidylate depletion has been produced in eukaryotic experimental models with inhibitors, e.g. MTX or 5-FdUrd, and, more recently, in cells with genetic defects in thymidylate synthase.[2,3] The mechanism for the toxicity has remained obscure but common features in many of the studies, in prokaryotes and eukaryotes, include loss of viability, within one generation time; requirement for a functioning DNA replication apparatus; and associated fragmentation of the DNA.

Our work on the toxicity of thymidylate deprivation grew out of studies on the mechanism by which Ura is excluded from DNA. Our initial studies were carried out with an in vitro system for DNA synthesis using isolated nuclei from cultured animal cells.[4] When dUTP was introduced in this system it was rapidly degraded by the enzyme dUTPase,[5] which is primarily responsible for preventing entry of Ura into DNA. During the brief period during which dUTP was present it was incorporated into DNA substituting for dTTP. The second mechanism for the exclusion from DNA then became evident; the incorporated Ura was promptly removed by Ura-DNA glycosylase[6] leaving an apyrimidimic (AP) site for each Ura (base) excision. This was followed by additional (non-Ura) base removal (after incision by AP endonuclease), a general mechanism for excision/repair of AP sites.[6,7] Analysis of the DNA under denaturing conditions showed marked fragmentation due to the many interruptions at AP sites undergoing excision/repair.[4] The appearance closely resembled previous results with E. coli mutants in which dUTPase was defective, allowing dUTP to accumulate.[8] Interestingly, in the in vitro mammalian system the defects in DNA resulting from high levels of dUMP incorporation were not repaired suggesting an irreversible lesion.

Our interest in the consequences of stressing the dUTPase and Ura-DNA glycosylase mechanisms in vivo led us to reconsider an old observation on the effects of thymidylate deprivation. It had been found a number of years earlier, in studies on the intracellular nucleotides of Lactobacillus casei starved for thymidylate (by limiting folate and supplementing with purines and amino acids), that not only was intracellular dTTP markedly depressed but there was greater than 30-fold increase in dUMP.[9] Some years later it was observed in several laboratories that animal cells treated with MTX or 5-FUra also accumulate dUMP, up to ~10x normal levels,[10-12] and in one of the later reports dUMP rose to 300x normal.[13]

We used a model of thymidylate deficiency produced by treatment of cultured animal cells (human lymphoblasts) with MTX in presence of purine and amino acid supplements. Analysis of acid soluble extracts by HPLC showed that dUMP from the drug treated cells was at least 1000-fold greater than in untreated cells.[14] This encouraged an effort to detect dUTP, which had not previously been measured in cells. The procedure used for this was very simple in principle (Fig. 1): the dUrd nucleotide pools were labeled under steady state conditions with tracer amounts of ^{3}H-dUrd; unlabeled dUTP was added at the time of acid extraction of intracellular pools and re-isolated free of contaminants; radioactivity associated with the re-isolated dUTP was converted to moles by independently determined specific activity. The major task in this procedure was the purification of dUTP free from traces of contaminants of much greater abundance e.g., dTTP, dUMP, dUDP, and dUDP-GlcNAc (see below). In the initial studies three consecutive paper chromatographic steps were used;[14] in a later modification ion exchange chromatography replaced two of the paper chromatographies.[15] In the final step the purified dUTP was dephosphorylated and analyzed as dUrd by reverse phase HPLC; the optical density of dUrd provided a correction for losses during the multiple purification steps (Fig. 1). Specific activity for dUrd nucleotide pools from MTX treated cells was determined by direct measurement by HPLC of optical density and radioactivity of intracellular dUMP. Measurements of both dUrd and dThd nucleotides were made on untreated cell extracts by the same procedure except that specific activity was measured on dTTP.

Using this method dUTP was easily measured in the deoxythymidylate-depleted cells whereas it could not be detected in untreated cells, even when additional measures (Fig. 1) were taken to reduce background and further increase sensitivity (Table 1).[14] The amount of dUTP in the drug treated cells was at least 700-fold greater than in normal cells. This, together with the marked fall in dTTP with drug treatment, resulted in an increase in the ratio of dUTP/dTTP from <1/100,000 in untreated cells to 1/5 in cells starved for deoxythymidylate (Table 1).[14]

2×10^8 cells[a] ± MTX (+Hx) + ^{3}H-dUrd (0.1 µM)
[10^9 cells][b]

TCA extract (+ unlabeled dUTP, internal standard)

DEAE column (de-salt)

←------------------[IO_4^-][b]

paper chromatography (NH_4 isobutyrate) x 3
(or DEAE column + paper chromatography x 1)[c]

phosphatase

-------→[paper chromatography (n-butanol, NH_3)][b]

HPLC (RP-18) ←------------

Fig. 1. Isolation of cellular dUTP[17,18]

[a]human lymphoid cells, suspension culture
[b]modified procedure for greater sensitivity[17]
[c]alternative procedure[18]

Table 1. Intracellular nucleotides (pmol/10^6 cells)[17]

	untreated	MTX-treated
dTTP	39	1
dUMP	0.85	1200
dUTP	< 0.0003	0.2
$\dfrac{dUTP}{dTTP}$	$< \dfrac{1}{100,000}$	$\dfrac{1}{5}$

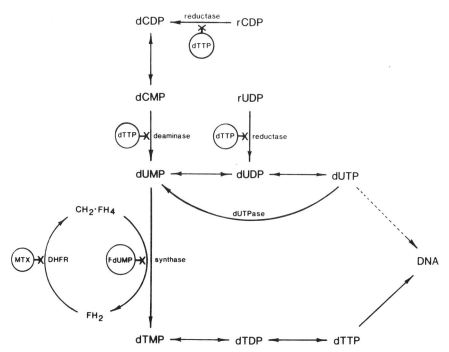

Fig. 2. Interrelationships between dUrd and dThd nucleotide pathways,
$CH_2 \cdot FH_4$ methylene tetrahydrofolate; FH_2, dihydrofolate;
reductase, ribonucleoside diphosphate reductase; deaminase, dCMP
deaminase; synthase, dTMP synthase; DHFR, dihydrofolate reductase.

The increase in dUMP is explained as activation of the normal pathways
for synthesis of dUMP in response to the fall in dTTP (Fig. 2).[12]
Approximately 2/3 of the dUMP is made by deamination of dCMP, the source of
which is dCDP; the remaining 1/3 of dUMP comes directly from dUDP. dTTP
normally inhibits both dCMP deaminase, and the synthesis of dCDP and dUDP
from CDP and UDP by ribonucleotide reductase; with the fall in dTTP, these
enzymes are activated.[12] In addition, it is likely that the dUMP pool is
further augmented by enhanced salvage of dUrd by dThd kinase, which is also
normally inhibited by dTTP.

Since DNA polymerases do not discriminate between dTTP and dUTP, the
above results implied that in these cells, for every five dTMPs
incorporated into DNA, one dUMP should have been inserted, as well. In
searching for evidence of this MTX treated and untreated cells were labeled
with radioactive dUrd, and the DNA isolated and analyzed for dUMP. As
predicted, dUMP was found in the DNA from deoxythymidylate depleted cells
(0.8 pmol/μmol DNA nucleotide), the first such demonstration in eukaryotic
cells.[16] No dUMP was detected in DNA from normal cells (<0.0001 pmol/μmol
DNA nucleotide).

Similar experiments, measuring intracellular dUrd and dThd nucleotides
and dUrd in DNA, were also carried out on cells made thymidylate deficient
with 5-FdUrd instead of MTX (+Hx), with entirely similar results.[15] The
principal difference was that 5-FdUMP, the intracellular derivative of 5-
FdUrd which directly inhibits thymidylate synthase, also followed the fate
of dUMP. The pools of 5-FdUMP enlarged to levels similar to dUMP; 5-FdUTP,
which is also susceptible to dUTPase,[15] accumulated along with dUTP; and
both 5-FdUMP and dUMP were incorporated into DNA, and then removed by the
mechanism described below.[15,17]

The amounts of dUMP in DNA from thymidylate deprived cells was far less than the 1/5 ratio of dUTP/dTTP; however, this is exactly what would be expected from the function of the Ura-DNA glycosylase mechanism, referred to above. This enzyme promptly removes Ura residues from DNA and what was detected represented a minute reflection of the very large amounts that were actually being incorporated (and removed). This was indicated by the several-fold increase in amount of dUrd in DNA of the thymidylate deficient cells when they were also treated with Ura, which inhibits Ura-DNA glycosylase (albeit weakly).[16] In further support of this dynamic process is the rapid disappearance of dUrd from DNA when the ^{3}H-dUrd was "chased" with an excess of unlabeled dThd.[16]

This previously unsuspected process of Ura insertion and removal could well account for the fragmentation of DNA that had long been observed with thymidylate deprivation. It is likely that the resulting nicks and gaps in DNA are also responsible for the enhanced rate of genetic recombination in thymine deficient cells.[2,3] Similarly, these effects on DNA structure may increase the rate of gene amplification, e.g., through effects on recombination or initiation of replication; and it may be no coincidence that the first demonstration of gene amplification in mammalian cells was with the gene for dihydrofolate reductase in cells resistant to its inhibitor, MTX.[18]

The apparent magnitude of the process of dUMP insertion and removal called attention to the possibility that it might have a part in the toxicity of the thymineless state, as well. Our first approach to testing this was to observe the effects of raising the level of dUMP in cells without lowering dTTP. It was not possible to do this by simply adding dUrd to cell growth medium; dTTP levels rose and, we believe, limited further utilization of dUrd (via dThd kinase). However, in "HAT" medium (containing Hx, MTX and dThd) dUrd caused elevation of intracellular dUMP, with normal or increased dTTP.[19] At the highest levels of dUrd in medium dUTP equaled dTTP. Interestingly, in these cells (hamster lung fibroblasts) MTX alone (+Hx) resulted in a ratio dUTP/dTTP of 8/1, much higher than in the previous studies, with human lymphoid cells. When viability was tested by measuring plating efficiency it was found that both the dUrd treatment (in HAT medium) and MTX (+Hx) resulted in death of most of the cells (>90%).[19] Likewise, analysis by sedimentation velocity of the DNA in alkaline sucrose gradients showed marked reduction in single strand length from both types of treatment.[19] We assume that this apparent fragmentation is a consequence of the many nicks and gaps in newly synthesized DNA at sites of Ura removal.

The results indicate that a high intracellular concentration of dUMP is toxic to cells even in the presence of normal concentration of dTTP. Thus, the toxicity of MTX (+Hx) could be accounted for by the effects of dUMP (and dUTP) pool expansion; however, because of the markedly diminished dTTP in drug treated cells it is not possible to be certain of the role of dUMP from this evidence, alone.

We have tried to resolve this by reducing dUTP without affecting dTTP, and observing the effects on cell survival. Our approach to this called upon an old observation that addition of GlcN to cells causes rapid depletion of intracellular UTP by forming UDP-GlcNAc.[20] In the course of purifying dUTP from MTX (+Hx) treated cells labeled with ^{3}H-dUrd, we had identified an unknown ^{3}H-labeled compound as dUDP-GlcNAc.[21] This deoxyribosyl analog of UDP-GlcNAc was found only in drug treated cells and we assumed that it is formed by dUTP competing with UTP for the UDP-GlcNAc pyrophosphorylase. We hoped that, in analogous fashion, GlcN would reduce dUTP, as well. The experiments were carried out during the S phase of cells (hamster lung fibroblasts) synchronized by serum starvation/repletion. GlcN added to MTX (+Hx) treated cells reduced dUTP by 95% and increased survival 10-fold (to ∼1/3 normal) without affecting dTTP.[22] Levels of dTTP were very low in the drug treated cells and were not affected by the presence of GlcN. In cells not treated with MTX GlcN had no effect on dTTP level, viability, or DNA synthesis rate (dThd incorporation). These studies also showed that for at least some cell lines (e.g. hamster lung fibroblast) treatment with MTX causes such a large increase in dUTP that, since the usual assay for dNTPs (with DNA polymerase) does not distinguish dUTP and dTTP, this may account for some reports of only modest depression of dTTP in MTX-treated cells.[19,22,23,24]

Missing at this time is an assessment of the effects of low dTTP alone, i.e. without expansion of deoxyuridylate. However, our work thus far indicates that expansion of dUMP pools, and thereby dUTP, plays a significant part in the effects from thymidylate starvation of cells treated with MTX (+Hx) and, we assume, in thymineless death, in general.

How does the incorporation of dUMP into DNA result in death of cells? The normal, and inevitable, response to the presence of Ura in DNA is its removal by Ura-DNA glycosylase followed by excision/repair of the resulting AP sites. One possibility is that the many single-strand gaps that are associated with this process are susceptible to nuclease(s), which would result in a DNA lesion that is difficult to repair, e.g., double-strand breaks. Another aspect of the process may further increase the risk of this or some other form of irreversible damage: dUMP may be re-inserted in the gap fill-in (repair) phase of the excision-repair process thus resulting in a self-defeating cycle of insertion, removal and re-insertion. The probability of re-insertion of dUMP at a particular AP repair site will depend on the ratio dUTP/dTTP and gap size; with high dUTP/dTTP, as, for example, with the hamster fibroblast line (dUTP/dTTP = 8-20/1), the probability of re-insertion of one or more dUMP per gap becomes very high, even for short gaps. Thus, it is possible that the net effect of this normal cellular response will be not removal and repair of dUMP-containing sites but accelerating rate of entry of dUMP into DNA and progressively widening single strand gaps (Fig. 3).

In support of the importance (in bacteria) of Ura removal and/or AP site excision/repair in the toxicity, both from thymine deprivation and from absence of dUTPase, is the protection from toxicity by loss of Ura-DNA glycosylase.[25,26] Also pertinent to note is a recent report of double strand breaks in DNA associated with thymidylate depletion in animal cells.[3]

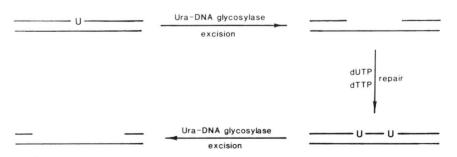

Fig. 3. Hypothetical scheme for the consequences of DNA excision/repair of dUMP-containing sites in DNA in presence of high ratio dUTP/dTTP.

In summary, our studies indicate that limiting deoxythymidylate unleashes the biosynthetic pathway for synthesis of deoxyuridylate, overcoming the dUTPase mechanism that ordinarily prevents Ura from entering DNA. This leads to the incorporation of large amounts of deoxyuridylate into DNA, which, in turn, causes the normal mechanism for removal of Ura from DNA to run amok resulting in lesions in DNA that account for a significant part of the pathology that characterizes thymineless death.

ACKNOWLEDGEMENTS

This work was supported by USPHS grants CA 11705 and CA 25440, and USPHS traineeship (HL 07107 to H.A.I.)

REFERENCES

1. S.S. Cohen and H.D. Barner, Proc. Natl. Acad. Sci. USA 40:885 (1954).
2. B.J. Barclay, B.A. Kunz, J.G. Little, and R.H. Haynes, Can J. Biochem. 60:172 (1982).
3. D. Ayusawa, K. Shimizu, H. Koyama, K. Takeishi, and T. Seno, J. Biol. Chem. 258:12448 (1983).
4. R.H. Grafstrom, B.Y. Tseng, and M. Goulian, Cell 15:131 (1978).
5. E. Bertani, A. Häggmark, and P. Reichard, J. Biol. Chem. 236:PC67 (1963).
6. T. Lindahl, Ann. Rev. Biochem. 51:61 (1982).
7. P.C. Hanawalt, P.K. Cooper, A.K. Ganesan, and C.A. Smith, Ann. Rev. Biochem. 48:783 (1979).
8. B-K. Tye, J. Chien, I.R. Lehman, G.K. Duncan, and H.R. Warner, Proc. Natl. Acad. Sci. USA 75:233 (1978).
9. M. Goulian, and W.S. Beck, Biochim. Biophys. Acta 129:336 (1966).
10. M.H.N. Tattersall, R.C. Jackson, T.A. Conners, and K.R. Harrap, Eur. J. Cancer 9:733 (1973).
11. C.E. Myers, R.C. Young, and B.A. Chabner, J. Clin. Inv. 56:1231 (1975).
12. R.C. Jackson, J. Biol. Chem. 253:7440 (1978).
13. R.G. Moran, C.P. Spears, and C. Heidelberger, Proc. Natl. Acad. Sci. USA 76:1456 (1979).
14. M. Goulian, B. Bleile, and B.Y. Tseng, J. Biol. Chem. 255:10630 (1980).
15. H.A. Ingraham, B.Y. Tseng, and M. Goulian, Mol. Pharm. 21:211 (1982).
16. M. Goulian, B. Bleile, and B.Y. Tseng, Proc. Natl. Acad. Sci. USA 77:1956 (1980).
17. H.A. Ingraham, B.Y. Tseng, and M. Goulian, Ca. Res. 40:998 (1980).
18. P.C. Hanawalt, p.257; and R.T. Schimke, p.317; in Gene Amplification, R.T. Schimke, ed. Cold Sring Harbor Laboratory, N.Y. (1982).
19. H.A. Ingraham, L.M. Dickey, and M. Goulian, submitted for publication.
20. C. Scholtissek, Biochim. Biophys. Acta, 277:459 (1972).
21. M.S. Peterson, H.A. Ingraham, and M. Goulian, J. Biol. Chem. 258:10831 (1983).
22. M. Goulian, L. Dickey, and S. Neynaber, to be published.
23. M.H.N. Tattersall and K.R. Harrap, Ca. Res. 33:3086 (1973).
24. D.W. Roberts and C. Peck, Ca. Res. 41:505 (1981).
25. F. Makino and N. Munakata, J. Bact. 134:24 (1978).
26. H.R. Warner, B.K. Duncan, C. Garrett, and J. Neuhard, J. Bact. 145:687 (1981).

ENHANCEMENT OF METHOTREXATE CYTOTOXICITY BY URACIL ANALOGUES THAT INHIBIT

DEOXYURIDINE TRIPHOSPHATE NUCLEOTIDOHYDROLASE (dUTPase) ACTIVITY

William R. Beck, George E. Wright, Neil J. Nusbaum,
James D. Chang, and Eric M. Isselbacher

Department of Medicine, Harvard Medical School, and
Hematology Research Laboratory, Massachusetts General
Hospital, Boston, Massachusetts 02114

INTRODUCTION

Impairment of DNA synthesis by lack of thymine nucleotides was long considered the critical cytotoxic event in cells exposed to methotrexate (MTX) (1). Recent work has shown, however, that the thymineless state is associated with increases in intracellular dUMP and dUTP with resulting uracil misincorporation into DNA (2-4). Subsequent unrepaired excisions of uracils by uracil-DNA-glycosylase (5,6) lead to DNA fragmentation (7-11).

We have found that these effects are more easily produced in some cell types, e.g., human lymphoblasts 8866, than in others, e.g., phyto-hemagglutinin (PHA)-stimulated lymphocytes (12). This accords with the known fact that cell lines differ in their sensitivity to MTX cytotoxicity (9). If uracil misincorporation into DNA is a critical event in the death of cells in these circumstances, it seemed likely that a cell's susceptibility to this state might be influenced by its levels of deoxy-uridine triphosphate nucleotidohydrolase (dUTPase) (EC 3.6.1.23), which tends to exclude dUTP from DNA synthesis. In recent work we found that the levels of these enzymes do show differences in a variety of diverse cell types (12,13). We also have evidence correlating sensitivity to MTX cytotoxicity with dUTPase levels in some cell lines (14).

This paper presents evidence that certain uracil derivatives (here termed analogues), which inhibit dUTPase activity in intact cells or soni-cates, may enhance MTX cytotoxicity as judged by (a) cell counts or clon-ing efficiency; (b) MTX-induced increases in [dUTP]/[dTTP] ratios; (c) uracil misincorporation into DNA; and (d) DNA strand breakage as assayed by the alkaline filter elution method (15) and end-labeling of new 3'-OH termini (16). Other methods used are discussed elsewhere (13,14).

RESULTS

Assays of dUTPase and Uracil-DNA-Glycosylase Levels in Various Cell Lines

An extensive survey showed wide variations in dUTPase and uracil-DNA-glycosylase levels in different cell lines (Fig. 1). Since dUTPase levels

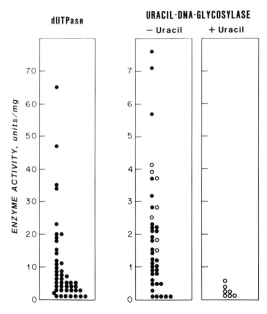

Fig. 1. Scattergrams showing dUTPase levels (left graph) and uracil-DNA-glycosylase levels (middle graph) observed in 46 separate cell lines or cell types as described elsewhere (13). Right graph shows uracil-DNA-glycosylase activity observed in seven cell lines (those in middle graph with open points) when assayed after 30 min incubation of cells with 15 mM uracil.

vary in the course of the cell cycle, we regard these assay results as average levels of enzyme activity in actively growing nonsynchronized cultures. The glycosylase results with open points represent cell lines in which glycosylase was also assayed in the presence of 15 mM uracil. The results in right graph confirm that free uracil is a powerful inhibitor of glycosylase (5,6).

Seven cell lines that differ widely in their dUTPase levels were selected. These were studied further in an effort to determine whether higher dUTPase levels could be correlated with lower MTX sensitivity and vice versa.

Uracil Misincorporation into DNA: Apparent Influence of dUTPase Levels

Measurements were made of actual rates of misincorporation of uracil from [5-³H]dUrd into DNA under standard conditions. In all cell types studied, observed rates of uracil misincorporation into DNA (in pmoles per 100 pmoles DNA nucleotide) in control cells exposed to no MTX were <0.1, both in cultures lacking uracil in the medium and in ones to which 15 mM uracil had been added 24 hr before harvest. Rates in cells exposed to 10 μM MTX for 5 hr in the absence and presence of added uracil were as shown in Table 1. The data confirm that added uracil enhances recovery of misincorporated uracil in DNA by inhibiting its excision by glycosylase. They also suggest (but do not prove) an inverse correlation between MTX-induced uracil misincorporation and dUTPase level in all cell lines tested.

Additional Evidence Associating dUTPase Level with MTX Cytotoxicity

Correlation of Cell Count Effects of MTX with dUTPase Levels. Next, experiments were performed in which MTX cytotoxicity, as judged by effects of MTX on cell counts or cloning efficiency in standardized cultures, was compared with the average dUTPase level of each cell line. The results in Fig. 2 show an apparent inverse correlation between MTX cytotoxicity and dUTPase levels in all cell lines tested but two (PC-4 cells and 3T3-R500 cells), which demonstrated MTX resistance that appears unrelated to dUTPase level. The latter is known to be MTX-resistant (17).

Table 1. Apparent Inverse Relation Between Level of dUTPase and Degree of Uracil Misincorporation Due to MTX in Seven Cell Lines

Cell line	dUTPase (units/mg)	Uracil-DNA-glycosylase (units/mg)	dUMP in DNA −Ura (pmol/100 pmol DNA nucleotides)	dUMP in DNA +Ura (pmol/100 pmol DNA nucleotides)
T cells (HSB-2)	2.1	<0.1	---	4.85
Lymphoblasts (8866)	3.0	1.3	0.72	1.98
Fibroblasts, SV-80 transformed	6.9	1.8	0.37	0.99
Lymphocytes, PHA-stimulated (72 hr)	28.0	2.8	<0.10	0.38
HL-60 cells	29.2	5.9	<0.10	0.14
CEM cells	33.2	5.7	<0.10	0.19
Smooth muscle cells, bovine aorta	61.0	2.9	<0.10	0.15

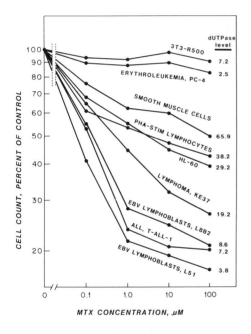

Fig. 2. Cell counts per cu mm in 72-96 hr cultures of various cell lines after incubation with indicated concentration of MTX for last 24 hr of culture period. Results are expressed as percent of control incubated with no MTX. Right column shows dUTPase levels.

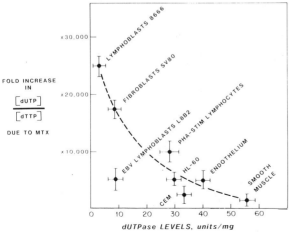

Fig. 3. Plot of -fold increase due to MTX (10 μM, 24 hr) in intracellular [dUTP]/[dTTP] in various cultured cell lines vs. dUTPase levels. Points and bars are ratios ± SDs.

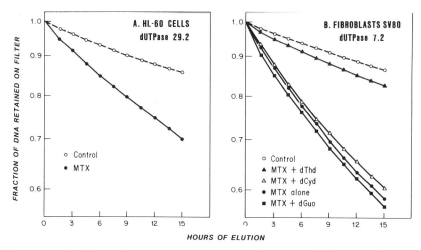

Fig. 4. Alkaline elution patterns in cells with different dUTPase levels. Demonstration of reversal by dThd. MTX (10 μM) was added 24 hr before harvest. A, HL-60 cells. B, fibroblasts SV80. In B, nucleosides (2 μM) were added 24 hr before harvest. Control cells were incubated without MTX.

Effects of MTX on [dUTP]/[dTTP] Ratios. Studies were then carried out on the same cell lines in which -fold increases in the [dUTP]/[dTTP] ratio due to MTX were compared with dUTPase levels. The results, summarized in Fig. 3, indicate that high dUTPase levels tend to minimize increases in [dUTP]/[dTTP] due to MTX. This suggests that lower rises in [dUTP]/[dTTP] result in lower rates of uracil misincorporation into DNA and lower cytotoxicity of MTX.

Alkaline Filter Elution Studies of DNA Strand Breakage. Curves in Fig. 4 indicate a greater effect of MTX in inducing strand breakage in fibroblasts SV80 cells, which have relatively low dUTPase levels, than in HL-60 cells, which have relatively high dUTPase levels. Fig. 4B shows that added dThd (but not dGuo or dCyd) reverses or prevents strand breakage.

Studies of Uracil Analogues: Search for Useful dUTPase Inhibitors

In view of these indications that low cellular dUTPase levels may be associated with relative sensitivity to MTX cytotoxicity we considered the possibility that exposure of a cell to a putative dUTPase inhibitor (along

Table 2. Uracil Analogues Selected For Further Study

Number		Chemical Name
I	GW-17B	6-(3',4'-trimethyleneanilino)uracil
II	NSC-72561C	1-β-D-(2'-deoxy-5'-thioribofuranosyl)thymine
III	NSC-97433N	5-amino-β-D-(2'-deoxyribofuranosyl)uracil
IV	NSC-986760	5-trifluoromethyl-1-β-D-(3',5'-di-O-acetyl-2'-deoxyribofuranosyl)uracil
V	NSC-225110S	3,5-dimethyl-1-β-D-(3',5'-di-O-methyl-2'-deoxyribofuranosyl)uracil
VI	NSC-259915Y	5-(iodacetamidomethyl)-1-β-D-2'-(deoxyribofuranosyl)uracil

with MTX) would enhance induced rises in [dUTP]/[dTTP] and thereby increase uracil misincorporation into DNA. If the basic premise is correct--that killing is due to uracil excision followed by inadequate repair--this maneuver should amplify the cytotoxicity of MTX.

We obtained 44 uracil analogues from G.F. Wright, Department of Pharmacology, University of Massachusetts, and from V.L. Narayanan, Drug Synthesis & Chemistry Branch, NCI. The approach was to test these compounds in various concentrations for their effects on (a) dUTPase; (b) uracil-DNA-glycosylase; (c) cell growth in culture in the presence and absence of MTX; and (d) uracil misincorporation into DNA. From a mass of data we selected for further testing compounds that had all of the following effects: (a) active inhibition of dUTPase in both cells and sonicates in concentrations of 0.1 to 0.25 mM; (b) little or no effect on uracil-DNA-glycosylase; (c) enhancement of cytotoxic effects of MTX on cell growth; (d) minor cytotoxicity in the absence of MTX; (e) enhancement of uracil incorporation into DNA compared to control cells exposed to MTX alone; and (f) little or no inhibition of other enzymes of deoxyribonucleotide metabolism (e.g., ribonucleotide reductase, thymidine kinase, thymidylate synthetase). The compounds listed in Table 2 satisfied these criteria.

None of these compounds were inhibitory when tested directly with purified dUTPase from leukemic blast cells (kindly provided by Dr. Y.-C. Cheng). Since they do inhibit the enzyme in intact cells and sonicates it appears likely that in these circumstances they are first converted to another form which is the active inhibitor. Presumably the active form is a nucleotide, probably a deoxynucleoside triphosphate. This suspicion is supported by data showing that the inhibitory potency of compound I is enhanced in sonicates by addition of 1 mM ATP (see below).

Effects of Uracil Analogues on MTX Cytotoxicity

Some of the results of this study are summarized in Fig. 5, which shows effects of certain uracil analogues on MTX cytotoxicity as judged by cell counts under standard conditions (Fig. 5A) and on DNA strand breakage (Fig. 5B). Effects of these analogues on uracil misincorporation into DNA are summarized in Table 3. None had inhibitory or stimulatory effects on uracil-DNA-glycosylase activity in concentrations up to 3 mM.

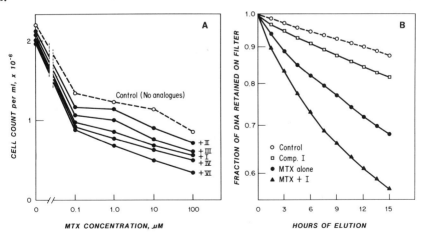

Fig. 5. Effects of selected uracil analogues on sensitivity of PHA-stimulated lymphocytes to indicated concentrations of MTX, as judged by (A) cell counts in 72 hr cultures, and (B) alkaline elution patterns. Analogues (0.1 mM) were added with MTX 24 hr before harvest.

Table 3. Effects of Uracil Analogues (0.1 mM) on Uracil Misincorporation
into DNA due to MTX (10 μM) in PHA-stimulated Lymphocytes

Uracil Analogue	Inhibition of:[a] dUTPase (per cent)	Glycosylase	Analogue Effect on Cell Count (per cent)	dUMP in DNA −Anal +MTX	+Anal +MTX	+Anal −MTX
				(pmol/100 pmol DNA nucleotide)		
I	67.3[b]	− 8.2	−40	2.2	4.6	1.1
II	78.7	0.0	−11	2.1	3.6	1.0
III	59.3	1.9	−30	2.6	10.4	0.9
IV	61.1	−12.6	−20	1.9	14.1	1.4
V	87.9	− 9.5	−18	1.8	13.1	1.3
VI	86.4	− 4.4	−50	2.3	3.0	0.5

[a] In sonicate after 30 min preincubation.
[b] Inhibition 90.4 per cent when 1 mM ATP also present.

Since excision of misincorporated uracils by uracil-DNA-glycosylase
creates new 3'-OH termini, assay of their concentration in the DNA of
variously treated cells should provide added perspective on how many
uracils were incorporated and then excised. This figure when added to
the number of misincorporated uracils still present should give a more
complete picture of the patterns of DNA synthesis occurring in MTX-treated
cells.

When free 3'-OH termini were assayed by end labeling of isolated DNA
with [α-^{32}P]ATP using terminal deoxynucleotidyl transferase, an increase
in their number was invariably caused by exposure to MTX (in the absence
of exogenous uracil) (Table 4). They were increased further by the presence of Compound I. This also suggests that dUTPase inhibition enhances
uracil misincorporation and MTX cytotoxicity.

DISCUSSION

Additional analogues are being screened in attempts to find compounds
active at lower concentrations. Ingram and Goulian have shown that several dUTP analogues are only moderate inhibitors of purified dUTPase (18).
We plan to convert the analogues discussed above to deoxynucleoside monophosphates and then to their di-, and triphosphate derivatives for further
studies of their effects on purified dUTPase. Early experiments had only
limited success in converting a few uracil analogues to deoxyribonucleosides by incubation with deoxyadenosine and trans-N-deoxyribosylase of
Lactobacillus leichmannii (19). We are attempting also to establish
whether analogues are converted to nucleotides in vivo by seeking evidence

Table 4. Effect of MTX and dUTPase Inhibitor (Compound I) on
Concentration of 3'-OH Termini in DNA

Additions to cell cultures[a]	3'-OH Termini in DNA (pmol per 100 pmol DNA nucleotide)
None	<0.1
MTX (10 μM, 24 hr)	2.9
Compound I (0.1 mM, 24 hr)	0.3
MTX plus Compound I	5.4

[a] PHA-stimulated lymphocytes

in HPLC analyses of acid-soluble extracts and DNA deoxyribonucleotides as to whether analogue derivatives are present. UV absorption spectra of these compounds are sufficiently distinctive to permit the identification of unusual HPLC peaks.

It is recognized that an inverse relation between dUTPase level and MTX resistance is but one of several relevant factors influencing response to MTX, especially when cell types of diverse origin are compared. Nonetheless, the data do suggest such an association and, if substantiated, they have three implications: (a) the level of dUTPase activity in cells is not grossly in excess of need and its activity threshold can be exceeded when dUTP levels are increased; (b) compounds that inhibit dUTPase activity can further elevate dUTP levels and thereby enhance uracil misincorporation into DNA and MTX cytotoxicity; and (c) to the several recognized mechanisms of MTX resistance (20), which are known or presumed to operate in 3T3-R500 cells and PC4 cells, may be added high dUTPase activity.

Acknowledgement. This work was supported by a research grant from the American Cancer Society (CH-282).

REFERENCES

1. R.C. Jackson and G.B. Grindley, The biochemical basis for methotrexate cytotoxicity, in: "Folate Antagonists as Therapeutic Agents, Vol. 1. Biochemistry, Molecular Actions, and Synthetic Design," F.M. Sirotnak, J.J. Burchall, W.B. Ensminger, and J.A. Montgomery, eds., Academic Press, Orlando (1984) pp. 290-315.
2. R.H. Grafstrom, B.Y. Tseng, and M. Goulian, The incorporation of uracil into animal cell DNA in vitro, Cell 15:131-140 (1978).
3. M. Goulian, B. Bleile, and B.Y. Tseng, Methotrexate-induced misincorporation of uracil into DNA, Proc. Natl. Acad. Sci. USA 77:1956-1960 (1980).
4. M. Goulian, B. Bleile, and B.Y. Tseng, The effect of methotrexate on levels of dUTP in cells, J. Biol. Chem. 255:10630-10637 (1980).
5. T. Lindahl, New class of enzymes acting on damaged DNA, Nature 259:64-66 (1976).
6. T. Lindahl, S. Ljungquist, B. Nyberg, and B. Sperens, DNA N-glycosidases, J. Biol. Chem. 252:3286-3294 (1977).
7. B.-K. Tye, J. Chien, I.R. Lehman, B.K. Duncan, and H.R. Warner, Uracil incorporation: a source of pulse-labeled DNA fragments in the replication of the Escherichia coli chromosome, Proc. Natl. Acad. Sci. USA 75:233-237 (1978).
8. E. Wist, O. Unhjem, and H. Krokan, Accumulation of small fragments of DNA in isolated HeLa cell nuclei due to transient incorporation of dUMP, Biochim. Biophys. Acta 520:253-270 (1978).
9. S.N. Wickramasinghe and A.V. Hoffbrand, Defective DNA synthesis in megaloblastic anemia. Studies employing velocity sedimentation in alkaline sucrose gradients, Biochim. Biophys. Acta 563:46-58 (1979).
10. K. Brynolf, R. Eliasson, and P. Reichard, Formation of Okazaki fragments in polyoma DNA synthesis caused by misincorporation of uracil, Cell 13:573-580 (1978).
11. J.C. Li and E. Kaminskas, Accumulation of DNA strand breaks and methotrexate cytotoxicity, Proc. Natl. Acad. Sci. USA 81:5694-5698 (1984).
12. T.-T. Pelliniemi and W.S. Beck, Prevention of uracil misincorporation into DNA in megaloblastic cells, Blood 54, Suppl. 1:43a (1979).
13. W.S. Beck, J.A. Vilpo, and J.D. Chang, Differences in deoxyuridine triphosphate nucleotidohydrolase (dUTPase) and uracil-DNA-glycosylase levels in various cell lines, Cancer Res. submitted for publication.

14. W.S. Beck, J.D. Chang, N.J. Nusbaum, E.M. Isselbacher, T.-T. Pelli-niemi, and J.A. Vilpo, Role of uracil misincorporation into DNA in the cytotoxicity of methotrexate: importance of deoxuridine triphos-phate nucleotidohydrolase (dUTPase) levels, Cancer Res., submitted for publication.

15. K.W. Kohn, R.A.G. Ewig, L.C. Erickson, and L.A. Zwelling, Measurement of strand breaks and cross-links by alkaline elution, in: "DNA Repair: A Laboratory Manual of Research Procedures, Vol. 1, Part B," E.C. Friedberg and P.C. Hanawalt, eds., Dekker, New York (1981), pp. 379-401.

16. U.W. Hagen, Measurement of strand breaks by end labeling, in: "DNA Repair: A Laboratory Manual of Research Procedures, Vol. 1, Part B," E.C. Friedberg and P.C. Hanawalt, eds., Dekker, New York (1981), pp. 431-445.

17. P.C. Brown, S.M. Beverly, and R.T. Schimke, Relationship of amplified dihydrofolate reductase genes to double minute chromosomes in unstably resistant mouse fibroblast cell lines, Mol. Cell. Biol. 1:1077-1083 (1981).

18. H.A. Ingraham and M. Goulian, Deoxyuridine triphosphatase: a potential site of interaction with pyrimidine nucleotide analogues. Biochem. Biophys. Res. Commun. 109:746-752 (1982).

19. W.S. Beck ad M. Levin, Purification, kinetics and repression control of bacterial trans-N-deoxyribosylase. J. Biol. Chem. 238:702-709 (1963).

20. A.M. Albrecht and J.L. Biedler, Acquired resistance of tumor cells to folate antagonists, in: "Folate Antagonists as Therapeutic Agents, Vol. 1 Biochemistry, Molecular Actions, and Synthetic Design," F.M. Sirotnak, J.J. Burchall, W.B. Ensminger, and J.A. Montgomery, eds., Academic Press, Orlando (1984), pp. 317-353.

INFLUENCE OF METHOTREXATE ON PURINE AND PYRIMIDINE POOLS AND ON CELL PHASE DISTRIBUTION OF CULTURED HUMAN LYMPHOBLASTS

Ronney A. De Abreu, Jos P. M. Bökkerink, Marinka A. H. Bakker, Tilly W. Hulscher, John M. van Baal, Chris H. M. M. de Bruijn and Egbert D. A. M. Schretlen

Centre for Pediatric Oncology SE Netherlands, Dept. of Pediatrics, University Hospital St. Radboud, Nijmegen The Netherlands

INTRODUCTION

Methotrexate (MTX) and 6-mercaptopurine (6MP) are among the most commonly used agents in the maintenance treatment of acute lymphoblastic leukemia (ALL) in children.[1-3] There are biochemical considerations for an increased efficacy of a combination of both drugs in maintenance therapy of ALL. MTX is a strong inhibitor of dihydrofolate reductase. As a result tetrahydrofolates are reduced. Tetrahydrofolate coenzymes are required for one-carbon transfer reactions in purine de novo synthesis and thymidine biosynthesis. As a consequence a purine-less and a thymidylate-less state will occur, ultimately resulting in inhibition of DNA biosynthesis.[4-9]

Previously, we have demonstrated in MOLT 4 and Raji cells an increase of intracellular 5-phospho-1-pyrophosphate (PRPP) levels due to inhibition of purine de novo synthesis by MTX.[10] Elevation of PRPP levels induced by MTX is also associated with enhanced intracellular incorporation of 6MP.[10]

In the present study we are investigating the effects of MTX on purine and pyrimidine concentrations in MOLT 4 and Raji cells, under conditions by which 6MP incorporation is enhanced after pretreatment of cells with MTX. Biochemical pertubations produced by MTX are correlated with effects on cell phase distribution and cytotoxicity.

METHODS

Conditions for cell culture are described previously.[10] At various time-intervals after addition of MTX, samples were taken to measure several data simultaneously: e.g. PRPP levels, purine and pyrimidine nucleotides, flow cytometry, cell growth and cell viability. Analysis of purine and pyrimidine nucleotides were performed by high performance liquid chromatography.[11] Deoxyribonucleotide triphosphates were measured after periodate oxydation of ribonucleotides.

RESULTS AND DISCUSSION

    Intracellular deoxyribonucleotide concentrations are decreased when
MOLT 4 cells are exposed to 0.2 μM MTX (Fig.1). A rapid decrease of dTTP
is observed. Thus, dihydrofolate reductase must be markedly inhibited by
a concentration of 0.2 μM MTX in these cells. The consequence of the inhibition of dihydrofolate reductase will be a lack of tetrahydrofolates,
resulting in reduced purine de novo synthesis and decrease of ATP and
GTP levels (Fig.1). Presumably, as a result of decreased DNA synthesis
also the concentrations of dATP and dGTP are decreased initially.

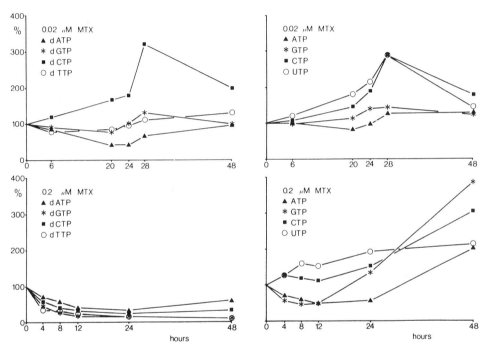

Fig.1. Effects of MTX on nucleotide concentrations in MOLT 4 cells.
        The concentrations (pmol/10[6] viable cells) are expressed as %
        of untreated cells.

    Treatment of MOLT 4 cells with 0.02 μM MTX does not result in signi-
ficant decrease   of dTTP levels (Fig.1), thus inhibition of dihydrofolate reductase is not complete under these conditions. Furthermore, purine
de novo synthesis is more resistant to MTX inhibition than thymidylate
synthesis.[9] Therefore, the inhibition of purine de novo synthesis can
only be partial in MOLT 4 cells treated with 0.02 μM MTX, so no decrease
in ATP and GTP is observed under these conditions (Fig.1).

Evidence for a partial inhibition of purine de novo synthesis was
obtained from data of ^{14}C-glycine incorporation studies and from measurements of PRPP and amidoimidazole carboxamide ribonucleotide (AICAR) levels.
After pretreatment of MOLT 4 with 0.2 µM MTX for 24 hours, ^{14}C-glycine incorporation into purine nucleotides is inhibited completely. Thus, purine
de novo synthesis is inhibited. After exposure of MOLT 4 cells to 0.02 µM
MTX, ^{14}C-glycine incorporation is still present suggesting a still active
de novo synthesis. Furthermore, when MOLT 4 cells are treated with 0.2 µM
MTX and 0.02 µM MTX, PRPP levels are elevated in these cells.[10] However,
with 0.02 µM MTX also AICAR levels are increased, whereas after pretreatment with 0.2 µM MTX no detectable AICAR levels are found. Therefore,
if MOLT 4 cells are exposed to 0.2 µM MTX, GAR formyltransferase is inhibited completely and PRPP levels are increased but no detectable amount
AICAR will be formed. However, if MOLT 4 cells are exposed to 0.02 µM MTX,
GAR formyltransferase and AICAR formyltransferase are inhibited partially,
since PRPP and AICAR levels are increased and ^{14}C-glycine can still be
incorporated into purine nucleotides.

In Raji cells, a significant decrease of dTTP concentrations is observed with both concentrations of MTX (Fig.2). The results indicate that
inhibition of thymidylate biosynthesis is more pronounced in Raji cells
during treatment with MTX. Although dihydrofolate reductase seems to be
markedly inhibited in Raji cells, DNA synthesis seems to be still active
since dATP, dGTP and dCTP concentrations are not affected significantly.
The results of glycine incorporation studies and of PRPP and AICAR measurements are similar to those of MOLT 4 cells. After exposure of Raji cells
to 0.2 µM MTX, ^{14}C-glycine incorporation is inhibited almost completely
and PRPP levels are enhanced. After treatment of Raji cells with 0.02 µM
MTX, ^{14}C-glycine incorporation is partially inhibited and PRPP as well
as AICAR levels are increased.

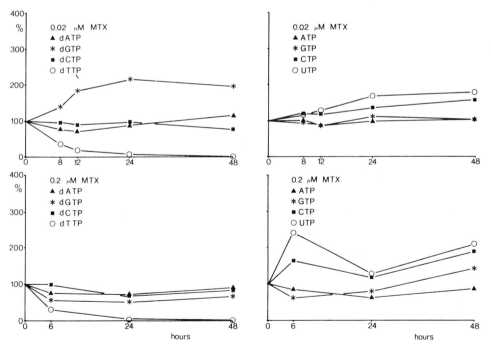

Fig.2. Effects of MTX on nucleotide concentrations in Raji cells.
The concentrations (pmol/10[6] viable cells) are expressed as
% of untreated cells.

As a result of the inhibition of thymidylate synthetase and the increased availability of PRPP by MTX, pyrimidine ribonucleotide concentrations are increased in both cell lines (Fig.1, 2 and 3). An exceptional increase of UMP levels can be observed in both lines, after exposure to 0.2 µM and 0.02 µM MTX, respectively.

UMP (pmol/10^6 viable cells)

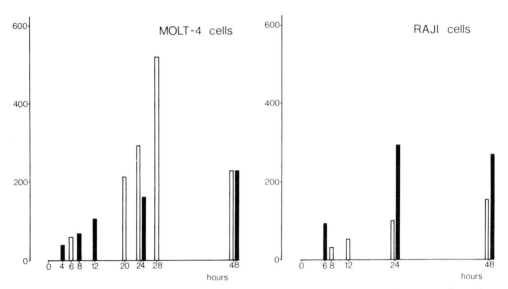

Fig.3.   Enhanced UMP concentrations after treatment of MOLT 4 cells (left) and Raji cells (right).
Open bars   : treatment with 0.02 µM MTX.
Closed bars: treatment with 0.2 µM MTX.

The results of flowcytometry studies are demonstrated in Fig.4. During treatment with 0.2 µM MTX, MOLT 4 cells are arrested in late G1-phase, followed by a complete prevention of cell progression through the cell cycle. With 0.02 µM MTX the cells are arrested predominantly in early S-phase. Accumulation of dCTP in early S-phase has been described earlier.[12] This may explain why dCTP levels are increased in MOLT 4 cells after pretreatment with 0.02 µM MTX (Fig.1). During treatment of Raji cells with 0.2 µM MTX a progressive decrease of amounts of cells in S-phase and G_2+M can be observed (Fig.4). However, in contrast with MOLT 4 cells, cell proliferation is still present in Raji cells. In these cells no accumulation of dCTP is observed during exposure to 0.02 µM MTX, because the distribution of cells shows an accumulation from late G1 to late S-phase (Fig.4).

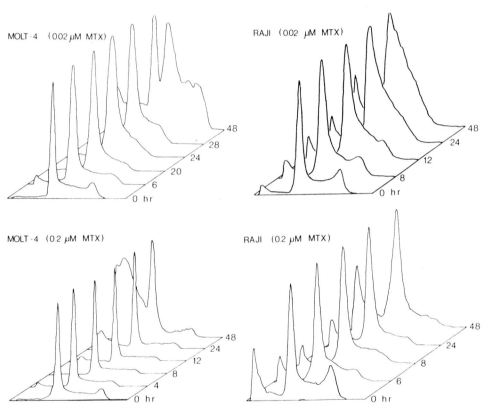

MOLT·4 (0.02 μM MTX)

RAJI (0.02 μM MTX)

MOLT·4 (0.2 μM MTX)

RAJI (0.2 μM MTX)

Fig.4. DNA histograms of MOLT 4 (left) and Raji (right) cells, before and at indicated times after addition of MTX.

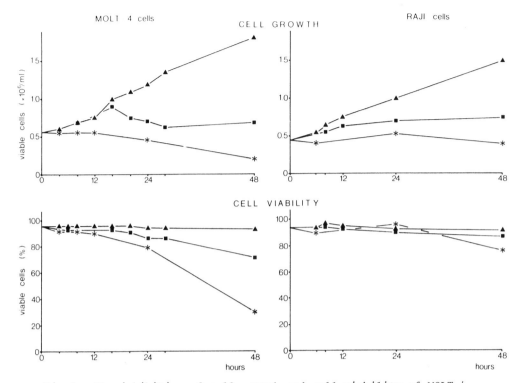

Fig.5. The inhibition of cell growth and cell viability of MOLT 4
        (left) and Raji (right) cells following treatment of MTX
        Triangles : untreated cells
        Squares   : cells treated with 0.02 μM MTX
        Asterisks : cells treated with 0.2 μM MTX.

The differences between MOLT 4 cells and Raji cells are also reflected in the effects of MTX on cell growth and cell viability (Fig.5).
0.2 μM MTX causes a cytostatic effect in Raji cells (after 48 hours 80%
of the cells are viable), whereas in MOLT 4 cells it causes a cytotoxic
effect (30% of the cells are viable).

In conclusion MTX exhibits a more cytotoxic effect on MOLT 4 cells,
whereas in Raji cells the disturbance of thymidylate biosynthesis is more
profound. Moreover, the data of our flowcytometry studies indicate that
MTX causes different effects on cell phase distribution of MOLT and Raji
cells and that the effects on the cell phase distribution in these cells
are dependent on the MTX concentrations. This may have consequences for
combinations of MTX and 6MP, since 6MP is preferentially incorporated
in RNA in G1-phase cells and preferentially in DNA in S-phase cells.

ACKNOWLEDGEMENT
        This work was supported by the Queen Wilhelmina Fund for Cancer
Research of the Netherlands, grant SNUKC 82-3.

REFERENCES

1. E. Frei III, E. J. Freireich, E. Gehan, D. Pinkel , J. F. Holland, O. Selawry, F. Haurani, C. L. Spurr, D. M. Hayes, G. W. James, H. Rothberg, D. B. Sodee, R. W. Rundles, L. R. Schroeder, B. Hoogstraten, I. J. Wolman, D. G. Traggis, T. Cooper, B.R. Gendel, F. Ebaugh and R. Taylor., Studies of sequential and combination antimetabolite therapy in acute leukemia: 6-mercaptopurine and methotrexate., Blood 18:431 (1961)
2. E. Frei III and S. E. Sallan, Acute lymphoblastic leukemia: treatment., Cancer 42:828 (1978)
3. D. Pinkel, The ninth annual David Karnofsky Lecture. Treatment of acute lymphocytic leukemia., Cancer 43:1128 (1979)
4. A. C. Sartorelli and G. A. LePage, A methopterin on the purine biosynthesis of susceptible and resistant TA3 ascites cells., Cancer Res. 18:1336 (1958)
5. W. C. Werkheiser, The biochemical, cellular and pharmacological action and effects of the folic acid antagonists., Cancer Res. 23:1277 (1963)
6. J. Borsa and G. F. Whitmore, Studies relating to the mode of action of methotrexate. II Studies in sites of action on L-cells in vitro., Mol. Pharmacol. 5:303 (1969)
7. R. G. Moran, M. Mulkins and C. Heidelberger, Role of thymidylate synthetase activity in development of methotrexate cytotoxicity., Proc. Natl. Acad. Sci. USA 76:5924 (1979)
8. W. M. Hryniuk, L. W. Brox, J. F. Henderson and T. Tamaoki, Consequences of methotrexate inhibition of purine biosynthesis in L5178Y cells., Cancer Res. 35:1427 (1975)
9. J. C. White, S. Loftfield and I. D. Goldman, The mechanism of action of methotrexate. III Requirement of free intracellular methotrexate for maximal suppression of ^{14}C-formate incorporation into nucleic acids and protein., Mol. Pharmacol. II:287 (1975)
10. J. P. M. Bökkerink, R. A. De Abreu, J. P. R. M. van Laarhoven, M. A. H. Bakker, T. W. Hulscher, E. D. A. M. Schretlen and C. H. M. M. de Bruijn, Increased availability of phosphoribosyl pyrophosphate as the basis for enhanced 6-mercaptopurine incorporation by methotrexate, in cultured human lymphoblasts., Submitted for publication in Purine Metabolism in Man-V
11. R. A. De Abreu, J. M. van Baal, J. A. J. M. Bakkeren, C. H. M. M. de Bruijn and E. D. A. M. Schretlen, High-performance liquid chromatographic assay for identification and quantitation of nucleotides in lymphocytes and malignant lymphoblasts., J. Chromatogr. 227:45 (1982)
12. T. W. Taylor, P. Slowiaczek, P. R. Francis and M. H. N. Tattersall, Biochemical and cell pertubations in methotrexate treated cells., Mol. Pharmacol. 21:204 (1982)

MODULATION OF CYTOTOXICITY AND METABOLISM OF 5-FLUOROURACIL

IN TWO INTESTINE CELL LINES

G.J. Peters, E. Laurensse, A. Leyva, and H.M. Pinedo

Department of Oncology, Free University Hospital
P.O. Box 7075, 1007 MB Amsterdam
The Netherlands

INTRODUCTION

The antitumor activity of 5-fluorouracil (5FU) is dependent on its conversion to active nucleotides. The rate of direct conversion to FUMP, catalyzed by orotate phosphoribosyl transferase (OPRT), depends on the availability of the co-substrate phosphoribosylpyrophosphate (PRPP). The rate of the indirect conversion to FUMP or FdUMP via fluorouridine (FUR) or fluorodeoxyuridine (FUdR), respectively, catalyzed by a pyrimidine nucleoside phosphorylase (PNP), depends on the availability of the cosubstrates ribose-1-phosphate (Rib-1-P) and deoxyribose-1-phosphate (dRib-1-P). Nucleosides can act as ribose donors. Especially inosine and deoxyinosine are good sources since the major pathway of their metabolism involves phosphorolysis to hypoxanthine (1). Furthermore, these nucleosides are relatively non-toxic (2). In a variety of systems either inosine or deoxyinosine have been shown to increase growth inhibition by 5FU (3,4,5) and to enhance antitumor activity (6). It has also been shown that deoxyinosine or inosine can protect cells against 5FU (4,7).

The mechanism of the synergistic effect on 5FU cytotoxicity is different for inosine or deoxyinosine. Inosine enhances the incorporation of 5FU into RNA (5), while deoxyinosine can enhance the accumulation of FdUMP (3) leading to inhibition of thymidylate synthase. A lack of synergism with inosine and deoxyinosine can be due to interference by the nucleosides with 5FU anabolism leading to decreased 5FU incorporation into RNA (4). By studying 5FU metabolism, the inhibition of thymidylate synthase and the incorporation of 5FU into RNA can be quantitated, but the contribution of each parameter to cytotoxicity may remain unclear. The use of purine nucleosides as modulators of 5FU metabolism may give more insight into specific pathways of 5FU anabolism responsible for 5FU cytotoxicity. In order to examine these pathways we compared the effect of deoxyinosine on 5FU metabolism and cytotoxicity in two human colonic cell lines differing in their sensitivity to 5FU and 5'-deoxy-5-fluorouridine (Doxifluridine, 5'dFUR) a precursor of 5-fluorouracil.

MATERIALS AND METHODS

The origins of the cell lines, culture media, fetal bovine serum

have been described previously (8) as well as the culturing
conditions and calculations for the growth inhibition experiments
(9). For these experiments cells were grown in dialyzed serum.
For the synergism studies nucleosides (or dIMP) and 5FU were added
simultaneously to the cells. Cells were cultured for 24 and 48 hr
after the addition of these compounds. The media in which the cells
were cultured were collected after 24 and 48 hr, deproteinized
and frozen until analysis. The concentrations of nucleosides and
nucleotides were measured using standard HPLC methods, nucleosides
on a RP-18 column and nucleotides on a Partisil SAX column.

The effect of nucleosides and dIMP on the PRPP concentration
in cells was determined as described previously (10,11). The
effect of nucleosides and dIMP on the concentration of the pentose
phosphates was determined by a method analogous to that used for
PRPP. The concentration of the pentose phosphates was measured by the
method of McIvor et al. (12) with slight modifications.

All enzyme assays were performed using labeled 5FU as a
substrate as described previously (9). Enzyme activities were
measured with the co-substrates Rib-1-P and dRib-1-P and with the
nucleoside precursors inosine and deoxyinosine. The activity of
thymidylate synthase was measured at 1 and 10 μM dUMP as described
(9). Measurement of the incorporation of tritiated 5FU into RNA and
DNA was performed according to methods slightly modified from
those described previously (8,9). Briefly, cells were precipitated on
a filter with trichloroacetic acid. One half of the filters were
counted directly, while the other half of the filters were incubated
with RNAse to remove RNA. Thereafter, DNA was reprecipitated with
trichloroacetic acid and the filters were counted.

RESULTS

Sensitivity to drugs of the WiDr and Intestine 407 cell lines
was determined after 24 and 48 hr exposure to 5FU or the prodrug
5'dFUR. The IC 50 values after 24 hr exposure are summarized in
Table 1. WiDr cells are more sensitive than Intestine 407 cells to
5FU as well as 5'dFUR. In order to study the effect of deoxyinosine
on sensitivity to 5FU, low relatively non-toxic concentratons of
5FU were used in combination with increasing concentrations of
deoxyinosine. Deoxyinosine itself did not inhibit cell growth at low
concentrations, but at a high concentration growth inhibition was
observed in WiDr cells and to a lesser extent in Intestine 407 cells
(Fig. 1). At 10 μM deoxyinosine a synergism with 0.5 μM 5FU in WiDr
and with 1.0 μM 5FU in Intestine 407 cells was found only after 24
hr exposure of the cells to the combination but not after 48 hr
exposure. At higher concentrations of deoxyinosine a synergism was also

Table 1. Sensitivity of WiDr and Intestine 407 cells to 5FU and 5'dFUR.

Cell line	IC 50 (μM)	
	5FU	5'dFUR
Intestine 407	1.7 $\pm$ 0.6	82 $\pm$ 23
WiDr	0.7 $\pm$ 0.1	18 $\pm$ 5

Values are means $\pm$ SEM from 3-4 separate experiments. Cell number
was determined after 24 hr exposure of the cells to the drugs, IC 50
values (the concentration that causes 50% growth inhibition) were
calculated from the individual growth inhibition curves.

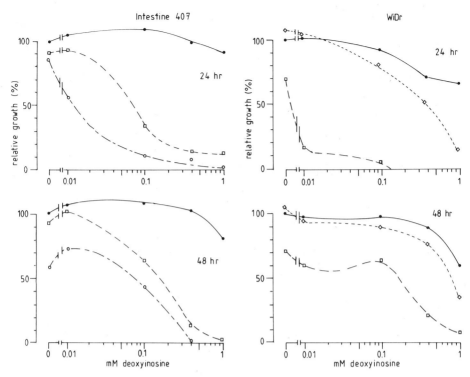

Fig. 1. Growth inhibition by 5FU in combination with deoxyinosine in Intestine 407 and WiDr cells. Values are means of 3-5 separate experiments. 5FU was present at 0.1 μM (◇---◇), 0.5 μM (□——□) and 1 μM (o— -o) or not present (●——●).

observed with 0.1 μM 5FU in WiDr and with 0.5 μM in Intestine 407 cells. In both cell lines the synergistic effect of the combination appeared to be lower after 48 hr. This was more pronounced in the WiDr cells.

Synergistic studies were also performed with dIMP which can serve as a precursor for deoxyinosine after its dephosphorylation. No growth inhibition by dIMP was observed in either cell line. In Intestine 407 cells a marked synergism with the various combinations of 5FU and dIMP was observed (Fig. 2). The synergism was comparable to that observed with deoxyinosine. However, no synergism with dIMP was observed in WiDr cells (data not shown).

In order to investigate the mechanism of growth inhibition in both cell lines, thymidine was added to the cultures simultaneously with the combinations of 5FU with deoxyinosine and dIMP (Table 2). Thymidine prevented the growth inhibition by 5FU and deoxyinosine, and by 5FU and dIMP in Intestine 407 cells. However, this was not the case in WiDr cells. Thymidine even enhanced the inhibition by 5FU and deoxyinosine or dIMP.

Examination of the medium in which the cells were cultured in the presence of deoxyinosine demonstrated that in both cell lines a considerable part of deoxyinosine was phosphorolyzed to hypoxanthine (Table 3). After 48 hr more deoxyinosine appeared to be converted to hypoxanthine in Intestine 407 cells than in WiDr cells. Both cell lines

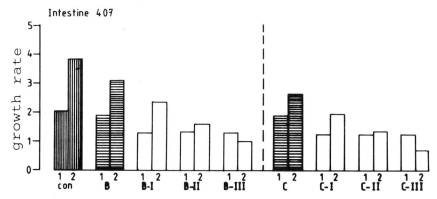

Fig. 2. Synergistic effect of 5FU and dIMP on growth of Intestine 407 cells. Growth rate was determined after one day (1) and two days (2). Bars are means of 2-4 separate experiments. Con, growth of control cells in the absence of any additions; B, 0.5 μM 5FU; C, 1.0 μM 5FU; I-III, 5FU and additional dIMP at 0.1, 0.4 and 1.0 mM, respectively.

Table 2. Effect of thymidine (TdR) on growth inhibition by 5FU in the presence of deoxyinosine or dIMP.

Cell line	Time (hr)	Control	5FU					
			No TdR		2 μM TdR		10 μM TdR	
			dino	dIMP	dino	dIMP	dino	dIMP
WiDr	24	1.76	1.09	1.42	1.23	1.35	0.92	0.9
	48	3.36	1.61	2.26	2.60	2.54	1.89	2.1
Intestine 407	24	1.96	1.19	1.27	1.46	1.75	1.71	1.7
	48	3.83	1.19	1.36	3.08	2.80	3.54	3.0

Growth rates are means of 3-6 separate experiments. In WiDr cells 5FU was present at 0.5 μM and in Intestine 407 cells at 1.0 μM. The concentration of deoxyinosine (dino) and dIMP was 0.4 mM. All compounds were added simultaneously to the cultures.

Table 3. Concentrations of metabolites of deoxyinosine and dIMP in culture medium of WiDr and Intestine 407 (I 407) cells.

Metabolite (μM)	Time (hr)	Addition (1 mM)			
		Deoxyinosine		dIMP	
		WiDr	I 407	WiDr	I 407
dIMP	24	–	–	647	650
	48	–	–	630	290
Deoxyinosine	24	750	727	64	198
	48	700	542	102	305
Hypoxanthine	24	243	249	73	144
	48	284	415	168	308
Inosine	24	6	22	–	–
	48	14	34	–	–

Table 4. Activity of PNP in WiDr and Intestine 407 cells with various co-substrates.

Co-substrate	WiDr	Intestine 407
Rib-1-P	1.3 ± 0.2	0.9 ± 0.1
Inosine	0.6 ± 0.2	0.85 ± 0.2
dRib-1-P	2.1 ± 0.3	1.3 ± 0.2
Deoxyinosine	0.1 ± 0.05	0.3 ± 0.13

Enzyme activities (in nmol/hr per 10^6 cells) are means ± SEM of 3-4 separate experiments and were measured with 5FU as substrate; the various co-substrates were present at 2.5 mM final concentration.

Table 5. Effect of inosine and deoxyinosine and dIMP on the concentration of Rib-1-P and dRib-1-P in WiDr and Intestine 407 (I 407).

Addition	Cell line	Rib-1-P	dRib-1-P
None	WiDr	0.72	-
	I 407	0.86	-
Inosine	WiDr	2.22	-
	I 407	2.54	-
Deoxyinosine	WiDr	0.72	1.80
	I 407	0.86	2.23
dIMP	WiDr	0.72	0.78
	I 407	0.86	2.97

Values (pmol/10^6 cells) are means of 3-5 separate experiments. Inosine, deoxyinosine and dIMP were present at 1 mM for 1 hr.

were able to convert hypoxanthine to inosine. Both WiDr and Intestine 407 cells converted dIMP to deoxyinosine and hypoxanthine. After 48 hr dephosphorylation was higher with the Intestine 407 cells than with WiDr cells. Inosine could not be detected under these circumstances.

The activities of enzymes responsible for 5FU activation were measured with 5FU as a substrate and with various co-substrates. The activity of OPRT, for which PRPP is the cosubstrate, was 2.07 in WiDr cells and 2.22 nmol/hr per 10^6 cells in Intestine 407 cells. The activity of PNP was slightly higher in the WiDr cell line (Table 4). PNP not only converts 5'dFUR to 5FU, but also 5FU to FUR and FUdR. PNP activity was also measured with inosine and deoxyinosine as precursors for the co-substrates Rib-1-P and dRib-1-P. In Intestine 407 cells PNP activity with inosine was comparable to that with Rib-1-P, but with deoxyinosine PNP activity was about 20% of that with dRib-1-P. In WiDr cells PNP activity with inosine and deoxyinosine was much lower than with Rib-1-P and dRib-1-P.

In both WiDr and Intestine 407 cells, inosine was able to increase the Rib-1-P concentrations 2- to 3-fold (Table 5). No dRib-1-P was detectable in untreated cells; after incubation with deoxyinosine the

Table 6. Inhibition of thymidylate synthase by FdUMP.

Cell line	FdUMP (μM)	Enzyme activity at	
		1 μM dUMP	10 μM dUMP
WiDr	0	110 $\pm$ 14	380 $\pm$ 41
	0.01	16 $\pm$ 2	144 $\pm$ 17
	0.1	4 $\pm$ 2	28 $\pm$ 12
Intestine 407	0	430 $\pm$ 51	1358 $\pm$ 57
	0.01	39 $\pm$ 10	586 $\pm$ 25
	0.1	9 $\pm$ 2	116 $\pm$ 36

Values (in pmol/hr per 10^6 cells) are means $\pm$ SEM from 3-5 separate experiments.

concentration of dRib-1-P was comparable to that of Rib-1-P obtained after incubation with inosine. Incubation with dIMP resulted in much higher concentrations of dRib-1-P in Intestine 407 cells than in WiDr cells. In both cell lines the concentration of PRPP was reduced after a 2 hr incubation with inosine, deoxyinosine or dIMP (data not shown). Since the increase of the dRib-1-P pool due to deoxyinosine would result in an elevation of FdUMP levels (3) we measured the inhibition of thymidylate synthase by FdUMP (Table 6). Inhibition was measured at two concentrations of the substrate dUMP to examine for any kinetic differences between the two cell lines. The activity of thymidylate synthase was 3-4 times higher in Intestine 407 cells than in WiDr cells at both dUMP concentrations. The apparent Km value for dUMP was about 4 μM in both cell lines. The relative inhibition by 0.01 μM FdUMP was comparable in the two cell lines. At 0.1 μM FdUMP enzyme activity was inhibited more than 90% in both cell lines. The stimulation of 5FU anabolism towards deoxynucleotides produced at most a slight increase in the incorporation into DNA (Table 7). However, incorporation of 5FU into RNA was inhibited in both cell lines.

DISCUSSION

Modulation of the metabolism of 5FU with purine nucleosides can result in a specific enhancement of particular pathways of 5FU metabolism (3,5,13). This type of biochemical modulation will provide more information into the relative contribution of these pathways to the cytotoxicity of 5FU. Since inosine did not enhance the sensitivity to 5FU of WiDr and Intestine 407 cells (data not shown), we concentrated on the effect of deoxyinosine. Deoxyinosine appeared to be synergistic with 5FU, as was also demonstrated in other systems (3,6). Analogous to the similar effect of guanosine and GMP on potentiation of 5FU (13,14), the more soluble dIMP also potentiated the effect of 5FU. dIMP might also prevent a depletion in Pi which can occur when deoxyinosine is given as a precursor. The synergistic effect of dIMP and deoxyinosine with 5FU was comparable in Intestine 407 cells but not in WiDr cells. Analysis of the medium demonstrated that dephosphorylation of dIMP by WiDr is relatively low, which correlates with the lower ecto-phosphatase and ecto-nucleotidase activity in this cell line compared to Intestine 407 (9). As a consequence the expansion of the dRib-1-P pool by dIMP in WiDr cells is relatively low (Table 5). This means that less dRib-1-P is available to enhance the conversion of 5FU to FUdR and thus to FdUMP. This might account for the relatively little synergism found in WiDr cells for dIMP and 5FU.

Table 7. Effect of deoxyinosine on the incorporation
of 5FU into RNA and DNA.

	WiDr		Intestine 407	
	RNA	DNA	RNA	DNA
Control	8.2	0.58	7.2	0.75
Deoxyinosine	3.7	0.84	3.0	0.87

Values (pmol/hr per 10^6 cells) are means of 2-3
experiments. Deoxyinosine (at 0.4 mM) was added
simultaneously with 5FU to the incubation mixture.

The enhancement of sensitivity to 5FU by deoxyinosine indicates
that inhibition of thymidylate synthase by FdUMP is an important
mechanism for cytotoxicity of 5FU in both WiDr and Intestine 407 cells.
FdUMP is a potent inhibitor of thymidylate synthase in both cell lines
(Table 6) and is formed during incubation with 5FU (data not shown) and
may inhibit thymidylate synthase in vivo. However, it is not clear
whether the complex thymidylate synthase-FdUMP is stable in WiDr cells.
It appears from the growth-inhibition experiments that inhibition is
most pronounced at 24 hr but that between 24 and 48 hr growth rate is
normal. The thymidylate synthase activity after 24 hr may be sufficient
to support growth. In Intestine 407 the growth inhibition is more long-
lasting since in the presence of 5FU and deoxyinosine or dIMP growth is
still inhibited between 24 and 48 hr. This difference in the duration
of growth inhibition might also account for the lack of reversal of
growth inhibition by thymidine in WiDr cells. It is possible that WiDr
only requires low amounts of dTMP to support growth since the activity
of thymidylate synthase is relatively low in these cells. So a small
recovery of thymidylate synthase would generate enough dTMP to support
growth. Higher thymidine concentrations might even result in elevated
dTTP levels which may lead in combination with purines to an enhanced
growth inhibition (15).

The contribution to cytotoxicity from 5FU incorporation into DNA,
which is even enhanced in WiDr cells by deoxyinosine is not clear. 5FU
incorporated into DNA will be excised (16). The amount of 5FU which is
incorporated into DNA may account for a steady state level; the actual
amount of 5FU that has been incorporated into DNA will be higher.
Fractionation of DNA of cells incubated with 5FU showed that lesions
occur which have pronounced effects on both synthesis and stability of
DNA (17). The higher amount of 5FU incorporated into DNA of WiDr in the
presence of deoxyinosine may reflect a higher proportion of lesions and
might be associated with toxicity. Since it may be possible that in the
presence of thymidine these lesions will not be repaired, this would
also account for the lack of reversal by thymidine of cytotoxicity.

In both cell lines 5FU is incorporated into RNA but deoxyinosine
decreases this incorporation. Thus, incorporation of 5FU into RNA may
have little or no impact on cells. However, in the presence of thymidine 5FU incorporation into RNA can be enhanced (18), which may also
account for the lack of reversal of cytotoxicity of thymidine.

In conclusion, deoxyinosine increased the sensitivity to 5FU of
WiDr and Intestine 407 cells. An acceleration of the metabolism of 5FU
to FUdR and subsequently to FdUMP, supported by the increased dRib-1-P
levels, appears to be responsible for this synergism. Inhibition of
thymidylate synthase by FdUMP plays a major part in the cytotoxic ef-

fect of 5FU in both cell lines, but in WiDr cells an additional mechanism such as 5FU incorporation into RNA or DNA may also be responsible for the growth inhibition.

ACKNOWLEDGEMENTS

This work was supported by the Dutch Cancer Foundation "Koningin Wilhelmina Fonds", by grant IKA 83-16. We also thank Hoffman-La Roche, Mijdrecht, The Netherlands for financial support.

REFERENCES

1. Peters, G.J., Oosterhof, A. and Veerkamp, J.H. (1983). Biochim. Biophys. Acta, 755: 127-136.
2. Henderson, J.F., Scott, F.W. and Lowe, J.K. (1980). Pharmac. Ther. 8: 573-604.
3. Evans, R.N., Laskin, J.D. and Hakala, M.T. (1981). Cancer Res. 41: 3288-3295.
4. Beltz, R.E., Waters, R.N. and Hegarty, T.J. (1983). Biochem. Biophys. Res. Comm. 112: 235-241.
5. Cory, J.G. and Carter, G.C. (1982). Biochem. Pharmacol. 31: 2841-2844.
6. Santelli, G. and Valeriote, F. (1980). J. Natl. Cancer Inst. 64: 69-72.
7. Ullman, B. and Kirsch, J. (1979). Mol. Pharmacol. 15: 357-366.
8. Peters, G.J., Laurensse, E., Lankelma, J., Leyva, A. and Pinedo, H.M. (1984). Eur. J. Cancer Clin. Oncol. 20: 1425-1431.
9. Peters, G.J., Laurensse, E., Leyva, A., Lankelma, J. and Pinedo, H.M. (1985). Cancer Res., in press.
10. Peters, G.J., Laurensse, E., Leyva, A. and Pinedo, H.M. (1984). FEBS Lett. 170: 277-280.
11. Peters, G.J., Laurensse, E., Leyva, A. and Pinedo, H.M. (1985) Int. J. Biochem. 17: 95-99.
12. McIvor, R.S., Wohlhueter, R.M. and Plagemann, P.G.W. (1982). Anal. Biochem. 127: 150-154.
13. Iigio, M., Kuretani, K. and Hoshi, A. (1983). Cancer Res. 43: 5687-5694.
14. Iigio, M. and Hoshi, A. (1984). Eur. J. Cancer Clin. Oncol. 20: 411-415.
15. Leyva, A., Appel, H. and Pinedo, H.M. (1982). Leuk. Res. 6: 483-490.
16. Ingraham, H.A., Tseng, B.Y., Goulian, M. (1980). Cancer Res. 40: 998-1001.
17. Lönn, U. and Lönn, S. (1984). Cancer Res. 44: 3414-3418.
18. Spiegelman, S., Nayak, R., Sawyer, R., Stolfi, R., Ritzi, E. and Martin, D. (1980). Cancer Res. 45: 1129-1134.

TOXICITY AND ANTITUMOR EFFECT OF 5-FLUOROURACIL

AND ITS RESCUE BY URIDINE

G.J. Peters, J. van Dijk, C. van Groeningen,
E.J. Laurensse, A. Leyva, J. Lankelma, and H.M. Pinedo

Department of Oncology, Free University Hospital
P.O. Box 7075, 1007 MB Amsterdam
The Netherlands

INTRODUCTION

The pyrimidine analog 5-fluorouracil (5FU) has been used in the treatment against several solid carcinomas, mainly of the colon, for several decades (1). However, overall response rate is only about 20%, while gastrointestinal and myeloid toxicity are dose-limiting (1). Research on 5FU has been directed towards enhancing 5FU anabolism by combination with e.g. methotrexate (2) or thymidine (3) and towards the development of several analogs of 5FU. One recent analog, Doxifluridine (5'-deoxy-5-fluorouridine, 5'dFUR) showed an improved therapeutic index against a broad range of murine and rat tumors (4,5) and therapeutic activity in human advanced rectosigmoid adenocarcinoma (6).

Martin et al. (7) and Klubes et al. (8) used delayed uridine admini-stration to "rescue" mice from 5FU toxicity. Martin et al. (7) com-bined 5FU with repeated i.p. injections of high-dose uridine, while Klubes et al. (8) combined 5FU with long-term subcutaneous moderate-dose uridine infusions. Both schedules permitted the use of high doses of 5FU, which would be lethal in the absence of uridine. Antitumor activity could be enhanced against several 5FU resistant tumors such as Colon 26 (7) and B16 melanoma (9) but not against L1210 leukemia (9). However, it was not clear whether the response rate of 5FU-sensitive tumors would be affected by the combination with uridine. This is of special interest since in clinical trials with 5FU followed by delayed uridine, it is not known if the tumor will be sensitive or resistant. Therefore, we evaluated the antitumor effect of 5FU and uridine on the murine colon carcinomas Colon 38 (sensitive) and against the murine colon carcinoma Colon 26, (less sensitive) (10). The results were compared with the antitumor effect of 5'dFUR. In C57Bl mice we investi-gated the time course of the hematological toxicity of 5FU and its "rescue" by uridine. In addition we investigated the toxic effects of uridine itself.

EXPERIMENTAL

For all experiments 2-month-old female mice were used. Subcu-taneous transplants of Colon 38 and Colon 26 were maintained in both flanks of C57Bl/6 and Balb-c mice, respectively. The size of the tumor was determined by caliper measurement (11) every 3-4 days. Treatment was started when tumor volume was between 50 and 150 mm^3. Mice were treated with 5FU or 5'dFUR (Hoffman-La Roche, Mijdrecht, Netherlands) by intraperitoneal injection once a week. Uridine was administered at 2 and 20 hr after 5FU. Tumor size was calculated relative to that of the first day of treatment (11). Therapeutic effectiveness was evaluated by calculations of T/C values (tumor size of treated animals divided by tumor size of untreated controls) at various days after treatment. In each experiment the control group consisted of 4-6 animals, and the treated group consisted of 5-6 animals.

Toxicity of 5FU and of the combination of 5FU and uridine were evaluated in C57Bl/6 mice. Blood samples were obtained weekly by retro-orbital bleeding and the number of leucocytes and thrombocytes and the hematocrit value were determined. Body temperature of the mice was monitored rectally with a thermosensitive probe. Statistical analyses were performed by Student's t-test.

RESULTS

Colon 38 and Colon 26 are both chemically induced murine colon carcinomas; Colon 38, an adenocarcinoma and Colon 26, an undifferentiated carcinoma with local fibrosarcoma structures (12). Colon 38 reached a volume of 100 mm^3 at about 20 days after transplant and the doubling time was 5.2 days. The take rate was about 90%. Mice tolerated a large tumor load and the median life-span was longer than 40 days. Colon 26 reached a volume of 100 mm^3 after about 13 days and the doubling time was considerably shorter, 1.9 days. The take rate was almost 100%. The tumor appeared to be aggressive since the median life span was only 18 days.

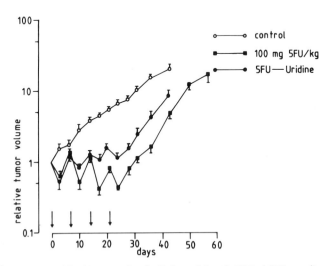

Fig. 1. Antitumor activity against Colon 38 of 5FU (100 mg/kg) and of 5FU (100 mg/kg) followed by uridine (3500 mg/kg) after 2 and 20 hr. Values are means ± SEM of 8-12 tumors. Arrows indicate the day of 5FU administration.

Colon 38 was sensitive to 5FU. At 60 mg/kg a significant tumor growth delay was observed (data not shown) when 5FU was administered for 4 weeks. Toxicity was moderate. At 100 mg 5FU/kg, also administered for 4 weeks, the tumor did not grow during treatment. A representative experiment is shown in Fig. 1. Administration of uridine after 5FU did not significantly alter the tumor growth rate (Fig. 1).

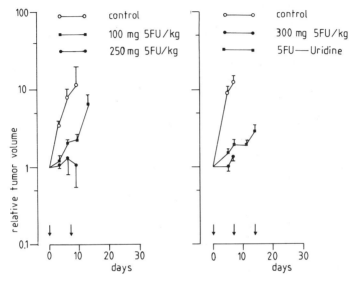

Fig. 2. Antitumor activity against Colon 26 of 5FU at various doses and of 5FU followed by uridine (3500 mg/kg) after 2 and 20 hr. values are means ± SEM of 8-12 tumors. Arrows indicate the day of 5FU administration.

Colon 26 appeared to be relatively resistant to 5FU at 100 mg/kg (Fig. 2). At 250 and 300 mg/kg tumor growth was arrested, but the mice died due to toxicity of 5FU. Administration of uridine increased the life-span of the mice but the tumors did not regress.

Doxifluridine (5'dFUR), a prodrug of 5FU, is considered to be less toxic than 5FU. With mice bearing Colon 38 and treated with 800 and 1000 mg/kg, a significant tumor growth delay could be observed (Fig. 3). The antitumor effect was comparable to that observed with 100 mg 5FU/kg. With mice bearing Colon 26 a significant tumor growth delay was also observed at 800 and 1000 mg 5'dFUR/kg. The antitumor effect was again comparable to that observed with 100 mg 5FU/kg, but treatment with 5'dFUR resulted in a considerable increase in life-span. The results are summarized in Table 1. With Colon 26 high T/C values were obtained after treatment with high, very toxic doses of 5FU. At 300 mg/kg a mean weight loss of almost 10% was observed. With Colon 38 maximal T/C values were obtained with 100 mg 5FU/kg, but with 5'dFUR the weight loss was less. None of these mice died due to toxicity from these drugs.

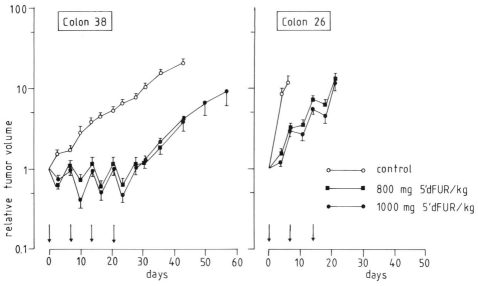

Fig. 3. Antitumoractivity of 5'dFUR against Colon 38 and Colon 26.
Values are means ± SEM of 8-12 tumors. Arrows indicate the day
of 5'dFUR administration.

Table 1. Antitumor activity against Colon 38 and Colon 26.

Drug	Dose (mg/kg;ip)	Days of treatment	Maximal T/C % (day)[a]	Median life span[b]	Weight loss[c]
Colon 26					
5FU	100	0,7	17.1 (9)	13	1.8
	250	0,7	9.2 (7)	8	4.3
	300	0,7	10.6 (7)	7	9.8
5FU-UR	300-3500	0,7	14.9 (7)	13	7.1
5'dFUR	800	0,7,14	24.6 (7)	24	5.4
5'dFUR	1000	0,7,14	23.3 (7)	17	6.2
Colon 38					
5FU	100	0,7,14,21	6.9 (24)	> 40	2.4
5FU-UR	100-3500	0,7,14,21	17.1 (24)	> 40	3.0
5'dFUR	800	0,7,14,21	9.0 (24)	> 40	0.8
5'dFUR	1000	0,7,14,21	7.2 (24)	> 40	0.7

[a] The day at which difference between T and C was maximal. All values
are significantly different (p < 0.05) from control values.
[b] Days after first treatment.
[c] Mean percentage weight loss, one day after treatment.
UR; uridine.

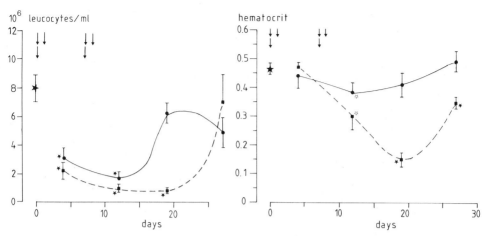

Fig. 4. Leucocyte count and hematocrit value in C57Bl/6 mice after treatment with 5FU (100 mg/kg; broken line) and in mice treated with 5FU followed by uridine (3500 mg/kg; solid line). Value at day 0 is of 18 mice, other values are of 5-7 mice and represent mean ± SEM. Closed asterisk, $p < 0.01$; open asterisk, $0.02 < p < 0.05$; differences are that between treated and control group.

Myeloid toxicity was studied by weekly monitoring of several hematological parameters. As expected 5FU caused a severe leukopenia which was more pronounced after the second treatment (Fig. 4). The leukopenia reached its nadir 18 days after the first administration of 5FU. One week thereafter the leucocyte count was within the normal range. With the combined 5FU-uridine treatment the nadir of the leucopenia was found after 12 days and was less severe than with 5FU alone. Furthermore, the mice recovered much more rapidly from the treatment than with 5FU alone. 5FU also seriously affected the hematocrit value. Again the nadir was after 18 days which is 11 days after the last treatment (Fig 4). With the combined treatment only a moderate effect on the hematocrit value was observed shortly after the second treatment. A modest thrombocytopenia was only observed in the 5FU treated group 4 days after the first treatment. On subsequent days thrombocyte count was in the normal range, but after 27 days an overshoot was observed (about twice the normal values), which was long-lasting, up to 68 days after the first treatment (data not shown).

The toxicity of uridine was limited to a pronounced effect on body temperature. At a dose of 3500 mg/kg temperature decreased very rapidly (Fig. 5). Hypothermia was accompanied by spasms, solitary behavior, discolored fur, etc. However, as soon as temperature increased, their behavior normalized. After 6 hr all mice recovered. At a lower dose the symptoms were less severe. At 1000 mg/kg no hypothermia was observed, but at 500 mg/kg a small but significant fall in body temperature occurred.

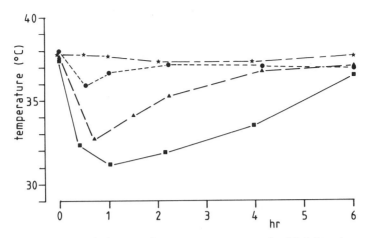

Fig. 5. Effect of uridine on body temperature of C57Bl/6 mice. Values are means of 6-12 mice. ■————■, 3500 mg/kg; ▲———▲, 2000 mg/kg; *————-*, 1000 mg/kg; and ●--●, 500 mg/kg.

DISCUSSION

In previous reports (7-9) it has been demonstrated that 5FU in combination with delayed uridine administration can be used for the treatment of 5FU resistant tumors. Our results demonstrated that 5FU sensitive tumors can also be treated with 5FU combined with delayed uridine, without affecting the antitumor effect significantly. The 5FU resistant tumor Colon 26 can be treated by 5FU in combination with uridine, resulting in an increased life-span. The main advantage of the delayed uridine administration actually appears to be a prevention of toxicity, rather than a rescue (7,8).

The antitumor effects of 5FU on Colon 38 and Colon 26 are comparable to that described previously (12,13) for these tumors. Delayed administration of uridine does not affect the differences in sensitivity between the tumors. The Colon 26 is relatively resistant to 5FU treatment and mice still die due to toxic effects of 5FU, although uridine increased the life-span. Doxifluridine, a precursor of 5FU, appeared to be as effective against Colon 38 as 5FU. Although a higher T/C value was found for the 5FU-uridine treatment, the difference with Doxifluridine was not significant. However, toxicity of Doxifluridine measured by weight loss, was less severe than for 5FU or 5FU-uridine. The T/C value for Doxifluridine in Colon 26 was higher than for 5FU alone or 5FU combined with uridine, but with Doxifluridine a significant increase in life-span was observed and the mice only died when the tumor load was comparable to that of untreated mice at the time of death. It appears that Doxifluridine has a better therapeutic index in the treatment of Colon 26 and to a lesser extent also for Colon 38 than treatment with 5FU alone. However, in clinical trials it appears that toxicity of Doxifluridine is more severe than expected (6,14). The possibility remains that the use of longterm infusions or other modes of administrations may enhance the therapeutic index of this new fluoropyrimidine.

The time course and extent of leukopenia is comparable to that described previously (15). However, the toxic effects of 5FU on the hematocrit were not reported (15), nor the protecting effect of uridine on the hematocrit value (7,8). 5FU did not cause severe thrombocytopenia, but the overshoot indicates that synthesis of thrombocytes was affected during 5FU therapy. The absence of an overshoot in the combination therapy, therefore, indicated that uridine also prevented toxic effects of 5FU on thrombocytes.

The mechanism of uridine rescue of 5FU toxicity is not totally clear, but there is evidence that toxicity in bone marrow is mainly mediated by incorporation of 5FU into RNA (7). Expansion of the uridine nucleotide pool by uridine administration would enhance the replacement of FUTP in RNA by UTP, preventing a further disturbance in the proliferation of hematopoietic progenitor cells. The selectivity of the rescue for bone marrow could be related to the relative large expansion of the uridine pool in blood compared to tissues (16).

The effect of uridine on body temperature indicates that uridine or its metabolites directly or indirectly interfere with the thermoregulatory function of the hypothalamus. This effect of uridine on body temperature has not been described previously. The mechanism is not clear but uracil or another catabolite might be involved since inhibition of uridine phosphorylase partially prevented the fall in body temperature (data not shown). Furthermore, the lowest temperature was observed after 30-60 min when plasma concentrations were highest (15-20 mM); temperature increased when uridine concentrations decreased. In contrast, high-dose uridine administration resulted in fever in both rabbits and patients (17). In patients this fever was only observed after long-term continuous infusions, but not with one-hr infusion (18) or with an intermittent administration schedule. With the latter schedule a reversal of 5FU induced toxicity was observed in two patients up to now (19).

In conclusion, delayed uridine administration prevents 5FU induced myelosuppression. The sensitivity of Colon 38 to 5FU is not or only slightly affected by delayed uridine administration. Colon 26 is relatively resistant to 5FU, but a higher dose of 5FU can be administered in combination with uridine, resulting in an increased life-span. However, Doxifluridine appears to have a better therapeutic effectiveness against Colon 26. Uridine has severe side-effects on body temperature, but they are not long-lasting or lethal. Preliminary clinical trials indicated that uridine can be given safely to patients by adjusting the administration schedule.

ACKNOWLEDGEMENTS

This work was supported by the Dutch Cancer Foundation, grant IKA 83-16. We acknowledge the Department of Hematology for their assistance and Ms C. Vanderlans for typing the manuscript. We also thank Hoffmann-La Roche, Mijdrecht, The Netherlands, for financial support.

REFERENCES

1. B.A. Chabner, Pyrimidine antagonists, in: Pharmacologic principles of cancer treatment, B.A. Chabner, ed., pp. 183-212, Saunders, Philadelphia (1982).
2. E. Mim and J.R. Bertino, Chemotherapia II: 147-162 (1985).
3. C.W. Young, T.M. Woodcock and D.S. Martin, Cancer Treatm. Rep., 65, (suppl. 3): 83-87 (1981).
4. W. Bollag, H.R. Hartmann, Eur. J. Cancer, 16: 427-432 (1980).
5. R.D. Armstrong and R.B. Diasio, Cancer Res. 40: 3333-3338 (1980).
6. R. Abele, P. Alberto, S. Kaplan, P. Siegenthaler, V. Hoffman, H.J. Ryssel, D. Hartman, E.E. Holdener and F. Cavalli, J. Clin. Oncol. 1: 750-754 (1983).
7. D.S. Martin, R.L. StolfiL, S.C. Sawyer, S. Spiegelman and C.W. Young, Cancer Res. 42: 3964-3970 (1982).
8. P. Klubes, I. Cerna and M.A. Meldon, Cancer Chemother. Pharmacol. 8: 17-21 (1982).
9. P. Klubes and I. Cerna, Cancer Res. 43: 3182-3186 (1983).
10. T.H. Corbett, D.P. Griswold, B.J. Roberts, J.C. Peckham and F.M. Schabel, Cancer Res. 35: 2434-2439 (1975).
11. E. Boven, W.J.F. van der Vijgh, M.M. Nauta, H.M.M. Schlüper and H.M. Pinedo, Cancer Res. 45: 86-90 (1985).
12. P.J. van Kranenburg-Voogd, H.J. Keizer and L.M. van Putten, Eur. J. Cancer 14 (suppl.): 153-157 (1978).
13. T.H. Corbett, D.P. Griswold, B.J. Roberts, J.C. Peckham and F.M. Schabel, Cancer 40: 2660-2690 (1977).
14. E.A. de Bruyn, A.T. van Oosterom, U.R. Tjaden, H.J.E.M. van Reeuwijk and H.M. Pinedo, Cancer Res. in press (1985).
15. A.M. Yeager, J. Levin and F.C. Levin, Exp. Hematol. 11: 944-952 (1983).
16. J.W. Darnofsky and R.E. Handschumacher, submitted (1985).
17. C.J. van Groeningen, A. Leyva, G.J. Peters, I. Kraal and H.M. Pinedo, Proc. AACR 35: 169 (abstract 670) (1984).
18. A. Leyva, C.J. van Groeningen, I. Kraal, H. Gall, G.J. Peters, J. Lankelma and H.M. Pinedo, Cancer Res. 44: 5928-5933 (1984).
19. C.J. van Groeningen, A. Leyva, I. Kraal, G.J. Peters and H.M. Pinedo, submitted (1985).

SYNERGY OF METHOTREXATE AND 6-MERCAPTOPURINE ON CELL GROWTH AND CLONOGENICITY OF CULTURED HUMAN T-LYMPHOBLASTS

Ronney A. De Abreu, Jos P.M. Bökkerink, Marinka A.H. Bakker, Tilly W. Hulscher, Gerard A.M. de Vaan, Chris H.M.M. de Bruijn and Egbert D.A.M. Schretlen

Centre for Pediatric Oncology SE Netherlands, Dept.of Pediatrics, University Hospital St.Radboud, Nijmegen The Netherlands

INTRODUCTION

Methotrexate (MTX) is a well known inhibitor of purine de novo synthesis and thymidylate synthesis. In cells with an active purine de novo synthesis 5-phosphoribosyl-1-pyrophosphate (PRPP) levels are increased significantly, when the cells are exposed to MTX.[1,2] Enhanced incorporation of 6-mercaptopurine (6MP) is observed, when 6MP is added at points of time at which PRPP levels are maximal.[3] These observations indicate to a possible synergistic action of MTX and 6MP under these circumstances. Therefore, we measured the effects of MTX and 6MP on cell growth, cell viability and clonogenic activity of MOLT 4 cells under conditions at which 6MP incorporation is enhanced in these cells after pretreatment with MTX.

MATERIALS AND METHODS

MOLT 4 cells were grown at 37°C in a water-saturated atmosphere containing 2.5% CO_2 in RPMI 1640 medium Dutch modification (Flow Laboratories, Ltd Irvine, Scotland), sodium pyruvate, penicillin and streptomycin in plastic culture flasks. During the experiments glutamine was added every 24 hours to a final concentration of approximately 2mM. Logarithmically growing cells were suspended in fresh medium in a concentration of 0.3 x 10^6 cells/ml 24 hours before each experiment.

Cell viability was measured by Trypanblue exclusion. Soft agar clonogenic assays were performed at various intervals of time after incubation with MTX, 6MP or a combination of both drugs. Prior to the assay, cells were washed three times and resuspended in medium. The cells were counted and plated in agar (final concentration 0.3%) containing the RPMI medium mentioned above. Colonies were allowed to grow for ten days at 37°C in a water-saturated atmosphere containing 5% CO_2.

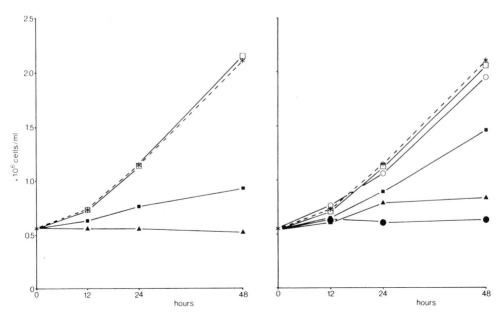

Fig.1. Effects of various concentrations of MTX and 6MP on cell growth
        of MOLT 4 cells.
        Left: (✳) untreated cells: (▫) 0.002 μM MTX; (▪) 0.02 μM MTX;
              (▲) 0.2 μM MTX.
        Right:(✳) untreated cells; (▫) 0.5 μM 6MP; (○) 2.0 μM 6MP;
              (▪) 5.0 μM 6MP; (▲) 10 μM 6MP; (●) 50 μM 6MP.

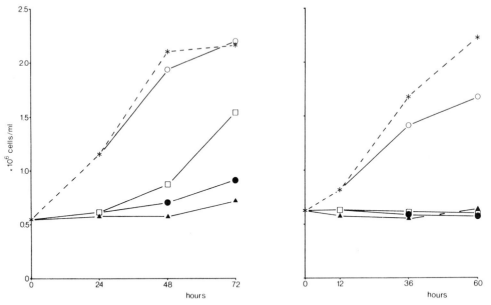

Fig.2. Effects of combinations of MTX and 6MP on cell growth of MOLT 4
        cells.
        Left: (✳) untreated cells; (▫) 0.02 μM MTX added at t=0;
              (○) 2.0 μM 6MP at t=24; (▲) 0.02 μM MTX and 2.0 μM 6MP
              at t=0; (●) 0.02 μM MTX at t=0 and 2.0 μM 6MP at t=24.
        Right:(✳) untreated cells; (▫) 0.2 μM MTX at t=0;
              (○) 2.0 μM 6MP at t=12; (▲) 0.2 μM MTX and 2.0 μM 6MP
              at t=0; (●) 0.2 μM MTX at t=0 and 2.0 μM 6MP at t=12.

RESULTS

The effect of various concentrations of MTX (0.002, 0.02 and 0.2 micromolar) and of 6MP (0.5, 2, 5, 10 and 50 micromolar) was measured on cell growth of MOLT 4 cells (Fig.1).
Incubations with 0.002 μM MTX or with 6MP concentrations of 0.5 and 2 μM, respectively, did not show significant differences with untreated cells. After incubation with 0.02 μM MTX cell growth was inhibited partially and with 0.2 μM MTX inhibition was completely. After exposure to a combination of 2 μM 6MP and 0.02 μM MTX, the cell growth of MOLT 4 cells was inhibited more pronounced than after exposure to 2 μM 6MP or 0.02 μM MTX alone (Fig.2), indicating synergism of both drugs.
Since cell growth of MOLT 4 cells is completely inhibited by 0.2 μM MTX, the synergistic action of MTX and 6MP could not be followed by this concentration of MTX (Fig.2).

We investigated the synergy of both drugs in more detail by cell viability studies (Fig.3) and by a soft agar clonogenic assay (Fig.4). A combination of 2 μM 6MP and 0.02 μM MTX was also more effective in these studies than either 2 μM 6MP or 0.02 μM MTX. The results of cell viability studies revealed a higher synergy when 6MP and MTX are added simultaneously compared with 6MP addition 24 hours after pretreatment with MTX.

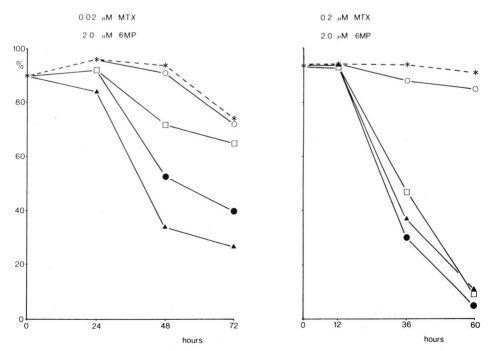

Fig. 3. Effects of 6MP and MTX and of combinations of both drugs on cell viability of MOLT 4 cells.
The symbols are explained by the legends of Fig.2.

Soft agar experiments revealed also a greater synergy of MTX and 6MP when these drugs are added simultaneously in concentrations of 0.02 µM and 2.0 µM, respectively (Fig.4).

In contrast to the experiments with 0.02 µM MTX, no synergy was observed when 0.2 µM MTX is added simultaneously with 2.0 µM 6MP, whereas synergy was clearly observed when 2 µM 6MP was added 12 hours after pretreatment with 0.2 µM MTX.

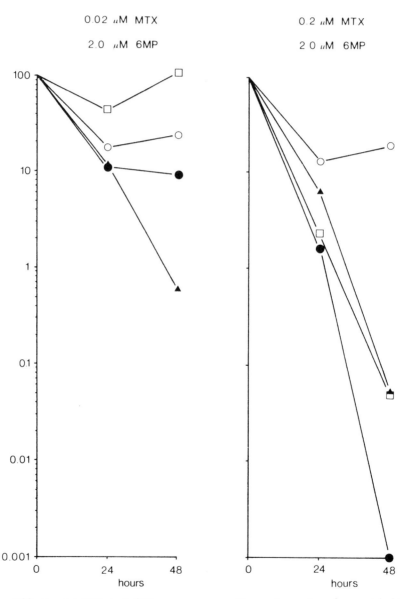

CLONOGENIC ASSAY

Fig.4. Effects of MTX and 6MP exposure on the colony forming activity of MOLT 4 cells. The symbols are explained by the legends of Fig.2. The time-scale indicates the duration of exposure to 6MP.

DISCUSSION

In previous studies we have observed that pretreatment with 0.02 μM and 0.2 μM MTX, respectively, results in elevated intracellular PRPP levels[3],[4], with concomitant enhancement of 6MP incorporation.[4] These observations indicate that combinations of MTX and 6MP will result in an optimal synergistic cytotoxicity at maximum PRPP levels. The experiments with 0.2 μM MTX and 2 μM 6MP on cell viability and colony forming activity of MOLT 4 cells are consistent with this statement.

Addition of 2 μM 6MP, 24 hours after pretreatment with 0.02 μM MTX, also results in a synergistic cytotoxicity of both drugs. However, when MOLT 4 cells are exposed to 0.02 μM MTX and 2 μM 6MP, a higher cytotoxic effect can be demonstrated when both drugs are added simultaneously instead off in sequence. There are biochemical circumstances which can abrogate the proposed mechanism of a maximal synergy at optimal availability of PRPP:

(1) In MOLT 4 cells, a concentration of 0.02 μM MTX induces a slow increase of PRPP levels to an optimum at 24 hours, followed by a decrease to normal values at 48 hours after addition of MTX. Presumably the slow but progressively higher PRPP concentrations will result in a higher incorporation of 6MP, when 0.02 μM MTX and 2 μM 6MP are added simultaneously to MOLT 4 cells. Despite optimal incorporation, addition of 6MP at 24 hours after MTX-pretreatment may be less effective, since PRPP levels drop to normal values the next 24 hours.

(2) To enhance 6MP incorporation in RNA and DNA, purine de novo biosynthesis should be inhibited by MTX so that competition between guanine nucleotides and thioguanine nucleotides for incorporation into polynucleotides will be in favour of the latter. We have experimental evidence from ^{14}C-glycine incorporation studies and AICAR measurements that inhibition of purine de novo synthesis is incomplete in MOLT 4 cells, after exposure to 0.02 μM MTX.[5] Therefore 0.02 μM MTX is not sufficient to provoke a purineless state in these cells.

Exposure of MOLT 4 cells to 0.2 μM MTX causes a complete inhibition of purine de novo synthesis[5] and a rapid increase of PRPP levels to a maximum at 8-12 hours.[4] This may be the reason why addition of 6MP, 12 hours after pretreatment with 0.2 μM MTX is more effective than a simultaneous addition of both drugs.

In summary, studies on cell growth, cell viability and clonogenic activity of MOLT 4 cells reveal synergism of methotrexate and 6-mercaptopurine. The sequence in which both drugs should be added, i.e. simultaneously or addition of 6MP (at optimal PRPP level), following pretreatment with MTX, is dependent on the MTX concentration.

ACKNOWLEDGEMENTS

We wish to thank Dr.J.A.J.M.Bakkeren for his stimulating suggestions during the study.

This work was supported by the Queen Wilhelmina Fund for Cancer Research in The Netherlands.

REFERENCES

1. J. M. Buesa-Perez, A. Leyva and H. M. Pinedo, Effect of methotrexate on 5-phosphoribosyl-1-pyrophosphate levels in L1210 leukemia cells in vitro; Cancer Res. 40:139 (1980).
2. E. Cadman, R. Heimer and L. Davis, Enhanced 5-fluorouracil nucleotide formation after methotrexate administration: explanation for drug synergism, Science 205:1135 (1979).
3. J. P. M. Bökkerink, T. J. Schouten, R. A. De Abreu, R. J. Lippens, G. A. M. de Vaan, C. H. M. M. de Bruijn en J. P. R. M. van Laarhoven, 6-Mercaptopurine en methotrexate, zicht op een rationeel gebruik na 35 jaar?, T.Kindergeneesk. 52:118 (1984).
4. J. P. M. Bökkerink, R. A. De Abreu, J. P. R. M. van Laarhoven, M. A. H. Bakker, T. W. Hulscher, E. D. A. M. Schretlen and C. H. M. M. de Bruijn, Increased availability of phosphoribosyl pyrophosphate as the basis for enhanced 6-mercaptopurine incorporation by methotrexate, in cultured human lymphoblasts, submitted for publication in: Purine Metabolism in Man-V.
5. R. A. De Abreu, J. P. M. Bökkerink, M. A. H. Bakker, T. W. Hulscher, J. M. van Baal, C. H. M. M. de Bruijn and E. D. A. M. Schretlen, Influence of methotrexate on purine and pyrimidine pools and on cell phase distribution of cultured human lymphoblasts, Submitted for publication in: Purine Metabolism in Man-V.

INCREASED AVAILABILITY OF PHOSPHORIBOSYL PYROPHOSPHATE AS THE BASIS FOR
ENHANCED 6-MERCAPTOPURINE INCORPORATION BY METHOTREXATE, IN CULTURED
HUMAN LYMPHOBLASTS

Jos P.M. Bökkerink, Ronney A. de Abreu, Jan P.R.M. van
Laarhoven, Marinka A.H. Bakker, Tilly W. Hulscher, Egbert
D.A.M. Schretlen and Chris H.M.M. de Bruijn

Centre for Pediatric Oncology SE Netherlands, Dept. of
Pediatrics, University Hospital St.Radboud, Nijmegen
The Netherlands

INTRODUCTION

In response to inhibition of purine de novo biosynthesis by methotrexate (MTX), an increase in the availability of 5-phosphoribosyl-1-
pyrophosphate (PRPP) occurs. As a result of this increased availability
of PRPP, potentiation of 6-mercaptopurine (6MP) incorporation can be expected in cells with an active purine de novo synthesis (Fig.1).

In this study we examined the enhancement of 6MP incorporation in
MOLT 4 and Raji cells, pretreated with MTX.

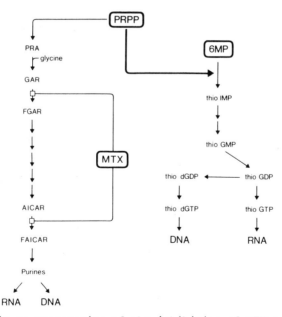

Fig.1. Schematic representation of the inhibition of MTX on purine
de novo synthesis. The block of MTX induces an elevation of
PRPP levels. The result of the elevation of PRPP is an increased
incorporation of 6MP to thio IMP and ultimately in DNA and RNA.

METHODS

Molt 4 cells (T-lymphoblasts)and Raji cells (B-lymphoblasts) are allowed to grow at 37°C in a water-saturated atmosphere containing 2.5 % CO_2 in RPMI medium 1640 Dutch modification, supplemented with 10 % fetal calf serum (v/v), sodium pyruvate, penicillin and streptomycin in plastic culture flasks. During the experiments, glutamine was added every 24 hours to a final concentration of ± 2 mM. Logphase cells were exposed to 0.02 µM and 0.2 µM MTX, respectively. Aliquots of 0.5 ml of cell culture were taken at various intervals of time after incubation with MTX for PRPP measurements and for 8-$[^{14}C]$-6MP incorporation studies.

RESULTS

A rapid accumulation of PRPP is observed when MOLT 4 cells are exposed to 0.2 µM MTX (Fig.2A). The PRPP levels reach an optimum 8 hours after addition of MTX, followed by a decrease to values of untreated cells at 48 hours. When MOLT 4 cells are exposed to 0.02 µM MTX, the increase of PRPP levels is much slower and an optimum is reached at 24 hours after pretreatment with MTX (Fig.2A).

The increased availability of PRPP, after pretreatment of MOLT 4 cells with MTX, can be used for enhanced incorporation of 6MP (Figs.2B and 2C). At indicated times after pretreatment with MTX, aliquots were taken and the cells were incubated during 20 minutes with 10 µM 8-$[^{14}C]$ -6MP, and 100 µM 8-$[^{14}C]$-6MP, respectively. As shown in Figs.2B and 2C, the enhanced incorporation of 6MP follows the increase of PRPP levels (Fig.2A).

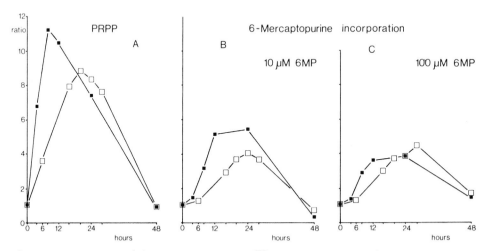

Fig.2.   Effects of (■) 0.2 µM MTX and (□) 0.02 µM MTX on intracellular PRPP levels and intracellular incorporation of 6MP in MOLT 4 cells.
A: Enhancement of PRPP levels by MTX.
B: Effect of MTX on the incorporation of 10 µM 6MP.
C: Effect of MTX on the incorporation of 100 µM 6MP.

After exposure of Raji cells to 0.02 µM MTX, accumulation of PRPP is very rapid. An optimum is reached within 6 hours and continues for a prolonged period of time (Fig.3A).
Pretreatment with 0.02 µM MTX also causes a more rapid increase of PRPP

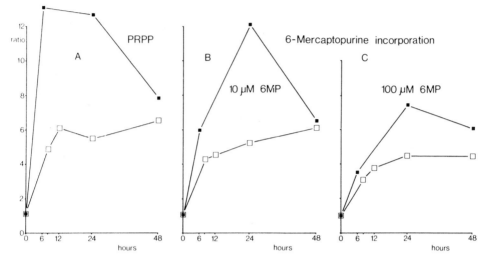

Fig.3. Effects of (■) 0.2 µM MTX and (▢) 0.02 µM MTX on intracellular PRPP levels and intracellular incorporation of 6MP in Raji cells.
   A: Enhancement of PRPP levels by MTX.
   B: Increased incorporation of 6MP (10 µM) after pretreatment with MTX.
   C: Increased incorporation of 6MP (100 µM) after pretreatment with MTX.

levels, compared to MOLT 4 cells. An optimum is reached after 12 hours and the PRPP concentration is still high 48 hours after addition of MTX (Fig.3A).

Differences between MOLT 4 cells and Raji cells are also observed by 6MP incorporation studies. Figs. 3B and 3C show that 6MP incorporation can be enhanced by MTX-induced increased availability of PRPP. However, optimal incorporation is at a later time point than the PRPP optimum, after pretreatment with 0.2 µM MTX. With both MTX concentrations the increase of 6MP incorporation is still high after 48 hours.

DISCUSSION

Methotrexate is a known inhibitor of dihydrofolate reductase, the enzyme that catalyses the reduction of dihydrofolate to tetrahydrofolate. As such activities of tetrahydrofolate dependent enzymes are reduced, e.g. in purine de novo synthesis.
PRPP is an important early intermediary compound for purine de novo synthesis and PRPP levels will be elevated by MTX concentrations which inhibit purine de novo syntheses. [1,2]

The present study demonstrates a correlation between the period of time after which PRPP levels are optimal and the concentration of MTX used. In MOLT 4 T-lymphoblasts and Raji B-lymphoblasts, maximal PRPP concentrations are reached at a later point of time by lower concentrations of MTX. In previous experiments with MOLT 4 cells[3], we found that 0.002 μM MTX did not result in any increase of PRPP during a period of 48 hours, 0.02 μM resulted in an optimum of PRPP after approximately 24 hours, 0.1 μM MTX after 18 hours and 0.2 μM MTX after 8 hours. Thus higher concentrations of MTX induced earlier peak levels of PRPP.

Elevated PRPP levels are induced by methotrexate only in cells with an active purine de novo synthesis. MOLT 4 and Raji cells have a high capacity of purine de novo synthesis, this is reflected by high PRPP levels in untreated cells (50-150 pmol/10^6 cells). This is in contrast with low PRPP levels found in normal peripheral blood lymphocytes (± 10 pmol/10^6 cells)[4].
We could not demonstrate an increase of PRPP in normal peripheral blood lymphocytes after exposure of cells to MTX. Moreover Becher et al[5] demonstrated a lack of purine the novo synthesis in bone marrow cells. These data suggest that human T- and B-lymphoblasts are more susceptible to biochemical pertubations caused by MTX, resulting in inhibition of purine de novo synthesis.

Increased availability of PRPP induced by MTX can be used for enhanced incorporation of 6-mercaptopurine (Figs. 2B, 2C, 3B and 3C). The data show a potentiation of 6MP incorporation after pretreatment of cells with MTX, indicating synergy of MTX and 6MP on cytotoxicity of human malignant T- and B-lymphoblasts. Synergistic action of MTX and other purine and pyrimidine analogues has been demonstrated in many in vitro studies, e.g. MTX and Fluorouacil[2,6,8] and MTX and thioguanine[9]. However, correlations between the period of time after which PRPP levels are optimal (i.e. maximal incorporation of the analogues) and concentrations of MTX, have not been referred. This may be an explanation for the lack of evidence for synergy in some clinical trials with combination of MTX with purine and pyrimidine analogues[10,11], while others are more succesfull[12].

At present we are investigating the effects of the sequence, time and dose dependency of combinations of MTX and 6MP in various human malignant lymphoblastic cell lines with respect to purine de novo synthesis, availability of PRPP and cell toxicity. The data may lift the veil of the increased efficacy of the combination of MTX and 6MP in the maintenance therapy of childhood acute lymphoblastic leukemia.

ACKNOWLEDGEMENT

This work was supported by the Queen Wilhelmina Fund for Cancer Research in the Netherlands.

REFERENCES

1. J.M.Buesa-Perez, A. Leyva, and H.M. Pinedo, Effect of methotrexate
    on 5-phosphoribosyl-1-pyrophosphate levels in L1210 leukemia cells
    in vitro, Cancer Res. 40:139 (1980).

2. E.Cadman, R.Heimer and L.David, Enhanced 5-fluorouracil nucleotide
    formation after methotrexate administration: explanation for drug
    synergism, Science 205:1135 (1979).

3. J.P.M.Bökkerink, T.J.Schouten, R.A.De Abreu, R.J.J.Lippens, G.A.M.
    de Vaan, C.H.M.M.de Bruyn en J.P.R.M.van Laarhoven, 6-Mercaptopu-
    rine en methotrexate, zicht op een rationeel gebruik na 35 jaar?,
    T.Kindergeneesk. 52:118 (1984).

4. G.J.Peters, A.Oosterhof and J.H.Veerkamp, Metabolism of phosphori-
    bosyl pyrophosphate in peripheral and phytohemagglutinin-stimulated
    mammalian lymphocytes., Int.J.Biochem. 13:577 (1981).

5. H.Becher, M.Weber and G.W.Lohr, Purine nucleotide synthesis in normal
    and leukemic blood cells., Klin.Wschr., 56:275 (1978).

6. J.R.Bertino, W.L.Sawicki, C.Linquist and V.S.Gupta, Schedule-depen-
    dent antitumor effects of methotrexate and 5-fluorouracil, Cancer
    Res. 37:327 (1977).

7. D.W.Kufe and M.Egan, Enhancement of 5-fluorouracil incorporation
    into human lymphoblast ribonucleic acid., Biochem.Pharmacol. 30:129
    (1981).

8. C.Benz and E.Cadman, Modulation of 5-fluorouracil metabolism and
    cytotoxicity by antimetabolite pretreatment in human colorectal
    adenocarcinoma HCT-8, Cancer Res. 41:994 (1981).

9. R.D.Armstrong, R.Vera, P.Snyder and E.Cadman, Enhancement of 6-thio-
    guanine cytotoxic activity with methotrexate, Biochem.Biophys.Res.
    Comm. 109:595 (1982).

10. G.P.Browman, Clinical application of the concept of methotrexate plus
    5-FU sequence-dependent "synergy": How good is the evidence?,
    Cancer Treat.Rep. 68:465 (1984).

11. S.B.Kaye, G.Sangster, A.Hutcheon, T.Habeshaw, F.Crossling, C.Ferguson,
    C.McArdle, D.Smith, W.D.George and K.C.Calman, Sequential metho-
    trexate plus 5-FU in advanced breast and colorectal cancers: a phase
    II study, Cancer Treat.Rep. 68:547 (1984).

12. R.Hermann, C.Manegold, M.Schroeder, F.J.Tigges, H.Bartsch, F.Jungi
    and D.Fritze, Sequential methotrexate and 5-FU in breast cancer
    resistant to the conventional application of these drugs, Cancer
    Treat.Rep. 68:1279 (1984).

REGULATION OF DEOXYADENOSINE AND NUCLEOSIDE ANALOG

PHOSPHORYLATION BY HUMAN PLACENTAL ADENOSINE KINASE

Mary C. Hurley, Bertha Lin, and Irving H. Fox

Human Purine Research Center, Departments of Internal Medicine
and Biological Chemistry, Clinical Research Center, The
University of Michigan, Ann Arbor, Michigan

SUMMARY

The enzymes responsible for phosphorylation of adenosine and
nucleoside analogs are important in the pathogenesis of adenosine deaminase
deficiency and for the activation of specific anticancer and antiviral
drugs. We examined the role of adenosine kinase in catalyzing these
reactions using an enzyme purified 4000-fold (2.1 umol/min/mg) from human
placenta. The Km values of adenosine and ATP are 135 uM and 4 uM,
respectively. Adenosine kinase phosphorylates adenine arabinoside with an
apparent Km value of 1 mM using adenosine kinase assay conditions. The Km
values for 6-methylmercaptopurine riboside and 5-iodotubercidin, substrates
for adenosine kinase, are estimated to be 4.5 uM and 2.6 nM, respectively.

These data indicate that dadenosine phosphorylation by adenosine
kinase is primarily regulated by its Km, and the concentrations of Mg^{2+},
ADP and AMP. The high Km values for phosphorylation of dadenosine and
adenine arabinoside suggest that adenosine kinase may be less likely to
phosphorylate these nucleosides in vivo than other enzymes with lower Km
values. Adenosine kinase appears to be important for adenosine analog
phosphorylation where the Michaelis constant is in the low micromolar
range.

INTRODUCTION

Nucleoside kinases are an important series of intracellular enzymes
which phosphorylate endogenous nucleosides and nucleoside analogs active as
anticancer and antiviral drugs.[1-8] The enzymes responsible for the
phosphorylation of dadenosine and adenine arabinoside have been studied to

141

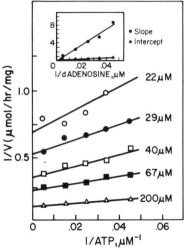

Fig. 1. Double reciprocal lot of
initial velocity studies
with variable ATP
concentrations ranging from
2 to 10 uM. Mg^{2+}
concentration was in 5 mM
excess of the ATP. Inset
shows secondary plot of the
slope and the intercept
versus the inverse of
dadenosine concentration in
uM. The Km for dadenosine
is 135 uM.

determine their relative importance in this activation step. Studies
indicate that both adenosine kinase and dcytidine kinase have a role in
adenosine analog phosphorylation, but no definitive pattern is evident.[9-15]

In an effort to clarify whether adenosine kinase has an important role
in adenosine analog phosphorylation, we have tested the hypothesis that
adenosine kinase phosphorylates dadenosine, adenine arabinoside and other
nucleoside analogs. Our studies allow specific conclusions to be made
concerning the regulation of nucleoside phosphorylation by this enzyme.

METHODS

Adenosine kinase was purified 4000-fold from human placental cytosol
to a specific activity of 2.1 umol/min/mg as described previously.[16-18]
Radiochemical enzyme assays for adenosine and dadenosine phosphorylation

were performed in triplicate in a total reaction volume of 50 ul using a modification of previously described methods.[18,19] Data from kinetics experiments were fit to a hyperbola using Cleland's program[20] on a Vax 11/730 minicomputer. Linear regression analysis of slope and intercept replots were performed on a Hewlett-Packard 9825-A programmable calculator. Ki and Ki' were designated for the K values derived from the slope and intercept replots, respectively.[21] The true Ki for nucleoside analogs was estimated from the apparent Ki using the method of Chung and Prusoff for competitive inhibitors.[22]

RESULTS

Initial Velocity Studies

Double reciprocal plots of initial velocity data at varying concentrations of $MgATP^{2-}$ and fixed concentrations of dadenosine with Mg^{2+} in 5 mM excess of ATP levels yield a series of intersecting lines (Figure 1). The intersection occurs to the left of the vertical axis and below the horizontal axis. Replots of the slope and intercept yield Km values of 135 and 4 uM for dadenosine and ATP, respectively. K^+ 150 mM is required to maintain linear double reciprocal plots. Mg^{+2} 5 mM in excess of the total nucleotide present is required for the Km for dadenosine to be as low as 135 uM (Table 1).

Phosphate Donor

Guanosine triphosphate is almost as effective a phosphate donor as ATP for dadenosine phosphorylation (Table 1), but promotes adenosine phosphorylation at about three times the rate of ATP. Both dATP and ITP are able to phosphorylate adenosine and dadenosine at 50% of the ATP rate. Other nucleotides tested (dGTP, CTP, dCTP, dITP, TTP, UTP, XTP) are less effective phosphate donors.

pH Effect

The rate of dadenosine phosphorylation is greatest at pH 7.5 but is essentially constant between pH 6.5 and 8.0. At pH 6.0 the rate of dadenosine phosphorylation is only 55% of the maximum rate; however, adenosine is phosphorylated at its maximum rate at this pH. The rate of adenosine phosphorylation at pH 7.5 is only 40% of the rate at pH 6.0. The ratio of dadenosine kinase to adenosine kinase activity is 0.11 at pH 6.0 and 0.47 at pH 7.5.

Inhibition by Nucleotides

ADP and AMP are the only nucleotides tested that significantly

Table 1. Kinetic Constants for Adenosine Kinase
Phosphorylation of dAdenosine and Nucleoside Analogs

Compound Studied (Conditions Used)		Kinetic Constant (uM)
dAdenosine Phosphorylation		
dAdenosine	Km (5 mM MgCl$_2$)	135
	Km (0.4 mM MgCl$_2$)	1500
ATP	Km (5 mM MgCl$_2$)	4
ADP	Ki (Variable MgATP^{2-})	13
	Ki,Ki' (Variable dAdenosine)	600,92
AMP	Ki,Ki' (Variable MgATP^{2-})	17,27
	Ki,Ki' (Variable dAdenosine)	177,15
Adenine Arabinoside Phosphorylation		
Adenine Arabinoside	Km (apparent)	1000
ATP	Km (apparent)	9
AMP	Ki,Ki' (MgATP^{2-} 1.2 mM, ara-A variable)	150,520
Adenosine Phosphorylation		
6-methylmercaptopurine riboside Ki		16
5-iodotubercidin Ki		0.003

decrease the rate of dadenosine phosphorylation. The rate of dadenosine
phosphorylation was 50% of control value when ATP and ADP were both present
in the assay at 1.2 mM and Mg^{+2} was 2.8 mM. ADP and AMP inhibition was
studied using two substrate kinetics[19] (Table 1). The other product of the
dadenosine phosphorylation, dAMP, did not inhibit the enzyme activity
unless it was present at unphysiological levels of 5 mM and above. Only a
30% reduction in catalytic rate was observed with 5 mM dAMP.

Nucleoside Inhibitors and Substrates

Thirty nucleosides were screened to determine their effects on
dadenosine phosphorylation. The nucleosides were held at 100 uM, while
dadenosine and ATP were 50 uM. Adenosine, 5-iodotubercidin, 6-
dimethylaminopurine riboside, 6-methylmercaptopurine riboside and

tubercidin caused 100% inhibition of dadenosine phosphorylation. Nucleosides causing 50 to 80% inhibition were 3'-dadenosine, 5'-deoxy-5'-methylthioadenosine, N^6-phenylisopropyladenosine and S-adenosylhomocysteine (Table 2).

The kinetic characteristics of several of these nucleosides were studied. Initial velocity studies of [2-^{3}H]adenine arabinoside phosphorylation were performed using the assay conditions for dadenosine phosphorylation. The apparent Km values for adenine arabinoside and ATP are 1 mM and 9 uM, respectively. AMP inhibits adenine arabinoside phosphorylation (Table 1).

Inhibition studies were performed with two adenosine analogs, previously shown to be substrates for adenosine kinase.[16] Assay conditions were those used for adenosine kinase and included 1 uM adenosine, 1.2 mM ATP and 1.6 mM $MgCl_2$. The apparent Ki values determined for the two adenosine analogues are 16 uM and 9 nM for 6-methylmercaptopurine riboside and 5-iodotubercidin, respectively. The true Ki values were estimated to

Table 2. Inhibition of dAdenosine
Phosphorylation by Nucleosides

Inhibitory action of nucleosides was determined by incubating 0.4 ug of adenosine kinase with 50 uM [8-^{3}H]dadenosine, 50 uM ATP and 5 mM $MgCl_2$ under standard assay conditions. The activity with no additional nucleoside was 1.54 umol/hr/mg.

Compound (100 uM)	Inhibition (%)
Adenosine	100
5-iodotubercidin	100
6ɣɣ-dimethylaminopurine riboside	100
6-methylmercaptopurine riboside	100
Tubercidin	100
N-ethylcarboxyamidoadenosine	95
Cyclohexyladenosine	85
N^6-phenylisopropyladenosine	78
3'-deoxyadenosine	78
5'-deoxy-5'-methylthioadenosine	70
S-adenosylmethionine	53
S-adenosylhomocysteine	52
2-Chloroadenosine	44
Ethyladenosine 5'-carboxylate	42
Aminoimidazolecarboxamide ribonucleoside	33
2',3'-isopropylidene adenosine	32

Compounds causing 20% or less inhibition:

Adenine arabinoside, EHNA, deoxycoformycin, cytidine, dcytidine, dguanosine, dinosine, guanosine, inosine, duridine, ribivarin, thymidine, uridine, xanthosine.

be 4.5 uM and 2.6 nM. Since these compounds are substrates of adenosine kinase, their Km values should be similar to the Ki values.

DISCUSSION

Our studies have systematically examined the phosphorylation of dadenosine and other adenosine analogs by adenosine kinase. Using conditions carefully optimized for dadenosine phosphorylation,[18,23] we have observed an activity as high as 49% of the specific activity for adenosine phosphorylation. The true Michaelis constant for dadenosine phosphorylation is 135 uM for the human placental enzyme. Previous studies have found apparent Michaelis constants from 360 to well above 1000 uM[24-26]. While is it possible to account for phosphorylation of dadenosine with a Michaelis constant of 135 uM, this could potentially be accomplished more effectively by an enzyme with a lower Michaelis constant for dadenosine. One enzyme with this latter property has recently been discovered.[27]

Adenosine analog phosphorylation is the focus of great interest as this reaction activates many potential and actual antiviral and anticancer compounds[1,2] and may form toxic intermediates of adenosine receptor agonists. In particular, the phosphorylation of adenine arabinoside has prompted extensive studies with variable conclusions.[6-8,12,14,24-26] Both adenosine kinase and dcytidine kinase appear to be responsible for phosphorylation of this compound.[6-8] Our data indicate that adenosine kinase phosphorylates adenine arabinoside with incubation conditions optimized for dadenosine phosphorylation. The apparent Michaelis constant for adenine arabinoside is 1.0 mM. This value is extremely high and it is reasonable to conclude that this activity may not be the major route for its phosphorylation in vivo.

A number of other nucleoside analogs inhibit dadenosine phosphorylation[17,24] (Table 2). Some of these compounds, including 5-iodotubercidin, tubercidin, 6-methylmercaptopurine riboside, and 3'-dadenosine are proven substrates of adenosine kinase.[17] To indirectly measure the Michaelis constants of two compounds, we estimated their Ki values using inhibition studies. Both iodotubercidin and 6-methylmercaptopurine riboside have low Ki values, indicating that they can be phosphorylated at relatively low concentrations. The Ki value for 6-methylmercaptopurine riboside of 4.6 uM is in close agreement with the Km value of 10 uM observed in Ehrlich ascites tumor cells.[28]

Our studies provide support for the concept that adenosine kinase is an important enzyme for the phosphorylation of dadenosine and other adenosine analogs. These phosphorylations are regulated by the levels of Mg^{2+}, ADP and AMP, and the Michaelis constants for the nucleosides. It is clear from our experiments and the work of others that there is a need for further understanding of the phosphorylation of dadenosine and other adenosine analogs. This will require careful observation of other nucleoside kinase enzymes, especially those with lower Michaelis constants for nucleoside phosphorylation.

ACKNOWLEDGEMENTS

We wish to thank Steve Schmaltz for assistance in using the Vax 11/730 computer, the nurses and patients in the Delivery Rooms of Womens Hospital, The University of Michigan for supplying fresh placenta and Sharon Demorest for typing the manuscript. This work was supported by grants from the United States Public Health service 1 R01 GM 32837 and 5 M01 RR00042.

REFERENCES

1. E. P. Anderson, in: "The Enzymes," P. D. Boyer, ed., Vol. 9, pp. 49-96. Academic Press, Inc., New York (1973).

2. J. F. Henderson, F. W. Scott, and J. K. Lowe, Toxicity of naturally occurring purine deoxyribonucleosides, Pharmacol. Ther. Part A Chemother. Toxicol. Metab. Inhibitors 8:573-604 (1980).

3. M. S. Coleman, J. Donofrio, J. J. Hutton, L. Hahn, A. Daoud, B. Lampkin, and J. Dyminski, Identification and quantitation of adenine deoxynucleotides and erythrocytes of a patient with adenosine deaminase deficiency and severe combined immunodeficiency, J. Biol. Chem. 253:1619-1626 (1978).

4. A. R. Cohen, R. Hirschhorn, S. D. Horowitz, A. Rubinstein, S. H. Polmar, R. Hong, and D. W. Martin, Jr., Deoxyadenosine triphosphate as a potentially toxic metabolite in adenosine deaminase deficiency, Proc. Natl. Acad. Sci. USA 75:472-476 (1978).

5. M. F. E. Siaw, B. S. Mitchell, C. A. Koller, M. S. Coleman, and J. J. Hutton, ATP depletion as a consequence of adenosine deaminase inhibition in man, Proc. Natl. Acad. Sci. USA 77:6157-6161 (1980).

6. N. R. Cozzarelli, The mechanism of action of inhibitors of DNA synthesis, Ann. Rev. Biochem. 46:641-668 (1977).

7. W. Plunkett, S. Chubb, L. Alexander, and J. A. Montgomery, Comparison of the toxicity and metabolism of 9-β-D-arabinofuranosyl-2-fluoro-adenine and 9-β-D-arabinofuranosyladenine in human lymphoblastoid cells, Cancer Res. 40:2349-2355 (1980).

8. W. Plunkett, and S. S. Cohen, Metabolism of 9-β-D-arabinofuranosyl-adenine by mouse fibroblasts, Cancer Res. 35:418-422 (1975).

9. B. Ullman, L. J. Gudas, A. Cohen, and D. W. Martin, Jr., Deoxy-adenosine metabolism and cytotoxicity in cultured mouse T-lymphoma cells: A model for immunodeficiency disease, Cell 14:365-375 (1978).

10. E. R. Pfefferkorn, and L. C. Pfefferkorn, The biochemical basis for resistance to adenine arabinoside in a mutant of toxoplasma gondii, J. Parasitol. 64:486-492 (1978).

11. B. Ullman, B. B. Levinson, M. S. Hershfield, and D. W. Martin, Jr., A biochemical genetic study of the role of specific nucleoside kinases in deoxyadenosine phosphorylation by cultured human cells, J. Biol. Chem. 256:848-852 (1981).

12. V. Verhoef, J. Sarup, and A. Fridland, Identification of the mechanism of activation of 9-β-D-arabinofuranosyladenine in human lymphoid cells using mutants deficient in nucleoside kinases, Cancer Res. 41:4478-4483 (1981).

13. M. S. Hershfield, J. E. Fetter, W. C. Small, A. S. Bagnara, S. R. Williams, B. Ullman, D. W. Martin, Jr., D. B. Wasson, and D. A. Carson, Effect of mutational loss of adenosine kinase and deoxycytidine kinase on deoxy ATP accumulation and deoxyadenosine toxicity in cultured CEM human T-lymphoblastoid cells J. Biol. Chem. 257:6380-6386 (1982).

14. C. E. Cass, M. Selner, and J. R. Phillips, Resistance to 9-β-D-arabi-nofuranosyladenine in cultured leukemia L1210 cells, Cancer Res. 43:4791-4798 (1983).

15. P. P. Saunders, and M. M. Lai, Nucleoside kinase activities of Chinese hamster ovary cells. Biochim. Biophys. Acta 761:135-141 (1983).

16. C. M. Andres, and I. H. Fox, Purification and properties of human placental adenosine kinase, J. Biol. Chem. 254:11388-11393 (1979).

17. T. D. Palella, C. M. Andres, and I. H. Fox, Human placental adenosine kinase kinetic mechanism and inhibition, J. Biol. Chem. 255:5264-5269 (1980).

18. M. C. Hurley, T. D. Palella, and I. H. Fox, Human placental deoxyadenosine and deoxyguanosine phosphorylating activity, J. Biol. Chem. 258:15021-15027 (1983).

19. M. C. Hurley, B. Lin, and I. H. Fox, Regulation of deoxyadenosine and nucleoside analog phosphorylation by human placental adenosine kinase. (1985). In preparation.

20. W. W. Cleland, The statistical analysis of enzyme kinetic data, Adv. Enzymol. 25:1-32 (1967).

21. M. Dixon, and E. C. Webb, Enzyme inhibition and activation, in "Enzymes," Academic Press, New York (1979).

22. Y. Cheng, and W. H. Prusoff, Relationship between the inhibition constant (Ki) in the concentration of inhibitor which causes 50% inhibition (I50) of an enzymatic reaction, Biochem. Pharmacol. 22:3099-3108 (1973).

23. M. C. Hurley, and I. H. Fox, Measurement of nucleoside kinases in crude tissue extracts, Biochem. Med. 30:89-100 (1983).

24. R. L. Miller, D. L. Adamczyk, W. H. Miller, G. W. Koszalka, J. L. Rideout, L. M. Beacham, III, E. Y. Chao, J. J. Haggerty, T. A. Krenitsky, and G. B. Elion, Adenosine kinase from rabbit liver. II. Substrate and inhibitor specificity, J. Biol. Chem. 254:2346-2352 (1979).

25. Y. Yamada, H. Goto, and N. Ogasawara, Purification and properties of adenosine kinase from rat brain, Biochim. Biophys. Acta 616:199-207 (1980).

26. Y. Yamada, H. Goto, and N. Ogasawara, Adenosine kinase from human liver, Biochim. Biophys. Acta 660:36-43.(1981).

27. Y. Yamada, H. Goto, and N. Ogasawara, Purine nucleoside kinases in human T- and B-lymphoblasts, Biochim. Biophys. Acta 761:34-40 (1983).

28. J. F. Henderson, A. Mikoshiba, S. Y. Chu, and I. C. Caldwell, Kinetic studies of adenosine kinase from Ehrlich ascites tumor cells, J. Biol. Chem. 247:1973-1975 (1972).

THE BIOSYNTHESIS OF DEOXYGUANOSINE TRIPHOSPHATE IN HERPES SIMPLEX TYPE-1 INFECTED VERO CELLS TREATED WITH ACYCLOVIR AND HYDROXYUREA

Catherine U. Lambe, Donald J. Nelson and
Phillip A. Furman

Wellcome Research Laboratories
Research Triangle Park, N.C. 27709

INTRODUCTION

Herpes simplex virus type 1 (HSV-1) infection of cells causes induction of virally specified thymidine kinase, DNA polymerase, ribonucleotide reductase and deoxyribonuclease[1-3]. Also accompanying infection is a rise in deoxyribonucleoside triphosphates (dNTPs), particularly dTTP[2,4]. Acyclovir (ACV) is an acyclic deoxyguanosine analog which, in its triphosphate form, inhibits viral DNA polymerase[5,6]. ACV treatment dramatically increased the pool sizes of dATP and dGTP in infected Vero cells compared to infected untreated cells[4]. The source of these elevated purine deoxyribonucleotide pools could be from either enhanced de novo synthesis, salvage of deoxyribonucleosides derived from host cell DNA, or possibly a combination of the two pathways.

Ribonucleotide reductase is essential to de novo dNTP synthesis both in uninfected[7] and HSV-1 infected cells[3,8]. Blocking ribonucleotide reductase should thus serve to separate the relative contributions of both pathways.

METHODS

Vero cells were grown for several generations in the presence of [1',2'-^{3}H]deoxyguanosine (36 Ci/mmole, 0.27 µM). Additional [1',2'-^{3}H]-deoxyguanosine was added with fresh media on days 1, 3, and 5. This

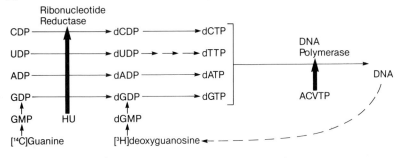

Fig. 1 DNA Precursor Synthesis.

tritium label was then chased into cellular DNA by four days of incubation in unlabeled medium.

The confluent cells were then infected or mock-infected with HSV-1 (Patton). One hour post-infection [8-^{14}C]guanine (51 mCi/mmol, 3.9 μM) was added to follow de novo formation of dGTP; ACV and hydroxyurea (HU) were also added at this time. Eight hours post-infection, cells were extracted and analyzed by high-pressure liquid chromatography[5,9].

RESULTS AND CONCLUSIONS

Uninfected Vero cells have very low levels of dNTPs (Table 1). As has been reported previously[4], HSV-1 infection causes elevation of dNTPs, peaking at 6 to 8 hours post-infection. Elevation of dTTP is particularly pronounced, probably due to induction of viral thymidine kinase.

This virally induced thymidine kinase catalyzes the initial phosphorylation of ACV. Cellular enzymes then complete the formation of acyclovir triphosphate (ACVTP), a DNA polymerase inhibitor with marked selectivity for HSV DNA polymerase[5,6].

Effect of ACV

Addition of ACV (100 μM) to infected cells caused slight decreases in dCTP and dTTP pools and 7-fold increases in dATP and dGTP pools (Table 1). ACVTP levels at 6 to 8 hours were 25% of dGTP levels, much more than necessary for DNA polymerase inhibition. Similar large increases in dATP and dGTP, with little change in dCTP and dTTP, have also been shown with phosphonoacetic acid, another HSV DNA polymerase inhibitor[10]. Thus inhibition of DNA polymerase is sufficient to cause these dramatic pool expansions. These elevated dNTP levels should be inhibitory to cellular ribonucleotide reductase, but are not inhibitory to the HSV-1 ribonucleotide reductase[1].

Table 1. Effects of Acyclovir (ACV) and Hydroxyurea (HU) on Deoxyribonucleoside Triphosphate Pools in Herpes Simplex Type I Infected Vero Cells

Cells	Treatment	dCTP	dTTP	dATP	dGTP	ACVTP
				$pm/10^6cells$		
Uninfected		19	46	6.1	n.d.	
		19	64	9.1	6.9	
		8	18	n.d.	n.d.	
		15	57	n.d.	n.d.	
Infected		38	384	25	81	
		23	354	6.9	30	
Infected	100 μM ACV	13	222	122	459	126
		7	170	99	382	88
Infected	100 μM ACV	14	291	11	15	1126
	+10[-2]M HU	30	285	11	5.0	973

Effect of HU

The addition of HU (10 mM), a ribonucleotide reductase inhibitor[1], to infected ACV treated cells caused essentially no change in pyrimidine deoxynucleoside triphosphates but drastically decreased purine dNTP levels (Table 1). DeoxyATP and dGTP decreased to 10% or less of infected treated levels, and below infected untreated levels. Concomitant with the drop in dATP and dGTP was a 10-fold increase in ACVTP levels. The combined changes in the pools resulted in an ACVTP/dGTP ratio of 103, which could greatly increase the antiviral potency of ACV by increased competition of ACVTP for DNA polymerase. Another less toxic ribonucleotide reductase inhibitor has indeed shown significant antiviral potentiation of ACV accompanying dGTP decrease and ACVTP increase[11]. The addition of HU alone to infected cells caused essentially no change in dNTP pools (unpublished data).

Ribonucleoside Triphosphates

In infected cells, dNTP levels rise with time, peaking at 6 to 8 hours instead of remaining constant, as in uninfected cells. The same time course of elevation is seen with ribonucleotides in infected cells (unpublished data). Peak pool sizes of ribonucleotides of approximately 2-fold uninfected levels are reached by 6 to 8 hours post-infection, the time chosen for this experiment. This is presumably due both to induction of viral enzymes and stimulation of cellular enzymes. These changes could be important in regulation of many enzymes, particularly ribonucleotide reductase. Both ATP and GTP have significant allosteric effects on mammalian ribonucleotide reductase[7].

Addition of ACV to infected cells decreased all four ribonucleoside triphosphates to levels below those found in infected untreated cells (Table 2). This is probably related to decreased viral stimulation of the cells, due to ACV inhibition of viral replication. With the addition of hydroxyurea as well as ACV, ribonucleotide levels were again similar to infected treated cells.

Table 2. Ribonucleoside Triphosphate Levels in Vero Cells $\pm$ HSV-1 Infection and Treatment

Cells	Treatment	CTP	UTP	ATP	GTP
			pm/10^6cells		
Uninfected		605	1449	4813	1131
		1344	3628	7956	2213
		258	884	2324	529
		538	1866	5157	1056
Infected		1142	5425	10819	2079
		1123	5136	10367	1956
Infected	100 μM ACV	349	1371	4547	966
		349	1389	4115	817
Infected	100 μM ACV + 10^{-2}M HU	1611	5117	8354	2324
		1351	4464	7598	2025

Source of dGTP

DeoxyGTP in uninfected cells contained a small amount of tritium label salvaged from cellular DNA and a larger amount from de novo synthesis (Table 3). Both [^{3}H]deoxyguanosine and [^{14}C]guanine labeling of dGTP increased during infection (Table 3). Although both de novo and salvage pathways contributed to the increased dGTP pool, the de novo contribution was the largest.

In ACV treated cells, a 7-fold rise in total dGTP was observed, but essentially no elevation in [^{3}H]dGTP derived from host cell DNA occurred (Table 3). The greatly increased ^{14}C/^{3}H ratio over that seen in infected untreated cells shows that almost all of this further increase in dGTP came from de novo synthesis through ribonucleotide reductase.

When HU was added to ACV treated infected cells, dGTP decreased to a level below that found in infected untreated cells. This decrease was paralled by a sharp drop in [^{14}C]dGTP, derived from [^{14}C]guanine, but the [^{3}H]dGTP was virtually unchanged (Table 3). Therefore, blocking ribonucleotide reductase is sufficient to totally inhibit the dGTP elevations caused by infection and magnified by ACVTP inhibition of DNA

Table 3. Effects of Acyclovir and Hydroxyurea on Radioactivity Recovered in Deoxyguanosine Triphosphate Isolated from HSV-1 Infected Vero Cells

Cells	Treatment	dGTP	[^{3}H]pm/10^6	[^{14}C]pm/10^6	[^{14}C]/[^{3}H]
Uninfected		n.d.	.001	none**	none**
		6.9	.002		
		n.d.	.001	.017	17.1
		n.d.	.002	.165	69.5
			(.002)*	(.091)	(43.3)
Infected		80.8	.016	.546	33.9
		29.5	.018	.386	21.2
		(55.2)	(.017)	(.466)	(27.5)
Infected	100 μM ACV	459	.015	16.3	1113
		381	.022	14.8	662
		(420)	(.019)	(15.6)	(887)
Infected	100 μM ACV + 10^{-2}M HU	15.3	.015	.109	7.2
		5.0	.015	.128	8.3
		(10.2)	(.015)	(.119)	(7.8)

n.d. = none detected (<5 pm/10^6 cells)
*Numbers in parentheses are averages of the duplicate experiments shown
**No ^{14}C-guanine used in these two experiments

polymerase. There was no increase in salvage of [^{3}H]deoxyguanosine derived from cell DNA.

These results with dGTP are in contrast with results reported with dTTP in HSV-2 infected HeLa cells[12]. In that case, inhibition of ribonucleotide reductase with HU resulted in no decrease in dTTP, as we also observed. However, those authors stated that contributions from host cell DNA via salvage increased to make up for de novo inhibition and maintain the dTTP pool. Hydroxyurea seems to have specific effects in decreasing the purine deoxynucleotide pools in rapidly growing cells and has also been reported to decrease turnover of dTTP without decreasing pools of dTTP[13,14]. These differences in radiolabeling of dGTP versus dTTP may reflect differences in cellular control mechanisms for pyrimidine and purine deoxyribonucleotide pools.

REFERENCES

1. D. R. Averett, C. Lubbers, G. B. Elion, and T. Spector, Ribonucleotide reductase induced by herpes simplex Type 1 virus, J. Biol. Chem. 258:9831 (1983).
2. Y. C. Cheng, B. Goz, and W. H. Prusoff, Deoxyribonucleoside metabolism in herpes simplex virus infected HeLa cells, B.B.A. 390:253 (1975)
3. K. Leary, J. Bratton, and B. Francke, Replication of herpes simplex virus Type 1 on hydroxyurea resistant baby hamster kidney cells, J. Virol. 47:224 (1983).
4. P. A. Furman, C. U. Lambe, and D. J. Nelson, Effect of acyclovir on the deoxyribonucleoside triphosphate pool levels in Vero cells infected with herpes simplex virus Type I, in: "Acyclovir Symposium", Am. J. Med., D. H. King and G. Galasso, eds., Technical Publishing, New York (1982).
5. G. B. Elion, P. A. Furman, J. A. Fyfe, P. de Miranda, L. Beauchamp, and H. J. Schaeffer, Selectivity of action of an antiherpetic agent, 9-(2-hydroxyethoxymethyl)guanine, Proc. Natl. Acad. Sci. USA 74:5716 (1977).
6. P. A. Furman, M. H. St. Clair, and T. Spector, Acyclovir triphosphate is a suicide inactivator of the herpes simplex virus DNA polymerase, J. Biol. Chem. 259:9575 (1984).
7. R. G. Hards and J. A. Wright, Regulation of ribonucleotide reductase activity in intact mammalian cells, Arch. Biochem. Biophys. 231:17 (1984).
8. B. Roller and G. H. Cohen, Deoxyribonucleoside triphosphate pools in synchronized human cells infected with herpes simplex virus Types 1 and 2, J. Virol. 18:58 (1976).
9. C. Garrett and D. V. Santi, A rapid and sensitive high pressure liquid chromatography assay for deoxyribonucleoside triphosphates in cell extracts, Anal. Biochem. 99:268 (1976).
10. P. A. Furman, C. U. Lambe, and D. J. Nelson, abstr. 199; Alteration of deoxyribonucleoside triphosphate pools by acyclovir and hydroxyurea in HSV-1 infected Vero cells, presented at Internatl. Workshop on Herpesviruses, Bologna, Italy (1981).
11. T. Spector, D. R. Averett, D. J. Nelson, C. U. Lambe, R. W. Morrison, M. H. St. Clair, and P. A. Furman, Potentiation of antiherpetic activity of acyclovir by ribonucleotide reductase inhibition, Proc. Natl. Acad. Sci. USA 82:4254 (1985).

12. L. Nutter, S. Grill, and Y. C. Cheng, abstr. 431; The sources of deoxynucleotides for virus DNA synthesis in herpes simplex virus Type 2 infected HeLa cells, presented at Federation of American Societies for Experimental Biology (1985).

13. L. Skoog and B. Nordenskjold, Effects of hydroxyurea and 1-β-arabinofuranosyl-cytosine on deoxyribonucleotide pools in mouse embryo cells, Eur. J. Biochem. 19:81 (1971).

14. B. Nicander and P. Reichard, Relations between synthesis of deoxyribonucleotides and DNA replication in 3T6 fibroblasts, J. Biol. Chem. 260:5376 (1985).

5-AZA-2'-DEOXYCYTIDINE SYNERGISTIC ACTION WITH THYMIDINE ON

LEUKEMIC CELLS AND INTERACTION OF 5-AZA-dCMP WITH dCMP DEAMINASE

R.L. Momparler, M. Rossi, J. Bouchard, S. Bartolucci,
L.F. Momparler, C.A. Raia, R. Nucci, C. Vaccaro, and S. Sepe

Dept. pharmacologie,Univ. de Montréal,Montreal,Quebec, Canada
Dept. Chim. Org. Biol.,Univ. of Naples,IIGB,CNR,Naples, Italy

INTRODUCTION

5-Aza-2'-deoxycytidine (5-AZA-CdR) is a potent antileukemic agent whose inhibitory effects are blocked by deoxycytidine (1-3). In order to be active, 5-AZA-CdR must first be phosphorylated by the enzyme deoxycytidine kinase (4). Leukemic cells deficient in the enzyme are resistant to 5-AZA-CdR (5). The lethal action of 5-AZA-CdR is related to its incorporation into DNA (6,7). The incorporation of 5-azacytosine analogs into DNA has been shown to induce the expression of differentiated phenotypes (8). This induction of differentiation by 5-AZA-CdR appears to be related to the inhibition of DNA methylation produced by these analogs (8,9).

In mammalian cells specific cytosine residues are modified by enzymatic transfer of methyl group from S-adenosyl-methionine to the 5 position in a reaction catalyzed by DNA methylase following DNA replication (10). This postreplicative enzymatic modification of DNA maintains the specific pattern of 5-methylcytosine residues in DNA and the inhibition of DNA methylation can induce inheritable changes in the expression. For some genes DNA methylation suppresses expression whereas demethylation results in gene activation (11,12). Since 5-AZA-CdR is a potent inhibitor of DNA methylation, it has the potential to induce the differentiation of cancer cells so that they lose their malignant properties making it an interesting therapeutic agent (13).

It is important to understand the nucleotide metabolism of 5-AZA-CdR because changes in the intracellular pool of the naturally occurring deoxynucleotides may modulate the action of nucleoside analogues. For example, cells whith an increased pool of dCTP are resistant to deoxycytidine analogue, cytosine arabinoside (14). Also, treatment of cells with thymidine produces enhancement of the antineoplastic action of 5-AZA-CdR (15). Thymidine after its conversion to dTTP, produces a decrease of the intracellular pool of dCTP (16) by feedback inhibition of ribonucleotide reductase (17). dTTP is also a feedback inhibitor of dCMP deaminase (18) an nnzyme which plays an important role in deoxynucleotide metabolism by modulating the intracelluar pool size of dCTP and dTTP (19). Cells defecient in this enzyme have been shown to have a reduced dTTP pool (19) and an increased rate of spontaneous mutation (20). The pharmacological activity of deoxycytidine analogues, such as cytosine arabinoside, can be modified by dCMP deaminase because they are often substrates for the enzyme or their activity may be modulated by pool size of dCTP or dTTP (21).

In this report we have observed a synergistic antineoplastic effect between 5-AZA-dCR and dTR on leukemia cells in culture. In order to understand the mechanism behind this interaction we have investigated the effets of dTTP on the deamination of 5-aza-2'-deoxycytidine-5'-monophosphate (5-AZA-dCMP) by dCMP deaminase. In addition we have also studied the effects of 5-AZA-dCTP on this enzyme.

EXPERIMENTAL METHODS

Cell Culture Leukemic cells were grown in suspension cultures in minimal essential medium containing non-essential amino acids and 5% fetal calf serum. Cell survival after drug exposure was determined by colony assay in 0.15% agar (2).

Biochemical Assays The incorporation of ^{3}H-5-AZA-CdR (Moravek Biochemicals) into DNA of leukemic cells was performed in medium containing 5% dialyzed serum. The amount of radioactivity incorporated into DNA was determined by trapping the cells on GF/C glass fiber filters and washing with cold TCA (7).

DNA methylation was assayed by the incorporation of 6-^{3}H deoxycytidine into cytosine and 5-methylcytosine of DNA after chemical hydrolysis and separation by HPLC as described previously (23).

Pool size measurements of deoxynucleoside triphosphates were performed by the DNA polymerase assay (24).

dCMP Deaminase Assay 5-AZA-dCMP and 5-AZA-dCTP were synthesized enzymatically from 5-AZA-CdR using purified deoxycytidine kinase, nucleoside monophosphate kinase and nucleoside diphosphokinase and purified by anion exchange chromatograpy (9). Homogeneous spleen dCMP deaminase had a specific activity of 920 units/mg (22). One unit of enzyme activity was defined as the amount of enzyme that catalyzed the deamination of 1 µmole of dCMP per min at 38ℓ C. The enzyme was assayed spectrophotometrically as described previously (22).

RESULTS

The cytoxic effects of 5-AZA-CdR and/or thymidine on murine L1210 leukemic cells and human RAJI lymphoid leukemic cells as determined by a colony-forming assay are shown in Table 1. The data clearly show that thymidine produces an enhancement of the cytotoxic action of 5-AZA-CdR on both murine and human leukemic cells.

Table 1. In Vitro Cytotoxic Action of 5-AZA-CdR and/or Thymidine on Leukemic Cells

Leukemic cell line	Addition[a]	Concentration (mg/ml)	Cell Kill (%)
L1210	5-AZA-CdR	0.005	28 ± 2[b]
L1210	TdR	5	0
L1210	5-AZA-CdR + TdR	0.005 + 5	48 ± 3
RAJI	5-AZA-CdR	0.005	24 ± 10
RAJI	TdR	5	19 ± 4
RAJI	5-AZA-CdR + TdR	0.005 + 5	96 ± 1

[a]Exposure times were 6 hr (L1210) and 48 hr (RAJI)
[b]Mean ± S.D.

158

In Table 2 is shown the effects of different concentrations of thymidine on the incorporation of radioactive 5-AZA-CdR into DNA of L1210 leukemic cells. Thymidine produced a marked stimulation of the incorporation of radioactive 5-AZA-CdR into DNA of L1210 leukemic cells, even at concentrations as low as 0.1 µg/ml.

Table 2. Effect of Thymidine on Incorporation of [^{3}H]-5-AZA-CdR into DNA of L1210 Leukemic Cells

Thymidine concentration (mg/ml)	Incorporation into DNA[a] [^{3}H]-5-AZA-CdR (cpm)
0	1,200 ± 230[b]
0.1	2,330 ± 280
1.0	3,650 ± 350
10.0	6,930 ± 200

[a]Cells were incubated with Thymidine and isotope for 4 hr.
[b]Mean ± S.D.

The effect of 5-AZA-CdR and/or thymidine on DNA methylation in L1210 leukemic cells as determined by the incorporation of radioactive deoxycytidine into cytosine and 5-methylcytosine of DNA is shown in Table 3. Thymidine alone had a minimal effect on DNA methylation. However, thymidine markedly increased the inhibitory effect of 5-AZA-CdR on DNA methylation.

Table 3. Effect of 5-AZA-CdR and/or Thymidine on DNA Methylation in L1210 Leukomic Cells[a]

Addition	Concentration (mg/ml)	5-MC / 5 Mc + C (%)	Inhibition (%)
None		4.20	0
5-AZA-CdR	0.005	3.64	13
TdR	5	3.98	5
5-AZA-CdR + TdR	0.005 + 5	2.99	29

[a]L1210 leukemic cells (4 X 10^6 cells/20 ml) were incubated for 6 hr in presence of indicated concentrations of drugs and subsequently incubated for 12 hr with 10 µCi of [6-^{3}H] deoxycytidine (4 Ci/mmol). DNA was isolated, hydrolyzed, analysed by HPLC and assayed for radioactivity in 5-methylcytosine (5-MC) and cytosine (C)

The effect of different concentrations of thymidine on the intracellular pool size of dCTP and dTTP is shown in Table 4. Thymidine at 10 μg/ml concentration produced no change in the pool of dCTP.

Table 4. Effect of Thymidine on the Pool Size of
dCTP and dTTP in L1210 Leukemic Cells

Concentration Thymidine[a] (mg/ml)	Pool Size[b]	
	dCTP	dTTP
	(% control)	
0	100 ± 3	100 ± 6
10	100 ± 6	135 ± 13
25	68 ± 13	490 ± 107

[a]Cells were incubated with indicated concentration of thymidine for 4 hr.
[b]Pool size (nmol/10^9 cells) of control were 33 and 25 for dCTP and dTTP, respectively.

In a previous study we observed that 5-AZA-dCMP was a substrate for dCMP deaminase (22). In Table 5 the effects of different concentrations of dTTP on the deamination of this nucleotide analog by dCMP deaminase are shown. dTTP at concentration as low as 4 μM produced significant inhibition of the deamination of 5-Aza-dCMP.

Table 5. dTTP induced Inhibition of Deamination of
5AZA-dCMP by dCMP Deaminase[a]

Concentration dTTP (μM)	5-AZA-dCMP deaminated (nmol/min)	Inhibition (%)
0	31	0
4	25	19
10	20	35
20	10	68
30	3	90

[a]Reaction mixture (0.3 ml) contained 98 μM 5-AZA-dCMP and 1.6 μg homogeneous spleen dCMP deaminase (22); incubation at 38° C.

The effects of dCTP and 5-AZA-dCTP on the dTTP-induced inhibition of dCMP deaminase when 5-AZA-dCMP was used as the substrate are shown in Table 6. Both dCTP and 5-AZA-dCTP could reverse completely the inhibitory effects of dTTP on dCMP deaminase.

160

Table 6. Effect of dCTP or 5-AZA-dCTP on dTTP
induced Inhibition of Deamination of
5-AZA-dCMP by dCMP Deaminase[a]

Additions (µM)	5-AZA-dCMP deaminated (nmol/min)	Relative Activity
None	18	100
dTTP 30	1	0
dTTP 30 + dCTP 0.1	7	39
dTTP 30 + dCTP 1.0	20	111
dTTP 30 + 5-AZA-dCTP 0.1	11	61
dTTP 30 + 5-AZA-dCTP 1.0	20	111

[a]Reaction mixture (0.3 ml) contained 41 µM
5-AZA-dCMP and 1.6 µg homogeneous spleen dCMP deaminase incubation at 38° C.

DISCUSSION

Thymidine produces an enhancement of the antineoplastic activity of
5-AZA-CdR against leukemic cells (Table 1, ref. 15). Molecular studies
have shown that thymidine increases the amount of 5-AZA-CdR incorporated
into DNA (Table 2) and augments the inhibition of DNA methylation produced
by this nucleoside analog (Table 3). One of the objectives of this study
was to elucidate the molecular mechanism by which thymidine produces a
synergistic interaction with 5-AZA-CdR with respect to the antileukemic
action of this analog.

One possible mechanism to explain the action of thymidine is the inhibition of ribonucleotide reductase by its nucleotide form (17) producing a
reduction in the intracellular pool of dCTP (16). Previous studies have
shown that dCTP competes with 5-AZA-dCTP for the catalytic site of DNA polymerase alpha (9) and acts as a feedback inhibitor of deoxycytidine kinase, the enzyme that phosphorylates 5-AZA-CdR (4). One argument against
this mechanism is that high concentrations of thymidine are required to reduce the pool size of dCTP (16). The studies in this report show that low
concentrations of thymidine can enhance the action of 5-AZA-CdR (Tables
1-3). These concentrations of thymidine do not reduce the pool size of
dCTP (Table 4). It is possible that thymidine may act by a different mechanism.

Since dTTP is a potent feedback inhibitor of dCMP deaminase (18), we
have investigated the interaction of this enzyme with the nucleotide forms
of 5-AZA-CdR. We have observed that 5-AZA-dCMP is a substrate for dCMP
deaminase and its deamination inhibited by different concentrations of dTTP
(Table 5; ref. 22). Deamination of this nucleotide analog occurs with
hyperbolic kinetics with an apparent Km of 0.1 mM, in contrast to the natural substrate, dCMP, where the kinetics are sigmoidal with an apparent Km
of 0.49 mM. However, the Vmax for the deamination of dCMP is about
100-fold greater than that of 5-AZA-dCMP. In addition, 5-AZA-dCTP was
observed to be an allosteric activator of this enzyme when dCMP was the

substrate, but not when 5-AZA-dCMP was the substrate. In both case 5-AZA-dCTP was able to reverse the inhibitory effects of dTTP (Table 6; ref. 22).

A possible mechanism by which low concentrations of thymidine can increase the antileukemic action of 5-AZA-CdR is by the inhibition produced by dTTP on the deamination of 5-AZA-dCMP by dCMP deaminase (Table 5). As shown in Table 4, low concentrations of thymidine do not decrease the pool size of dCTP, which supports our hypothesis. Therefore, the inhibition of ribonucleotide reductase by dTTP producing a reduction of the dCTP pool appears not to be the mechanism to explain the synergistic interaction between 5-AZA-CdR and thymidine under the experimental conditions used in this study. The modulation of the antineoplastic activity of deoxycytidine anlogs by allosteric effectors such as dTTP may have the potential to increase the effectiveness of the chemotherapy for acute leukemia.

Acknowledgements This work was supported by Grant 6356 of the Medical Research Council of Canada and by Contract 84-00429-44 of the Progetto Finalizzato Oncologia of the Italian CNR.

REFERENCES

1- F. Sorm and J. Vesely, Effect of 5-AZA-2'-deoxycytidine against leukemic and hemopoietic tissues in AKR mice. Neoplasma (Bratisl.) 15:339 (1968).

2- R.L. Momparler and F.A. Gonzales, Effect of intravenous infusion of 5-AZA-2'-deoxycytidine on survival time of mice with L1210 leukemia. Cancer Res., 38: 2673 (1978).

3- R.L. Momparler and J. Goodman, In vitro cytotoxic and biochemical effects of 5-AZA-2'-deoxycytidine. Cancer Res., 37:1636 (1977).

4- R.L. Momparler and D. Derse, Kinetics of phosphorylation of 5-AZA-2'-deoxycytidine by deoxycytine kinase. Biochem. Pharmacol., 28:1448 (1979).

5- J. Vesely, A. Cihak and F. Sorm, Characteristics of mouse leukemic cells resistant to 5-azacytidine and 5-AZA-2'-deoxycytidine. Cancer Res., 28:1995 (1968).

6- J. Vesely and A. Cihak, Incorporation of a potent antileukemic agent, 5-AZA-2'-deoxycytidine, into DNA of cells from leukemic mice. Cancer Res., 37:3684 (1977).

7- R.L. Momparler, J. Vesely, L.F. Momparler and G.E. Rivard, Synergistic action of 5-AZA-2'deoxycytidine and 3-deazauridine on L1210 leukemic cells and EMT6 tumor cells. Cancer Res., 39:3822 (1979).

8- P.A. Jones, S.M. Taylor, Cellular differentiation, cytidine analogs, and DNA methylation. Cell, 20:85 (1980).

9- J. Bouchard and R.L. Momparler, Incorporation of 5-AZA-2'-deoxycytidine-5'triphosphate into DNA interactions with mammalian DNA polymerase and DNA methylase. Mol. Pharmacol., 24:209 (1983).

10- A. Razin and A.D. Riggs, DNA methylation and gene function. Science, 210:604 (1980).

11- J.D. McGhee and G.D. Grinder, Specific DNA methylation sites in the vicinity of the chicken β-globin genes. Nature, 280:419 (1979).

12- M. Busslinger, J. Hurst and R.A. Flavell, DNA methylation and the regulation of globin gene expression. Cell, 34:197 (1983).

13- L. Sachs, The differentiation of meloid leukemic cells. New possibilities for therapy. Br. J. Haematol., 40:509 (1978).

14- R.L. Momparler, M.Y. Chu and G.A. Fischer, Studies on a new mechanism of resistance of L5178Y murine leukemic cells to cytosine arabinoside Biochim. Biophys. Acta, 161:481 (1968).

15- S. Grant, F. Rauscher, J. Margolin and E. Cadman, Dose and schedule depedent activation and drug synergism between thymidine and 5-AZA-2'-deoxycytidine in a human promyelocytic leukemia cell line. Cancer Res., 42:519 (1982).

16- G. Bjursell and P. Reichard, Effect of thymidine in deoxyribonucleoside triphosphate pools and deoxyribonucleic acid systhesis in Chinese hamster ovary cells. J. Biol. Chem., 218:3904 (1973).

17- E.C. Moore and R.B. Hurlbert, Regulation of Mammalian deoxyribonucleotide biosynthesis by nucleotides as activators and inhibitors. J. Biol. Chem., 241:4802 (1966).

18- F. Scarano, G. Geraci and M. Rossi, Deoxycytidylate aminohydrolase. II. Kinetic properties. The activatory effect of deoxycytidine triphosphate and the inhibitory effect of deoxythymidine triphosphate. Biochemistry, 6:192-201 (1967).

19- B.R. de Saint-Vincent and M. Dechamps, and G. Buttin, The modulation of the thymidine triphosphate pool of Chinese hamster cells by dCMP deaminase and UDP reductase: thymidine auxotrophy induced by CTP in dCMP deaminase-deficient line. J. Biol. Chem., 255:162 (1980).

20- G. Weinberg, B. Ullman and D.W. Martin, Mutator phenotypes in mammalian cell mutants with distinct biochemical defects and abnormal deoxyribonucleoside triphosphate pools. Proc. Natl. Acad. Sci. (USA), 78:2447 (1981).

21- W.R. Mancini and Y.C. Cheng, Human deoxycytidylate deaminase substrate and regulator specificities and their chemotherapeutic implications. Mo. Pharmacol., 23:159-164 (1983).

22- R.L. Momparler, M. Rossi, J. Bouchard, C. Vaccaro, L.F. Momparler and S. Bartolucci, Kinetic Interaction of 5-AZA-2'-deoxycytidine-5'-monophosphate and Its 5'-triphosphate with deoxycytodilate deamirase. Mol. Pharmacol., 25:436 (1984).

23- R.L. Momparler, L.F. Momparler and J. Samson, Comparison of the antileukemic activity of 5-AZA-2'-deoxycytidine, 1-β-D-arabinofuranosyl-cytosine and 5-azacytidine against L1210 leukemie. Leukemie Res.,8:1043 (1984).

24- J. Bouchard, Pharmacologie moléculaire de la 5-AZA-2'-deoxycytidine. Ph.D Thesis., Univ. Montréal (1983).

SELECTION OF PURINE NUCLEOSIDE ANALOGS BASED ON MULTIPLE BIOLOGICAL

AND BIOCHEMICAL PARAMETERS

M. Jean Noujaim, George Zombor and J. Frank Henderson

Cancer Research Group
University of Alberta
Edmonton, Alberta
Canada T6G 2H7

INTRODUCTION

Thousands of purine and purine nucleoside analogs have been synthe-sized in attempts to identify new anticancer, antiviral or immunosuppres-sant agents or other cytotoxic or cytocidal drugs. Because they are similar in chemical structure and in the pathways by which they are anabolized and catabolized, purine analogs are sometimes also perceived as rather homogeneous in terms of their biological and biochemical effects and mechanisms of action; such a generalization is prejudicial to detailed preclinical and clinical evaluation of individual analogs.

This study, therefore, was carried out to assess and demonstrate the degree of individual variation in biochemical and biological effects in a group of purine analogs when a number of such agents were studied in detail under exactly the same conditions. Furthermore, and in contrast to most previous attempts to do "biochemical screening" using intact cells (e.g., ref. 1, 2, 3), the present studies were done under condi-tions of growth inhibition by the analogs.

METHODS

Human lymphoblast RPMI 6410 cells were grown in RPMI 1640 medium containing 5% dialyzed fetal calf serum with and without concentrations of each analog that gave 20% of control growth at 24 h. Unless otherwise indicated, all measurements of drug effects were carried out after 5 h treatment; measurements of radioactive precursor metabolism were carried out for 1 h, beginning at 5 h of treatment.

RESULTS

Seventy-four purine and purine nucleoside analogs were initially assessed; only 36 were growth inhibitory in the system chosen and were carried further for detailed studies of their effects. The total number of biological and biochemical parameters studied was 29; effects of drugs on only a few of these are shown here. Furthermore, only the analogs that had the greatest effect on each parameter are listed.

Table 1. Inhibition of Thymidine Incorporation into DNA

	% of Control
6-Methylthiopurine ribonucleoside	0.240
Thiazole-4-carboxamide 2-ribonucleoside	1.37
Adenine arabinonucleoside	5.35
6-Thiopurine	6.11
6-Dimethylallylaminopurine ribonucleoside	7.45
6-Thiopurine ribonucleoside	7.76
7-Deazaadenine xylonucleoside	9.25
6-Chloropurine ribonucleoside	10.3
6-Dimethylaminopurine ribonucleoside	10.9
Adenine xylonucleoside	11.6
6-Ethanoylaminopurine ribonucleoside	23.1
2-Amino-6-thiopurine	24.9

Table 2. Inhibition of Uridine Incorporation into RNA

	% of Control
6-Methylthiopurine ribonucleoside	1.36
Thiazole-4-carboxamide 2-ribonucleoside	4.42
6-Thiopurine ribonucleoside	7.52
6-Thiopurine	10.0
6-Dimethylaminopurine ribonucleoside	11.0
6-Ethanoylaminopurine ribonucleoside	17.8
5,6-Dichlorobenzimidazole ribonucleoside	21.3

Table 3. Inhibition of Formate Incorporation into ATP + GTP

	% of Control
6-Hydroxylaminopurine ribonucleoside	0
6-Methoxypurine ribonucleoside	0.300
Adenine 3'-amino-3'-deoxyribonucleoside	2.26
Adenine xylonucleoside	3.00
6-Methylthiopurine ribonucleoside	3.37
Adenine 3'-deoxyribonucleoside	5.40
6-Dimethylallylaminopurine ribonucleoside	6.30
7-Deazaadenosine	12.0
6-Chloropurine ribonucleoside	13.5
Purine ribonucleoside	13.5
2-Amino-6-thiopurine	20.2
1-Oxyadenosine	21.2
6-Dimethylaminopurine ribonucleoside	23.8

Table 4. Inhibition of Nucleotide Synthesis from Hypoxanthine
          or of Guanylate Synthesis from Inosinate

	% of Control	
	GTP	ATP
6-Methoxypurine ribonucleoside	2.88	5.35
6-Hydroxylaminopurine ribonucleoside	0	7.74
Adenine 3'-amino-3'-deoxyribonucleoside	22.7	25.2
Purine ribonucleoside	23.1	31.6
7-Deazaadenosine	18.6	35.0
2-Amino-6-chloropurine ribonucleoside	0.990	56.1
6-Chloropurine ribonucleoside	5.05	59.0
Thiazole-4-carboxamide 2-ribonucleoside	2.01	64.3

Table 5. Lowered and Elevated GTP or ATP Concentrations

	% of Control	
	GTP	ATP
6-Ethanoylaminopurine ribonucleoside	34.8	31.0
6-Dimethylallylaminopurine ribonucleoside	26.7	37.6
6-Methylthiopurine ribonucleoside	20.7	37.8
6-Dimethylaminopurine ribonucleoside	32.9	44.4
6-Thiopurine ribonucleoside	26.3	75.7
6-Thiopurine	27.0	78.3
6-Chloropurine ribonucleoside	18.5	85.5
Thiazole-4-carboxamide 2-ribonucleoside	33.3	87.6
6-Methoxypurine ribonucleoside	170	237

Table 6. Lowered and Elevated UTP or CTP Concentrations

	% of Control	
	UTP	CTP
6-Methoxypurine ribonucleoside	10.9	13.6
1-Oxyadenosine	64.0	14.7
6-Dimethylaminopurine ribonucleoside	134	100
6-Thiopurine ribonucleoside	137	134
2-Amino-6-thiopurine ribonucleoside	147	122
Thiazole-4-carboxamide 2-ribonucleoside	151	124
6-Thiopurine	160	108
6-Ethanoylaminopurine ribonucleoside	170	128
6-Methylthiopurine ribonucleoside	175	105

Twelve analogs inhibited thymidine incorporation more than 75% (Table 1), and seven analogs inhibited uridine incorporation more than 75% (Table 2). No analog produced substantial inhibition of leucine incorporation.

Thirteen analogs inhibited purine biosynthesis de novo (measured as formate incorporation into ATP + GTP) by more than 75% (Table 3), whereas Table 4 shows that five analogs inhibited hypoxanthine incorporation into ATP by more than 65%. However, another three analogs preferentially inhibited hypoxanthine incorporation into GTP, indicating a block between inosinate and guanylate.

Table 5 lists eight purine analogs that produced marked decreases in the concentration of ATP, GTP or both. In addition, 6-methoxypurine ribonucleoside produced marked increases in the concentrations of ATP and GTP.

Two purine analogs produced marked decreases in the concentrations of UTP or CTP (Table 6). In addition, substantial increases in UTP and/or CTP concentrations were found in cells treated with another seven analogs.

The toxicity of seven purine analogs was markedly potentiated by low concentrations of guanine; these are 6-dimethylallylaminopurine ribonucleoside, 6-thiopurine ribonucleoside, 1-oxyadenosine, 6-hydroxylaminopurine ribonucleoside, 6-chloropurine ribonucleoside, 6-thiopurine, and 5,6-dichlorobenzimidazole ribonucleoside.

Chromatographic profiles were examined for analog nucleotide formation. Table 7 shows that this route of metabolism of analogs was detected in 16 cases; the putative triphosphate derivative was the predominant metabolite in 11 cases.

Table 7. Formation of Drug Nucleotides

	Tentative Identification	Concentration (fmoles/cell)
6-Ethanoylaminopurine ribonucleoside	mono-P	1.80
1-Oxyadenosine	mono-P	2.20
	tri-P	5.24
6-Methylthiopurine ribonucleoside	mono-P	1.10
6-Dimethylaminopurine ribonucleoside	mono-P	2.00
7-Deazaadenosine	di-P	0.91
	tri-P	1.90
Adenine arabinonucleoside	mono/di-P	0.43
6-Dimethylallylaminopurine ribonucleoside	mono/di-P	2.10
2,6-Diaminopurine	tri-P	2.60
6-Methylpurine	tri-P	2.70
6-Methylpurine ribonucleoside	tri-P	4.10
Adenine 3'-amino-3'-deoxyribonucleoside	tri-P	1.40
6-Chloropurine ribonucleoside	tri-P	0.39
6-Methoxypurine ribonucleoside	tri-P	0.14
Purine ribonucleoside	tri-P	10.0
6-Hydroxylaminopurine ribonucleoside	tri-P	3.20
Adenine 3'-deoxyribonucleoside	tri-P	3.20

Treatment with seven purine analogs led to substantial cell lysis and/or increased membrane permeability, as judged by trypan blue uptake, fluorescein diacetate exclusion, and accumulation of cell debris. These are adenine 3'-amino-3'-deoxyribonucleoside, 6-dimethylallylaminopurine ribonucleoside, 1-oxyadenosine, 7-deazaadenosine, 2-amino-6-thiopurine, 3-deazaadenosine, and 6-methylpurine ribonucleoside.

Some of the other parameters studied, which are not presented here for lack of space, include reversal of inhibition by adenine and hypoxanthine; potentiation of growth inhibition by deoxycoformycin; cell volume; cell cycle parameters; and the formation of drug catabolites found in the culture medium.

These results show that almost every analog has a unique pattern of biological and biochemical effects. Furthermore, they identify several previously unstudied analogs that deserve further investigation; these include 6-dimethylallylaminopurine ribonucleoside, 7-deazaadenine xylonucleoside, 6-dimethylaminopurine ribonucleoside, 6-ethanoylaminopurine ribonucleoside, 6-hydroxylaminopurine ribonucleoside, 6-methoxypurine ribonucleoside, adenine 3'-amino-3'-deoxyribonucleoside, and 1-oxyadenosine. Finally, they may guide decisions regarding choice of compounds for pharmacological studies and for clinical trials, and may guide chemists in the synthesis of new agents.

Future studies may include effects of drugs in the following additional biochemical parameters: deoxyribonucleotide concentrations, metabolism of deoxyribonucleosides, ribonucleotide reduction in intact cells, DNA strand breaks, synthesis of PP-ribose-P, PRPP amidotransferase activity in intact cells, metabolism of adenine, guanine, adenosine, uridine, cytidine, orotate and bicarbonate, concentrations of UDP-sugar coenzymes, adenosylhomocysteine hydrolase activity, methylthioadenosine phosphorylase activity, and ATP catabolism.

REFERENCES

1. J. F. Henderson, A. R. P. Paterson, I. C. Caldwell, P. Paul, M. C. Chan, and K. F. Lau, Inhibitors of nucleoside and nucleotide metabolism, Cancer Chemother. Rep. (Part 2) 3:71 (1972).
2. L. L. Bennett, Jr., M. H. Vail, S. Chumley, and J. A. Montgomery, Activity of adenosine analogs against a cell culture line resistant to 2-fluoroadenine, Biochem. Pharmacol. 15:1719 (1966).
3. R. E. Parks, Jr. and P. R. Brown, Incorporation of nucleosides into the nucleotide pools of human erythrocytes. Adenosine and its analogs, Biochemistry 12:3294 (1973).

EFFECT OF m-AMSA ON THE CELLULAR PHARMACOLOGY OF Ara-CTP IN HUMAN

LEUKEMIC CELLS DURING THERAPY WITH HIGH-DOSE Ara-C

William Plunkett,[1] and Michael Keating[2]

Departments of Chemotherapy Research[1] and Hematology
The University of Texas M.D. Anderson Hospital and
Tumor Institute at Houston, Houston, Texas 77030

INTRODUCTION

The pyrimidine nucleoside analog, 1-β-D-arabinofuranosylcytosine
(ara-C), is the most effective drug for the successful treatment of adult
acute myelogenous leukemia (1). The cytotoxic activity of ara-C is
dependent upon its intracellular conversion to the active 5'-triphosphate
ara-CTP. Ara-CTP competes with dCTP in the inhibition of DNA polymerase,
but its incorporation into DNA indicates that it also serves as a
substrate for the enzyme. These actions are associated with the
cytotoxicity of the drug. However, ara-CTP also participates as an analog
in the intermediary metabolism of phospholipids and is glycoprotein
synthesis. In fact, ara-CTP is required for all of the known mechanisms
of ara-C-induced cell killing.

While ara-C has conventionally been administered by continuous
infusion, new regimens using large (3 g/m^2) intermittent doses have shown
promising antileukemic activity. Cellular pharmacology studies of ara-CTP
during such "high-dose ara-C" therapy of relapsed acute leukemia have
demonstrated strong correlations between clinical response and the
pharmacokinetics of ara-CTP in circulating leukemic blasts (2). These
high-dose ara-C regimens are currently being combined with other drugs in
clinical trials. Preclinical investigations have demonstrated that when
used in combination with ara-C, drugs such as hydroxyurea and 3-
deazauridine, may enhance the total intracellular exposure to ara-CTP. We
have presented evidence that the 3-deazauridine effect on ara-CTP may also
be seen during therapy of patients with acute leukemia with this
combination (3). On the other hand, drugs such as methotrexate and
epipodophyllotoxins may limit ara-CTP accumulation, either by metaboliic
antagonism or interference with transport.

Another drug with significant antileukemic activity, 4'-(9-
acridinylamino)methansulfon-m-anisidide (m-AMSA), is currently undergoing
clinical trials in combination with ara-C (4). The present investigation
was designed to evaluate the effect of m-AMSA on the cellular
pharmacokinetics of ara-CTP in leukemic blasts during therapy of adult
leukemia with high-dose ara-C and m-AMSA.

PATIENTS AND METHODS

Patients. Adults with acute leukemia were treated with either of two therapeutic protocols. In the first, patients with acute leukemia in relapse received ara-C alone in the high-dose regimen. This consisted of 4-12 doses of 3 g/m^2 of ara-C infused over a 2-hr duration at 12-hr intervals (2). A second protocol, consisting of a two-arm combination chemotherapy regimen, was used for patients with previously untreated acute leukemia. Patients received either six doses of high-dose ara-C at 12-hr intervals or 2 doses of 3 g/m^2 at 12-hr intervals followed by a continuous infusion of ara-C (70 mg/m^2) for 7 days. Each treatment arm required the administration of m-AMSA (30 mg/m^2) infused intravenously over 1 hr simultaneously with the second infusion of ara-C. The design of these protocols made it possible to study the cellular ara-CTP pharmacokinetics after serial doses of ara-C alone (protocol 1) and after consecutive doses of ara-C where the second dose was accompanied by a simultaneous infusion of m-AMSA (protocol 2).

Cellular Pharmacology Studies. Samples of peripheral blood (10 ml) were obtained at 1-2 hr intervals for 24 hr during the first two doses of ara-C. Leukemic blasts were isolated by Ficoll-Hypaque density centrifugation. Cells from each sample were then counted and their mean cell volume was determined using an electronic particle counter that was equipped with a cell sizing apparatus (Coulter Electronics, Inc.). Cellular nucleotides were extracted with HC10$_4$ and separated using anion-exchange high-pressure liquid chromatography as described (5). Quantitation of absorbance at 280 nm was determined by electronic integration and referenced to preprogrammed response factors. Determination of the absolute quantity of ara-CTP contained in the HC10$_4$-soluble material extracted from a given number of cells of known mean cell volume permitted the expression of cellular contents of ara-CTP in terms of concentration (μM). The total intracellular exposure to ara-CTP was calculated as the area under the curve (AUC) of ara-CTP accumulation and elimination (2). This calculation summed the area under the linear accumulation phase with that under the mono-exponential elimination phase. The ara-CTP AUC values were expressed in units of μM-hr. In calculating the AUC values for the second dose of ara-C, the ara-CTP μM-hr value calculated to remain in the cells after the first dose of ara-C was subtracted from the ara-CTP AUC value determined after the second dose of ara-C.

RESULTS

Serial sampling of peripheral blood during and following ara-C infusions gave a pharmacokinetic picture of ara-CTP metabolism in leukemic cells , permitted comparison of intrapatient variability and enabled a search for correlations with clinical response (2). Substantial heterogeneity was seen among patients with respect to the ability of the circulating leukemic cells to accumulate and to retain ara-CTP. Some characteristics of the cellular pharmacokinetics of ara-CTP in circulating blasts of patients with relapsed acute leukemia were strongly correlated with clinical response to high-dose ara-C therapy (2). While the peak ara-CTP values were not significantly different between patients who achieved complete remission and for treatment failures, significant relationships were found between response and the rate of ara-CTP elimination, the trough ara-CTP concentration, and the area under the ara-CTP accumulation and elimination curve (2).

Even without this compelling evidence, the importance for achieving maximal intracellular exposure to ara-CTP in leukemic cells is clear. The design of the two treatment protocols permitted us to ask whether m-AMSA administered simultaneously with ara-C would affect the cellular metabolism of ara-CTP. To provide a frame of reference for this investigation, we first determined the variability of the ara-CTP AUC in the circulating blasts of patients who received treatment with ara-C alone. The cellular pharmacokinetics of these patients were determined after serial doses of ara-C (Table 1). While these determinations revealed both increases and decreases, the AUC values differed by more than 17% in only a single patient. The median change in the ara-CTP AUC values between the two doses of ara-C was found to be a decrease of 5%. When the ara-CTP AUC was determined after administration of ara-C alone and then with m-AMSA, the median decrease in intracellular ara-CTP AUC was found to be 20% (Table 1). Four of the 6 patients studied on this protocol showed decreased ara-CTP AUC values during the simultaneous infusion of m-AMSA. While the cellular pharmacokinetics of patients treated on the two protocols were not significantly different, the decreases in ara-CTP AUC that accompanied infusion of m-AMSA tended to be greater than those recorded after consecutive doses of ara-C.

TABLE 1

Effect of m-AMSA on the area under the ara-CTP accumulation and elimination curve (AUC) in leukemic blasts. Separate groups of patients with acute leukemia were treated with either intermittent doses of ara-C (3 g/m^2) at 12-hr intervals or with this regimen accompanied by m-AMSA (30 mg/m^2) with the second dose of ara-C. The intracellular AUC of ara-CTP was determined after each dose of ara-C.

	ara-CTP AUC in leukemic cells (μM-hr)						
	ara-C				ara-C plus m-AMSA		
ara-C dose	1	2	Δ%		1	2	Δ%
Patient				Patient			
1	1670	1940	17	6	1180	1690	43
2	5240	3400	-54	7	3920	3520	-11
3	580	600	4	8	1340	1470	9
4	6370	6040	-5	9	5300	4080	-30
5	6120	5450	-12	10	5870	4470	-31
				11	1540	1100	-40
Median difference		-5%					-20%

DISCUSSION

The results in Table 1 indicate that the ara-CTP AUC in leukemic cells after serial doses of ara-C showed only a 5% decrease after the second dose. This relative constancy was the basis for using intracellular ara-CTP pharmacokinetic parameters to direct the intervals between doses and the dose levels of ara-C on separate protocols (2). Given the complexity associated with the repeated sampling required to construct the pharmacokinetic profiles in the cells of each patient and the likely possibility of ongoing cellular damage, the level of variation seen in the cells of the patients who received ara-C alone was considered to be remarkably small.

Following simultaneous infusions of m-AMSA with the second dose of ara-C the intracellular exposure to ara-CTP decreased by a median value of 20%. It is possible that this may reflect ara-CTP-induced damage to a specific cohort of leukemic cells in the population that is immediately susceptible to its action. The variation of the ara-CTP AUC values in each patient population was sufficient to rule out finding significant effects of ara-CTP AUC associated with infusion of m-AMSA. However, there was a trend in the direction of decreased intracellular exposure to ara-CTP when m-AMSA was administered with ara-CTP. The clinical evidence that the combination of ara-C and m-AMSA is more effective in inducing antileukemic response than either agent alone compels the conclusion that cells exposed to the combination sustain greater damage than if treated only with ara-C (4). We have recently provided presumptive biochemical evidence for this likelihood by demonstrating DNA strand breaks in leukemic cells during therapy with m-AMSA (6).

The approach that we have described to the study of drug-drug interactions in tumor cells during combination chemotherapy builds upon our earlier efforts (3,4,7). We plan to extend this approach to both investigations of the biochemical modulation of ara-C activity in hematological malignacies and to inquires into the significance of drug-induced DNA damage in leukemic cells.

ACKNOWLEDGEMENTS

This work was supported by grants CA28513 and CA32839 from the National Cancer Institute, DHHS. The authors are grateful for the excellent technical assistance of Theresa Adams, Sherri Chubb, Rodney Croft, and Billie Nowak.

REFERENCES

1. M.J. Keating, K.B. McCredie, G.P. Bodey, T.L. Smith, E. Gehan, and E.J. Freireich, Improved prospects for long-term survival in adults with acute myelogenous leukemia, J. Amer. Med. Assoc., 248:2481, 1982.
2. W. Plunkett, S. Iacoboni, E. Estey, L. Danhauser, J.O. Liliemark, and M.J. Keating, Pharmacologically directed ara-C therapy for refractory leukemia, Sem. Oncol. 12:(Suppl 3), 20, 1985.
3. B. Barlogie, W. Plunkett, M. Raber, J. Latreille, M. Keating and K. McCredie, In vivo cellular kinetic and pharmacological studies of 1-β-D-arabinofuranosylcytosine and 3-deazauridine chemotherapy for relapsing acute leukemia, Cancer Res., 41:1227, 1981.
4. M. Keating, T. Smith, K. McCredie, E. Estey, and E. J. Freireich, Successful introduction of a new m-AMSA containing regimen in acute leukemia, Proc. Amer. Soc. Clin. Oncol. 2:174, 1983.
5. W. Plunkett, S. Chubb, and B. Barlogie, Simultaneous determination of 1-β-D-arabinofuranosylcytosine 5'-triphosphate and 3-deazauridine 5'-triphosphate in human leukemia cells by high-performance liquid chromatography. J. Chromatogr., 221:445, 1980.
6. R. Meyn, W. Plunkett, and E. Estey, Quantitation of DNA strand breaks in circulating blasts from leukemia patients during therapy with m-AMSA, Proc. Amer. Assn. Cancer Res., 26:161, 1985.

7. W. Plunkett, R.S. Benjamin, M.J. Keating and E.J. Freireich,
    Modulation of 9-β-D-arabinofuranosyladenine 5'-triphosphate and
    deoxyadenosine triphosphate in leukemic cells by 2'-deoxycoformycin
    during therapy with 9-β-D-arabinofuranosyladenine, Cancer Res.
    42:2092, 1982.

5-FLUORO-5'-DEOXYURIDINE IS AN INHIBITOR OF

URIDYLATE NUCLEOTIDASE IN L1210 LEUKEMAEMIA

C. Roobol, G.B.E. De Dobbeleer, and J.L. Bernheim

Laboratory for Experimental Chemotherapy
Dept. of Hematology, Immunology and Oncology
Free University of Brussels
Laarbeeklaan 103/E, B - 1090 Brussels, Belgium

INTRODUCTION

Fluoropyrimidines are widely used in anti-cancer chemotherapy. Of all cytotoxic drugs, their mechanism of action is propably the best understood.[1] Essentially, two processes are known to be involved in the cytotoxic action of fluoropyrimidines. First, the incorporation of fluoropyrimidines into nascent RNA, leading to the formation of defective RNA species, has been reported and has been correlated to their cytotoxic effect [2,3]. Second, dTMP depletion has been shown to occur via the inhibition of thymidylate synthetase by FdUMP, leading to an arrest of DNA synthesis[4,5] A third potential target has been suggested to be the incorporation of fluoropyrimidines into nascent DNA, but the contribution of this phenomenon to the final cytotoxic effect is less understood[6].

Recently, a novel fluoropyrimidine, 5-fluoro-5'-deoxyuridine (dFUrd) has been synthesised [7]. It has been shown that dFUrd is activated intracellularly by a pyrimidine nucleoside phosphorylase mediated conversion to FUra [8,9]. However, several observations have raised questions about this proposed mechanism of action. First, for several experimental tumours, dFUrd showed a significantly better in vivo effect as compared to the in vitro effect on the same cells[10,11]. Second, the intracellular activity of pyrimidine nucleoside phosphorylase correlates only poorly with the effect of dFUrd on tumour cell lines [10]. Third, dFUrd and FUra have been reported to follow different routes of intracellular activation in Novikoff hepatoma cells [12].

For these reasons, the mechanism of activity of dFUrd was studied in L1210 leukaemia and compared with those of FdUrd and FUra. It is shown that FdUrd has thymidylate synthetase as the decisive target for its cytotoxic action. FUra is converted to 5-fluorouridine-5'-triphosphate (FUTP) via a series of consecutive phosphorylation reactions and incorporated into nascent RNA, leading to the formation of defective RNA species. Contrary to current concepts which consider dFUrd as a prodrug for FUra, in L1210 leukaemia dFUrd is not degraded to FUra by pyrimidine nucleoside phosphorylase. It is shown that dFUrd did not interfere with the de novo synthesis of pyrimidine nucleotides, but inhibited the formation of uridine derived from the orotic acid pathway. This activity does not require metabolic activation of 5'-dFUrd.

RESULTS

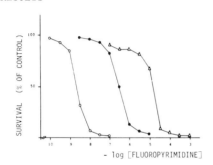

Fig. 1

3 x 10³ L1210 cells were incubated in
150μl of medium RPMI 1640 , supplemented with 10% (v/v) of foetal calf
serum, in the continous presence of
varying concentrations of FdUrd (o-o),
FUra (●-●) or dFUrd (Δ-Δ). After an
incubation of 3 days at 37°C in a
humidified 5% CO_2 atmosphere, the
cell numbers were determined and survival was expressed as a percentage
of the untreated control.

The antiproliferative effect of fluorinated pyrimidines on L1210
leukaemia was demonstrated via an incubation of cells in the continuous
presence of varying concentrations of the different fluoropyrimidines
(figure 1). The cytotoxic effects were quantitated as the inhibitory
concentration 50 (IC 50), i.e. the concentration of antimetabolite that
leads to a 50% inhibition of cell growth. The observed IC 50 values were
2.10^{-9} M for FdUrd, 3.10^{-7} M for FUra and 2.10^{-5}M for dFUrd.

In order to obtain a first indication with respect to the targets
for the action of the respective antimetabolites, the effects of thymidine
(figure 2A) and uridine (figure 2B) on the cytotoxicity of the different
fluoropyrimidines were studied, where the IC 50 values for the respective
fluoropyrimidines were determined as described for the experiment shown
in figure 1. Rescue by exogenous thymidine would imply synthesis of
thymidine-5'-monophosphate via the salvage pathway, thereby circumventing
a blockade of thymidylate synthetase. As shown in figure 2A, thymidine
has a pronounced effect on the cytotoxicity of FdUrd, without affecting the
antiproliferative effects of FUra and dFUrd. This finding suggests that
thymidylate synthetase is involved in the cytotoxic action of only FdUrd,
but is not the decisive target for either FUra or dFUrd. When uridine is
added (figure 2B), rescue is observed only from dFUrd toxicity, but not
from FdUrd or FUra toxicity. This phenomenon has been explained by the
observation that uridine competitively inhibits the pyrimidine nucleoside
phosphorylase mediated degradation of dFUrd to FUra in murine tumours [8].

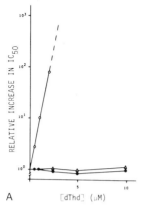

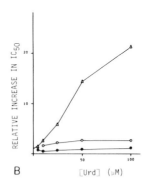

Fig. 2

The effects of varying concentrations of thymidine (panel A) or
uridine (panel B) on the cytotoxicities of FdUrd (o-o), FUra
(●-●) or dFUrd (Δ- Δ) were studied as described in the legend
to figure 1. The relative increase in the IC 50 value is
defined as IC 50 (+ nucleoside)/IC 50 (- nucleoside)

In order to investigate the role of thymidylate synthetase in the cytotoxic action of fluoropyrimidines in more detail, the effects of FdUrd, FUra and dFUrd on the de novo pathway of dTMP synthesis were studied. In situ thymidylate synthetase activity can be measured via the incorporation of ^{3}H-dThd or ^{3}H-dUrd into nascent DNA. In the case of ^{3}H-dUrd incorporation , the radioactive precursor will be converted to ^{3}H-dUMP, the substrate for thymidylate synthetase. Hence, a blockade of thymidylate synthetase by fluoropyrimidines will lead to a complete inhibition of the incorporation of ^{3}H-dUrd into nascent DNA [13] . In the case of ^{3}H-dThd incorporation, DNA thymine will be in part unlabelled, derived from de novo synthesis, and in part labelled, derived from exogenous ^{3}H-dThd that is incorporated via the salvage pathway. If de novo thymidylate synthesis is blocked , all DNA thymine will be derived from the salvage pathway, hence resulting in an increase in the level of ^{3}H-dThd incorporation [13].

As shown in figure 3, incubation of L1210 cells with FdUrd leads to an inhibition of in situ thymidylate synthetase activity. More important, this inhibition coincides with the inhibitory effect of FdUrd on cell proliferation (figure 4A), thus showing that thymidylate synthetase is the decisive target for FdUrd. This finding is in agreement with the observation that L1210 cells can be rescued from FdUrd toxicity by addition of exogenous dThd, as demonstrated by the experiment shown in figure 2.

Incubation of L1210 cells with excessive concentrations of FUra does lead to an inhibition of in situ thymidylate synthesis as measured via ^{3}H-dThd incorporation (figure 3). However, the inhibition of thymidylate synthetase does not coincide with the antiproliferative activity of FUra (figure 4B). This experiment shows that intracellular conversion of FUra to FdUMP has occurred, but that this conversion is not related to the cytotoxic action of FUra. Therefore, thymidylate synthetase is not the decisive target for FUra.

Incubation of L1210 cells with dFUrd leads to an inhibition of the incorporation of ^{3}H-dUrd into nascent DNA, which, however, does not coincide with the cytotoxic activity of dFUrd (figure 4C). More important, the inhibition of ^{3}H-dUrd incorporation does not occur with a concomitant increase in the incorporation of ^{3}H-dThd, as confirmed by the experiment shown in figure 3. This finding suggests that the inhibition of ^{3}H-dUrd incorporation by dFUrd does not occur at the level of thymidylate synthetase. The observation that in the case of FUra, inhibition of de novo thymidylate synthesis did occur (figure 4B) suggests that in L1210 cells dFUrd does not act as a prodrug for FUra.

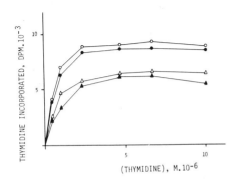

THYMIDINE INCORPORATED, DPM.10^{-3}

(THYMIDINE), M.10^{-6}

Fig. 3

5×10^4 L1210 leukaemia cells were incubated in the presence of varying concentrations of ^{3}H-dThd and in the presence of either 10^{-6} M FdUrd (o-o), 10^{-4} M FUra (●-●), 10^{-3} M dFUrd (▲-▲) or in the absence of drugs (Δ-Δ). After an incubation of 2 hours at 37°C, the cells were collected on glass fiber filter and the amount of radioactivity retained was determined

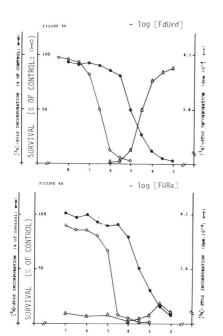

Fig. 4A

5 x 10^4 L1210 leukaemia cells were incubated in the presence of either 10μM
of ^{3}H-dThd (Δ-Δ) or 10μM of ^{3}H-dUrd (●-●)
and the effect of varying concentrations
of FdUrd on the level of incorporation
was determined as described for figure
3 and compared with the antiproliferative effect of FdUrd (o-o) measured as
described for figure 1.

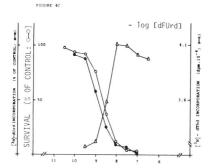

Fig. 4B

5 x 10^4 L1210 leukaemia cells were incubated in the presence of either 10μM
of ^{3}H-dThd (Δ-Δ) or ^{3}H-dUrd (●-●) and
the effect of varying concentrations of
FUra on the incorporation was determined as described for figure 3 and
compared with the antiproliferative
effect of FUra, (o-o) measured as
described for figure 1.

Fig. 4C

5 x 10^4 L1210 leukaemia cells were
incubated in the presence of either
10μM of ^{3}H-dThd (Δ-Δ) or ^{3}H-dUrd (●-●)
and the effect of varying concentrations of dFUrd on the incorporation
was determined as described for figure
3 and compared with the antiproliferative effect of dFUrd (o-o) measured as
described for figure 1.

        In order to obtain more information concerning the possible absence
of a prodrug function of dFUrd, the pyrimidine nucleoside phosphorylase
mediated degradation of different nucleosides was studied in cell free
extracts of L1210 leukaemia. As shown in figure 6 , even in the presence
of excessive amounts of L1210 protein, only uridine is converted to its
free base uracil, but no degradation of dThd or dFUrd can be detected.
However, if the effect of dFUrd on the pyrimidine nucleoside phosphorylase
mediated degradation of ^{3}H-Urd to ^{3}H-Ura is studied using initial rate
kinetics, competitive inhibition is observed (figure 5). This proves that,
although not being degraded by pyrimidine nucleoside phosphorylase, dFUrd
has a low but detectable affinity for the enzyme.

        The affinity of pyrimidine nucleoside phosphorylase for different
nucleosides and bases was determined by their effect on the degradation
of ^{3}H-Urd to ^{3}H-Ura, as has been described for figure 5. The combined
results are summarised in table I, and lead us to the following conclusions.
First, with respect to the inhibition of the degradation of 3H-Urd to
^{3}H-Ura , L1210 pyrimidine nucleoside phosphorylase has equal affinities
for Urd and dThd. However, despite this comparable effinity, dThd does
not act as a substrate for pyrimidine nucleoside phosphorylase, as shown
in figure 6.

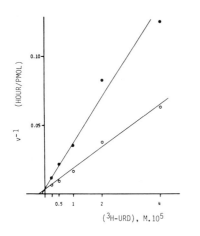

Fig. 5

20μg of L1210 protein were incubated in the presence of varying concentrations of ^{3}H-Urd, both in the absence (o-o) and presence (●-●) of 10^{-3} M dFUrd and the initial rate of uracil formation was determined as described in the legend to figure 6.

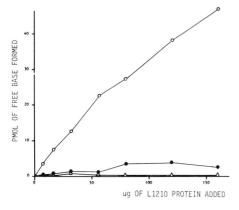

Fig. 6

Varying amounts of L1210 protein were incubated in the presence of 30μM of ^{3}H-Urd (o-o). ^{3}H-dThd (●-●) or ^{14}C-dFUrd (Δ-Δ) in 50μl of a buffer containing 20mM Tris-Hcl, pH 7,5; 100mM Kcl; 1,5mM Mg-acetate,0,1mM EDTA; 10mM 2-mercaptoethanol and 25mM K2HPO4 for 1 h. at 37°C. The reaction was stopped by addition of 50μl of ethanol, the mixture was analysed by thin layer chromatography and the amount of free base formed was determined.

The effect of pyrimidine nucleosides and bases on the pyrimidine nucleoside phosphorylase mediated degradation of ^{3}H-Urd to ^{3}H-Ura

nucleoside/base	Km-value	Ki-value
Urd	$6,0 \times 10^{-5}$	$1,7 \times 10^{-4}$
dThd	n.d.	$1,1 \times 10^{-4}$
dFUrd	n.d.	$1,2 \times 10^{-3}$
Ura	n.d.	$2,0 \times 10^{-4}$
FUra	n.d.	$2,3 \times 10^{-5}$

Second, FUra has a pronounced inhibitory effect on the degradation of ^{3}H-Urd to ^{3}H-Ura and has an affinity for pyrimidine nucleoside phosphorylase that largely exceeds that of the native substrate, Urd, or that of the native product, Ura. If dFUrd were able to act as a substrate, a severe product inhibition would have occurred. Third, pyrimidine nucleoside phosphorylase has a very low affinity for dFUrd but, as is the case for dThd, dFUrd does not act as a substrate for the enzyme (figure 6). In summary, these data eliminate the possibility of a prodrug function of dFUrd in L1210 leukaemia.

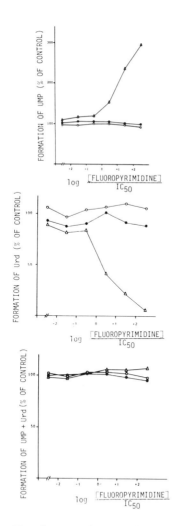

Fig. 7A

Th effect of varying concentrations of FdUrd (o-o), FUra (o-o) or dFUrd (Δ-Δ) on the formation of ^{3}H-UMP, derived from ^{3}H-orotate , was determined as described in the legend to figure 7C.

Fig. 7B

The effect of varying concentrations of FdUrd (o-o), FUra (o-o) or dFUrd on the formation of ^{3}H-Urd, derived from ^{3}H-orotate was determined as described in the legend to figure 7C.

Fig. 7C

5µg of L1210 protein were incubated with 400 pmol of ^{3}H-orotate in 50µl of a buffer containing 20mM Tris-Hcl, pH 7,5, 100mM Kcl, 5mM Mg-acetate, 0,26mM 5-phosphoribosyl-1-pyrophosphate, 10mM 2-mercaptoethanol and varying concentrations of FdUrd (o-o), FUra (o-o) or dFUrd (Δ-Δ). After an incubation of 1h. at 37°C, the reaction was stopped by addition of 10µl of 100mM EDTA and the mixtures were analysed by paper chromatograpghy and the amount of orotate metabolised was determined.

    Thusfar, we have only shown how dFUrd does not act in L1210 leukaemia. The question that comes now is which step in the cellular metabolism may represent the actual target for the action of dFUrd in L1210 leukaemia. After unsuccessfully exploring several potential pathways, an effect of dFUrd on the de novo synthesis of pyrimidine nucleotides via the orotic acid pathway could be demonstrated. When ^{3}H-orotate is incubated in the presence of 5-phosphoribosyl-1-pyrophosphate and varying concentrations of the different fluoropyrimidines, the amount of orotic acid that is metabolised appeares to be unaffected (figure 7C), thereby showing that both orotate phosphoribosyl transferase and orotidylate decarboxylase are not inhibited by dFUrd, FUra or FdUrd. However, if the formation of uridine-5'-monophosphate is measured separately, a strong increase is observed in the presence of dFUrd, whereas FdUrd and FUra have no effect (figure 7A). The increase in the formation of UMP occurs with a concomittant decrease in the formation of Urd (figure 7B), showing that the process which is inhibited in this experiment is the degradation of UMP to Urd. More important, the fact that this inhibition is observed with dFUrd, but not with FUra shows that in this respect dFUrd is active in itself and doens not require a pyrimidine nucleoside phosphorylase mediated degradation of dFUrd to FUra. Furthermore, when the effect of dFUrd on the formation of Urd derived from orotic acid is compared with the effect of dFUrd on the degradation of tritiated UMP, it appears that the latter phenomenon is only moderately inhibited by dFUrd (figure 8).

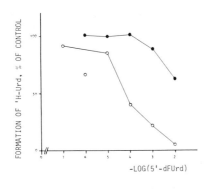

Fig. 8

5µg of L1210 protein were incubated with 400 pmol of ^{3}H-UMP in 50µl of a buffer containing 20mM Tris-Hcl,pH 7,5; 100mM Kcl; 5mM Mg-acetate, 10mM 2-mercaptoethanol and varying concentrations of dFUrd. After an incubation of 1 hour at 37°C the reaction was stopped by addition of 50µl ethanol and the extent of Urd formation was determined by paper chromatography (o-o). Alternatively, the effect of dFUrd on the extent of Urd-formation derived from orotate was determined as described for figure 7B.

Therefore, it is postulated that dFUrd exerts a specific inhibitory effect on a uridylate nucleotidase mediated formation of free uridine, derived from the orotic acid pathway. This metabolism can be distinguished from the phosphatase mediated degradation of UMP (figure 8). Finally, it should be emphasised that in L1210 leukaemia, uridylate nucletidase has been identified as a potential target for dFUrd, which is not necessarily the decisive target for this drug. The role of uridylate nucleotidase inhibition in the mechanism of action of dFUrd remains to be elucidated.

REFERENCES

1. C. Heidelberger, in:Handbook of Experimental Pharmacology,Eds Sartorelli & Johns,Berlin Springer-Verlag, 38-2:193 (1975)
2. R. Glazer & M. Legraverend, The effect of 5-fluorouridine-5'-triphosphate on RNA transcribed in isolated nuclei in vitro, Molec Pharmacol. 17: 279 (1980)
3. R. Glazer & A.L. Peale, The effect of 5-fluorouracil on the synthesis of nuclear RNA in L1210 cells in vitro, Molec Pharmacol. 16:270 (1979)
4. P.V. Danenberg, Thymidylate synthetase, a target enzyme in cancer chemotherapy, Biochem. Biophys. Acta 473: 73 (1977)
5. C.P. Spears, A.H. Shahinian, R.G. Moran, C. Heidelberger & T.H. Corbet, In vivo kinetics of thymidylate synthetase inhibition in 5-fluorouracil sensitive and resistant murine colon adenocarcinomas, Cancer Res 42: 450 (1982)
6. P. Major, E. Egan, D. Herrick & D.W. Kufe, 5-fluorouracil incorporation in DNA of human breast carcinoma cell lines, Cancer Res 42: 3005 (1982)
7. A.F. Cook, M.J. Holman, M.J. Kramer & P.W. Trown, Fluorinated pyrimidine nucleosides. Synthesis and antitumor activity of a series of 5'-deoxy-5-fluoropyrimidine nucleosides, J. Med. Chem. 22: 1330 (1979)
8. H. Ishitsuka, M. Miwa, K. Takemoto, K. Fukuoka, A. Itoga & H.B. Maruyama, Role of uridine phosphorylase for antitumor activity of 5'-deoxy-5-fluorouridine, Gann 71:112 (1980)
9. A. Kono, Y. Hara, S. Sugata, Y. Karube, Y. Matsushima & H. Ishitsuka, Activation of 5'-deoxy-5-fluorouridine by thymidine phosphorylase in human tumours, Chem. Pharm. Bull. 31:175 (1981)
10. H.R. Hartman & A. Matter, Antiproliferative action of a novel fluorinated uridine analog, 5'-deoxy-5-fluorouridine, measured in vitro and in vivo on four different murine tumor lines, Cancer Res.42:2412 (1982)
11. A.Rosowski, S.H. Kim, D. Trites & M. Wick, Synthesis and in vivo antitumor activity of potential 5-fluorouracil prodrugs,J.Med.Chem 25:1034 (1982)
12. J.G. Corey & G.L. Carter, Evidence that 5'-deoxy-5-fluorouridine may not be activated by the same mechanism as 5-fluorouracil, Biochem.Pharmacol 31:2841 (1982)
13. C. Roobol, G.B.E. De Dobbeleer & J.L. Bernheim, 5-fluorouracil and 5-fluoro-2'-deoxyuridine follow different metabolic pathways in the induction of cell lethality in L1210 leukaemia, Br.J.Cancer 49:739 (1984)

THE EFFECT OF PHOSPHONOFORMIC ACID ON

WILD TYPE AND MUTANT S49 CELL LINES

Daniel A. Albert and Lorraine J. Gudas

Department of Medicine, University of Chicago, Chicago, IL
60637, and Department of Pharmacology, Dana Farber Cancer
Institute, Harvard Medical School, Boston, MA 02115

Phosphonoformic acid was synthesized in 1924 by Nylen (1) and has
been investigated recently as an antiviral agent because of its
inhibition of viral DNA polymerase (2,3,4). Herpes simplex virus
polymerase activity is quite sensitive to PFA inhibition (K_i=3.5 uM) (2).
Partially purified mammalian (calf thymus) DNA polymerase is less
sensitive to PFA inhibition, and this sensitivity depends on the enzyme
concentration (K_i>3.5 uM) (2).

In contrast to the results with mammalian DNA polymerases in cell
free systems, mammalian cellular DNA synthesis is very resistant to PFA
inhibition, requiring two orders of magnitude higher concentrations of
PFA to achieve 50% inhibition (K_i 1 to 2 mM) (2). We studied the
mechanism of inhibition of cellular replication by PFA by selecting
mutant cell lines resistant to growth inhibition by phosphonoformic acid,
and compared them with wild type S49 cells that are more sensitive to PFA
inhibition.

METHODS

Cell Growth. Cells were cultured in Dulbecco's modification of
Eagle's medium containing 10% heat-inactivated horse serum at 37° in 10%
CO_2 (in air) atmosphere. Growth curves were performed by plating 1×10^5
cells/ml in increasing drug concentrations. Cells were harvested and
counted at 72 hours after plating.

Mutant Selection Procedure. 1.5×10^8 cells were exposed to N-
methyl-N'-nitro-N-nitrosoguanidine (4 mg/ml) for 4 hours, and then
plated out five to seven days later in soft agar containing various
concentrations of phosphonoformic acid, over a mouse embryo fibroblast
feeder layer. Clones were picked by aspirating a visible colony of
cells under a dissecting microscope and culturing these cells in growth
media at one to three weeks after plating. 19 colonies were picked from
plates containing 3.0 mM PFA. Plates with higher concentrations of PFA
had no colonies whereas plates with lower concentrations had numerous
colonies.

Determination of Nucleotide Pools. Nucleotide pools were measured
by the method of Garrett and Santi (5). 2.5 to 5×10^7 cells were

washed in phosphate-buffered saline and an aliquot was counted on a Coulter Model ZBI. Cells were centrifuged, and then resuspended in 200 ul ice cold 0.1 M potassium phosphate buffer (pH 7.2). 1.5 N perchloric acid (40 ul) was added, and 60 seconds later, 80 ul of 1.2 N KOH (final pH = 7.0) was added. The tubes were left on ice for 5 to 10 additional minutes. For deoxyribonucleotide pools, ribonucleotides were removed by periodation with 0.02 M sodium periodate for 90 minutes at 37°. The reaction was terminated by the addition of .03 M rhamnose. Extracts were then analyzed on a Beckman high performance liquid chromatography system using a SAX Partisil Column and elution buffer consisting of 0.38 M ammonium phosphate buffer with 1.5% acetonitrile at pH 3.45. Peaks were measured at 254 and 280 nm and concentration was proportional to peak height above base-line (6).

Ribonucleotide Reductase Assay. Ribonucleoside diphosphate reductase activity was measured by the conversion of CDP to deoxy CDP by permeabilized cells using the method of Lewis and Wright (7). $1-2 \times 10^7$ cells were permeabilized by exposure to 1% Tween 80, and the assay mixture consisted of 39 mM HEPES, 6.6 mM ATP, 8 mM $MgCl_2$, 22 mM dithiothreitol (DTT), 50 mM CDP and 0.42 uCi $[^{14}C]$-CDP, in a final volume of 300 ul after addition of a 5-fold concentrated mixture. Reactions were run for 20 and 40 minutes and were linear with time. The reactions were terminated by boiling for 4 minutes. Crotalus atrox venom was then added (6.0 mg per assay) and samples were incubated for 4 h at 37^OC. Samples were diluted with 500 ul H_2O, and then loaded on 1 ml borate Dowex AG-1-X-8 columns. The $[^{14}C]$ deoxycytidine product was eluted from the column with 2 ml H_2O and counted by liquid scintillation.

DNA Polymeraseα Assay. DNA polymeraseα was partially purified and assayed by the method of Baril et al (8). Briefly, 5×10^8 cells were centrifuged at 1500 rpm at 4° and lysed in hypotonic 10 mM Tris. The supernatant was dialyzed against 50 mM Tris, 1 mM DTT, 1 mM EDTA (pH 7.4) (TDE buffer). PMSF (0.1 uM) was added as a proteinase inhibitor. After centrifugation for 60 min. at 35 K rpm (150,000 xg) (SW41 rotor in a Beckman Model L5-75 Ultracentrifuge), the material was concentrated on a sucrose bed, loaded onto a 2.5 ml DEAE column, and eluted with a 0.3 M KCl; the preparation was subsequently dialyzed against TDE buffer overnight prior to assay. The assay mixture contained 0.2 mM dCTP, 0.2 mM dGTP, 0.2 mM dATP, 0.8 mM $MgCl_2$, 20 mM Tris, 1 mM DTT, 100 uM TTP, 0.27 uCi ^{3}H-TTP, and a variable amount of protein (1 to 100 ug) in 100 ul final volume. Samples were incubated at 35° for 60 minutes. Subsequently, samples were processed by adding carrier bovine serum albumin, and the ^{3}H-labelled polynucleotide product was precipitated onto sodium pyrophosphate treated glass fiber filters in 10% cold trichloroacetic acid. These filters were counted by liquid scintillation.

RESULTS

Growth Curves. We initially characterized the mutant lines with respect to their growth in the presence of a variety of different drugs which affect DNA synthesis. Most of the PFA-resistant lines (3-1,3-2, etc.) exhibited moderate resistance to acyclovir, and slight resistance to thymidine, deoxyadenosine, and aphidicolin, suggesting that deoxyribonucleoside metabolism may play a role in resistance to PFA. In addition, the TK⁻, dCK⁻ and dATP ribonucleotide reductase feedback-resistant lines all showed moderate resistance to PFA (data not shown), again suggesting an association of deoxyribonucleotide levels with resistance to phosphonoformic acid. The variability in the patterns of resistance suggested that there might be more than one mechanism for achieving partial resistance to phosphonoformic acid.

Nucleotide Pools. Measurement of the deoxyribonucleoside triphosphate pools in exponentially growing PFA-resistant mutant lines revealed some differences from wild type levels. dGTP pools in most mutants were within normal limits, whereas the dATP and dTTP pools were often slightly elevated. The dCTP level was dramatically elevated in one mutant line, PFA-3-9. Ribonucleotide pools were not significantly different in the mutants, as compared to wild type S49 cells.

DNA Polymerase α Assay. The elevated pools of deoxycytidine triphosphate found in the mutant line PFA 3-9 could be the result of an alteration in the Km of DNA polymerase α for dCTP. Thus, we partially purified DNA polymerase according the method of Baril et al (8) and assayed the partially purified enzyme for activity in the absence of dCTP, and in the presence of small increments of dCTP. Calculation of a true Km is complicated by the presence of about 20% to 30% of maximal activity in the absence of any added dCTP. However, in four separate assays the calculated Km was within two-fold of the calculated Km for wild type cells. In addition, a phosphonoformic acid inhibition curve of DNA polymerase activity done under V_{max} conditions was virtually identical for the wild type and the mutant 3-9 line cells. The K_is for PFA of DNA polymerase from wild type and from PFA 3-9 cells were 7.0 and 7.5 uM, respectively.

Ribonucleotide Reductase Assay. An elevated dCTP pool could be associated with a mutation in the enzyme ribonucleotide reductase. As shown in Figure 1, the mutant PFA 3-9 exhibits about a 3-fold greater CDP reductase activity in permeabilized cells than wild type. Both wild type and PFA 3-9 ribonucleotide reductase activities are sensitive to PFA when PFA is added to the reaction mixture (Fig. 1). Inhibition of CDP reductase activity in the presence of PFA, as a percent of control, was similar for wild type and PFA 3-9 mutant cells; thus, the K_i for PFA in both cell types is approximately 3.3 mM. Moreover, PFA inhibited ribonucleotide reductase activity in partially purified preparations from wild type S49 and PFA 3-9 mutant cells, except that the K_i was 1.4 mM (data not shown).

Deoxyribonucleotide Pools During PFA Treatment. PFA (4 mM) diminished the dCTP pools in wild type S49 cells and PFA 3-9 cells (Table 1). Deoxycytidine (10 uM) repleted the dCTP pools in both wild type, but this did not reverse the toxicity of PFA. The mutant is inhibited almost to the same extent as wild type cells (18.9% vs. 11.4% survival at 72 hr in 4 mM PFA). Thus, in spite of the increased ribonucleotide reductase activity in PFA 3-9 cells associated with a large dCTP pool, the mutant PFA 3-9 cells are only about 2-fold resistant to the growth inhibitory effects of PFA.

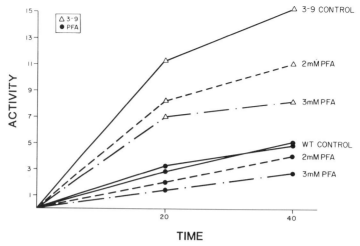

CDP reductase assay in wild type and mutant PFA 3-9 permeabilized cells. X axis: assay time; Y axis: ^{14}C-deoxycytidine cpm per 10^5 cells. (●——●), WT control; (●— —●), WT plus 2 mM PFA; (●—·—·—●) WT plus 3 mM PFA in assay. (△——△), PFA 3-9 mutant control; (△ — — △), mutant plus 2 mm PFA; (△ —·—·—· △), mutant plus 3 mm PFA in the assay.

Table 1. PFA Inhibition

	Growth at 72 h	dCTP	dTTP	dATP	dGTP
WT	100%	30.8±1.6	46.0±5.4	31.9±6.2	14.0±1.1
4mM PFA (4h)	11.4%	11.4±4.6	27.5±1.8	25.3±3.3	7.3±1.0
4mM PFA (4h) 10 uM CdR	12.8%	28.3±3.2	44.5±2.8	28.4±2.2	17.4±2.1
4mM PFA (8h)	--	4.4	15.6	20.3	7.9
PFA 3-9	100%	187.8	59.9	37.5	20.1
4mM PFA (4h)	18.9%	104.7	45.3	30.8	10.9
4mM PFA (8h)	--	70.0	35.1	23.6	8.5

Growth inhibition (in percent viable cells after 72h) and deoxyribonucleoside triphosphate pool size in wild type (WT) and mutant PFA 3-9 cell lines in the presence and absence of phosphonoformic acid, with or without deoxycytidine (CdR) at concentrations indicated (after 4h or 8h). These results were obtained with a Waters HPLC system and an Altex Ultrasil-SAX column.

DISCUSSION

Several conclusions can be drawn from these studies. First, it is clear that phosphonoformic acid inhibits mammalian DNA polymerase α. In fact, partially purified DNA polymerase α appears to be more sensitive to inhibition (K_i <10 uM) than cell growth inhibition studies would suggest $EC_{50} \sim 1.3$ mM). This difference could result from poor transport of PFA into cells.

Second, phosphonoformic acid appears to inhibit ribonucleotide reductase activity (CDP reduction) in permeabilized cells, and in cell extracts at concentrations similar to those required for cell growth inhibition.

Third, it appears from the preliminary characterization of the PFA-resistant clones that a spectrum of phenotypes results in a resistance to PFA. This result suggests that PFA may inhibit cell function at a variety of sites.

The direct inhibition of ribonucleotide reductase activity by PFA in permeabilized cells, and the corresponding decrease in the dCTP pool when PFA is added to exponentially growing cells suggest that PFA inhibits CDP reductase activity. However, the failure to reverse PFA toxicity with deoxycytidine in wild type cells, even though this treatment replenishes the dCTP pool, suggests that the inhibition of ribonucleotide reductase, as determined by dCTP pool depletion, is not the only mechanism of PFA toxicity. This conclusion is supported by observations in the mutant PFA 3-9 cell line. These mutant cells have significantly higher CDP reductase activity and an increase in their dCTP pool, but the cells were only slightly more resistant to PFA toxicity.

REFERENCES

1. P. Nylen, Chem. Ber. 57B:1023 (1924).
2. E. Helgstrand et al. Science 201:819 (1978).
3. D. Derse, K. F. Bastow, and Y. Cheng, J. Biol. Chem. 257:10251 (1982).
4. S. Shin, J. Donovan, and M. Nonoyama, Virology 124:196 (1983).
5. C. Garrett and D. Santi, Anal. Biochem. 99:268 (1979).
6. D. A. Albert and L. J. Gudas, J. Biol. Chem. 260:679 (1985).
7. W. H. Lewis, B. A. Kuzik, and J. A. Wright, J. Cell Physiol. 94:287 (1978).
8. P. Lamothe, B. Baril, A. Chu,. L. Lee, and B. Baril, Proc. Natl. Acad. Sci. USA 78:4723 (1981).

MECHANISMS OF DEOXYGUANOSINE TOXICITY FOR HUMAN T AND B LYMPHOCYTES

J.G.M. Scharenberg, L.J.M. Spaapen, G.T. Rijkers, S.K. Wadman+, G.E.J. Staal* and B.J.M. Zegers

Departments of Immunology and Metabolic disorders+, University Hospital for Children and Youth 'Het Wilhelmina Kinderziekenhuis', Nieuwe Gracht 137, 3512 LK Utrecht, The Netherlands; * Dept. of Medical Enzymology, University Hospital, Utrecht, The Netherlands

INTRODUCTION

The discovery of a causal association between a deficiency of one of the enzymes of the purine degradative pathway and a selective cellular immunodeficiency has initiated studies on the relation of the purine/pyrimidine metabolism and lymphocyte function.

Out of the metabolites which accumulate in purine nucleoside phosphorylase (PNP) deficiency, deoxyguanosine (dGuo) appears to be especially toxic to lymphocytes. In vivo and in vitro studies in PNP deficient patients have shown that T lymphocytes are sensitive to dGuo intoxication while B lymphocytes are relatively unaffected (1,2,3). T lymphocyte functions which are proliferation-independent (such as helper cell activity) are also not affected by dGuo (2).

The mechanism by which dGuo is thought to interfere with lymphocyte proliferation is inhibition of ribonucleotide reductase, mediated by dGTP, the end-product of dGuo phosphorylation (4). Indeed, high levels of dGTP are found in erythrocytes and lymphocytes of PNP deficient patients (5,6). We addressed the question of the differential sensitivity for dGuo of various lymphocyte functions by analyzing the effect of dGuo on the in vitro function of T cells and B cells obtained from the peripheral blood of adult donors. It is shown that in the effect of dGuo on the in vitro proliferation of T cells both the pathway which leads to phosphorylation of dGuo as well as the salvage pathway which starts with degradation of dGuo by PNP are operative and do contribute to toxicity. In B cell proliferation as well as B cell differentiation the only route by which dGuo exerts its toxic effect is through generation of GTP, a pathway which requires the action of PNP and HGPRT.

MATERIALS AND METHODS

Isolation and culture of the lymphocytes.

Mononuclear cells (MNC) from peripheral blood of healthy donors and patients with a deficiency of PNP or HGPRT were isolated and cultured according to methods described in detail elsewhere (7). To analyse the effect of dGuo on T cell proliferation, MNC were cultured with phytohaemagglutinin (PHA). Nucleosides, purine bases and the PNP-inhibitor 8-aminoguanosine (8-NH$_2$Guo) were added at the initiation of the cultures. The analysis of the effect of dGuo on B cell function was performed using purified B cells from normal donors, isolated and characterized as described (8). B cell activation was achieved in two different ways: 1. B cells were cultured with a T cell replacing factor produced by T cells cultured with PWM for 24 hours (8). 2. B cells were cultured with formalinized Staphylococcus aureus Cowan I (STA) during 18 hours, and the preactivated B cells were cultured for another 120 hours with the supernatant of Con A-activated T cells (Con A sup; 9) In the first system most probably only B cell differentiation takes place (8). The second approach gives rise to B cell proliferation as measured by ^{14}C-thymidine incorporation and B cell differentiation, measured by enumerating immunoglobulin-containing plasmacells respectively.

Table 1

Inhibition of in vitro T cell and B cell proliferation by guanosine and deoxyguanosine

	Proliferation (% of control ^{14}C-TdR incorporation)			
	T cells [a]		B cells [b]	
μM	dGuo	Guo	dGuo	Guo
—	100	100	100	100
25	103		107	87
50	76	73	52	56
100	46	53	27	41
150	35	43	16	35
200	21	30	13	30
ID$_{50}$	80 ± 8 μM	104 μM	68 ± 5 μM	63 ± 13 μM

(a) MNC cultured with PHA in RPMI-1640 supplemented with 10% heat-inactivated (2 hrs, 64°C) fetal calf serum in the presence of various concentrations deoxyguanosine (dGuo) of guanosine (Guo).
(b) B cells were precultured with STA and subsequently with 10% v/v Con A sup in various concentrations of dGuo or Guo. Proliferation was measured as ^{14}C-TdR incorporation, control cultures generated 11,900 ± 1,400 cpm (PHA) and 2,341 ± 1,335 cpm (STA/Con A sup). Data are arithmetic means of 2-6 independent experiments.

Table 2

Inhibition of B cell differentiation by deoxyguanosine

culture	μM dGuo causing 50% inhibition of plasmacell generation
MNC + PWM	87 ± 8
B + STA/Con A sup	54 ± 4
B + TRF	62 ± 5

Cultures without dGuo generated 24.3 x 10^3 (PWM), 24 ± 6 x 10^3 (STA/Con A sup) and 77 ± 13 x 10^3 (TRF) plasmacells/well. For further explanation see Materials and Methods.

In the second system nucleosides, purine bases and $8-NH_2Guo$ were only present in the B cell culture with Con A sup.

Measuring of intracellular ribonucleotides and deoxyribonucleotides.

Concentrations of intracellular ribonucleotides were measured in extracts of B cells cultured with STA and the Con A sup, and in extracts of PBL stimulated with PHA in the presence or absence of dGuo and/or other metabolites. The amounts of purine and pyrimidine bases, ribonucleosides and ribonucleotides were measured with high performance liquid chromatography (HPLC; 7) and deoxyribonucleotide pools were measured with a DNA polymerase assay (9).

RESULTS AND DISCUSSION

The in vitro proliferation of normal MNC, as induced by mitogens, is inhibited by deoxyguanosine (dGuo). Guanosine is capable of causing a similar effect, provided that sera used to supplement culture media contain no PNP activity (e.g. fetal calf serum heated for 2 hrs at 64^oC). Mitogens and antigens used in MNC cultures induce T cell proliferation. In order to investigate whether B cell proliferation is also sensitive to inhibitory effects of dGuo and Guo, we used a system in which B cells are preactivated with STA and subsequently cultured in a Con A sup (see below). B cell proliferation occurring in this system is inhibited by dGuo as well as by Guo in a dose-dependent way (Table 1). The concentrations of dGuo or Guo needed to cause a 50% inhibition (ID_{50}) of proliferation of T cells or B cells are about the same indicating that T and B cells obtained from the peripheral blood of normal donors are equally sensitive to dGuo and Guo.

Culturing MNC with the T cell dependent polyclonal B cell activator PWM leads to B cell differentiation which can be analysed by quantitating the number of plasmacells generated and the levels of immunoglobulins secreted into culture supernatants. dGuo as well as Guo interfere with B cell differentiation as measured by both parameters (Table 2 and data not shown). The interpretation of this observation is complicated by the fact that the inhibition of B cell differentiation might be the result of the action of dGuo and Guo on T cells. Furthermore, we do not know whether B cell proliferation (which is also sensitive to (d)Guo inhibition) is a prerequisite for B cell differentiation.

We therefore have preactivated purified B cell preparations with STA which renders them receptive for growth and differentiation factors. When these B cells are cultured with a Con A sup containing those factors cell proliferation and differentiation into plasmacells takes place. Alternatively, purified B cells cultured with a T cell-derived helper factor differentiate into plasmacells without any detectable proliferation. In both these systems dGuo and Guo exert their inhibitory effect (Table 2 and data not shown), indicating that the differentiation process itself is sensitive to the toxic effect of (d)Guo.

It thus appears that dGuo inhibits the proliferation of normal T cells and the proliferation and differentiation of normal B cells. Furthermore, these phenomena are also observed when using Guo instead of dGuo. This immediately addresses the issue of the mechanism(s) of the observed toxicity since Guo cannot be converted into dGTP, the recognized inhibitor of lymphocyte proliferation.

In order to obtain more insight in the mechanism(s) by which (d)Guo inhibits lymphocyte proliferation, a number of purine and pyrimidine metabolites have been tested for their ability to antagonize

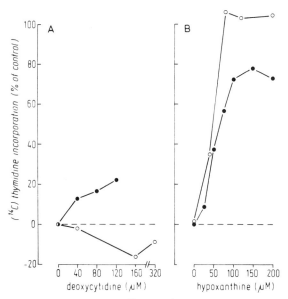

Figure 1
The effect of deoxycytidine (A) and hypoxanthine (Hx) on dGuo-mediated inhibition of T cell proliferation (•—•) and B cell proliferation (○—○). T and B cells were incubated with dGuo, 80 µM and 100 µM respectively, the ^{14}C-TdR incorporation of these cultures was considered zero percent. Control value for T cells stimulated with PHA was 11,600 ± 1,250 cpm and for B cells cultured with STA and Con A sup was 1,700 ± 160 cpm.

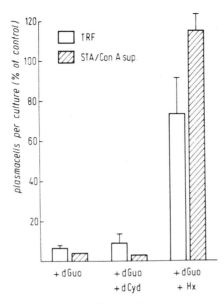

Figure 2
Antagonizing effect of hypoxanthine (Hx) on
deoxyguanosine-(dGuo-)mediated inhibition
of B cell differentiation. Purified B cells
were cultured with a T cell derived helper
factor (TRF; open bars) or with Staphylococcus aureus/Con A sup (hatched bars).
dGuo (100 μM), deoxycytidine (dCyd, 40 μM)
and Hx (100 μM) were added at initiation of
the culture. 100% plasmacells is 77 ± 13 x
10^3 plasmacells for TRF and 24 ± 6 x 10^3
plasmacells per culture for STA and Con A
sup.

the effects of (d)Guo. Deoxycytidine (dCyd), which can compete with
dGuo for phosphorylation by deoxycytidine kinase, and hypoxanthine
(Hx), which competes with guanine for conversion by HGPRT, are useful
in this respect. dCyd antagonizes the dGuo-mediated inhibition of
PHA-induced proliferation of normal lymphocytes (Fig. 1A). The
protective effect of dCyd is only partial: at the highest concentration
tested the proliferative response is restored for 20-30%. Moreover, at
high dGuo concentrations dCyd has no effect at all (data not shown).
While dCyd at least partially antagonizes T cell proliferation, it is
without any effect in dGuo-mediated inhibition of STA/Con A sup-induced
B cell proliferation (Fig. 1A). Hx, on the other hand, allows both T
and B cell proliferation at otherwise inhibitory dGuo concentrations
(Fig. 1B). The inhibition by dGuo of B cell differentiation, induced by
either STA/Con A sup or TRF, can effectively be reversed by Hx; dCyd is
without any effect (Fig. 2).

Both the observation that Guo inhibits in vitro lymphoycte
functions as well as the fact that Hx is capable to antagonize Guo- but
also dGuo-toxicity point towards a role for the purine
degradative/salvage pathway in (d)Guo toxicity. We therefore extended
our studies to patients with a deficiency of either PNP or HGPRT.

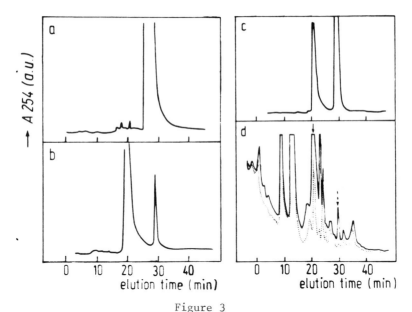

Figure 3

HPLC analysis of cellular uptake and metabolism of 8-amino-
guanosine (8-NH$_2$Guo). Purified 8-NH$_2$Guo (a); 8-NH$_2$Guo
incubated with PNP in 0.2 M KH$_2$PO$_4$ (pH 7) for 4 hours at
37°C (b); 8-NH$_2$Guo incubated in 0.5 M HCl (pH 1) for 1 hour
at 80°C (c); extracts of mononuclear cells incubated for
36 hours with PHA without (broken line) or with 500 μM
8-NH$_2$Guo (solid line; d). Products are separated on two
serially connected reverse phase columns (LiChromsorb,
10RP18; Chrompack). Absorbance at 254 nm is expressed in
arbitrary units (a.u.).

Lymphocytes of these patients differ from normal cells in two respects.
First, PNP$^-$ and HGPRT$^-$ cells are less sensitive to dGuo and
(apparently) insensitive to Guo intoxication. Second, dCyd but not Hx
antagonizes dGuo toxicity (7).

The elucidation of the mechanisms of lymphocytotoxicity in ADA
deficiency has been greatly facilitated by the availability of an
effective inhibitor of this enzyme namely deoxycoformycin. Recently a
competitive inhibitor of PNP has been described, 8-NH$_2$guanosine (10).
We have synthesized this compound and used it to study the role of PNP
in the mechanism of lymphocytotoxicity of dGuo. HPLC analysis of
8-NH$_2$Guo shows that it elutes with a retention time (RT) of 28 minutes
(Fig. 3a). The product of acid hydrolysis of 8-NH$_2$Guo i.e. 8-NH$_2$guanine
(Fig. 3c) and the product obtained after incubation of 8-NH$_2$Guo with
PNP (Fig. 3b) both elute at an RT of 22 minutes. Chromatograms of
extracts of lymphocytes cultured for 48 hours with PHA in the presence
of 500 μM 8-NH$_2$Guo show a major additional peak at 22 minutes and a
minor one at 28 minutes as compared to control cultures without
8-NH$_2$Guo (Fig. 3d). It thus appears that, at the concentrations used,
8-NH$_2$Guo is able to pass the cell membrane and to inhibit intracellular
PNP. Moreover, 8-NH$_2$guanine is also an effective inhibitor of PNP (10).
Fig. 4 shows that at high concentrations 8-NH$_2$Guo the dGuo-mediated
inhibition of proliferation of T cells and B cells is prevented. The
inhibition of B cell differentiation by dGuo or Guo can also be
antagonized by 8-NH$_2$Guo (8,9).

All data presented thusfar suggest a central role for PNP and the subsequent salvage pathway in (d)Guo toxicity for both proliferation of T cells and differentiation of B cells. In order to directly test this hypothesis we analyzed intracellular purine and pyrimidine ribonucleotide pools in T and B lymphocytes under conditions of (d)Guo inhibition and the effects of dCyd and Hx. Significant effects were observed in (d)GXP levels; other metabolites, notably ATP and ADP, remained constant. In both T and B cells addition of dGuo results in an increase of dGTP and GTP (Table 3). The addition of dCyd next to dGuo abrogates the increase of dGTP while GTP remains elevated. Hx (and 8-NH$_2$Guo) prevents dGuo-induced GTP accumulation without significantly affecting the (increased) dGTP levels.

We conclude that the pathway leading to generation of dGTP contributes only marginally to the observed inhibitory effect of dGuo on the proliferation of normal T lymphocytes. A more important route in T cells and the only one in B cells for dGuo in order to interfere with lymphocyte function is through generation of GTP, a pathway which requires the action of PNP and HGPRT. This is one of the explanations why B cells in PNP deficiency escape (d)Guo intoxication: because PNP is lacking, (d)Guo cannot be converted into GTP. The addition of (d)Guo to in vitro lymphocyte cultures not only causes a rise in intracellular GTP, but GDP and GMP also increase (data not shown). The mechanism by which one or more of these guanine ribonucleotides interfere with lymphocyte proliferation and differentiation is currently unknown (9,11).

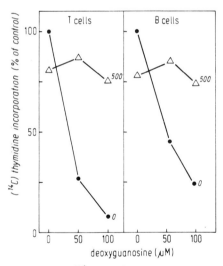

Figure 4

8-aminoguanosine (8-NH$_2$Guo) prevents deoxyguanosine-(dGuo-)mediated inhibition of lymphocyte proliferation. Mononuclear cells were cultured with 53 µg/ml pokeweed mitogen (PWM) for 6 days. Purified B cells were cultured with STA/Con A sup (see legend Table 1). dGuo without (●) or with (Δ) 500 µM 8-NH$_2$Guo was added at initiation of the cultures. 100% ^{14}C-TdR incorporation is 4,700 cpm for PWM; 2,341 ± 1,335 cpm for STA/Con A sup.

Table 3

Intracellular pools of GTP and dGTP in T and B cells: effect of deoxyguanosine, 8-NH$_2$guanosine, hypoxanthine and deoxycytidine

additions	T cells [a]		B cells [b]	
	GTP [c]	dGTP [c]	GTP [c]	dGTP [d]
none	365	< 2	500 ± 45	< 2
+ dGuo	720	68	960 ± 240	44 ± 17
+ dGuo + 8-NH$_2$Guo	n.d.	n.d.	490 ± 90	67 ± 19
+ dGuo + Hx	e	27	610 ± 115	26 ± 6
+ dGuo + dCyd	780	< 2	1020 ± 170	6 ± 1

MNC were stimulated with PHA for 24 hrs in culture medium supplemented with 10% FCS (a). B cells were precultured with STA and subsequently with Con A sup for 24 hrs (b). Intracellular pools of GTP and dGTP as determined by eiter HPLC (c) or DNA polymerase (d) are expressed in pmol/10^6 cells. Initial concentrations dGuo: 100-200 μM; 8-NH$_2$Guo: 500 μM; Hx: 100 μM; dCyd: 40 μM. n.d. = not done. e) In a parallel experiment Hx was found to prevent the dGuo-mediated increase in GTP.

We have indications that the early events occurring during lymphocyte activation are not affected by dGuo (9, J.G.M. Scharenberg et al., elsewhere in this volume), which contrasts the effects of adenosine and deoxyadenosine in T cell activation (12,13). Studies are in progress in our laboratory to unravel the guanine ribonucleotide-mediated toxicity for lymphocytes. In this respect human peripheral blood B lymphocytes are extremely useful because in these cells dGuo toxicity is mediated solely by GXP whereas T cells are sensitive for both GXP and dGTP.

ACKNOWLEDGEMENTS

We are indebted to Drs J.W. Stoop, M. Duran and G. Rijksen for stimulating discussions during design of the experiments and preparation of the manuscript. The expert technical assistance of Mrs Elly Toebes and secretarial assistance of Mrs Helene van Moorsel is gratefully acknowledged. Part of this study was supported by the Foundation for Medical Research (FUNGO), which is subsidized by the Netherlands Organization for the Advancement of Pure Research (ZWO).

REFERENCES

1. E.R. Giblett, Adenosine deaminase and purine nucleoside phosphorylase deficiency: how they were discovered and what they mean, in: Enzyme defects and immune dysfunction, Excerpta Medica, Amsterdam (1979).
2. J.W. Stoop, B.J.M. Zegers, G.F.M. Hendrickx, L.H. Siegenbeek van Heukelom, G.E.J. Staal, P.K. de Bree, S.K. Wadman, and R.E.

Ballieux, Purine nucleoside phosphorylase deficiency associated with selective cellular immunodeficiency, N.Engl.J.Med. 296:651 (1977).

3. D.W. Martin jr., and E.W. Gelfand, Biochemistry of diseases of immunodevelopment, Ann.Rev.Biochem. 50:845 (1981).

4. B. Ullman, L.J. Gudas, S.M. Clift, and D.W. Martin jr., Isolation and characterization of purine nucleoside phosphorylase deficient T-lymphoma cells and secondary mutants with altered ribonucleotidereductase: genetic model for immunodeficiency disease, Proc.Nat.Acad.Sci.USA 76:1074 (1979).

5. A. Cohen, L.J. Gudas, A.J. Ammann, G.E.J. Staal, and D.W. Martin jr., Deoxyguanosine triphosphate as a possible toxic metabolite in the immunodeficiency associated with purine nucleoside phosphorylase deficiency, J.Clin.Invest. 61:1405 (1978).

6. L.J.M. Spaapen, G.T. Rijkers, G.E.J. Staal, G. Rijksen, S.K. Wadman, J.W. Stoop, and B.J.M. Zegers, The effect of deoxyguanosine on human lymphocyte function. I. Analysis of the interference with lymphocyte proliferation in vitro, J.Immunol. 132:2311 (1984).

7. L.J.M. Spaapen, G.T. Rijkers, G.E.J. Staal, G. Rijksen, M. Duran, J.W. Stoop, and B.J.M. Zegers, The effect of deoxyguanosine on human lymphocyte function. II. Analysis of the interference with B lymphocyte function, J.Immunol. 132:2318 (1984).

8. J.G.M. Scharenberg, L.J.M. Spaapen, G.T. Rijkers, M. Duran, G.E.J. Staal, and B.J.M. Zegers, Functional and mechanistic studies on the toxicity of deoxyguanosine for the in vitro proliferation and differentiation of human peripheral blood B lymphocytes (submitted for publication).

9. J.D. Stoeckler, C. Cambor, V. Kuhns, S.-H. Chu, and R.E. Parks, Inhibitors of purine nucleoside phosphorylase, Biochem.Pharmacol. 31:163 (1982).

10. Y. Sidi, and B.S. Mitchell, 2'Deoxyguanosine toxicity for B and mature T lymphoid cell lines is mediated by guanine ribonucleotide accumulation, J.Clin.Invest. 74:1640 (1984).

11. D. Redelman, H.G. Bluestein, A.H. Cohen, J.M. Depper, and S. Wormsley, Deoxyadenosine inhibition of newly activated lymphocytes: blockade at the G0-G1 interface, J.Immunol. 32:2030 (1984).

12. S. Seto, C.J. Carrero, M. Kubota, D.B. Wasson, and D.A. Carson, Mechanism of deoxyadenosine and 2-chlorodeoxyadenosine toxicity to nondividing human lymphocytes, J.Clin.Invest. 75:377 (1985).

.

ROLE OF INTRACELLULAR DEOXYNUCLEOSIDE TRIPHOSPHATE LEVELS

IN DNA REPAIR IN HUMAN LYMPHOCYTES

Amos Cohen
Division of Immunology/Rheumatology
Research Institute
The Hospital for Sick Children
555 University Avenue
Toronto, Ontario, Canada M5G 1X8

INTRODUCTION

Resting peripheral blood lymphocytes (PBL) contain single strand breaks[1,2] and are very sensitive to DNA damaging agents[3]. Resting PBL contain very low levels of intracellular deoxynucleoside triphosphates which can be disrupted by the addition of extracellular deoxyadenosine in the presence of the adenosine deaminase inhibitor deoxycoformycin[4,5] thus creating imbalance between deoxyATP levels and other deoxynucleoside triphosphates. In the present work I have investigated the role of this deoxynucleoside triphosphate imbalance in the repair of DNA damage induced by deoxyadenosine[6,7] and other DNA damaging agents.

METHODS

Nucleoid sedimentation - Measurements of single strand breaks by nucleoid migration on sucrose gradient was performed according to Johnstone and Williams[1]. Cells were lysed inside a layer of 400 ul of 0.5% Triton x100, 10 mM Tris, 10 mM EDTA, 2 M NaCl pH 8.0 on top of 11.5 ml gradient of 5-20% (w/v) sucrose in 10 mM Tris, 10 mM EDTA, 2 M NaCl pH 8.0 in a Beckman SW41 centrifuge tube. After incubation for 15 min at 20°C the gradients were centrifuged at 25,000 rpm for 40 min at 20°C and eluted from the tube through the flow cell of a Gilford recording spectrophotometer set at 260 nm.

Fluorimetric analysis of DNA unwinding rate - single strand DNA breaks were also determined by ethidium bromide binding analysis as described by Jevcak and Birnboim[8]. DNA was treated in alkali (pH 12.8) for 1 hr at 15°C causing unwinding. The rate of unwinding is increased by the presence of DNA strand breaks. The pH of the cell lysate was adjusted to pH 11.0 and the fluorescent dye ethidium bromide, which binds selectively to double-stranded DNA, was added. The contribution to the fluorescence of material other than double-stranded DNA is estimated from a sample of cell extract sonicated to cause complete DNA unwinding (blank). Another aliquot of the same cells before exposure to alkali provides a measurement of the amount of double-stranded DNA in the extract (total). The percent double-stranded DNA remaining after1 hr unwinding at 15°C and pH 12.8 was calculated as follows: %D=[(sample-blank)/(total-blank)]x100.

Deoxynucleoside triphosphates detemination-dNTP levels were measured using the DNA polymerase technique described by Hunting and Henderson[9].

Incorporation of [H3]amino acids into proteins. The mitogenic response of lymphocytes was determined using [H3]amino acids incorporation into proteins since additions of extracellular deoxynucleosides did not allow direct measurement of DNA synthesis [H3] thymidine incorporation.

RESULTS AND DISCUSSION

The effect of deoxycoformycin and deoxyadenosine on single strand DNA breaks in resting PBL was studied using the method of nucleoid migration on sucrose density gradients[1]. Single strand breaks in the DNA decrease the constraint on the superoiled structure allowing it to expand and cause the nucleoids to migrate more slowly. Nucleoids from lymphocytes treated with the combination of deoxycoformycin and deoxyadenosine sedimented slower through the ensity gradients than those derived from either control lymphocytes or cells treated with deoxycoformycin alone (Table 1). Addition of any of the other deoxynucleosides, thymdine, deoxycytidine or deoxyguanosine, but not nucleosides such as guanosine, prevented the DNA damage caused by deoxyadenosine. Addition of thymidine, deoxycytidine or deoxyguanosine alone did not result in any change in the number of singe strand breaks.

Table 1:   Induction of DNA Single Strand Breaks by
            Deoxyadenosine Measured by Nucleoid Sedimentation

Deoxyadenosine (10 uM) + Deoxycoformycin (1 uM)	Deoxynucleosides (50 uM)	Migration distance (cm)
-	-	10.5
+	-	2.1
+	thymidine	7.8
+	deoxycytidine	8.0
+	deoxyguanosine	9.2
+	guanosine	2.4

Previous studies have shown that resting lymphocytes contain extremely low levels of dNTP,[4,5] and that addition of deoxyadenosine and deoxycoformycin results in accumulation of large amounts of dATP causing imbalance in intracellular dNTP pools. Addition of deoxycytidine prevents dATP accumulation[4,5]. Addition of either deoxyguanosine or thymidine does not interfere with dATP accumulation but replenish dGTP and dTTP pools respectively (data not shown) thus preventing the dNTP imbalance caused by the addition of deoxyadenosine.

Thymocytes contain much higher concentrations of intracellular dNTP as compared to peripheral blood T lymphocytes[4,5] and thus may be less sensitive to intracellular dNTP imbalance. In order to determine whether thymocytes will be also less sensitive to the induction of DNA breaks by deoxyadenosine we have measured the effect of deoxycoformycin and deoxyadenosine on single strand DNA breaks in thymocytes using the alkaline unwinding technique. The results summarized in Table 2 show indeed that deoxyadenosine in combination with deoxycoformycin did not induce DNA breaks in thymocytes.

Table 2:   Effect of Deoxyadenosine and Deoxycoformycin on
            Single Strand DNA Breaks in Thymocytes.

Additions	Percent Single stranded DNA
None	32.6
Deoxycoformycin (1 uM)	33.0
Deoxycoformycin (1 uM) + deoxyadenosine (10 uM)	35.2
deoxycoformycin (1 uM) + deoxyadenosine (30 uM)	34.1

The effect of deoxyadenosine on the repair of single-strand breaks induced by the alkylating agent N-methyl-N'-nitro-nitrosoguanidine (MNNG) is summarized in Table 3. Deoxyadenosine has no effect on the accumulation of single strand breaks during the first 1 hr in the presence or absence of MNNG. However, deoxyadenosine inhibits the repair of MNNG-induced single strand breaks after MNNG removal suggesting that deoxyadenosine-induced DNA damage is the result of inhibition of DNA repair rather than direct induction of DNA breaks.

Table 3:   Effect of Deoxyadenosine on the Repair of MNNG
            Induced Single Strand Breaks

Additions	Percent Single Stranded DNA	
	1 hr.	20 hrs.
None	19.7	22.8
Deoxyadenosine (10 uM) + deoxycoformycin (1 uMO	20.5	43.0
MNNG (5 ug/ml, 1 hr)	52.7	25.1
MNNG (5 ug/ml, 1 hr) + deoxyadenosine (10 uM) + deoxycoformycin (1 uM)	55.3	88.7

Deoxyadenosine inhibits the mitogenic response of peripheral blood lymhocytes when present during the first 24 hrs of culture[10]. On the other hand repair of spontaneous single strand breaks has been associated with early steps of mitogen stimulation of peripheral blood lymphocytes[1,2]. In order to determine whether the inhibition of mitogenic response by deoxyadenosine is related to inhibition of DNA

repair, lymphocytes were treated with the DNA alkylating agent
N-methyl-N'-nitro-N-nitrosoguanidine (MNNG) for two hours, causing
extensive DNA breaks, and then their mitogenic response was determined
in the presence or absence of deoxyadenosine added for the first 24
hrs. The mitogenic response was evaluated after 72 hrs by the ability
of lymphocytes to incorporate [H^3]amino acids into proteins (Table
4). Low concentrations of MNNG (0.1-2 ug/ml) alone did not inhibit the
incorporation of amino acids into protein in lymphocytes stimulated
with mitogen. The combination of deoxyadenosine (5 uM) with
deoxycoformycin (1 uM) only partially (40%) inhibited the mitogenic
response of lymphocytes whereas in the presence of MNNG deoxyadenosine
and deoxycoformycin caused complete inhibition of the mitogenic
response (> 99% inhibition at 2 ug/ml MNNG). Addition of the three
other deoxynucleosides completely prevented the effects of MNNG and
deoxyadenosine (data not shown).

Table 4: Effect of Deoxyadenosine on the Mitogenic Response of
MNNG Treated Lymphocytes

Deoxynucleosides added	[H^3]amino-acid incorporation (cpm x 10-3) MNNG added (ug/ml/1 hr)			
	0	0.1	0.5	2.0
None	33.5	28.7	30.2	32.7
Deoxyadenosine (5 uM) + deoxycoformycin (1 uM)	19.2	6.1	1.7	0.2

Imbalance in dNTP levels have been shown to be mutagenic to
mammalian cells in culture[11] and to increase the mutagenicity and
cytotoxicity of DNA-alkylating agents[12,13] but the mechanism of
action of these deoxynucleotide imbalances is unclear. The results
presented here suggest that nucleotide imbalance may act by interfering
with the repair of DNA single strand breaks.

ACKNOWLEDGEMENTS

This research was supported by the Medical Research Council of
Canada and the National Cancer Institute of Canada. Amos Cohen is a
Scholar of the National Cancer Institute of Canada. I am grateful to
Ellen Thompson and Chungyee Leung for expert technical assistance.

REFERENCES

1. A. P. Johnston and G. T. Williams, Role of DNA breaks and
   ADP-ribosyl transferase activity in eukaryotic differentiation
   demonstrated in human lymphocytes, Nature 300:368, (1982).
2. W. L. Greer and J. G. Kaplan, Regulation of repair of naturally
   occuring DNA strand breaks in lymphocytes, Biochem. Biophys. Res.
   Comm. 122:366, 1984.
3. D. K. Kwan and A. Norman, Radiosensitivity of human lymphocytes and
   thymocytes, Radiat. Res. 69:143, 1977.
4. A. Cohen, J. Barankiewicz, H. M. Lederman and E. W. Gelfand, Purine
   and pyrimidine metabolism in human T lymphocytes. Regulation of
   deoxyribonucleotide metabolism, J. Biol. Chem. 258:12334, 1983.
5. R. F. Kefford and R. M. Fox, Purine deoxynucleoside toxicity in
   nondividing human lymphoid cells, Cancer Res. 42:324, 1982.

6. L. Brox, A. Ng, E. Pollock and A. Belch, DNA strand breaks induced in human T-lymphocytes by the combination of deoxyadenosine and deoxycoformycin, Cancer Res. 44:934, 1984.

7. S. Seto, C. J. Carrera, M. Kubota, B. D. Wasson and D. A. Carson, Mechanism of deoxyadenosine and 2-chlorodeoxyadenosine toxicity to nondividing human lymphocytes, J. Clin. Invest. 75:377, 1985.

8. J. Jevcak and H. C. Birnboim, Fluorometric method for rapid detection of DNA strand breaks in human white blood cells produced by low doses of radiation, Cancer Res. 41:1889, 1981.

9. D. Hunting and J. F. Henderson, Methods for the determination of deoxyribonucleoside triphosphate concentrations, Meth. Cancer Res. 20:245, 1982.

10. J. Uberti, J. J. Lightbody and R. M. Johnson, The effect of nucleosides and deoxycoformycin on adenosine and deoxyadenosine inhibition of human lymphocyte activation, J. Immunol. 123:189, 1979.

11. M. Meuth, The genetic consequences of nucleotide precursor pool imbalance in mammalian cells, Mutation Res. 126:107, 1984.

12. A. R. Peterson, J. R. Landoph, H. Peterson and C. Heidelberger, Mutagenesis of Chinese hamster cells facilitated by thymidine and deoxycytidine, Nature 276:508, 1978.

13. M. Meuth, Role of deoxynucleoside triphosphate pools in the cytotoxic and mutagenic effects of DNA alkylating agents, Somat. Cell Genet. 7:89, 1981.

GENETIC ANALYSIS OF DEOXYADENOSINE TOXICITY
IN DIVIDING HUMAN LYMPHOBLASTS

Dennis A. Carson, Carlos J. Carrera, Masaru Kubota,
D. Bruce Wasson, and Taizo Iizasa

Dept. of Basic and Clinical Research
Scripps Clinic and Research Foundation
La Jolla, CA 92037

An inherited deficiency of adenosine deaminase (ADA) impairs specifically the development of the human lymphoid system (reviewed in 1).
In ADA deficient children, plasma deoxyadenosine (dAdo) concentrations
reach 1-2 μM. Micromolar concentrations of dAdo are toxic toward immature human T-lymphocytes, and toward T-lymphoblastoid cell lines grown in
medium supplemented with deoxycoformycin, a tight binding ADA inhibitor.[1]
Compared to other cell types, human T-lymphoblasts preferentially phosphorylate dAdo and accumulate dATP.

The metabolism of dAdo in dividing human lymphoblasts and the
mechanisms for dAdo toxicity have aroused considerable controversy. To
investigate this problem, we have used techniques of biochemical genetics. Three different stable mutant human lymphoblastoid cell lines have
been selected that resist the anti-proliferative effects of dAdo. The
dAdo resistant mutants differ from wild type cells in one of three ways:
(1) a decrease in deoxycytidine (dCyd) kinase activity;[2] (2) an increase
in cytoplasmic nucleotidase activity;[3] and (3) an increase in the activity of ribonucleotide reductase.[4] All three genetic changes cause a
secondary rise in de novo dCyd formation and excretion, and a reciprocal
inability to phosphorylate dAdo and to form dATP.

Deoxycytidine Kinase Deficient, Deoxyadenosine Resistant T Cells

In collaboration with Dr. Michael Hershfield and Dr. Buddy Ullman,
we selected and characterized a dCyd kinase deficient mutant of the human
CEM T-lymphoblastoid cell line.[2] The mutant lymphocytes, designated CEM-
AraC-8D, were selected following mutagenesis and growth in increasing
concentrations of cytosine arabinoside. Compared to wild type CEM cells,

the CEM-AraC-8D mutants had undetectable dCyd kinase activity. At dAdo concentrations of 10 μM or less, the dCyd kinase deficient mutants in-efficiently formed dATP. Compared to wild type CEM lymphoblasts, the dCyd kinase deficient variant was three-fold resistant to dAdo toxicity, and excreted three times more dCyd into the culture medium (Table 1). These studies established (1) that dCyd kinase primarily phosphorylates dAdo at the plasma concentrations found in ADA deficient children, and (2) that phosphorylation is required for the toxic effects of dAdo toward T-lymphocytes.

Deoxyadenosine Resistant T Cells with Increased Cytoplasmic Nucleotidase

The dCyd kinase deficient CEM T-lymphoblasts were selected after incubation in cytosine arabinoside. It was important to define the phenotypes that arose after selection of T cells directly in medium supplemented with cytotoxic concentrations of dAdo. To this end, muta-genized CEM cells were grown in medium supplemented with gradually in-creasing concentrations of dAdo. Cells resistant to 10 μM dAdo were selected and cloned. One dAdo resistant clone was studied in detail.[3] Compared to wild type CEM cells, the mutant lymphoblasts were ten-fold resistant to dAdo toxicity, and formed dATP inefficiently (Table 1). The mutants were also cross-resistant to cytosine arabinoside and to deoxy-guanosine. However, the dCyd kinase activities in extracts of mutant and wild type T-lymphoblasts were equivalent.

Further analysis showed that wild type CEM T-lymphoblasts produced dCyd de novo at the minimal velocity of 1 pmols/hr/10^6 cells. In con-trast, the dAdo resistant variant released dCyd at the markedly increased rate of 17 pmols/hr/10^6 cells. Indeed, the phenotype of the dAdo resis-tant CEM T-lymphoblasts was very similar to that of human B-lymphoblas-toid cell lines. The B cells are naturally resistant to dAdo and deoxyguanosine, form minimal dATP and dGTP, and excrete substantial quantitites of dCyd into the medium.

Human lymphoblasts contain at least three distinct nucleotidase activities.[1] Two of the three enzymes are located in the cytosol. One preferentially dephosphorylates deoxynucleotides, and is unaffected by ATP. The other cytosolic enzyme reacts better with ribonucleotides (particularly IMP) than deoxyribonucleotides, and requires ATP for opti-mal activity. The human CEM T-lymphoblastoid cell line contains minimal deoxynucleotidase activity, but has substantial cytosolic ribonucleo-tidase. In the T cells, the cytosolic ribonucleotidase is capable of dephosphorylating deoxynucleotides, albeit at a slow rate. Compared to wild type CEM T-lymphocytes, the dAdo deficient variant had 3-4 fold elevated activity of the ATP-activated cytosolic nucleotidase. This increase in nucleotidase activity was associated with a 17-fold rise in dCyd excretion, and a reduction in intracellular dCTP pools.

Deoxyadenosine Resistant Human Lymphoma Cells with Increased Ribonucleotide Reductase

The human histiocytic lymphoma cell line DHL-9 naturally lacks ADA. Dividing DHL-9 cells excrete dAdo, dCyd, and thymidine into the extra-cellular space.[4] We selected and analyzed a dAdo resistant clone, desig-nated DHL-9R, from a mutagenized population of wild type lymphoma cells. The dAdo resistant mutants excreted deoxynucleosides into the extra-cellular space at a 6-7 fold higher rate than wild type lymphoma cells. The deoxynucleoside overproduction was accompanied by a reduced ability to form dATP from exogenous dAdo. Compared to wild type cells, the dAdo resistant mutants had a 2-fold increase in ribonucleotide reductase activity, and had elevated deoxynucleoside triphosphate pools. Again, dCyd kinase levels were equivalent in extracts prepared from the wild type and dAdo resistant sublines. Furthermore, the relative rates of dATP degradation in intact cells were equivalent in DHL-9 and its dAdo resis-tant variant. The collective data strongly suggests that the augmented deoxynucleoside excretion in DHL-9R cells was caused by an increase in the steady-state rate of deoxynucleotide synthesis. The accumulation of dCyd, in turn, interfered competitively with the phosphorylation of dAdo and deoxyguanosine, and blocked the formation of dATP and dGTP.

Conclusions

Several different metabolic changes can impair dAdo-deoxyguanosine phosphorylating activity in mammalian cells (Table 2).[2-7] In addition to mutational loss of dCyd kinase, the enzyme may become functionally defi-cient as a result of changes in the rate of dCDP synthesis,[4,5] the pace of dCMP dephosphorylation,[3] or the velocity of dCMP or dCyd deamina-tion.[6,7] Deoxyadenosine and deoxyguanosine are inefficient substrates for dCyd kinase.[1,2] It is not suprising that an increase in dCyd-dCMP cycling is sufficient to impair dATP and dGTP formation.

These results have clinical implications. Several useful anti-cancer drugs are phosphorylated by dCyd kinase. These include cytosine arabinoside, guanine arabinoside, 2-chloro-dAdo, 2-fluoro-dAdo, and dAdo itself.[1] It seems likely that a malignant cell could achieve pleomorphic resistance to several deoxynucleoside analogs, simply by increasing the rate of de novo deoxynucleoside formation, relative to the rate of DNA synthesis. Preliminary experiments from our laboratory suggest that this phenomenon may be related to deoxynucleoside resistance in vivo.

Acknowledgements

This research was supported by grants GM23200, CA35048, AM07144 and RR00833 from the National Institutes of Health.

Table 1. Comparison of Three Deoxyadenosine Resistant Mutants of Human Lymphoblastoid Cell Lines

Parental Cell Line	Mutant Phenotype	Deoxycytidine Excretion	Deoxynucleotide Pools	dATP Formation
CEM	Decreased dCyd kinase	Increased	Normal	Decreased
CEM	Increased cytoplasmic nucleotidase	Increased	Decreased	Decreased
DHL-9	Increased ribonucleotide reductase	Increased	Increased	Decreased

Table 2. Metabolic Basis for Impaired Deoxyadenosine-Deoxyguanosine phosphorylating activity

1. Mutational loss of deoxycytidine kinase

2. Increased intracellular deoxycytidine $\rightarrow$ functional deoxycytidine kinase deficiency.

    a. Increased CTP synthesis
    b. Increased ribonucleotide reduction
    c. Increased dCMP dephosphorylation
    d. reduced dCMP or dCyd deamination

REFERENCES

1. C.J. Carrera, and D.A. Carson, Enzyme deficiencies associated with immunologic disorder, In: "Molecular Basis of Blood Diseases," G. Stamatoyannopoulos, A.W. Nienhuis, P. Leder, and P.W. Majerus, eds., W.B. Saunders Co., Philadelphia, in press (1985).

2. M.S. Hershfield, J.E. Fetter, M.C. Small, A.S. Bagnara, S.R. Williams, B. Ullman, D.W. Martin, D.B. Wasson, and D.A. Carson, Effects of mutational loss of adenosine kinase and deoxycytidine kinase on deoxyATP accumulation and deoxyadenosine toxicity in cultured CEM human T-lymphoblastoid cells, J Biol Chem 257:6380 (1982).

3. C.J. Carrera, T. Iizasa, D.B. Wasson, E.H. Willis, and D.A. Carson, Elevated cytoplasmic 5'-nucleotidase in deoxyadenosine-resistant human T-lymphoblasts, manuscript in preparation.

4. M. Kubota, C.J. Carrera, D.B. Wasson, and D.A. Carson, Deoxynucleoside overproduction in deoxyadenosine-resistant, adenosine deaminase-deficient human histiocytic lymphoma cells, Biochim et Biophys Acta, 804:37 (1984).

5. B. Aronow, T. Watts, J. Lassetter, W. Washtien, and B. Ullman, Biochemical phenotype of 5-fluorouracil resistant murine T-lymphoblasts with genetically altered CTP synthetase activity. J Biol Chem 259:9035 (1984).

6. T.-S. Chan, B.D. Lakhchaura, and T.-F. Hsu, Differences in deoxycytidine metabolism in mouse and rat, Biochem J 210:367 (1983).

7. S. Eriksson, S. Skog, B. Tribukait, and K. Jaderberg, Deoxyribonucleoside triphosphate metabolism and the mammalian cell cycle, Exp Cell Res 155:129 (1984).

INHIBITION OF ^{3}H-THYMIDINE INCORPORATION BY ADENOSINE AND DEOXYADENOSINE IN HUMAN PERIPHERAL LYMPHOCYTES AND MALIGNANT LYMPHOID CELL LINES

P.M. van der Krann, P.M. van Zandvoort, R.A. De Abreu,
J.A.J.M. Bakkeren, J.P.R.M. van Laarhoven and C.H.M.M. de Bruijn

Centre of Pediatric Oncology S.E. Netherlands, Dept. of Pediatrics and Human Genetics, Univ. of Nijmegen The Netherlands

INTRODUCTION

The purine converting enzyme adenosine deaminase (ADA, E.C. 3.5.4.4.) catalyzes the conversion of adenosine and deoxyadenosine into inosine and deoxyinosine respectively. The inherited deficiency of ADA is associated with an impairment of both cellular and humoral immunity[1,2]. Although all cells of patients with ADA deficiency lack a functional ADA enzyme, the expression of the deficiency seems to be restricted mainly to the immune system. This selective effect of ADA deficiency on lymphoid cells has drawn attention to the possibility of using ADA inhibitors as potential selective chemotherapeutic agents in treatment of lymphoproliferative disorders[3].

The use of deoxycoformycin (dCF), a potent ADA inhibitor, has already been investigated in phase 1 studies[4-12]. The use of dCF resulted in induction of partial and complete remissions in patients with T-ALL and in patients with nonBnonT-ALL[4-12]. Recently, succesful chemotherapy with dCF in a patient with adult T-cell lymphoma-leukemia has been reported[13].

Inhibition of ADA by dCF is not likely to be the direct cause of the observed lymphopenia, as in cultured cell systems ADA inhibitors are usually not toxic at concentrations required for complete inhibition of ADA activity [14-19]. Deoxycoformycin produces toxicity by potentiating the toxicity of adenosine and/or deoxyadenosine, which are normally detoxified by deamination [14-21].

We have investigated the toxicity of adenosine and deoxyadenosine, in presence of dCF, to normal human mitogen stimulated peripheral blood lymphocytes (PBL) and malignant lymphoid cell lines.

MATERIAL AND METHODS

Peripheral blood lymphocytes were isolated from buffy coats by Percoll gradient centrifugation followed by counterflow elutriation, in essence according to the method of De Mulder et al.[22]. Optimal mitogen concentrations and stimulation times were determined, PHA 92 hours, 0.50 μg/ml, Con A 116 hours, 25 μg/ml, PWM 92 hours, dilution 1:2400, SpA 116 hours, 15 μg/ml, formalinised Staphylococcus aureus Cowan I (StA) 130 hours, $4-8 \times 10^5$ bacteria/ ml. Lymphocytes and cultured lymphoid cells were grown at 37°C and 2.5% CO_2 in round bottom microtiter plates. Adenosine, deoxyadenosine and dCF were added at the initiation of the experiments. Twenty hours before cell harvesting ^{3}H-thymidine was added to the cell cultures. All experiments were carried out in quadruplicate.

RESULTS

In figure 1 it is shown that adenosine inhibits the mitogenic response of PBL, in presence of 10 μM dCF. Incubation with adenosine resulted in a rapid decrease of ^{3}H-thymidine incorporation at low concentrations (<30 μM). At higher adenosine concentrations ^{3}H-thymidine incorporation decreased less rapidly. Adenosine concentrations inhibiting the mitogenic response 50 per cent (ID50) were 18 ± 10 μM for PHA, 17 ± 7 μM for ConA, 13 ± 6 μM for PWM, 11 ± 7 μM for SpA and 7 ± 1 μM for StA (n=5). Peripheral blood lymphocytes stimulated with T cell dependent (PWM, SpA) or T cell independent (StA) B cell mitogens were at least as sensitive as lymphocytes stimulated with T cell mitogens (PHA, ConA) to adenosine. Lymphocytes stimulated with StA seemed to be the most sensitive cells.

Inhibition of the mitogenic response of PBL by deoxyadenosine (figure 2) was at low nucleoside concentrations comparable with the inhibition by adenosine. At nucleoside concentrations higher than 40 μM deoxyadenosine affected ^{3}H-thymidine incorporation in PBL to a greater extent than adenosine did. The ID50 values of deoxyadenosine were PHA 23 ± 10 μM, ConA 24 ± 3 μM, PWM 11 ± 4 μM, SpA 10 ± 6 μM and StA 3 ± 1 μM (n=6). Like adenosine deoxyadenosine was as toxic to PBL stimulated with B cell mitogens as to those stimulated with T-cell mitogens. Peripheral blood lymphocytes stimulated with StA appeared to be more sensitive to deoxyadenosine than PBL stimulated with other mitogens. Furthermore, incorporation of ^{3}H-thymidine in ConA stimulated PBL was greatly enhanced at low deoxyadenosine concentrations (1-5 μM), this was not observed when adenosine was used.

Figure 3 shows the effect of adenosine on ^{3}H-thymidine incorporation in malignant lymphoid cell lines. The nonBnonT cell line Reh was as sensitive as the T-cell line Molt 4 to the inhibitory effects of adenosine.

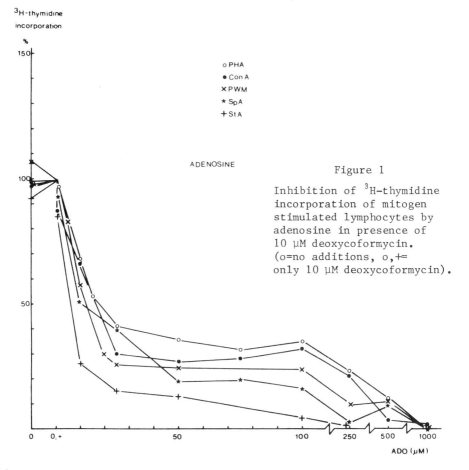

Figure 1

Inhibition of ^{3}H-thymidine incorporation of mitogen stimulated lymphocytes by adenosine in presence of 10 μM deoxycoformycin. (o=no additions, o,+= only 10 μM deoxycoformycin).

Inhibition of ^{3}H-thymidine incorporation in the malignant B cell line Raji
was less than in the Molt and Reh cell lines, especially at adenosine concentrations above 10 μM. The ID50 values calculated were Molt 4, 4.1±1.0 μM
(n=7), Reh 4.3±2.8 μM (n=3) and Raji 7.6±0.3 μM (n=7).

The inhibition of ^{3}H-thymidine incorporation in malignant lymphoid cell
lines by deoxyadenosine is shown in figure 4. The malignant T cell line Molt
4 was more sensitive to deoxyadenosine than to adenosine. The cell lines Reh
and Raji were affected less by deoxyadenosine than Molt 4. The sensitivity
of Reh to adenosine and deoxyadenosine was comparable. Incorporation of
labelled thymidine in the Raji cell line was affected less by deoxyadenosine
than by adenosine. The ID50 values were respectively Molt 4 0.7±0.2 μM (N=5),
Reh 10±2 μM (n=3), Raji 38±5 μM (n=5). Incubation of PBL or cultured lymphoid
cell lines with only 10 μM dCF had little effect on ^{3}H-thymidine incorporation.

DISCUSSION

It has been described that T lymphocytes were affected more by adenosine
and deoxyadenosine than B lymphocytes[16,23-27]. In some reports it was described that T- and B lymphocytes were affected to a comparable extent[28,29].
The greater sensitivity of T lymphocytes could not be confirmed by our experiments. Peripheral blood lymphocytes stimulated with T cell mitogens
appeared to be not more sensitive to the combinations dCF and adenosine or
deoxyadenosine than PBL stimulated with B cell mitogens. Hayward suggested
that the in vitro suppression of the mitogenic response of B cells was not due
to direct toxicity of deoxyadenosine but due to interference with T cell help
[27]. This does not seem likely, we have observed that lymphocytes stimulated
with a T cell independent B cell mitogen (StA) were even more sensitive to

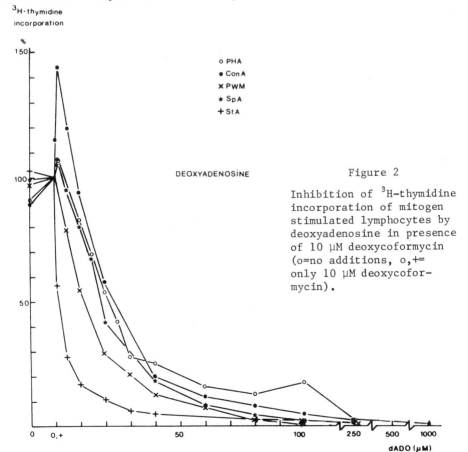

[3]H-thymidine
incorporation

%

o PHA
● Con A
× PWM
★ SpA
+ StA

DEOXYADENOSINE

dADO (μM)

Figure 2

Inhibition of ^{3}H-thymidine
incorporation of mitogen
stimulated lymphocytes by
deoxyadenosine in presence
of 10 μM deoxycoformycin
(o=no additions, o,+=
only 10 μM deoxycofor-
mycin).

deoxyadenosine than lymphocytes stimulated with other mitogens. The potent toxicity of adenosine and deoxyadenosine to stimulated B lymphocytes that we observed in our experiments seems to be in agreement with in vivo experiments in mice[30]. Ratech et al. reported that the effects of dCF administration on the lymphoid system of the mouse was rather an effect on B cells than on T cells[30].

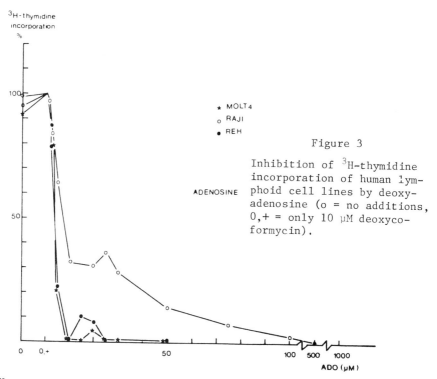

Figure 3

Inhibition of ^{3}H-thymidine incorporation of human lymphoid cell lines by deoxyadenosine (o = no additions, 0,+ = only 10 µM deoxycoformycin).

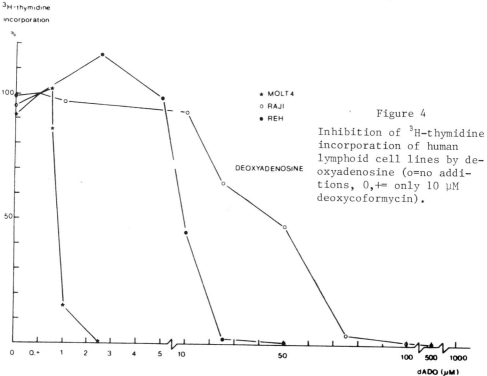

Figure 4

Inhibition of ^{3}H-thymidine incorporation of human lymphoid cell lines by deoxyadenosine (o=no additions, 0,+= only 10 µM deoxycoformycin).

The conflicting results obtained from in vitro experiments might be due to differences in experimental conditions. Our PBL isolation procedure was based on Percoll gradient centrifugation and counterflow elutriation while most of the other authors only used Ficoll gradient centrifugation. Counterflow elutriation makes it possible to obtain highly purified lymphocyte populations free of contamination with erythrocytes, granulocytes and thrombocytes which might influence the relative sensitivity of lymphocyte subpopulations to nucleosides. Also the use of different mitogen concentrations and batches might have resulted in stimulation of different lymphocyte subpopulations. Concanavalin A stimulated PBL showed an enhanced ^{3}H-thymidine incorporation at low deoxyadenosine concentrations, this in contrast to PHA stimulated PBL. It has been reported that PHA stimulates mainly T-helper cells and ConA both T-suppressor and T-helper cells[31,32]. One might speculate that the enhanced ^{3}H-thymidine incorporation is a result of a greater vulnerability to deoxyadenosine of T-suppressor cells than T-helper cells. This will result in net help activity at low deoxyadenosine concentrations. Ratech et al. found that administration of dCF to mice one day after antigen challenge resulted in enhancement of the immune response[30]. This would be due to destruction or inactivation of the activated suppressor cells[30]. Matsumoto et al. found comparable ID$_{50}$ values of deoxyadenosine whith freshly isolated leukemic cells as we found with cultured malignant cell lines[33]. The values he described were 0.8 uM for T-ALL, 8.1 uM for NULL-ALL and greater than 20 uM for B cell lymphoma. As mentioned before, our values were 0.7 uM for Molt 4 cells, 10 uM for Reh cells and 38 uM for Raji cells. Molt 4 was more affected than the other cell lines and mitogen stimulated PBL, this is confirmed by other authors[19,34]. Also in phase 1 studies T cell leukemias respond better to dCF therapy than other types of leukemia[4-12].

The nonBnonT cell line Reh was less sensitive to deoxyadenosine than Molt 4 but as sensitive to adenosine as Molt 4. Fox et al. observed in their experiments that the nonBnonT cell lines KM-3 and Reh were as sensitive to the inhibitory effects of adenosine [35] as the T cell lines CCRF-CEM, 8402, CCRF-HSB and Molt 4. The ultimate goal of anticancer chemotherapy is: to achieve killing of all malignant cells while non-malignant cells remain unharmed. In our experiments adenosine and deoxyadenosine appeared to be more toxic to leukemic nonTnonB and especially leukemic T cells than to mitogen stimulated PBL. These immature malignant cells are more vulnerable to these nucleosides than mature PBL. On this basis a certain selectivity of dCF to leukemic cells might be achieved.

REFERENCES

1. E.R. Giblett, J.E. Anderson, F. Cohen, B. Pollara and H.J. Meuwissen, Adenosine deaminase deficiency in two patients with severely impaired cellular immunity, Lancet 2, 1067 (1972).
2. H.J. Meuwissen, B. Pollara and R.J. Pickering, Combined immunodeficiency disease associated with adenosine deaminase deficiency, J. Pediatr. 86, 169 (1975).
3. J.P.R.M. van Laarhoven and C.H.M.M. de Bruijn, Purine metabolism in relation to leukemia and lymphoid cell differentiation, Leukemia Res. 7, 451 (1983).
4. H.G. Prentice, K. Ganeshaguru, K.F. Bradstock, A.H. Goldstone, J.F. Smyth, B. Wonke, G. Janossy and A.V. Hoffbrand, Remission induction with adenosine-deaminase inhibitor 2'-deoxycoformycin in thy-lymphoblastic leukemia, Lancet 1, 170 (1980).
5. N.H. Russel, H.G. Prentice, N. Lee, A. Piga, K. Ganeshaguru, J.F. Smyth and A.V. Hoffbrand, Studies on the biochemical seguelae of therapy in thy-acute lymphoblastic leukemia with the adenosine deaminase inhibitor 2'-deoxycoformycin, Br. J. Haematol. 49, 1 (1981).
6. H.G. Prentice, N.H. Russel, N. Lee, K. Ganeshaguru, H. Blacklock, A. Piga, J.F. Smyth and A.V. Hoffbrand, Therapeutic selectivity of and prediciton of response to 2'-deoxycoformycin in acute leukemia, Lancet 1, 1250 (1981).

7. C.A. Koller, B.S. Mitchell, M.R. Grever, E. Mejias, L. Malspeis and E.N. Metz, Treatment of acute lymphoblastic leukemia with 2'-deoxycoformycin: Clinical and biochemical consequences of adenosine deaminase inhibition, Cancer Treatm. Rep. 63, 11 (1979).

8. M.R. Grever, M.F.E. Siaw, W.F. Jacob, J.A. Neidhart, J.S. Miser, M.S. Coleman, J.J. Hotton and S.P. Miser, The biochemical and clinical consequences of 2'-deoxycoformycin in refractory lymphoproliferative disease, Blood 57, 406 (1981).

9. D.G. Poplack, S.E. Sallan, G. Rivera, J. Holcenberg, S.B. Murphy, J. Blatt, J.M. Lipton, P. Venner, D.L. Glaubiger, R. Ungerleider and D.G. Johns, Phase 1 study of 2'-deoxycoformycin in acute lymphoblastic leukemia, Cancer res. 41, 3343 (1981).

10. J.F. Smyth, R.M. Paine, A.L. Jackman, K.R. Harrap, M.M. Chassin, R.H. Adamson and D.G. Johns, The clinical pharmacology of the adenosine deaminase inhibitor 2'-deoxycoformycin, Cancer Chemother. Pharmacol. 5, 93 (1980).

11. P.P. Major, R.P. Agarwal and D.W. Kufe, Clinical Pharmacology of deoxycoformycin, Blood 58, 91 (1981).

12. R.F. Kefford and R.M. Fox, Deoxycoformycin induced response in chronic lymphocytes, Br. J. Haematol. 50, 627 (1982).

13. S. Daenen, R.A. Rojer, J.W. Smyth, M.R. Halic and H.O. Nieuweg, Succesful chemotherapy with deoxycoformycin in adult T-cell lymphoma-leukemia, Br. J. Haematol. 58, 723 (1984).

14. I.H. Fox and W.N. Kelley, The role of adenosine and 2'-deoxyadenosine in mammalian cells, Ann. Rev. Biochem. 47, 655 (1978).

15. J.F. Henderson, L. Brox, G. Zombor, D. Hunting and C.A. Lomax, Specificity of adenosine deaminase inhibitors, Biochem. Pharmacol. 26, 1967 (1977)

16. J.E. Seegmiller, T. Watanabe and M.H. Schreier, The effect of adenosine on lymphoid cell proliferation and antibody formation. In: "Purine and pyrimidine metabolism", CIBA foundation symposium 48, 249 (1977).

17. F.F. Snyder, J. Mendelshon and J.E. Seegmiller, Adenosine metabolism in phytohemagglutinin-stimulated human lymphocytes, J. Clin. Invest. 58, 654 (1976).

18. K.R. Harrap and R.M. Paine, Adenosine metabolism in cultured lymphoid cells, Adv. Enzyme Regul. 15, 169 (1976).

19. M.S. Herschfield, F.F. Snyder and J.E. Seegmiller, Adenine and adenosine are toxic to human lymphoblasts defective in purine salvage enzymes, Science 197, 1284 (1977).

20. D.A. Carson, J. Kaye and J.E. Seegmiller, Differential sensitivity of human leukemic T cell lines and B cell lines to growth inhibition by deoxyadenosine, J. Immunol. 121, 1726 (1978).

21. J. Uberti, J.J. Lightbody and R.M. Johnson, The effects of nucleosides and deoxycoformycin on adenosine and deoxadenosine inhibition of human lymphocyte activation, J. Immunol. 123, 189 (1979).

22. P.H.M. de Mulder, J.M.C. Wessels, D.A. Rossenbrand, J.R.J.M. Smulders, D.J.T. Wagener and C. Haanen, Monocyte purification with counterflow centrifugation monitored by continuous flow cytometry, J. Immun. Meth. 47, 31 (1981).

23. R.Hirschhorn and E. Sela, Adenosine deaminase and immunodeficiency: An in vitro model, Cell. Immunol. 32, 350 (1977).

24. J.H. Fox, E.C. Keystone, D.D. Gladman, M. Moore and D. Cane, Inhibition of mitogen mediated lymphocyte blastogenesis by adenosine, Immunol. Commun. 4, 419 (1975).

25. J. Uberti, J.J. Lightbody, J.W. Wolf, J.A. Anderson, R.H. Reid, R.M. Johnson, The effect of adenosine on mitogenesis of ADA-deficient lymphocytes, Clin. Immunol. Immunopath. 10, 446 (1978).

26. A.D.B. Webster and S. Pereira, Effects of deoxyadenosine on the activation of ADA inhibited T and B cells, Clin. Exp. Immunol. 50, 587 (1982).

27. A.R. Hayward, Resistance of pokeweed mitogen-stimulated B cells to inhibition by adenosine, Clin. Exp. Immunol. 41, 141 (1980).

28. D.A. Carson and J.E. Seegmiller, Effect of adenosine deaminase inhibition upon human lymphoblastogenesis, J. Clin. Invest. 57, 274 (1976).

29. L.W. Brox, E. Pollock and A. Belch, Adenosine and deoxyadenosine toxicity in colony assay systems for human T-lymphocytes, B lymphocytes and granulocytes, Cancer Chemother. Pharmacol. 9, 49 (1982).

30. H. Ratech, M.K. Bell, R. Hirschhorn and G.J. Thorbecke, Effects of deoxycoroformycin in mice. I suppression and enhancement of in vivo antibody response to thymus dependent and independent antigens, J. Immunol. 132, 3071 (1984).

31. M. Moretta, M.C. Ferrarini, A. Mingari, A. Moretta and S.R. Webb, Subpopulations of human T cells identified by receptors for immunoglobulins and mitogen responsiveness, J. Immunol. 117, 2171 (1976).

32. F.L. Shand, J.M. Orme and J. Ivanyi, The induction of suppressor T cells by Concanavalin A is independent of cellular proliferation and protein synthesis, Scand. J. Immunol. 12, 223 (1980).

33. S.S. Matsumoto, A.L. Yu, L.C. Bleeker, B. Bakay, F.H. Kung and W.L. Nyhan, Biochemical correlates of the differential sensitivity of subtypes of human leukemia to deoxyadenosine and deoxoadenosine and deoxycoformycin, Blood 60, 1096 (1982).

34. R.L. Wortmann, B.S. Mitchell, N.L. Edwards and H.J. Fox, Biochemical basis for differential deoxyadenosine toxicity to T and B lymphoblasts: Role for 5'-nucleotidase, Proc. Natl. Acad. Sci. USA 76, 2434 (1979).

35. R.M. Fox, E.H. Tripp, S.K. Piddington and M.H.N. Tattersal, Sensitivity of human leukemic null lymphocytes to deoxynucleosides, Cancer Res. 40, 3383 (1980).

S-ADENOSYLMETHIONINE METABOLISM AS A TARGET FOR ADENOSINE TOXICITY

E. Olavi Kajander, Masaru Kubota, Eric H. Willis, and
Dennis A. Carson

Scripps Clinic and Research Foundation
Department of Basic and Clinical Research
La Jolla, California 92037

INTRODUCTION

Adenosine exerts marked cytostatic and cytotoxic actions to
mammalian cells. The nucleoside has been reported to inhibit pyrimidine
nucleotide synthesis, to foster cyclic AMP accumulation and to cause the
accumulation of S-adenosylhomocysteine (AdoHcy).[1,2] Adenosine kinase (EC
2.7.1.20) deficient mammalian cells do not phosphorylate adenosine, but
adenosine still blocks their growth, and this is not reversed by addition
of uridine.[2,3] Thus, adenosine may exert toxicity at the nucleoside
level. Also, some adenosine analogs are cytotoxic without being converted to nucleotides.[4]

To approach directly the mechanism of toxicity adenosine exerts as a
nucleoside, we selected adenosine (plus deoxycoformycin) resistant clones
from an adenosine kinase negative mouse T lymphoma cell subline (R1.1
AK6). All mutant clones had elevated S-adenosylmethionine (AdoMet)
pools. They were resistant to adenosine, to adenosine plus L-homocysteine, and to diverse nucleosides that increased AdoHcy or formed AdoHcy
analogs in the cell. They also grow in medium supplemented with inhibitors of AdoMet synthesis. It appears that AdoMet over-producers are
protected against most, if not all agents that competitively inhibit
AdoMet synthesis and utilization. AdoMet metabolism is also a target for
adenosine toxicity in R1.1 wild type cells.

METHODS

Adenosine kinase deficient R1.1 cells and the adenosine resistant
mutant sublines were isolated as described elsewhere.[5] Both the parent
clone (AK_6) and the adenosine resistant clone (AKR4) presented here,
were devoid of measureable adenosine kinase activity (less than 1% of

wild type activity). Both sublines were totally resistant to >250 μM 6-methylmercaptopurine riboside.

Adenosine and methionine metabolites, and enzymes metabolizing the two compounds, were measured as described previously.[5] Growth-inhibitory properties of drugs were measured in several independent assays. The IC_{50} values indicate the concentration that reduced cell growth by 50% after 72 to 80 h.

RESULTS AND DISCUSSION

All of four isolated adenosine (plus deoxycoformycin) resistant clones had elevated AdoMet pools. Clone AKR4 was characterized in detail. The AKR4 clone contained 4 to 7-fold higher AdoMet levels and 1.5-fold higher AdoMet synthetase activity than the parent AK6 clone under usual culture conditions. When R1.1 cells were cultured in medium containing low methionine levels, AdoMet pools declined and AdoMet synthetase activity increased concomitantly. Cycloleucine, a competitive inhibitor of AdoMet synthesis, increased further the AdoMet synthetase activity in AKR4 cells, but not in the wild type or in AK6 cells in medium with 10 μM methionine (Table 1). This augmentation of AdoMet

Table 1. Comparison of AdoMet Metabolites and Properties of AdoMet Synthetase in Adenosine Kinase (AK) Negative Clones with the Kinase Positive Wild Type Cells.

| | AK positive wild type (WT) | Ratios over WT | |
		Adenosine Sensitive (AK6)	Adenosine Resistant (AKR4)
AdoMet	50 pmol/10^6 cells	1.1	7
AdoHcy	3.4 pmol/10^6 cells	1.4	1.4
AdoMet/AdoHcy	0.07	1.1	0.1
AdoMet synthetase*	22 pmol/min/10^6 cells	1	1.5
stimulated**	42 pmol/min/10^6 cells	1	3.4
AdoMet synthetase			
K_m methionine	13 μM	1	1
K_i AdoMet	160 μM	1	1

*AdoMet synthetase activities in cell lysates obtained from cultures in regular culture medium (RPMI-1640) were measured after removal of low-molecular weight inhibitors by gel filtration.
**Before assay, cells were cultured for 16 h in low-methionine medium (10 μM methionine) in the presence of 7.7 mM cycloleucine.

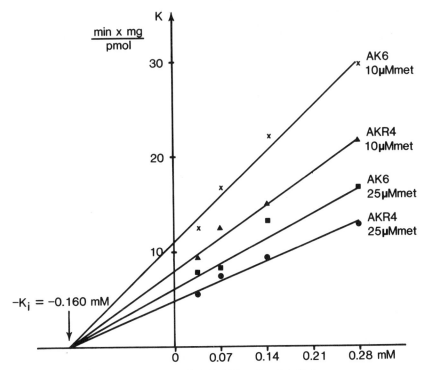

Figure 1. AdoMet Synthetase Inhibition by AdoMet.
The x-axel shows the concentration of AdoMet added to the
incubation mixtures.

synthesis was the only observed biochemical difference between the adenosine resistant and sensitive cells. All other enzymes (besides adenosine
kinase) engaged in adenosine metabolism were equivalent in the two cell
types.

The increased activity of AdoMet synthetase could not be attributed
to enzyme kinetic changes, since the K_m for the rate-limiting substrate,
methionine, as well as the K_I for AdoMet, were similar in adenosine-
resistant and adenosine sensitive cells (Table 1). As shown in Fig. 1,
both AK6 and AKR4 enzymes were noncompetitively inhibited by AdoMet with
a K_I of 160 μM. The intracellular concentrations of AdoMet in R1.1 cells
are of similar magnitude. Thus, the ability of AdoMet to inhibit its own
synthesis is an important regulator of cellular AdoMet pools, and probably prevented the AdoMet levels from increasing indefinitely in AKR4
cells.

AKR4 cells were more than 10-fold resistant to growth-inhibition by
adenosine (plus deoxycoformycin) alone or in combination with L-homocysteine. After treatment with adenosine, all R1.1 cells contained similarly increased levels of AdoHcy, a potent inhibitor of AdoMet-mediated
transmethylation reactions. AKR4 cells were protected from AdoHcy

accumulation, because the AdoHcy to AdoMet ratio did not rise much above 0.1. On the contrary, this ratio reached 1.0 in AK6 cells treated with 200 μM adenosine in the presence of deoxycoformycin.

AKR4 cells were cross-resistant (10 to 30-fold) to other compounds that are known inhibitors of AdoHcy hydrolase, and thus are presumed to block AdoMet-mediated transmethylation reactions (e.g. carbocyclic adenosine, carbocyclic 3-deazaadenosine, adenine, neplanocin A). AKR4 cells were also protected from the toxicity of 3-deazaadenosine, 8-azaadenosine and formycin A. These agents are capable of forming AdoHcy analogs in the cell.[6] Further, the adenosine resistant mutant lymphoblasts were resistant to the growth-inhibitory effects of the inhibitors of AdoMet synthesis cycloleucine and L-ethionine. AKR4 cells were also resistant to methylthioadenosine, methylthiotubercidin and S-isobutylthioadenosine, compounds that may block other pathways of AdoMet utilization.

An important question is, whether the results from adenosine kinase negative, deoxycoformycin treated cells are applicable to normal cells. First, it should be noted that inhibition of adenosine deaminase was unimportant for the toxicities of 3-deazaadenosine, carbocyclic adenosine, carbocyclic 3-deazaadenosine, methylthioadenosine and related analogs. AKR4 cells were resistant to adenosine toxicity even in the absence of deoxycoformycin, although IC_{50} values were 3-4-fold higher. Lack of adenosine kinase is not crucial for the toxicity of adenosine and several adenosine analogs. As shown in Table 2, a comparison of adenosine kinase positive R1.1 cells to the kinase negative AK6 and AKR4 mutants indicates that the growth inhibitory effects of adenosine, 3-deazaadenosine, carbocyclic 3-deazaadenosine, carbocyclic adenosine, neplanocin A and methylthioadenosine and its tubercidin analog are not much affected by presence or absence of adenosine kinase activity. One would predict that adenosine kinase positive AdoMet over-producers would be protected to some extent against nucleotides forming AdoMet analogs, when cultured in regular medium. On the contrary, 8-azaadenosine and formycin A form highly toxic nucleotides. The observed resistance of the adenosine kinase deficient AdoMet over-producer, is not relevant to the in vivo toxicity of these latter compounds.

In conclusion, the AdoMet over-producer mutant lymphoblasts have enabled us to implicate AdoMet metabolism as an important target for the toxic actions of adenosine, adenine, methylthioadenosine, and several synthetic adenosine congeners. The cell model provided a simple screening method to identify compounds that competitively block AdoMet synthesis or metabolism. For example, the model has confirmed that interference with AdoMet metabolism is the principle mechanism of neplanocin A toxicity.[7,8] More important, in malignant murine T-lymphoma cells these effects were observed at nM concentrations of the nucleoside. Further evaluation of the potential anti-lymphocyte properties of neplanocin A and related compounds are warranted.

Table 2. Comparison of Growth-Inhibitory Properties of Adenosine (Ado) and its analogs in Adenosine Kinase Negative with Kinase Positive Cells.

	wild type IC_{50}	AK6 WT	AKR4 WT
Ado	20 μM	1	10
3-deazaAdo	2-8 μM	1	10-20
carbocyclic 3-deazaAdo	0.3-0.5 μM	1	30
carbocyclic Ado	0.1-0.2 μM	1	30
neplanocin A	0.001 μM	1	30
tubercidin	0.05 μM	280	440
8-azaAdo	<0.1 μM	>66	>500
formycin A	<0.1 μM	>62	>480
methylthioAdo	40 μM	1	10
methylthiotubercidin	2-3 μM	1	100

REFERENCES

1. J. F. Henderson and F. W. Scott, Inhibition of animal and inverte-brate cell growth by naturally occurring purine bases and ribonucleosides, Pharmacol. Ther. 8:539 (1980).
2. N. M. Kredich and M. S. Hershfield, S-adenosylhomocysteine toxicity in normal and adenosine kinase-deficient lymphoblasts of human origin, Proc. Natl. Acad. Sci. USA 76:2450 (1979).
3. M. Kubota, N. Kamatani, P. E. Daddona, and D. A. Carson, Character-ization of an adenosine deaminase-deficient human histiocytic lymphoma cell line (DHL-9) and selection of mutants deficient in adenosine kinase and deoxycytidine kinase, Cancer Res. 43:2606 (1983).
4. J. A. Montgomery, S. J. Clayton, H. J. Thomas, W. M. Shannon, G. Arnett, A. J. Bodner, I.-K. Kion, G. L. Cantoni, and P. K. Chiang, Carbocyclic analogue of 3-deazaadenosine: A novel antiviral agent using S-adenosylhomocysteine hydrolase as a pharmacological target, J. Med. Chem. 25:626 (1982).
5. E. O. Kajander, M. Kubota, C. J. Carrera, J. A. Montgomery and D. A. Carson, Multiple drug resistance in mutant lymphoblasts with enlarged S-adenosylmethionine pools, submitted for publication.
6. T. P. Zimmerman, G. Wolberg, G. S. Duncan, and G. B. Elion, Adenosine analogues as substrates and inhibitors of S-adenosylhomo-cysteine hydrolase in intact lymphocytes, Biochem. 19:2252 (1980).

7. R. T. Borchardt, B. T. Keller, and U. Patel-Thombre, Neplanocin A.
    A potent inhibitor of S-adenosyhomocysteine hydrolase and of
    vaccinia virus multiplication in mouse L929 cells, J. Biol. Chem.
    259:4353 (1984).
8. R. I. Glazer and M. C. Knode, Neplanocin A. A cyclopentenyl analog
    of adenosine with specificity for inhibiting RNA methylation, J.
    Biol. Chem. 259:12964 (1984).

ETHANOL INDUCED NUCLEOTIDE CATABOLISM IN

MOUSE T LYMPHOBLASTOID CELLS IN VITRO

Jerzy Barankiewicz and Amos Cohen

Division of Immunology, Research
Institute
The Hospital for Sick Children, Toronto
Ontario, Canada

Impaired cellular immunity and gout are often prospects for alcoholic patients. Study of immunological status of alcoholic patients evaluated by both in vivo and in vitro parameters showed that in vivo reactivity measured by skin testing with dinitrochlorobenzene or tuberculin was not affected by alcohol intake[1]. However, cell mediated immunity was markedly affected. Thymus-derivative lymphocyte populations were impaired both qualitatively and quantitatively[1]. Transient granulocytopenia and lymphopenia without splenomegaly, cirrhosis, infection and hypocellular bone marrow with few mature granulocytes may suggest depressed granulopoetic activity in alcoholic patients[2]. It has been also found that ethanol administration results in functional impairment of granulocytes, macrophages and lymphocytes[2]. On the other hand, increased serum immunoglobulins in alcoholics with cirrhosis have been reported[1]. The significance of these immunoglobulins is unknown.

Ethanol in concentrations similar to those found in blood of alcoholic patients inhibited PHA- and STA-induced lymphocyte transformation in vitro[3] as well as inhibiting bone marrow granulocyte colony growth[3].

A direct relationship between alcohol consumption, gout and hyperuricemia has been reported[4], suggesting an inducing role for ethanol in the overproduction of uric acid. Ethanol increased serum lactate levels[5-7] and decreased renal urate excretion[8-9]. Oral or intravenous ethanol administration increased uric acid and oxypurine excretion and increased uric acid clearance[10]. Because ethanol induces overproduction of uric acid, it was suggested that the mechanism is associated with acceleration of adenine nucleotide degradation[10-12]. Recent studies by Puig and Fox[13] indicate that both ethanol and an acetate-product of ethanol oxidation increased purine nucleotide degradation by enhancing the turnover of the adenine nucleotide pool in patients.

To determine whether ethanol induced nucleotide degradation in cells of immune system and in consequence may impair cellular immunity, the effect of ethanol on a mouse T lymphoblastoid cell line (NSU-1) was studied. This purine nucleoside phosphorylase deficient cell line is convenient for catabolic study because the number of excreted products of nucleotide catabolism is reduced.

In T lymphoblastoid cells, ethanol induced ATP degradation which increased in parallel with ethanol concentrations (Table 1).

The major product of ATP catabolism excreted to the medium was inosine, however small amounts of adenosine were also detected. Simultaneously with increased excretion of nucleosides, increased ATP degradation with no significant changes in ADP or AMP levels were found (Table 1). Because no changes in adenosine excretion were found, inhibition of adenosine deaminase activity by preincubation with 40 μM deoxycoformycin had no significant effect on the ethanol-induced ATP catabolism pathways (results not shown). This indicates that ethanol-induced ATP catabolism proceeds mainly via AMP deamination and IMP dephosphorylation and adenosine deaminase is not significantly involved in the catabolic reactions.

In contrast to other nucleotide catabolism inducers[15], ethanol had no effect on the intracellular orthophosphate concentrations (results not shown).

Ethanol in low concentrations significantly inhibited salvage of purine and pyrimidine bases on nucleosides. Increasing concentrations of ethanol resulted in increasing inhibition of incorporation of adenine, uridine and thymidine into both nucleotides and nucleic acids. Inhibition of purine

Table 1.   Effect of ethanol on ATP catabolism

Ethanol (%)	ATP	ADP	AMP	Inosine	Adenosine
			(cpm/10^6 cells)		
0.0	8220	1222	241	845	162
0.2	6833	998	255	1551	251
0.4	6552	974	230	1882	248
1.0	5984	1101	258	3672	264
2.0	3822	982	182	4421	588

Mouse T lymphoblastoid (NSU-1) cells (20 x 10^6) were incubated with 4 μCi of radioactive adenine for 1 hr. The unicorporated precursor was washed out and 1 x 10^6 cells containing radioactive adenine nucleotides were incubated with different concentrations of ethanol for 20 min. Radioactivity in nucleotides of cell extracts and nucleosides excreted into the incubation medium was measured as described by Barankiewicz and Cohen[14].

and pyrimidine salvage and nucleotide incorporation into nucleic acids by ethanol together with induction of nucleotide catabolism may significantly reduce nucleotide availability not only for DNA and RNA synthesis but also for other cellular processes. It may therefore influence blastogenic transformation and lead to lymphopenia in acute alcoholism.

Table 2. Effect of ethanol on adenine, uridine and thymidine salvage

Substrate	Ethanol %	Nucleotides	Nucleic acids
		cpm/10^6 cell	
Adenine	0	25647	952
	0.2	24331	861
	0.5	22112	744
	1.0	20004	654
	2.0	15101	489
Uridine	0	16982	2544
	0.2	16302	2401
	0.5	12556	2002
	1.0	7210	1551
	2.0	6106	1361
Thymidine	0	1206	22461
	0.2	1197	21406
	0.5	1181	15144
	1.0	1100	9642
	2.0	802	8338

Mouse T lymphoblastoid (NSU-1) cells (1 x 10^6 cells) were incubated with different concentrations of ethanol and 1 μCi of radioactive adenine, uridine and thymidine respectively for 20 min. Incorporation of radioactivity into nucleotides and nucleic acids was measured according to Hunting et al[16].

References

1. J. Lundy, J.H. Raaf, S. Deakins, H.J. Wanebo, D.A. Jacobs, T. Lee, D. Jacobowitz, C. Spear, and H.F. Oettgen. The acute and chronic effects of alcohol on the human immune system. Surgery, Gynecology & Obstetrics, 141:212-218 (1975).
2. Y.K. Liu. Effect of alcohol on granulocytes and lymphocytes. Seminars in Hematology 17:130-136 (1980).
3. G. Tisman and V. Herbert. In vitro myelosuppression and immunosuppression by ethanol. J. Clin. Invest. 52:1410-1414 (1973).

4. J.B. Wyngaarden and W.N. Kelley. Drug induced hyperuricemia and gout. in: Gout and Hyperuricemia. Grunde & Stretton Inc., New York, 369-380 (1976).
5. T.F. You, J.H. Sirota, L. Berger, M. Halpern, A.B. Gutman. Effect of sodium lactate infusion on urate clearance in man. Proc. Soc. Exp. Biol. Med. 96:809-813 (1952).
6. M.J. Maclachlan and G.P. Rodnan. Effects of food fast and alcohol on serum uric acid and acute attacks of gout. Am. J. Med. 42:38-57 (1967).
7. L.H. Beck. Clinical disorders of uric acid metabolism. Med. Clin. N. Am. 65:401-411 (1981).
8. C.S. Lieber and C.S. Davidson. Some metabolic effects of ethanol accumulation. Am. J. Med. 33:319-327 (1962).
9. C.S. Lieber, D.P. Jones, M.S. Losowsky, and C.S. Davidson. Interrelation of uric acid and ethanol metabolism in man. J. Clin. Invest. 41:1863-1870 (1962).
10. J. Faller and I.H. Fox. Ethanol induced alteration of uric acid metabolism. Adv. Exp. Med. Biol. 165A:457-462 (1984).
11. J. Grunst, G. Dietze and M. Wicklmayr. Effect of ethanol on uric acid production in human blood. Nutr. Metab. (Suppl. 1) 138-141 (1982).
12. H. Hartmann and H. Förster. Studies of the influence of ethanol and lactic acid on uric acid metabolism. Adv. Exp. Med. Biol. 76A:509-518 (1977).
13. J.G. Puig and I.H. Fox. Ethanol induced activation of adenine nucleotide turnover. J. Clin. Invest. 74:936-941 (1984).
14. J. Barankiewicz and A. Cohen. Evidence for distinct catabolic pathways of adenine ribonucleotides and deoxyribonucleotides in human T lymphoblastoid cells. J. Biol. Chem. 259:15178-15181 (1984).
15. J. Barankiewicz and A. Cohen. Nucleotide catabolism and nucleoside cycles in human thymocytes. Biochem. J. 219:197-203 (1984).
16. D. Hunting, J. Horden, and J.F. Henderson. Quantitative analysis of purine and pyrimidine metabolism in chinese hamster ovary cells. Can. J. Biochem. 59:838-847 (1981).

INCREASED LEVEL OF RIBONUCLEOTIDE REDUCTASE IN DEOXYADENOSINE
RESISTANT ADENOSINE DEAMINASE DEFICIENT HUMAN HISTIOCYTIC
LYMPHOMA CELLS

Yvonne Dahbo, Dennis Carson* and Staffan Eriksson

Karolinska Institutet, Medical Nobel Institute, Department
of Biochemistry I, Stockholm, Sweden, *Scripps Clinic and
Research Foundation. Department of Basic and Clinical
Research, La Jolla, California

INTRODUCTION

Inherited deficiencies of the two enzymes of purine metabolism,
adenosine deaminase and purine nucleoside phosphorylase, cause
severe immunodeficiency disease in humans (1). The mechanisms
by which deoxynucleosides exert their cytotoxic effects are not
fully understood but the importance of the conversion of deoxyribonucleosides to their corresponding nucleotides stems from
the observation that it is the deoxyribonucleoside triphosphates
that are cytotoxic. The harmful effects of deoxyadenosine, deoxyguanosine and thymidine are apparently caused by inhibition of
ribonucleotide reductase (2-5). The accumulation of high intracellular pools of one of the effector substances, i.e. dATP,
dGTP and dTTP, will lead to interference with the de novo
synthesis of the other deoxyribonucleotides, essential for DNA-
synthesis.
A human histiocytic lymphoma, DHL-9 wild type, naturally devoided
of adenosine deaminase activity was isolated (6,7). From this
cell line a mutant, DHL-9 dAR-2, resistant to high concentrations
of deoxyadenosine was also selected (8). The deoxyadenosine
resistant cells excrete both deoxyadenosine and thymidine at a
six-seven fold higher rate than the wild type lymphoma cells.
The pace of adenosine excretion, the growth rate and the levels
of multiple other enzymes involved in deoxyadenosine and adenosine metabolism are equivalent in the two cell lines.
The purpose of this study was to clearify if the high resistance
of DHL-9 dAR-2 cells to deoxyadenosine was due to either a
higher level of ribonucleotide reductase or to an alteration
in the allosteric control of the nucleotide binding subunit, M1
of ribonucleotide reductase (9).

MATERIALS AND METHODS

Cell Lines

The naturally adenosine deaminase deficient human histiocytic lymphoma cell line, DHL-9, was kindly provided by Dr. Henry Kaplan (Stanford University). The cells were routinely cultivated in RPMI 1640 medium containing 10% heat-inactivated fetal bovine serum, penicillin (150U/ml), streptomycin (150μg/ml), puruvate (1mM), mycostatin (25U/ml,E.R. Squibb Ltd.) and L-glutamine (0.6mg/ml). The cultures were kept in logarithmic phase at 37°C in a humidified atmosphere of 93% air and 7% CO_2.

Preparation of Cellextracts

The wild type and the deoxyadenosine resistant cells were harvested at a density of 1 million cells per ml by centrifugation and thereafter homogenized in 20mM hypothon Hepes (4-(2-hydroxyethyl)-1-piperazinsulphonicacid), pH7.6, 10mM $MgCl_2$ and 2mM DTT (dithioerythritol). After centrifugation at 12.000 rpm for 30min the enzyme is found in the supernatant. The crude extracts were purified by affinity chromatography on a dATP-Sepharose column (see in 9) and eluted, after several washing steps, with ATP at different concentrations.

^{35}S-Labeling of the Cells

The cells were starved for methionine for 2hrs by letting them grow in DME-medium lacking methionine and containing 6% horse serum, whereafter ^{35}S-methionine (Amersham) was added (0.2mCi/10ml). After 3hrs the cells were harvested as described above.

Ribonucleotide Reductase Activity

The activity of ribonucleotide reductase was measured in both crude extracts and purified enzyme, with and without the addition of excess M1- and M2-protein, purified from calf thymus (9) and hydroxyurea resistant mouse cells (10), respectively. The reduction of ^{3}H-CDP to ^{3}H-dCDP was measured in the presence of ATP, Mg^{2+}, DTT and Fe^{3+} as described (9). The activity in the presence of different concentrations of dATP was also determined.

SodiumDodecylSulphate-PolyacrylAmideGelElectrophoresis and Western Blotting

The method used for electrophoresis was the one described in reference 9. In the cases where the proteins were not radioactive labeled they were blotted from the SDS-gel to a nitrocellulose sheet whereafter the M1-subunit were incubated with a mouse anti M1 antibody, AD203 provided by Y.Engström and L. Thelander (11). The M1-subunit was visualized with a second antibody, rabbit anti mouse (Sigma), either conjugated with alkaline phosphatase enzyme or detected with ^{125}I-Protein A and autoradiography. The band intensities on the gel and autoradiography were measured with a Joyce-Loebl microdensitometer.

RESULTS

The specific activity in crude extracts of ribonucleotide reductase was 2 to 5 times higher in dAR-2 cells compared to wild type cells, Table 1. Addition of excess purified M1 or M2 from calf thymus and mouse cells respectively did not stimulate the activity further. Incubation with different concentrations of dATP did not show significant effects on the activity, presumably due to rapid breakdown of dATP.

Table 1. Ribonucleotide reductase activity in crude cell extracts

Extract		Specific activity (pmols/min/mg)						
			+M1	+M2	50	100	200	400 μM dATP
DHL-9 wild type	a)	13	13	13	-	-	-	-
	b)	13	-	-	20	17	23	10
DHL-9 dAR-2	a)	60	40	47	-	-	-	-
	b)	33	-	-	37	23	17	13

a) crude extract b) after desalting of crude extract on Sephadex G-50,fine

Ribonucleotide reductase activity of protein M1 after dATP-Sepharose was not detectable, either due to inactivation during purification or to that mouse protein M2 was unable to crossreact with human protein M1.
The crude extracts were run on a 7.5% SDS-PAGE, blotted to nitro-cellulose, incubated with mouse anti M1 antibody followed by rabbit anti mouse antibody. Finally the bands were detected with ^{125}I-Protein A and autoradiographed. The result (Fig. 1A) showed that the deoxyadenosine resistant cell line contained 2-4 times more M1-protein than the wild type cells.
Both wild type and dAR-2 cells were labeled with ^{35}S-methionine as described above and then purified on two parallel dATP-Sepharose columns. The elutes were put on a 7.5% SDS-PAGE and after electrophoresis the gel was autoradiographed. The result (Fig. 1B) showed that the dAR-2 cell line contained 1.5-2.5 times more M1 than the wild type.
When wild type and dAR-2 extracts were purified on dATP-Sepharose and eluted with stepwise increasing concentrations of ATP, from 0.1 to 6.0 mM, no difference in elution pattern could be seen, indicating there is no difference in the binding of dATP between the two cell lines (data not shown).

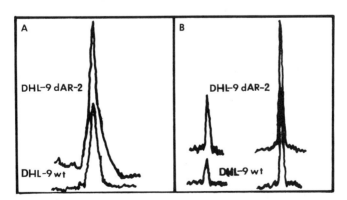

Fig.1. Microdensitometry scanning of the bands corresponding to the M1-protein. A. Crude extracts of 60µg DHL-9 wild type and 60µg DHL-9 dAR-2. B. Same volume of DHL-9 wild type and dAR-2 extracts purified on dATP-Sepharose and eluted with 0.1mM ATP (left peaks) and 6mM ATP, respectively.

DISCUSSION

A 2 to 5 fold higher ribonucleotide reductase activity was found in crude extracts of the deoxyadenosine resistant DHL-9 cells compared to wild type cells. The result from Western blotting of crude extracts as well as gel electrophoresis on dATP-Sepharose purified enzyme, showed a 2-3 fold increase of the M1 subunit in the mutant cells. Our results does not indicate any difference in binding to dATP-Sepharose between mutant and wild type, but this method may not be sensitive enough to detect subtile changes. The higher level of ribonucleotide reductase in the mutant cell line is sufficient to account for an increased excretion of deoxynucleosides since the same amount of deoxyribonucleotides are required for DNA synthesis. Elevated intracellular concentrations of deoxynucleosides and deoxynucleotides can be expected to interfere competetively with the phosphorylation and further metabolism of deoxyadenosine and related compounds derived from an extracellular source (12). Thus the overproduction of deoxynucleosides by DHL-9 dAR-2 cells can explain the decreased ability to form dATP (8) and the insensitivity of the deoxynucleoside-overproducing cell line to the toxic effects of deoxyadenosine. These result suggest that relatively small changes in the de novo synthetic rate of deoxynucleotides may significantly influence sensitivity not only to deoxyadenosine toxicity but also to the effects of several other nucleoside analouges used in cancer therapy.

ACKNOWLEDGEMENT

This work was supported by grants from the Swedish Medical Research Council, the Cancer Society, the Medical Faculty of the Karolinska Institute and also by NIH grant GM23200 and PHS grant CA31459 given to D. Carson.

REFERENCES

1. Giblett,E.R., Anderson,J.E., Cohen., Polara,B. and Meuwissen H.S. (1972) Lancet ii, 1067-1069
2. Martin,D.W.,Jr. and Gelfand,E.W. (1981) Annu.Rev.Biochem. 50, 845-847
3. Carson,D.A., Kaye,J. and Seegmiller,J.E. (1977) Proc.Natl. Acad.Sci. U.S.A. 74, 5677-5681
4. Ullman,B., Gudas,L.J., Cohen,A. and Martin,D.W.,Jr. (1978) Cell 14,365-375
5. Gudas,L.J., Ullman,B., Cohen,A. and Martin,D.W.,Jr. (1978) Cell 14, 531-538
6. Epstein,A.L., Levy,R.L., Kim,H., Henle,G. and Kaplan,H.S. (1978) Cancer (Phila) 42, 2379-2391
7. Kubota,M., Kamatani,N., Dadonna,P. and Carson,D.A. (1983) Cancer Research 43, 2606-2610
8. Kubota,M., Carrera,C., Wasson,D. and Carson,D. (1984) Biochem.Biophys,Acta 804,37-43
9. Engström,Y., Eriksson,S., Thelander,L. and Åkerman,M. (1979) Biochemistry 18, 2941-2948
10. Thelander,M., Gräslund,A. and Thelander,L. (1985) J.Biol. Chem. 260, 2737-2741
11. Engström,Y., Rozell,B., Hansson,H.A., Stemme,S. and Thelander,L. (1984) EMBO Journal 3, 863-867
12. Henderson,J.F., Scott,F.W. and Lowe,J.K. (1980) Pharm. Ther. 8, 573-604

INHIBITION OF T CELL CYTOTOXICITY BY CYCLOSPORINE (CSA), ADENOSINE (ADO) AND AN INHIBITOR OF ADENOSINE DEAMINASE (ADA)

Caliann T. Lum, Stephen R. Jennings, Frederick J. Wanner and Satvir S. Tevethia

The University of Texas Health Science Center at San Antonio, Department of Surgery, San Antonio, Texas and The Pennsylvania State University College of Medicine, Department of Microbiology and Cancer Research Center, Hershey, Pennsylvania USA

INTRODUCTION

Cyclosporine (CSA-previously known as cyclosporin A) is the prototype of a new class of polypeptide lipophilic immunosuppressive agents with cell membrane affinity.[1,2] The predominant immunosuppressive effects of this compound in vivo seem to result from depressed helper T cell function.[3] In vitro studies indicate that CSA inhibits proliferation of alloreactive cells in mixed lymphocyte culture and the generation of alloreactive cytolytic lymphocytes (CTL).[4] As shown by Orosz et al,[5] this inhibition of clonal expansion relates specifically to the alloantigen driven response of both helper and CTL. However, from current reports CSA does not appear effective against antigen-directed cytotoxic cells already committed to that function.[5,6]

Because of our ongoing interest in the relationship of adenosine metabolism to immune function, we had previously looked at the effect of CSA and adenosine on ATP levels in C57BL/6 mouse spleen cells and IL-2 dependent cloned CTL derived from the same mouse strain.[7,8] We observed that CSA and the adenosine deaminase (adenosine aminohydrolase, E.C. 3.5.4.4, ADA) inhibitor EHNA caused ATP depletion in unsensitized T cells isolated from mouse spleen cells but not in cloned murine CTL.

We postulated that this resistance to ATP depletion might be one reason that CTL cytolytic function is not inhibited by CSA. The current study was undertaken to test the effect of CSA, EHNA and adenosine on CTL function.

MATERIALS AND METHODS

Cells - IL-2 dependent cloned CTL directed against SV40 antigen were T cells derived from C57BL/6J male mice (The Jackson Laboratory,

---

Abbreviations used in this paper: SV40, simian virus 40; CTL, cytotolytic T lymphocytes; IL-2, human T cell growth factor; ATP, adenosine triphosphate; EHNA, erythro=9-(2-hydroxy-3-nonyl)adenine.

Bar Harbor, ME). Clones were established and maintained according to the methods of Brunner et al[9] and Braciale et al[10] and as previously described.[11] Specificity of these cells against SV40 was determined as previously described.[12] The T cell growth factor used was human interleukin-2 purchased from Amgen Biologicals, Thousand Oaks, CA. Target cells were derived by transformation of C57BL/6 embryo fibroblasts with wild-type SV40 as previously described.[12]

Cytotoxicity assay - A 5-hr ^{51}Cr-release cytotoxicity assay was performed as previously described using anti-SV40-CTL as effectors in a ratio of 10:1 and 3:1 against the syngeneic target cells bearing SV40 antigens.

Effector cell treatments - Anti-SV40-CTL effectors were adjusted to a concentration of 10^7 cells/ml and incubated for 10 min at 37°C with the following treatments. Media or drugs were added to the cells in a volume of 10 ul/ml of cell suspension. 1- RPMI 1640 (media control-Grand Island Biological Co., Grand Island, NY), 2- adenosine to a concentration of 10^{-3}M, 3- EHNA to a concentration of 10^{-3}M, 4- CSA to a concentration of 10 ug/ml and 5- CSA to a concentration of 25 ug/ml, 6- Ado (10^{-3}M) and CSA (25 ug/ml), 7- Ado (10^{-3}M) and EHNA (10^{-3}M), 8- EHNA (10^{-3}M) and CSA (25 ug/ml), 9- Ado (10^{-3}M) and EHNA (10^{-3}M) and CSA (25 ug/ml), and 10- 95% ethanol (used as a solvent for CSA in all groups). After treatment, effector cells were >90% viable when tested by Trypan blue dye exclusion.

Effector cells were then added in 100 ul volumes to ^{51}Cr labelled target cells (2×10^4 cells in 100 ul) in a ratio of 10:1 and 3:1 for the cytotoxicity assay. All samples were performed in quadruplicate.

Statistical analysis - statistical analysis of the results was conducted using Chi-square analysis calculated using the ethanol (treatment 10) as the control. CSA was a gift from Dr. David Winter of Sandoz, Inc., East Hanover, NJ; EHNA was provided by Wellcome Research Laboratories, Burroughs Wellcome Co., Research Triangle Park, NC.

RESULTS AND DISCUSSION

These data indicate that when the effector to target ratio (E:T) is

Table 1.

% specific ^{51}Cr release

Treatment	E:T 10:1		3:1	
1. No treatment	19.3 ± 2.0		31.1 ± 1.2	$p < 0.05$
2. Ado 10^{-3}M	21.6 ± 2.2	$p < 0.05$	24.7 ± 0.6	
3. EHNA 10^{-3}M	21.4 ± 0.7	$p < 0.05$	24.4 ± 0.4	
4. CSA 10 ug/ml	19.4 ± 0.9	$p < 0.05$	20.3 ± 0.8	$p < 0.05$
5. CSA 25 ug/ml	21.6 ± 1.3	$p < 0.05$	17.8 ± 0.9	$p < 0.05$
6. Ado + CSA 25 ug/ml	21.7 ± 1.7	$p < 0.05$	20.9 ± 1.3	$p < 0.05$
7. Ado + EHNA	17.2 ± 0.8	$p < 0.05$	18.5 ± 1.1	$p < 0.05$
8. EHNA + CSA 25 ug/ml	24.8 ± 1.0		19.7 ± 3.1	
9. Ado + EHNA + CSA 25 ug/ml	11.5 ± 0.6	$p < 0.05$	13.4 ± 0.8	$p < 0.05$
10. Ethanol	26.2 ± 0.8		23.7 ± 1.1	

low (3:1), if one controls for the toxicity of ethanol as the solvent/carrier for CSA, the combination of Ado + EHNA + CSA is most effective in hindering cytolytic function of cloned CTL. At the same E:T ratio, less effective but significant toxicity is also seen with CSA alone, and with combinations of Ado with CSA or EHNA. The effect of CSA alone appears to be dose related. However, Ado and EHNA alone have no effect in the concentrations studied.

At a higher E:T ratio of 10:1 there is a slight statistically significant reduction of cytotoxicity for all treatment groups except EHNA + CSA. However, again, the most reduction of cytolytic function is seen in treatment group 9 where Ado, EHNA and CSA are combined. Also of note is the loss of dose effect seen for the two doses of CSA.

The toxicity of Ado to mammalian cells has been described previously[13,14]. This effect can be potentiated by inhibitors of the enzyme adenosine deaminase in mouse T lymphoma cells[15]. Inhibition of lymphocyte-mediated cytolysis by adenosine was described by Wolberg et al in 1975.[16] They noted at that time that this effect was potentiated by the ADA inhibitor EHNA.

Cyclosporine has been reported to be ineffective in preventing the cytolytic activity of presensitized lymphocytes[4,6] although it appears to prevent the generation of alloreactive CTL[4,5]. Our data, however, show that CSA can inhibit cytolytic activity of committed CTL if the E:T ratio is low.

CONCLUSION

In conclusion, our study suggests that CSA does possess a modest ability to inhibit cytolytic activity of committed CTL although this has not been described previously. We postulate that in our system adenosine toxicity is enhanced by the addition of the adenosine deaminase inhibitor, EHNA, and that the addition of CSA then results in marked inhibition of the cytolytic activity of committed CTL. ATP does not appear directly related to the cytolytic function of these cloned T cells since even though ATP levels were preserved in previous studies, impaired function does occur in the presence of Ado + EHNA + CSA.

Because the clinical efficacy of CSA in treating established allograft rejection is marginal, we propose that Ado combined with the ADA inhibitor EHNA, may increase the cytotoxic range of CSA to include effector CTL and thus improve the results of organ transplantation rejection therapy.

ACKNOWLEDGMENTS

This work was supported by National Institutes of Health Grant R23 AM 31519, and National Cancer Institute Grant CA 25000.

REFERENCES

1. J.F. Borel, From our laboratories. Cyclosporin A, Triangle 20:97 (1981).
2. S.S. LeGrue, A.W. Friedman, and B.D. Kahan, Binding of cyclosporine by human lymphocytes and phospholipid vesicles, J. Immunol. 131:712 (1983).
3. C.T. Van Buren, R. Kerman, G. Agostino, W. Payne, S. Flechner, and B.D. Kahan, The cellular target of cyclosporin A action in humans, Surgery 92:167 (1982).

4. B.S. Wang, E.H. Heacock, K.H. Collins, I.F. Hutchinson, N.L. Tilney, and J.A. Mannick, Suppressive effects of cyclosporine A on the induction of alloreactivity in vitro and in vivo, J. Immunol. 127:89 (1981).

5. C.G. Orosz, R.K. Fidelus, D.C. Roopenian, M.B. Widmer, R.M. Ferguson, and F.H. Bach, Analysis of cloned T cells function I. Dissection of cloned T cell proliferative responses using cyclosporin A, J. Immunol. 129:1865 (1982).

6. T. Horsburgh, P. Wood, and L. Brent, Suppression of in vitro lymphocyte reactivity by cyclosporin A: existence of a population of drug-resistant cytotoxic lymphocytes, Nature 286:609 (1980).

7. C.T. Lum, F.J. Wanner, A.F. Tilberg, and C.A. Robertson, Cyclosporine-induced ATP depletion in murine T and B lymphocytes, Surgery 96:256 (1984).

8. C.T. Lum, F.J. Wanner, and S.S. Tevethia, Preservation of ATP in cyclosporine (CSA) treated cytotoxic T cells, Transplant, Proc. 17:1378 (1985).

9. K.T. Brunner, H.R. MacDonald, and J.C. Cerottini, Quantitation and clonal isolation of cytolytic T lymphocyte precursors selectively infiltrating murine sarcoma virus-induced tumors, J. Exp. Med. 154:362 (1981).

10. T.J. Braciale, M.E. Andrew, and V.L. Braciale, Heterogeneity and specificity of cloned lines of influenza-virus-specific cytotoxic T lymphocytes, J. Exp. Med. 153:910 (1981).

11. A.E. Campbell, F.L. Foley, and S.S. Tevethia, Demonstration of multiple antigenic sites of the SV40 transplantation rejection antigen by using cytotoxic T lymphocyte clones, J. Immunol. 130:490 (1983).

12. S.S. Tevethia, D.C. Flyer, and R. Tjian, Biology of simian virus 40 (SV40) transplantation antigen (TrAg), Virology 107:13 (1980).

13. I.H. Fox and W.N. Kelley, The role of adenosine and 21-deoxyadenosine in mammalian cells, Ann. Rev. Biochem, 47:655 (1978).

14. H. Green and T.-S. Chan, Pyrimidine starvation induced by adenosine in fibroblasts and lymphoid cells: role of adenosine deaminase, Science 182:836 (1973).

15. N.M. Kredich and D.W. Martin, Role of S-adenosylhomo-cysteine in adenosine-mediated toxicity in cultured mouse T lymphoma cells, Cell 12:931 (1977).

MECHANISM OF URATE PRODUCTION BY GUINEA-PIG ILEUM

D.R. Webster, G.D. Boston, and D.M. Paton

Department of Pharmacology and Clinical Pharmacology
University of Auckland School of Medicine
Auckland, New Zealand

INTRODUCTION

Previous studies have demonstrated considerable efflux of purine consisting mostly uric acid but with smaller quantities of xanthine, hypoxanthine, inosine and adenosine) from isolated guinea-pig ileal segments (Webster et al., 1985). The aim of the present study was to further investigate the origin of these purines. Our earlier studies had shown that the purine efflux was accompanied by considerable weight loss of the tissue and the appearance in the bathing medium of the enzymes of purine degradation, adenosine deaminase, purine nucleoside phosphorylase, and xanthine oxidase. As it has been suggested that muscle activity may be a cause of tissue enzyme release (Jones et al., 1983), the initial portion of the study investigated the effects of different stimulation conditions on muscle activity and purine release, while the second portion of the study was to investigate the biochemical origin of purines by the use of enzyme inhibitors.

METHODS

Tissue Preparation

Guinea-pigs of either sex were killed with 100% CO_2. The terminal ileum was removed and the contents gently washed out with a syringe containing warm physiological salt solution. Ileal sections of a defined weight (usually 100mg) were prepared. The preparation was suspended in a tissue bath containing 3.0ml physiological salt solution containing, in millimolar concentrations, NaCl 116, KCl 5.4, $CaCl_2$, $MgCl_2$ 1.2, NaH_2PO_4 1.2, Na_2EDTA 0.04, $NaHCO_3$ 22.0 and D-glucose 11.2. The baths were gassed continuously with 5% CO_2 in O_2. Phentolamine (1µM) and propranolol (1µM) were added to prevent actions at adrenoceptors.

Preparations were set up under a resting tension of 500mg and allowed to stabilise. Field stimulation (Grass stimulator, SD9) of the preparations at different frequencies (mostly 0.2Hz) was achieved by passing biphasic pulses of 1ms duration and supramaximal voltage between platinum electrodes located at the top and bottom of the organ baths.

Determination of purine metabolism

The tissues were allowed to stabilise for 80 min with changes of bath fluid every 20 min. When the effects of drugs were studied, these were added to the bath in the last rinse. Following equilibration sections were stimulated at 0.2Hz for 20 min. Sampling was begun after 5 min and $100\mu l$ samples of medium were taken at 0,1,2,4,8, and 15 min with a microsyringe.

Sample preparation

Medium samples were deproteinated with TCA (trichloroacetic acid) at a final concentration of 5% and extracted with water-saturated diethylether until the pH was greater than 4.0

Purine Analysis

Purines were separated on a Bondapak C-18 column (Waters Associates) and detected by their uv absorbance at 280 and 254nm simultaneously on a Waters M440 dual wavelength detector at 0.01 aufs. Purines were eluted with a linear gradient from 0.007mol/l KH_2PO_4 to 20% methanol in phosphate in 15min at 1.0ml/min. Purines were identified by retention times, ratio of absorbance at 280nm to that at 254nm, co-chromatography and enzyme peak shift. Quantitation was by comparison of peak height with external standard.

Chemicals

Purine standards were obtained from Sigma Chemical Co., St. Louis, Mo. HNBTGR (6-(2-hydroxy-5-nitrobenzylthio)-guanosine) was obtained from Calbiochem (San Diego, Ca.). Deoxycoformycin (dCF) was a generous gift from Dr. J.F. Henderson.

RESULTS AND DISCUSSION

The effects of the initial prestimulation, the frequency of stimulation and the muscle contraction on weight loss of the tissue and purine efflux are shown in Table 1. There was no significant effect on any of these parameters on purine efflux. Purine leakage and weight loss was not apparently only from the cut ends of the tissue, since the percentage weight loss was not dependent on the length of the tissue segment used (Table 2).

Table 1. Effect of different stimulation conditions on weight loss and purine efflux. After equilibration 150mg ileal segments were either rested or stimulated at 0.2Hz for 5 min, then the bath medium changed and the tissue stimulated or rested, in the presence or absence of atropine (10^{-6}M).

Pretreatment 5 min	Treatment 15 min	Muscle twitch	Weight loss, %	Purine efflux, nmol/150mG
none	none	no	47 ± 2	25.5 ± 7.2
stim 0.2Hz	none	no	47 ± 3	29.7 ± 4.8
stim 0.2Hz	stim 0.2Hz	yes	46 ± 2	31.8 ± 5.4
stim 0.2Hz	stim 0.2Hz	yes	47 ± 2	28.5 ± 6.6
stim 0.2Hz	stim 0.2Hz + atropine	no	47 ± 1	26.7 ± 3.0

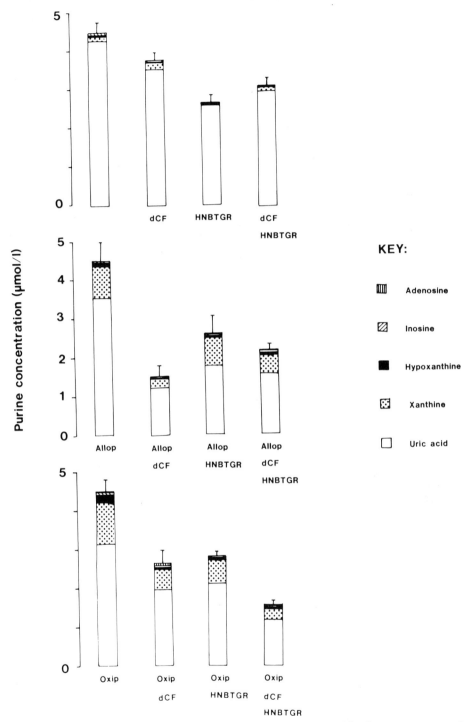

Figure 1. Net purine release from guinea-pig ileal segments during stimulation at 0.2Hz for 15 min. Drugs were added 20 min before the start of the experiment, at the following concentrations: allopurinol 10µM, oxipurinol 10µM, HNBTGR 1µM and dCF 2µM.

Table 2. Effect of length of ileal segment on tissue weight loss. Segments were equilibrated then remained suspended without stimulation for 5 min, the bath medium was changed and the tissue left a further 15 min before it was removed and weighed.

Starting mass mg	Loss %
100	50 ± 1
150	47 ± 2
250	46 ± 1

The biochemical basis for the purine efflux from ileal segments during stimulation was studied by the addition of deoxycoformycin (dCF, 2μM, HNBTGR (1μM), allopurinol (10 and 50μM) and oxipurinol (10μM) alone or in combination. The purine efflux was linear up to 15 min stimulation. The net purine efflux i.e. the concentration found at 15 min less the initial concentration is shown in Fig. 1.

Inhibition of xanthine oxidase with allopurinol or oxipurinol increased the amount of xanthine and hypoxanthine released but had no effect on total purine efflux. Increasing the allopurinol concentration to 50uM doubled the xanthine and hypoxanthine efflux but again did not affect total purine efflux.

Inhibition of adenosine deaminase with dCF slightly reduced purine efflux. This was further reduced by the concomitant inhibition of xanthine oxidase with a marked reduction in the uric acid component. This indicated that about one third of the urate was produced via adenosine. It has been shown (Webster et al., 1985) that there is considerable purine degradation products (inosine, hypoxanthine, xanthine and uric acid) present in tissue samples at the start of the experiment. Purine efflux may therefore be directly from this material or resulting from nucleotide breakdown during the experiment.

The total purine efflux was reduced in the presence of the nucleoside transport inhibitor HBNTGR. This indicated that purine metabolism is both intra- and extracellular in this system. It has previously been shown that purine metabolising enzymes are present in incubation medium from this preparation (Webster et al., 1985).

Inhibition of xanthine oxidase, nucleoside transport and adenosine deaminase together did not further decrease purine efflux.

In summary it appears that most of the uric acid appearing in incubation medium from stimulated guinea-pig ileal segments was produced via adenosine. Negligible IMP was present in either medium or tissue samples after these incubations, but a large amount of uric acid and other purine degradation products was present in the tissue immediately after it was prepared (Webster et al., 1985) and it is possible that some of this uric acid is released into the medium following tissue lysis. These results imply that the concentration of adenosine at adenosine receptors may bear little relationship to concentrations of adenosine measured in medium samples, since there is a continuing flux of endogenous purines through the degradative pathway via adenosine.

REFERENCES

Jones, D.A., Jackson, M.J., and Edwards, R.H.T. 1983, Release of intracellular enzymes from an isolated skeletal muscle preparation. Clin. Sci. 65:193-201.

Webster, D.R., Boston, D.G., and Paton, D.M. 1985. Measurement of adenosine metabolites and metabolism in isolated tissue preparations. J.Pharmacol. Meths. 13:339.

IN VITRO DEGRADATION OF PYRIMIDINE BASES: STUDIES OF RAT LIVER

DIHYDROPYRIMIDINE DEHYDROGENASE

Mendel Tuchman, Margaret L. Ramnaraine, and Robert F. O'Dea

Departments of Pediatrics and Pharmacology, University
of Minnesota Hospitals
420 Delaware St. S.E., Minneapolis, Minnesota 55455 U.S.A.

INTRODUCTION

There is considerable interest in the study of pyrimidine base degradation because this system's activity may be age dependent and is altered in neoplastic tissues (1,2). Pyrimidine base analogues (e.g. 5-fluorouracil, 5-FU) are degraded by the same enzymatic pathway (3) and changes in the function of pyrimidine base degrading enzymes may affect the elimination kinetics of these drugs as well as their adverse effects (4). Finally, four reports have been published indicating that genetic defects are present in this system (5-8) and it is likely additional information will become available on the nature of inborn errors of metabolism in the pyrimidine degradation system.

Pyrimidine bases are degraded primarily in the liver and to a lesser degree in the thymus, intestinal mucosa, spleen, kidney, brain, muscle and heart (9). The biological importance of this system is unclear. There is evidence that the rate of degradation of pyrimidine bases determines the degree of their reincorporation into RNA and DNA through "salvage" pathways. This concept is supported by the observation that incorporation of 5-FU into RNA is inversely proportional to the rate of its degradation (10). Furthermore, the activity of the pyrimidine degradation system and especially that of dihydropyrimidine dehydrogenase (DPD), the rate limiting enzyme, are low in growing liver cells, liver cancer cells or liver cells from young animals (9). This phenomenon is biologically logical in that rapidly growing cells depend on a readily available supply of pyrimidine nucleotides for protein synthesis and for replication. The mechanisms that control the degradation of pyrimidine bases are currently unknown. It is also unknown whether this activity can be detected in cultured cells such as fibroblasts or various cancer cell lines.

In this investigation we examined the effects of the nucleosides, uridine and thymidine, on the activity of DPD which was partially purified from rat liver using 5-FU, uracil and thymine as substrates.

METHODS

DPD was partially purified from rat liver using an ammonium sulfate precipitation procedure as described previously (11). One gram of liver yielded approximately 5 mg of enzyme protein. DPD activity was assayed spectrophotometrically by following the oxidation of added NADPH at 340

nm for 10 minutes as previously described (11). The blank reaction cell
contained all the reagents except the substrate. Results were expressed
as units/mg protein. One unit (U) of DPD activity is defined as nmoles
of NADPH oxidized per minute. Enzyme protein was assayed by the Lowry
method (12). 5-FU, uracil and thymine were used as substrates over concentrations from 0.5 to 100 μM. The reduction rate of 5-FU and uracil
by DPD was tested also in the presence of 1 μM uridine and 100 μM thymidine while reduction of thymine was tested in the presence of 0.1 μM
uridine and 10 μM thymidine. The possibility that uridine phosphorylase
activity was present as a contaminant in the DPD preparation was examined
by incubating uridine, DPD and NADPH. After 10 minutes the reaction was
terminated with perchloric acid and uracil was assayed by capillary gas
chromatography using a methodology previously described by this laboratory (8).

Km and Ki values were calculated according to the formulas:

$$Km = [S](\frac{Vmax}{V} - 1) \qquad Vapp = \frac{Vmax}{1 + \frac{[i]}{Ki}}$$ The data for the inhibition of 5-FU

reduction by uridine and thymidine were plotted graphically according to
the Lineweaver-Burk method. The statistically defined intercepts for
each line were determined using the Wilkinson regression analysis (13).

RESULTS

The spontaneous oxidation of NADPH over 10 minutes in the presence
of enzyme protein without substrate (blank) amounted to less than 6%
of the total NADPH in the incubation mixture. 5-FU, uracil and thymine
were all reduced to their dihydro derivatives by DPD in the presence of
NADPH. At saturating concentrations of each substrate, DPD showed the
highest specific activity for 5-FU (0.8 units/mg protein) whereas the
activities for uracil and thymine were 0.7 U/mg and 0.6 U/mg, respectively. The Km values were similar for uracil (2.3 μM) and thymine (2.2 μM)
while 5-FU had a Km of 3.5 μM. Reduction of 5-FU, uracil and thymine
was inhibited by the addition of uridine and thymidine as shown in
Table 1.

Table 1 Inhibition of DPD by Uridine and Thymidine

Substrate	Ki (μM) Uridine	Ki (μM) Thymidine
5-FU	0.7	24
Uracil	0.6	41
Thymine	0.2	9

Uridine was found to be a more potent inhibitor of DPD than thymidine.
Thymine reduction by DPD was more susceptible to inhibition by either
nucleoside than the reduction of 5-FU and uracil. Inhibition of DPD by
uridine and thymidine appears to occur in a noncompetitive manner(Fig.1).No
activity of uridine phosphorylase was detected in the DPD enzyme preparation used in this study.

DISCUSSION

The control mechanisms of pyrimidine base degradation are currently
unknown. It has been previously shown that uridine and thymidine can
inhibit the degradation of $2\text{-}^{14}C\text{-uracil}$ to $^{14}CO_2$ (14). The postulated

mechanism for this inhibition was the production of the competitive substrates, uracil and thymine arising from the degradation of uridine
and thymidine by nucleoside phosphorylase activity in the liver supernatant. A nucleoside phosphorylase contaminant in our enzyme preparation
could, if present, generate uracil and thymine from uridine and thymidine
resulting in a decrease in the inhibitor concentration and an increase
in the effective substrate concentration. However, no uridine phosphorylase activity was detected in the partially-purified hepatic DPD preparation used in our studies. Furthermore, the majority of the nucleoside
phosphorylase enzymes should precipitate in the 32% ammonium sulfate
fraction which was discarded during the DPD purification (15). Utilization of NADPH by uridine and thymidine is another possible explanation
for the inhibitory mechanism, however, such a reaction has not been demonstrated in tissues. It has been previously shown that the concurrent
administration of thymidine with 5-FU prolonged the plasma 5-FU half
life (16). We have shown that uridine is a more potent inhibitor of DPD
in vitro whereas thymidine is a much less potent inhibitor. In vivo
studies in humans have shown that administered isotopically labeled
uridine is mainly incorporated into nucleic acids and the elimination of
labeled CO_2 is very slow (more than 20 days). In contrast, labeled
uracil is largely degraded and rapidly eliminated (over a period of
hours) (17). It is possible that the effect of uridine on DPD plays a
role in regulating its incorporation into nucleic acids. It is also
possible that the administration of 5-FU with uridine may permit the
usage of smaller 5-FU doses, thereby avoiding central nervous system toxicity thought to be caused by fluorinated metabolites of 5-FU. Additional
detailed studies will be needed to characterize the theoretical bene-
ficial effects of uridine on 5-FU metabolism and toxicity.

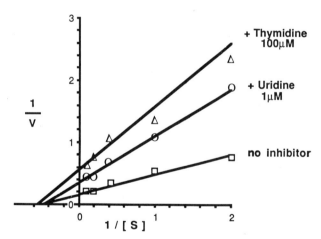

Figure 1. Lineweaver-Burk plot for the inhibition of 5-FU reduction
by uridine and thymidine. □-5-FU with no inhibitor, average of 15-19
experiments per point; O-5-FU + 1 μM uridine, average of 3-4 experiments per point; Δ-5-FU + 100 μM of thymidine, average 3-4 experiments per point. The intercepts for each line were determined using
the Wilkinson regression analysis (13).

REFERENCES

1. J.A. Ferdinandus, H.P. Morris, and G. Weber, Behavior of opposing
    pathways of thymidine utilization in differentiating, regenerating

and neoplastic liver, Canc Res 31:550-556 (1971).

2. G. Weber, S.F. Queener, and J.A. Ferdinandus, Control of gene expression on carbohydrate, pyrimidine and DNA metabolism, Advan Enzyme Regul 9:63-95 (1971).

3. N.K. Chaudhuri, K.L. Mukherjee, and C. Heidelberger, Studies on fluorinated pyrimidines. VII - the degradative pathway, Biochem Pharm, 1:328-341 (1958).

4. E. Matthes, D. Barwolff, and P. Langen, Inhibition by 6-aminothymine of the degradation of nucleosides (5-iododeoxyuridine, thymidine) and pyrimidine bases (5-iodouracil, uracil and 5-fluorouracil) in vivo, Acta Biol Med Germ 32:483-502 (1974).

5. G. Berglund, J. Greter, S. Lindstedt, G. Steen, J. Waldenstrom, and U. Wass, Urinary excretion of thymine and uracil in a two year old child with a malignant tumor of the brain, Clin Chem 25:1325-1328 (1979).

6. S.K. Wadman, F.A. Beemer, P.K. de Bree, et al, New defects of pyrimidine metabolism, in: "Purine Metabolism In Man". IV. Adv Exp Med Biol, C.H.M.M. De Bruyn, H.A. Simmonds, M.M. Muller, eds. 165A:109-114 (1984).

7. J.A.J.M. Bakkeren, R.A. De Abreu, R.C.A. Sengers, F.J.M. Gabreels, J.M. Maas, and W.O. Renier, Elevated urine, blood and cerebrospinal fluid levels of uracil and thymine in a child with dihydrothymine dehydrogenase deficiency, Clin Chim Acta 140:247-256 (1984).

8. M. Tuchman, J.S. Stoeckeler, D.T. Kiang, R.F. O'Dea, M.L. Ramnaraine, and B.L. Mirkin, Familial pyrimidinemia and pyrimidinuria associated with severe fluorouracil toxicity, N Engl J Med 313:245-249 (1985).

9. S.F. Queener, H.P. Morris, and G. Weber, Dihydrouracil dehydrogenase activity in normal differentiating and regenerating liver and in hepatomas, Canc Res 31:1004-1009 (1971).

10. K.L. Mukherjee, J. Boohar, D. Wentland, F.J. Ansfield, and C. Heidelberger, Studies on fluorinated pyrimidines. XVI Metabolism of 5-fluorouracil-2-^{14}C and 5-fluoro-2'deoxyuridine-2-^{14}C in cancer patients, Canc Res 23:49-66 (1963).

11. P. Fritzson, Properties and assay of dihydrouracil dehydrogenase of rat liver, J Biol Chem, 235:719-725 (1960).

12. O.H. Lowry, N.J. Rosebrough, A.L. Farr, and R.J. Randall, Protein measurement with the folin phenol reagent, J Biol Chem, 193:265-275 (1951).

13. G.N. Wilkinson, Statistical estimations in enzyme kinetics, Biochem J, 80:324-332 (1961).

14. H.W. Barrett, S.N. Munavalli, and P. Newmark, Synthetic pyrimidines as inhibitors of uracil and thymine degradation by rat liver supernatant, Biochim Biophys Acta, 91:199-204 (1964).

15. E.S. Canellakis, Pyrimidine catabolism II. Enzymatic pathways of uracil anabolism, J Biol Chem, 227:329-338 (1957).

16. T.M. Woodcock, D.S. Martin, L.A.M. Damin, N.E. Kemeny, and C.W. Young, Combination clinical trials with thymidine and fluorouracil: A phase I and clinical pharmacologic evaluation, Cancer 45:1135-1143 (1980).

17. S.M. Weissman, and L.M. Karon, The metabolism of isotopically labeled uracil and uridine in man, Metabolism, 12:60-67 (1963).

REGULATION OF ENZYMES BY LIGAND INDUCED CHANGE IN POLYMERIZATION

Thomas W. Traut, Nancy Cheng, and Margaret M. Matthews

Department of Biochemistry
University of North Carolina School of Medicine
Chapel Hill, North Carolina, USA

INTRODUCTION

Most enzymes form homopolymers, some of which are stable while others readily change their aggregation state by dissociation or association of subunits. This constitutes a mechanism for regulation if the different molecular weight species vary in their specific activity, and if the equilibrium between these species is readily perturbed by physiological effectors. Several enzymes in pyrimidine metabolism appear to be regulated by this mechanism: uridine kinase (EC 2.7.1.48); β-alanine synthase* (EC 3.5.1.6), the last enzyme in the catabolism of uracil; and the multifunctional protein UMP synthase which has 2 enzyme activities, orotate phosphoribosyltransferase (EC 2.4.2.10) and OMP decarboxylase (EC 4.1.1.23).

MATERIALS AND METHODS

The preparation of UMP synthase (1) and uridine kinase (2) from Ehrlich ascites cells have been described. β-alanine synthase was prepared from rat liver (3) using the fraction that precipitates between 40% and 50% of saturation with ammonium sulfate. The molecular size of enzymes was measured by various techniques: gel permeation chromatography using Ultrogel AcA 34 from LKB (UMP synthase, uridine kinase, and β-alanine synthase); equilibrium sedimentation with a Beckman airfuge (β-alanine synthase); and velocity sedimentation through 10-40% sucrose gradients (UMP synthase, uridine kinase, β-alanine synthase. In all molecular weight studies, the column or centrifuge tube contained varying concentrations of effectors, as indicated in the appropriate figures.

Enzyme assays were performed as previously described (1-3). Rapid initial kinetic studies with uridine kinase were done by a protocol similar to that used with UMP synthase (4). Standard reagents were

---

*This enzyme has been described under a variety of names: ureido-propionase, N-carbamoyl-β-alanine decarbamylase, NCβA amidohydrolase. Since β-alanine is important biosynthetically, and since the enzyme that synthesizes β-alanine appears to be allosterically regulated by its product, we feel that β-alanine synthase would be a better trivial name.

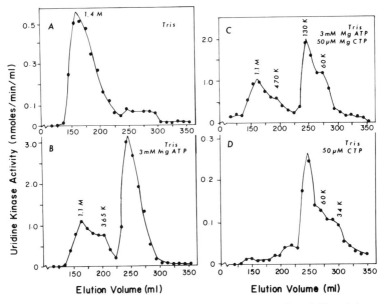

Figure 1. Various molecular weight species of uridine kinase
as a function of effectors. An Ultrogel AcA 34 column was
equilibrated with Tris buffer (50 mM, pH 8.1) plus effectors,
as shown. The enzyme sample was the 100,000 x g supernate
from an Ehrlich ascites cell homogenate.

obtained from Sigma Chemical Co., or from ICN Biomedicals. Radioactive 5-[^{14}C]-N-carbamoyl-β-alanine (NCβA) was synthesized from β-
alanine and [^{14}C] KCNO as described (3).

RESULTS

Uridine kinase has the fairly special attribute of existing as
a mixture of different aggregation states that only slowly equilibrate
with each other. These species therefore produce individual molecular
weight peaks by sedimentation studies, or by gel permeation chromatography (shown in Fig. 1); the subunit molecular weight by SDS gel
electrophoresis is 31,000 (2) and 34,000 by gel permeation chromatography.
The end product regulator, CTP, leads to the dissociation of larger polymers and to the appearance of dimers and monomers (Fig. 1). Since CTP
leads to dissociation of uridine kinase polymers, the activity of the
monomer was tested using rapid kinetic studies with enzyme preincubated
for 16 h with varying concentrations of CTP. Enzyme activity was measured
(at 4°) by the addition of substrates, and the appearance of product was
determined as shown in Fig. 2. The lag in the progress curve for enzyme
preincubated with higher concentrations of CTP (at which the enzyme should
exist predominantly as monomers) suggests that the monomer is inactive.
The addition of the substrate, ATP, leads to the formation of active
polymers (Fig. 1B, 1C) and a linear progress curve. It takes at least
30 s for this polymerization to occur after the addition of ATP.

Uridine kinase that has not been dissociated with CTP has no
lag in the progress curve, suggesting that polymers containing 4
or more subunits are intrinsically active. We have insufficient data
to decide if the dimeric species is catalytically active. The profile
in Fig. 1D might appear to suggest that both monomers and dimers are

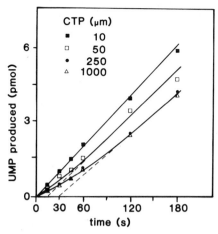

Figure 2. Initial rate studies with uridine kinase that has been pre-incubated with the inhibitor CTP. The enzyme sample had been purified 370 fold.

active. However, for the experiments of Fig. 1, activity is measured by taking an aliquot of an enzyme sample eluting as a monomer (for example) and then adding it to a reaction assay containing substrates. As shown by Figs. 1B and 1C the presence of ATP should cause aggregation of subunits to form polymeric species, and thereby result in measurable catalytic activity.

Is there an allosteric regulatory site to mediate dissociation/ inhibition by CTP? Since ATP, the normal phosphate donor, and the inhibitor CTP are sufficiently similar, both might exert their effects at the catalytic site. However, the experiments of Table 1 show that the catalytic site readily uses ribonucleotides and deoxyribonucleotides as phosphate donors, and while purine nucleotides are probably used *in vivo*, deoxypyrimidine nucleotides could be phosphate donors. Only with

Table 1

Nucleotides as Phosphate Donors for Uridine Kinase

Phosphate Donor	Vmax (μmole/min/mg)	Km (μM)	Vmax/Km (1/min/mg)
ATP	157	450	0.35
GTP	8.5	82	0.10
dATP	68	500	0.14
dGTP	19	87	0.22
dUTP	53	235	0.23
dTTP	23	200	0.12
dCTP	7.2	250	0.03
UTP	0	-	-
CTP	0	-	-

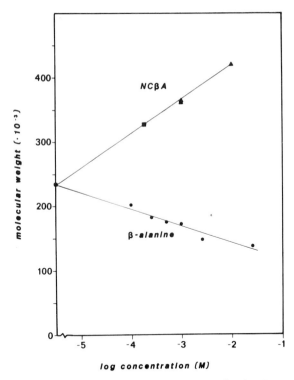

Figure 3. Change in the molecular size of β-alanine synthase
as a function of effectors. N-carbamoyl-β-alanine (NCβA) is
the substrate; β-alanine, the product, is also an inhibitor.
Experiments were done using an Airfuge (●), sucrose gradients
(▲), or column chromatography (■).

UTP and CTP is there no enzymatic activity. It is unlikely that
these two compounds would not bind productively at the catalytic site.
Since these are also the only two nucleotides that dissociate uridine
kinase, it seems reasonable that a regulatory site exists at which UTP
or CTP can bind to produce dissociation and render the enzyme inactive.

For β-alanine synthase the native enzyme in the absence of effectors has a molecular weight of 235,000 Da and is probably a tetram-
er. In the presence of effectors, the enzyme associates towards oct-
amers, or dissociates to dimers (Fig 3). Since a single molecular weight
peak is always observed, mixtures of different aggregation species must be
in rapid equilibrium. As shown in Fig. 3, the substrate, NCβA, promotes
association, while the product β-alanine is a negative effector that
produces dissociation and inhibition.

β-alanine is biosynthetically important for the synthesis of carn-
osine, a dipeptide (β-alanyl-histidine) that is abundant in muscle and
brain (5). The regulation of β-alanine synthase by its product is suggested by the observation that hyper β-alaninemia is associated with severe
neuropathology in humans (5). This enzyme can also decarbamoylate N-car-
bamoyl-β-aminoisobutyrate in the catabolism of thymine, to form an alternate product β-amino-isobutyrate (2-methyl β-alanine). The latter compound, although a normal physiological product, has no effect on the
polymer size of β-alanine synthase. Since β-aminoisobutyrate has no
known biosynthetic function, this suggests that the regulatory effects
produced by β-alanine 1) are physiologically significant, and 2) occur at

Table 2

Ligand Binding Constants with β-alanine Synthase

Ligand	Structure	Ki† (mM)	Kd* (mM)
NC-β-alanine	H₂N-C(=O)-NH-CH₂CH₂C(=O)-O⁻	0.091 ± 0.015	<0.065
ammonium	NH₄⁺	0.5	not effector
Urea	H₂N-C(=O)-NH₂	1.0	not effector
γ-aminobutyrate	H₃N⁺-CH₂CH₂CH₂-C(=O)-O⁻	1.8	not effector
β-aminoisobutyrate (product)	H₃N⁺-CH₂CH(CH₃)-C(=O)-O⁻	2.5	not effector
β-alanine	H₃N⁺-CH₂CH₂-C(=O)-O⁻	7.36 ± 0.11	3.5 ± 1.2

*Measured in polymerization studies (Fig.3).
†Km for NCβA, Ki for other compounds.

a site other than the catalytic site, where all substrates and products
can bind. In Table 2 the various equilibrium and dissociation constants
are compared. Compounds that are better inhibitors than β-alanine are not
effectors that produce changes in the enzyme's size. This suggests that
all compounds bind at the catalytic site, but that there is probably a
special regulatory site for β-alanine. At this point our data are insufficient to determine if a separate regulatory site exists for NCβA.

For UMP synthase, all the experimental data have been published
(1,4,5,6) and will only be summarized here, for comparison. In the
absence of effectors UMP synthase exists only as monomers; positive
effectors, of which OMP is physiologically most significant, produce
dimerization and a conformational change to an active species (1,4).
Orotate phosphoribosyltransferase activity is observed in monomers and
dimers; OMP decarboxylase activity appears only in the dimer. Since
this multifunctional protein catalyzes 2 sequential reactions
(orotate → OMP → UMP) it was suggested that the rationale for an active
dimer was the ability to channel OMP, the intermediate, and thereby
spare it from catabolic degradation by pyrimidine nucleotidases (5,6).

The physical and kinetic aspects of the different enzyme systems
are summarized and compared in Fig.4. For UMP synthase, species larger
than dimer are seldom observed and may not be significant. With β-alanine
synthase we have not observed subunits in the absence of denaturing agents.
For all three of these systems, positive effectors lead to larger polymers,
while negative effectors lead to dissociation. While this pattern is
common, there are enzymes where dissociation leads to the active species;
examples include amidophosphoribosyltransferase (8) and carbamoylphosphate
synthetase (9).

For UMP synthase and uridine kinase regulation is unimodal, being
positive for UMP synthase and negative for uridine kinase. More complex
bimodal regulation is observed with β-alanine synthase. Since dissociation and inhibition are both produced in the presence of β-alanine,

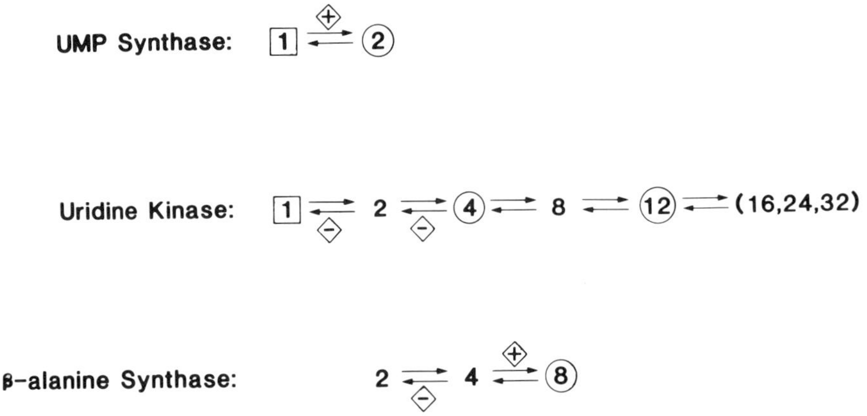

UMP Synthase: $\boxed{1} \overset{\oplus}{\underset{}{\rightleftharpoons}} ⟨2⟩$

Uridine Kinase: $\boxed{1} \underset{\ominus}{\rightleftharpoons} 2 \underset{\ominus}{\rightleftharpoons} ⟨4⟩ \rightleftharpoons 8 \rightleftharpoons ⟨12⟩ \rightleftharpoons (16,24,32)$

β–alanine Synthase: $2 \underset{\ominus}{\rightleftharpoons} 4 \overset{\oplus}{\rightleftharpoons} ⟨8⟩$

Figure 4. Regulation of enzyme activity by ligand induced change in polymerization. Numbers represent the quantity of subunits in a particular species; □ = inactive form of enzyme, 0 = active form of enzyme. The action of positive and negative effectors are included.

we assume that the dimer is inactive, but this has yet to be confirmed with rapid kinetic studies.

It may also be useful to distinguish between primary and secondary-effectors. Primary effectors bind at a special allosteric regulatory site (each enzyme probably has at least one regulatory site) and produce a change in polymerization and activity. Secondary effectors bind at the same sites, and act as blockers to displace the primary effectors. As examples: CTP is a primary negative effector on uridine kinase. ATP competes with CTP, and prevents dissociation/ inactivation. With UMP synthase OMP is the primary positive effector; P_i can displace OMP and prevent activation (1).

REFERENCES

1. Traut, T.W., Payne, R.C. and Jones, M.E. (1980) *Biochemistry* 19, 6062-6068.

2. Payne, R.C., Cheng, N., and Traut, T.W. (1985) *J. Biol. Chem.* 260 (in press).

3. Traut, T.W. and Loechel, S. (1984) *Biochemistry* 23, 2533-2539.

4. Traut, T.W. and Payne, R.C. (1980) *Biochemistry* 19, 6068-6074.

5. Scriver, C.S., Perry, T.L., and Nutzenadel, W. (1983) in *The Metabolic Basis of Inherited Diseases,* (Frederickson, D., Wyngaarden, J., and Stanbury, J., eds.) 4th ed., pp. 570-585.

6. Traut, T.W. (1980) *Arch. Biochem. Biophys* 200, 590-594.

7. Traut, T.W. (1982) *Trends Biochem. Sci.* 7, 255-257.

8. Holmes, E.W. (1981) *Adv. Enz. Regul.* 19, 215-231.

9. Lusty, C.J. (1981) *Biochemistry* 20, 3665-3675.

SYNTHESIS AND TURNOVER OF PURINE NUCLEOSIDE PHOSPHORYLASE IN HUMAN LYMPHOCYTES

Floyd F. Snyder, Kuldeep Neote, Eddie Kwan and Ellen R. Mably

Departments of Pediatrics and Medical Biochemistry
Faculty of Medicine, University of Calgary, Calgary
Alberta, Canada

The inherited deficiency of purine nucleoside phosphorylase results in severe T cell immunodeficiency disease[1]. We have studied the synthesis of purine nucleoside phosphorylase during phytohumagglutinin induced T cell transformation and have examined the turnover of purine nucleoside phosphorylase in the human lymphoblast WI-L2.

We have used a polyclonal antibody to purine nucleoside phosphorylase in the synthetic and turnover studies. Purine nucleoside phosphorylase was purified to homogeneity from human erythrocytes according to the method of Zannis et.al.[2], for the preparation of a rabbit antiserum. WI-L2 lymphoblast proteins were typically labelled by 16 hour exposure to [^{3}H] leucine. Purine nucleoside phosphorylase antiserum was added to labelled lysates and the antibody-antigen complex was precipitated with S. aureus cells, washed, and followed by SDS-polyacrylamide gel electrophoresis. Fluorography of a dried gel (Fig. 1) shows the total labelled lysate (lane 1), immunoprecipitate with control antiserum (lane 2) and immunoprecipitate with purine nucleoside phosphorylase antiserum (lane 3 and 4). The 32,500 molecular weight subunit of purine nucleoside phosphorylase is identified as a discrete band.

Turnover of purine nucleoside phosphorylase in WI-L2 lymphoblasts

The turnover of purine nucleoside phosphorylase was studied by labelling cells with [^{3}H] leucine as described for Figure 1 and after resuspension of labelled cells in non-radioactive medium, the decay of label in total protein and purine nucleoside phosphorylase was studied over a 15 hour time course. The half lives for total protein and purine nucleoside phosphorylase were essentially the same, 14.5 and 14.1 hours respectively. Lymphoblasts cultured in the presence of the non-toxic purine nucleoside phosphorylase substrate, inosine, 0.5 mM, had a half-life for purine nucleoside phosphorylase of 11.2 hours. Studies in erythrocyte lysates have shown that inosine can stabilize purine nucleoside phosphorylase in thermal inactivation studies[3]. In intact cells, however, inosine can also reduce the intracellular phosphate concentration through phosphorolysis to hypoxanthine and ribose-1-phosphate[4]. We have previously observed phosphate to increase the thermal stability of murine purine nucleoside phosphorylase[5].

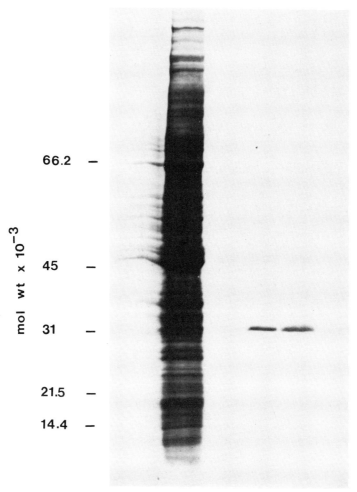

Figure 1. Fluorography of purine nucleoside phosphorylase immunoprecipitates from WI-L2 cells. Lymphoblast proteins were labelled during 16 hour culture with [4,5-³H] leucine, 0.2 mCi/15 ml. Cell lysates were electrophoresed on a 10% SDS-polyacrylamide gel and examined by fluorography after drying the gel. Samples were 20 μg of total protein (lane 1), immunoprecipitate of 200 μg protein with control serum (lane 2) and purine nucleoside phosphorylase antiserum (lanes 3 and 4).

We examined the effect of phosphate and formycin B, a C-C glycoside and inhibitor, on purine nucleoside phosphorylase stability. Thermal inactivation of purine nucleoside phosphorylase from WI-L2 cell lysates in the presence of 100 mM phosphate gave a half-life of 36.7 min versus 4.5 min in the absence of phosphate at 65°C (Fig. 2). In the absence of phosphate and presence of 5 mM formycin B the half-life was 20.9 min. Thus both phosphate and to a lesser degree the nucleoside substrate analog stabilize the enzyme. Thus it seems likely that the decreased half-life of purine nucleoside phosphorylase caused by inosine in the intact cell was due to a reduction in the intracellular phosphate concentration.

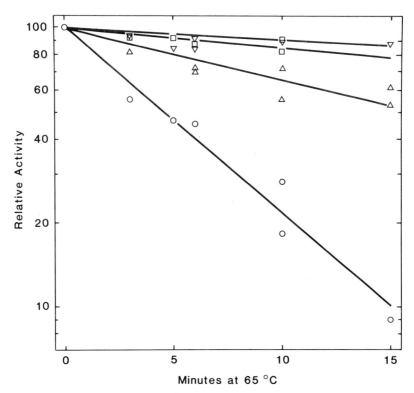

Figure 2. Stability of purine nucleoside phosphorylase. Lymphoblast lysates were incubated for various times at 65°C in the presence of 100 mM sodium phosphate and 5 mM formycin B (∇), 100 mM sodium phosphate ($\square$), 5 mM formycin B ($\triangle$) or no addition (0) at 0.25 mg/ml protein in 0.1 M Tris buffer pH 7.4. Lysates were diluted 5-fold and assayed in 100 mM phosphate, pH 7.4 and 0.4 mM [8-^{14}C] inosine.

Synthesis of purine nucleoside phosphorylase during T cell transformation.

The inherited deficiency of purine nucleoside phosphorylase appears to adversely affect some later stages of T cell differentiation. Proliferating T cells in in vitro studies are considerably more sensitive to the toxic effects of deoxyguanosine than non-dividing T cells[6,7]. We have examined the synthesis of purine nucleoside phosphorylase during mitogen induced transformation of human T cells.

T lymphocytes were isolated from peripheral blood by rosetting with sheep red blood cells, stimulated in culture with phytohemagglutinin and pulsed at various times with [^{35}S] methionine. The rate of total protein and purine nucleoside phosphorylase synthesis was measured. The results of four such experiments are given in Table 1 where the 0 hour rate was set at 1.0 and other time points were expressed relative to this value. The rate of total protein synthesis increased 10-fold over the time course with a 2 to 5-fold increase in the first 14 hours. There was an overall 30-fold increase in the rate of purine nucleoside phosphorylase synthesis, with greater than 10-fold increase during the first 14 hours. These experiments indicate there is some increased synthesis of purine nucleoside phosphorylase relative to total protein synthesis. In addition there is some preferential increase in the synthesis of purine nucleoside phosphorylase during the first 14 hours of T cell transformation.

Table 1   Purine Nucleoside Phosphorylase
Synthesis in T Lymphocytes

| Hours after Stimulation | Relative Protein Synthesis | |
	Total Protein	Purine Nucleoside Phosphorylase
0	1.0	1.0
14	3.4 ± 2.2	11.1 ± 6.2
22	5.1 ± 2.4	16.8 ± 7.7
40	9.0 ± 1.7	29.9 ± 4.8
62	10.6 ± 0.4	28.0 ± 4.6

At various times, phytohemagglutinin stimulated T lymphocyte cultures, 1×10^6 cells/ml, were pulsed with [^{35}S] methionine, 0.2 mCi per 12 ml culture. Cell lysates were immunoprecipitated with purine nucleoside phosphorylase antibody and processed as described in Fig. 1. The 32,500 molecular weight species was excised from dried gels and counted. Total labelled protein was obtained by collection of trichloroacetic acid precipitates on glass-fiber filters. Maximum labelling occurred between 40 and 60 hours and was typically 6-10 $\times 10^6$ dpm/mg protein for total protein and 1-3 $\times 10^5$ dpm/mg protein for purine nucleoside phosphorylase. The results of four experiments are expressed as the rate of protein synthesis relative to the zero hour culture.

Acknowledgements: This work was supported by the Medical Research Council of Canada, grant MT-6376.

REFERENCES

1. E.R. Giblett, A.J. Ammann, D.W. Wara, R. Sandman, and L.K. Diamond, Nucleoside phosphorylase deficiency in a child with severely defective T cell immunity and normal B-cell immunity, Lancet, 1: 1010 (1975).
2. V. Zannis, D. Doyle, and D.W. Martin Jr., Purification and characterization of human erythrocyte purine nucleoside phosphorylase and its subunits, J. Biol. Chem., 253: 504 (1978).
3. I.H. Fox, C.M. Andres, E.W. Gelfand, and D. Biggar, Purine nucleoside phosphorylase deficiency: Altered kinetic properties of a mutant enzyme, Science, 197: 1084 (1977).
4. G. Planet and I.H. Fox, Inhibition of phosphoribosylpyrophosphate synthesis by purine nucleosides in human erythrocytes, J. Biol. Chem., 251: 5839 (1976).
5. F.F. Snyder, F.G. Biddle, T. Lukey, and M.J. Sparling, Genetic variability of purine nucleoside phosphorylase activity in the mouse: Relationship to Np-1 and Np-2, Biochem. Genet., 21: 323 (1983).
6. D.A. Carson, J. Kaye, and J.E. Seegmiller, Lymphospecific toxicity in adenosine deaminase deficiency and purine nucleoside phosphorylase deficiency: Possible role of nucleoside kinase(s), Proc. Natl. Acad. Sci., 74: 5677 (1977).
7. A. Cohen, J.W.W. Lee, H.M. Dosch, and E.W. Gelfand, The expression of deoxyguanosine toxicity in T lymphocytes at different stages of maturation, J. Immunol., 125: 1578 (1980).

REVERSIBLE INTERCONVERSION BETWEEN SULFO AND

DESULFO XANTHINE DEHYDROGENASE

Takeshi Nishino

Department of Biochemistry
Yokohama City University School of Medicine
Minami-ku, Yokohama 232, Japan

In mammals oxidation of hypoxanthine to xanthine and xanthine to uric acid are steps in purine catabolism prior to excretion, uric acid being the end product in primates. These steps are catalyzed by xanthine dehydrogenase (oxidase), a complex molybdoflavoprotein.

It has been revealed in the last decade that the active and inactive forms of xanthine dehydrogenase exist in purified enzyme preparations (1-4): this is somewhat unique in that the inactive form lacks a specific sulfur atom at the molybdenum center of the functional enzyme (Mo=S), while other complements of molybdenum, FAD and iron sulfur prosthetic groups are contained fully in these two forms. The desulfo form has been considered to be merely preparation or storage artifacts (5). However almost half of the enzyme molecule in supernatant solution of rat liver homogenate was found to be desulfo form (6). Furthermore the molecular activity of chicken liver xanthine dehydrogenase changes with protein content in diet (7). These suggest that a mechanism exist in vivo whereby interconversion between sulfo and desulfo form may be effected.

These two forms can be separated by an affinity chromatography (8) and reversible interconversion between two forms can be done by the system of rhodanese, $Na_2S_2O_3$ and a sulfhydryl reagent (9). However, rhodanese may not be considered to catalyze sulfuration and desulfuration of xanthine dehydrogenase physiologically, because xanthine dehydrogenase is revealed to exist in cytosol while rhodanese is confined to mitochondria (10, 11). We have found that the alternate enzyme system of mercaptopyruvate sulfur transferase (3-mercaptopyruvate: cyanide sulfurtransferase, EC2, 8, 1, 2) can also reactivate the desulfo enzyme. Fairly high activity of this enzyme is found in 100,000 x g supernatant solution of rat liver homogenate. By the reaction system consisting of the purified mercapto-pyruvate sulfur transferase, mercaptopyruvate and sulfhydryl reagent, desulfo enzyme was activated and also sulfo enzyme was inactivated (Fig. 1). This suggests that this reaction is also reversible.

The reversible interconversion between sulfo and desulfo forms by these systems can be explained by the involvement of a common reaction intermediate. As we mentioned previously(8)possible reactive sulfur species such as S^{--} or $R-S-S^-$ may be involved in the reactions.

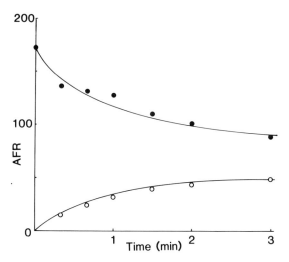

Fig. 1. Activation and inactivation of xanthine oxidase
by the system of mercaptopyruvate sulfurtransferase.
    Sulfo enzyme (AFR 172, ●) or desulfo enzyme (AFR 0, ○)
was incubated with 30 mM mercaptopyruvate, 5 mM dithiothreitol
and purified rat liver mercaptopyruvate sulfurtransferase in
0.1 M pyrophosphate buffer (pH 8.5) at 37 ° C.

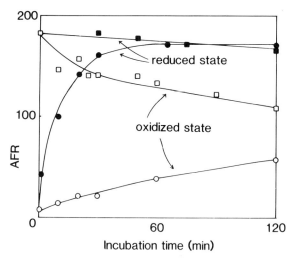

Fig. 2. Activation of desulfo enzyme and inactivation
of sulfo enzyme by sodium sulfide.
    Sulfo enzyme (AFR 175, □) or desulfo enzyme (AFR 7.4,
○) was incubated with 10 mM sodium sulfide at 37° C.
Sulfo enzyme (■) or desulfo enzyme (●) was previously
reduced with 5 mM dithionite under anaerobic conditions
before addition with sodium sulfide.

Rhodanese system:

$$Na_2S_2O_3 \quad + \quad RSH \quad \rightarrow \quad Na_2S_2O_3 \quad + \quad RSSH$$

$$RSSH \quad + \quad RSH \quad \rightarrow \quad RSSR \quad + \quad H_2S$$

Mercaptopyruvate sulfur transferase system:

$$HSCH_2\ COCOOH + RSH \quad \rightarrow \quad CH_3COCOOH + RSSH$$

$$RSSH \quad\quad + RSH \quad \rightarrow \quad RSSR \quad\quad + H_2S$$

Zn^{++} ion, which should precipitate sulfide as ZnS, was found to inhibit these reactivation while not inhibit mercaptopyruvate sulfurtransferase itself. This confirms that sulfide ion formed in these systems is incorporated into desulfo xanthine oxidase. On the contrary, in the desulfuration reaction of sulfo enzyme by these systems sulfide at the molybdenum of xanthine oxidase is assumed to be released by the attack of sulfide anion. In fact sodium sulfide was found to inactivate sulfo enzyme. This inactivation did not occur when the enzyme had previously been reduced with dithionite (Fig. 2).

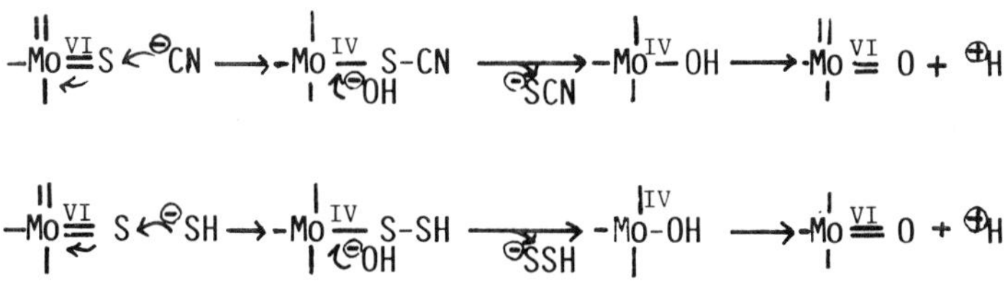

Fig. 3:   Possible mechanisms for the release of Mo=S complexes by cyanide and sulfide.

A mechanism of desulfuration of sulfo enzyme by HS^- is considered to be similar to that of desulfuration by CN^- (Fig. 3), because inactivation of sulfo enzyme by both HS^- and CN^- were protected when xanthine oxidase had been previously reduced with dithionite. Furthermore, both HS^- and CN^- are classified as soft base (12) and calculated E_{HOMO} value of HS^- is very close to that of CN^- (13).

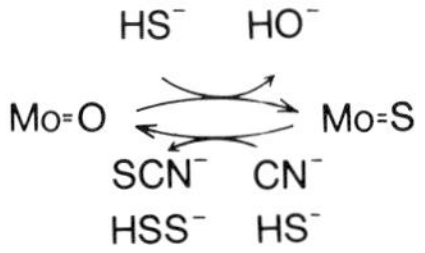

Physiologically, sulfide anion formed by an enzyme system may be involved in the interconversion between sulfo and desulfo xanthine dehydrogenase so that two forms exist in equilibrium. This may explain why desulfo enzyme exists in vivo. It is possible that reductants such as xanthine, hypoxanthine and NADH shift this equilibrium far to the formation of sulfo form.

Acknowledgements: The author would like to express his graditude to Prof.
K. Tsushima, Yokohama City University, for his helpful discussions and to
Mrs. C. Usami for her skillful technical assistance. This work was
supported in part by a Grant-in-Aid for Scientific Research from the
Japanese Ministry of Education, Science, and Culture (No. 17709 and No.
458081).

REFERENCES

1. V. Massey and D. Edmondson, On the mechanism of inactivation of
    xanthine oxidase by cyanide. J. Biol. Chem., 245, 6595-6598, (1970)
2. W.F. Cleere, and M. P. Coughlan, Turkey liver xanthine dehydrogenase;
    Reactivation of cyanide inactivated enzyme by sulphide and by selenide.
    Biochem. J. 143, 331-340, (1974)
3. T. Nishino, Purification of hepatic xanthine dehydrogenase from chicken
    fed a high protein diet. Biochim. Biophys. Acta, 341, 93-98, (1974)
4. T. Nishino, R. Itoh and K. Tsushima, Studies on chicken liver xanthine
    dehydrogenase with reference to the problem of non equivalence of FAD
    moieties. Biochim. Biophys. Acta, 403, 17-22, (1975)
5. R. C. Bray, Molybdenum hydroxylases, in "The enzymes", 3rd ed. 12, 299-
    419, Academic Press, New York (1975)
6. T. Ikegami and T. Nishino, The existence of desulfo-xanthine
    dehydrogenase in rat liver, in "Flavins and flavoproteins", 703-706,
    Walter de Gruyter, Berlin (1984)
7. T. Nishino, T. Nishino and K. Tsushima, On the nature of the
    cyanolysable sulfur in xanthine oxidase. in "Flavins and
    flavoproteins", Elsevier North Holland, Inc., pp. 792-795, 1982
8. T. Nishino, T. Nishino and K. Tsushima, Purification of highly active
    milk xanthine oxidase by affinity chromatography on Sepharose 4B/folate
    gel., FEBS Lett., 131, 369-372, (1981)
9. T. Nishino, C. Usami and K. Tsushima, Reversible interconver-conversion
    between sulfo and desulfo xanthine oxidase in a system containing
    rhodanese, thiosulfate, and sulfhydryl reagent. Proc. Natle. Acad. Sci.
    USA, 80, 1826-1829,(1982)
10. S. Ludewig and A. Chanutin, Distribution of enzymes in the livers of
    control and X-irradiated rats. Arch. Biochem., 29, 441-445, (1950)
11. C. De Duve, B.C. Pressman, R. Gianetto, R. Watteaux and F. Applemans,
    Tissue fractionation studies. Biochem. J. 60, 604-617, (1955)
12. R.G.Pearson,Hard and soft acids and bases.,J.Amer.Chem.Soc.,85, 3533-
    3539, (1963)
13. G. Klopman, Chemical reactivity and the concept of charge- and frontier-
    controlled reactions. J. Amer. Chem. Soc.,90, 223, (1968)

MAMMALIAN MUTANTS GENETICALLY ALTERED IN CTP SYNTHETASE ACTIVITY

Bruce Aronow and Buddy Ullman

Department of Biochemistry
University of Kentucky Medical Center
Lexington, Kentucky 40536-0084

SUMMARY

From wildtype mouse lymphoma cells, a clone (FURT-1A), was isolated by virtue of its resistance to 1μM 5-fluorouracil. In comparative growth rate experiments, FURT-1A cells were also less sensitive than parental cells to the growth inhibitory effects of thymidine, deoxyguanosine, 5-fluorouridine, and arabinosylcytosine. The altered growth sensitivity of FURT-1A cells to cytotoxic nucleosides was directly related to their decreased ability to accumulate the corresponding triphosphate from exogenous nucleoside. FURT-1A cells contained elevated cytidylate nucleotide pools which prevented normal growth sensitivity and interfered with the salvage of nucleosides. Metabolic flux experiments with [^{3}H]-uridine in situ indicated that FURT-1A cells had a 2-fold enhanced rate of conversion of UTP to CTP. Kinetic analyses indicated that the CTP synthetase activity in extracts of FURT-1A cells was refractory to inhibition by CTP. The genetic loss of normal allosteric inhibition of the CTP synthetase activity in FURT-1A cells could account for the unusual phenotypic properties of these cells.

INTRODUCTION

The enzyme CTP synthetase catalyzes the de novo synthesis of CTP from UTP. CTP synthetase is subject to feedback inhibition by its product. Buttin and co-workers have characterized mutant Chinese hamster fibroblast cells possessing a CTP synthetase activity which resists complete inhibition by CTP and consequently possess elevated CTP and dCTP pools (1). Moreover, Meuth and coworkers have described Chinese hamster V79 cells with a similar biochemical phenotype that have 5- to 300-fold elevated rates of spontaneous mutation to several independent genetic markers (2).

In our studies, we have used somatic cell genetic approaches to characterize mutant S49 cells with altered CTP synthetase activity. These cells had deranged cytidylate pools and were profoundly altered in their abilities to metabolize exogenous purine and pyrimidine nucleosides. The characterization of the mutant S49 cells with an altered CTP synthetase activity has important implications for understanding the regulation of purine and pyrimidine metabolism and specifically for determining the extent to which CTP synthetase activity is regulated in exponentially growing cells. These biochemical genetic studies may contribute to the rational design of chemotherapeutic protocols which utilize cytotoxic nucleoside or nucleobase analogs.

MATERIALS AND METHODS

Cell Culture The origins of wildtype (3) and FURT-1A (4) cells have been described. The abilities of S49 cell lines to survive and multiply in cell culture medium containing growth inhibitory and cytotoxic agents were deterined as described previously (4)

Measurement of Intracellular Nucleotide and Extracellular Nucleoside Concentrations Intracellular nucleotide pool levels were measured as described by Aronow et al. (4). Nucleoside triphosphates were separated on a Partisil-SAX high pressure column at a flow rate of 1.2 ml/min with ammonium phosphate buffers at pH = 3.4; isocratic elution at 0.3M (with respect to phosphate) for 12 ml, followed by a 24 ml linear gradient to 0.4M, and finally a 24 ml isocratic elution with the latter buffer.

Measurements of CTP Synthetase Activity In Situ and In Vitro The rate of flux of uridine nucleotides into cytidine nucleotides was determined on exponentially growing wildtype and FURT-1A cells as described previously in detail (4) The in situ rate of CTP synthetase activity was estimated by equation (1) derived by Maybaum and coworkers, taking into account the amount of radiolabel incorporated into CTP and the time average specific radioactivity of the UTP precursor pool (5). The in vitro CTP synthetase activity measurements were performed as previously described (4). The CTP and UTP were separated by the high pressure liquid chromatographic system described above. When radiolabel was employed the UTP and CTP peaks were collected individually and analyzed for radioactivity by liquid scintillation spectrometry. When CTP synthetase activity was measured in the absence of radiolabel, the CTP product was determined by cutting and weighing the appropriate peak at 280nm from the chart paper and comparing with that of a peak obtained with a known CTP standard.

RESULTS

Growth Rate Determination As the FURT-1A clone was isolated initially in semi-solid selective medium, we compared the sensitivities of wild type and FURT-1A cells growing in suspension culture to increasing concentrations of 5-fluorouracil. The results (Table I) indicated that FURT-1A cells were approximately 5-fold less sensitive to the growth inhibitory effects of 5-fluorouracil. The effective concentration of 5-fluorouracil which inhibited growth by 50%, i.e., the E.C.$_{50}$ value, was 0.45μM for wildtype cells and 2.3μM for FURT-1A cells. Unexpectedly, FURT-1A cells were also much less sensitive to the cytotoxic effects of high concentrations of a spectrum of cytotoxic nucleosides, Table I. The E.C.$_{50}$ values for other nucleosides on wildtype and mutant cells are listed in Table I. FURT-1A cells were as sensitive as parental cells to growth inhibition by adenosine-EHNA and deoxyadenosine-EHNA, but were less sensitive to thymidine, deoxyguanosine, 5-fluorouridine, 6-azauridine, 5-fluorodeoxyuridine and arabinosylcytosine (Table I).

264

Table I.

Comparison of growth sensitivities of wildtype and mutant cells to cytotoxic nucleoside analogs.

The effective concentration which inhibited growth by 50% (E.C.$_{50}$ value) is reported for each agent. Each growth rate determination was performed at least twice with each analog.

	Cell line	
Cytotoxic Agent	Wildtype	FURT-1A
	E.C.$_{50}$ Value	
	(μM)	
5-fluorouracil	0.45	2.3
5-fluorouridine	0.007	0.033
5-fluorodeoxyuridine	0.0005	0.0015
deoxyguanosine	30	120
thymidine	25	150
arabinosylcytosine	0.03	0.15
6-azauridine	0.4	3.0
arabinosyladenine	2.5	2.5
deoxyadenosine	7	4.5
adenosine	3.6	5

Quantitation of 5-Fluorouracil Metabolites   To ascertain the mechanism of 5-fluorouracil growth resistance in FURT-1A cells, the abilities of wildtype and FURT-1A cells to convert exogenous [6-^{3}H]fluorouracil into intracellular metabolites were examined.   The FURT-1A cells metabolized 5-fluorouracil into virtually all the same intracellular metabolites as did wildtype cells but at lower rates (data not shown).

Uptake of Deoxyribonucleosides   In order to elicit their growth inhibitory effects, most nucleosides require metabolism to their respective nucleoside triphosphate.   Since FURT-1A cells were relatively insensitive to growth inhibition by either thymidine or deoxyguanosine, but were sensitive to deoxyadenosine-EHNA, we examined the ability of these mutant cells to convert the three deoxyribonucleosides to the triphosphate level.   The results indicated that wildtype cells had a greater capacity to convert thymidine to TTP and deoxyguanosine to dGTP than did FURT-1A cells.   The capacity to accumulate dATP from exogenous deoxyadenosine-EHNA, however, was similar for both cell lines.   These data suggested that the relative growth resistance of FURT-1A cells to cytotoxic nucleosides or nucleoside analogs was at least partly due to their inability to efficiently convert these agents to phosphorylated derivatives.

Nucleotide Pool Determinations   Two potential mechanisms by which cells can become growth resistant to a multiplicity of cytotoxic nucleosides are by a deficiency in nucleoside transport or through an alteration in endogenous nucleotide concentrations.   Clonal isolates of nucleoside transport-deficient S49 cells have been shown to be cross-resistant to adenosine and deoxyadenosine and cannot incorporate even small amounts of exogenously added nucleosides into nucleotides (6).   We therefore

Table II.

Nucleotide pool levels in wildtype and mutant S49 cells.

Nucleotide pool sizes were determined by high performance liquid chromatography. Deoxyribonucleoside triphosphate pools were quantitated subsequent to periodate oxidation of the ribonucleotides.

Nucleotide	Cell Line	
	Wildtype	FURT-1A
	$(nmol/10^9$ cells)	
CTP	168	729
UTP	297	246
ATP	2070	2088
GTP	372	318
CDP	10.8	29.4
ADP	189.4	207
CMP	14.1	20.3
dCTP	15.8	57
dTTP	14.5	13.1
dATP	19.1	15.3
dGTP	9.7	8.8

measured the levels of nucleotides in wildtype and FURT-1A cells (Table II). FURT-1A cells possessed a 4-fold elevated CTP pool and a 2.5-fold elevated dCTP pool as compared to wildtype cells (Table II). CDP and CMP concentrations was also increased about 3.5-fold and 1.5-fold, respectively, over levels found in wildtype cells. Conversely, uridine triphosphate pools were lower in FURT-1A cells than in wildtype cells, while purine nucleotide concentrations were similar in both cell types (Table II).

Rate of Synthesis of CTP in Situ   To ascertain the mechanism by which CTP levels were elevated in FURT-1A cells, the rates of UTP conversion to CTP were determined in situ by the time average specific activity method of Maybaum and coworkers (12). Although the rate of incorporation of radiolabelled uridine into UTP pools was 4- to 5-fold faster in wildtype cells than in FURT-1A cells, a substantially greater proportion of the total [^{3}H]UTP is converted to [^{3}H]CTP by the mutant cells (Figure 1). Using time average specific activities, the calculated rates of CTP synthetase activity in situ for the wildtype and FURT-1A cell lines were 190±25 and 380±30 nmol/h/10^9 cells, respectively. If the CTP pool elevation in the mutant cells was due to a decreased rate of CTP degradation, the rate of conversion of UTP to CTP would have been identical for both cell lines.

CTP Synthetase Assays   The elevated concentration of intracellular CTP and the increased flux of UTP into CTP suggested that the CTP synthetase activity in FURT-1A cells was genetically altered. Therefore, we compared the CTP synthetase catalytic activities in vitro in gel-sieved extracts prepared from wildtype and FURT-1A cells. When measured in the absence of CTP, the specific catalytic

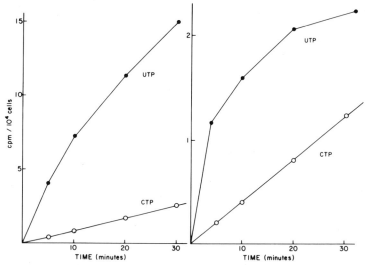

Figure 1. Comparison of the abilities of intact wildtype and FURT-1A
cells to convert [^{3}H]uridine to [^{3}H]CTP. The abilities of wildtype
(left panel) and FURT-1A cells (right panel) to metabolize exogenous 1μM
[^{3}H]uridine (1 Ci/mmol) were measured by the time average specific
activity method of Maybaum and coworkers (5). Cells were harvested at
various time intervals, and the radioactivity in and the magnitude of the
UTP (●) and CTP (o) pools quantitated by liquid scintillation
spectrometry and high performance liquid chromatography.

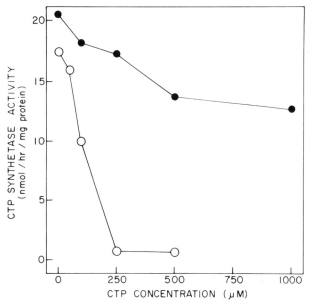

Figure 2. Effects of CTP on the CTP synthetase activities from wildtype
and FURT-1A cells. The abilities of extracts of wildtype (o) and FURT-1A
(●) cells to convert UTP to CTP were performed as described in Materials
and Methods. The CTP synthetase activities were always performed under
conditions which were linear with both time and protein. To demonstrate
sensitivity of CTP synthetase activity to varying CTP concentrations
[^{14}C]UTP was employed as substrate.

activities of the CTP synthetase activities were similar for both cell lines (Figure 2). Since CTP synthetase is normally negatively regulated by the CTP product (7), we also examined the relative sensitivities of the enzymes from wildtype and FURT-1A cell extracts to feedback inhibition by CTP. As shown in Figure 2, concentrations of CTP as low as 0.25mM virtually abolished the conversion of UTP to CTP by the wildtype enzyme. The enzyme from mutant cells, however, was inhibited by only 30% in the presence of CTP concentrations as high as 1.0mM. Thus, resistance of the CTP synthetase activity to feedback inhibition by CTP in FURT-1A cells could account for the increased levels of cytidylate nucleotides in FURT-1A cells.

DISCUSSION

The genetic lesion in the FURT-1A cell line was an alteration in an allele coding for CTP synthetase activity which caused the enzyme to lose its normal sensitivity to inhibition by CTP. This genetic alteration caused the CTP synthetase to overproduce CTP in mutant cells. The deranged nucleotide pool levels in FURT-1A cells markedly altered both their ability to metabolize nucleosides and 5-fluorouracil and their sensitivity to growth inhibition by these cytotoxic compounds. The elevated CTP and dCTP pools inhibited cellular nucleoside kinases, activated cytosolic nucleotide dephosphorylating enzymes, and ameliorated the normal in situ sensitivity of the CDP reductase activity of ribonucleotide reductase to allosteric effectors (4). The elevated CTP pools and the nearly complete loss of allosteric inhibition of the CTP synthetase activity in FURT-1A cells have important implications concerning the regulation of pyrimidine nucleotide synthesis in mammalian cells. First, since the Vmax activities of the CTP synthetase activities in extracts prepared from wildtype and FURT-1A cells were identical, synthesis of the mammalian enzyme is not repressed by elevated CTP concentrations. Second, since the metabolic flux through CTP synthetase was 2-fold greater in FURT-1 cells than wildtype cells, the wildtype enzyme must be normally inhibited by about 50%. Thus, the adequate and balanced supply of intracellular cytidylate nucleotides synthesized by CTP synthetase can be finely regulated by small changes in CTP levels.

Clearly, the rate of de novo pyrimidine biosynthesis in mammalian cells, unlike that in E. coli (8), is not regulated by CTP inhibition of the aspartate transcarbamylase activity, since CTP pools in FURT-1A cells were elevated 4-5 fold, whereas UTP levels were decreased only 20%. Since the total levels of pyrimidine nucleotides were elevated in mutant cells implying an increased rate of their synthesis, elevated CTP levels cannot have a negative regulatory effect on de novo pyrimidine biosynthesis. Rather, the slightly depleted UTP concentrations must release the pyrimidine pathway from normal inhibition. UTP is a known inhibitor of the carbamyl phosphate synthetase activity in mammalian cells (9). Thus, the rate of de novo synthesis of pyrimidines in situ appears to be tightly regulated by uridine nucleotides at the level of carbamyl phosphate synthetase and by CTP at the level of CTP synthetase.
Finally, the unusual phenotypic alterations caused by expanded CTP pools may have important clinical implications. Elevated CTP pools and increased CTP synthetase activity have been associated with rapid tumor growth and cell proliferation (10). Tumor cell populations with a genetic alteration in CTP synthetase activity and increased CTP levels might be generated in chemotherapeutic protocols which employ agents such as 5-fluorouracil, arabinosylcytosine, or any nucleoside to which FURT-1A cells are growth insensitive. To eradicate tumor cells with a FURT-1A phenotype, chemotherapeutic strategies might include a requirement for

efficient nucleoside salvage or an effective UDP reduction component unencumbered by excess CDP concentrations. Identification of the biochemical nature of the genetic lesion in FURT-1A cells and the elucidation of the mechanisms by which expanded cytidylate pools contribute to the mutant growth phenotype may provide a biochemical rationale for the design of chemotherapeutic regimens designed to eliminate specific drug-resistant malignant cells.

ACKNOWLEDGEMENTS

The work described in this manuscript was supported by grant R01 CA32580 from the National Institutes of Health. B.U. is a recipient of a Research Career Development Award from the National Institutes of Health.

REFERENCES

1   de Saint Vincent, R.B., Dechamps, M., and Buttin, G. (1980) J. Biol. Chem. 255, 162-167.
2. Meuth, M., L'Heureux-Huard, N., and Trudel, M. (1979) Proc. Natl. Acad. Sci. USA 76, 6505-6509.
3. Horibata, K. and Harris, A.W. (1970) Exp. Cell Res. 60, 61-77.
4. Aronow, B., Watts, T., Lasseter, J., Washtien, W., and Ullman, B. (1984) J. Biol. Chem. 259, 9035-9043.
5. Maybaum, J., Ullman, B., Mandel, H.G., Day, J.L., and Sadee, W. (1980) Cancer Res. 40, 4209-4215.
6. Cohen, A., Ullman, B., and Martin, D.W. Jr. (1979) J. Biol. Chem. 254, 112-116.
7. Long, C.W. and Pardee, A.B. (1967) J. Biol. Chem. 242, 4715-4721.
8. Gerhart, J.C. and Pardee, A.B. (1962) J. Biol. Chem. 237, 891-896.
9. Mori, M. and Tatibana, M. (1978) Methods in Enzymol. 51, 111-121.
10. Weber, G., Shiotani, T., Kizaki, H., Tzeng, D., Williams, J.C., and Gladstone, N. (1978) Adv. Enz. Regul. 16, 3-19.

INCREASED INOSINATE DEHYDROGENASE ACTIVITY IN

MYCOPHENOLIC ACID RESISTANT NEUROBLASTOMA CELLS

Stephen D. Hodges, Ernest Fung, C.C. Lin and Floyd F. Snyder

Departments of Pediatrics and Medical Biochemistry
Faculty of Medicine, University of Calgary
Calgary, Alberta, Canada, T2N 4N1

By incremental increase in the concentrations of inhibitors of essential biosynthetic pathways, a number of specific genes have been amplified in cultured mammalian cells. We have attempted to obtain an amplified phenotype by the progressive selection of resistance to mycophenolic acid, a potent inhibitor of the final two enzymes in guanine ribonuleotide synthesis, IMP dehydrogenase[1,2] and GMP synthetase[2]. In previous studies the inhibition of IMP dehydrogenase by mycophenolic acid has been shown to result in marked overproduction of purines in both human lymphoblasts[3,4] and mouse lymphoma cells[5]. Mycophenolic acid induced changes in ribonucleotide[3,5,6] and deoxyribonucleotide[6] pools are well characterized. A mouse lymphoma line resistant to 20 µM mycophenolic acid has been obtained by sequential mutagenesis[5] and single step mutagen induced resistant hamster V79 cells have also been obtained[7].

In this report we describe the isolation and characterization of neuroblastoma clones selected for resistance to 200 µM mycophenolic acid without mutagenesis. Mouse neuroblastoma cells were cultured as previously described[8] with at least 10% incremental increases in mycophenolic acid concentration with each passage. The acquired and progressive resistance of these cells is illustrated in Figure 1 with wild type cells showing 50% growth inhibition at 0.1 µM mycophenolic acid. At passage 109, these cells were clones in the presence of 200 µM mycophenolic acid and the growth properties of clones Myco-0.2A and Myco-0.2B are shown in Figure 2. Approximately 500 µM mycophenolic acid was required to produce 50% growth inhibition in the resistant clones, representing a 5000-fold decrease in sensitivity from wild type cells.

The result of the progressive selection to mycophenolic acid resistance on IMP dehydrogenase activity is shown in Table 1. The final increase in specific activity of the 200 µM resistant clones was 10 or 11-fold greater than that of wild type cells. In kinetic studies of IMP dehydrogenase the apparent k_m for IMP was found to be 10 µM for wild type, versus 85 µM for Myco-0.2A and 99 µM for Myco-0.2B. In addition the apparent k_i for mycophenolic acid was markedly increased from 6×10^{-9} M for wild type to 5×10^{-6} M for Myco-0.2A and 1×10^{-5} M for Myco-0.2B. These findings are in several respects remarkably similar to those of Ullman, who selected for 20 µM mycophenolic acid resistance after mutagenesis, and obtained cells with 10-15-fold increased IMP dehydrogenase activity and a 10-fold increase in the k_m for IMP[5].

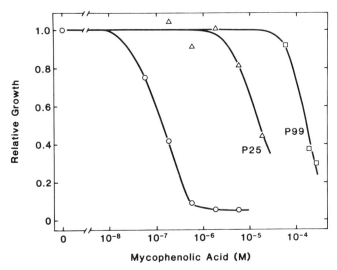

Figure 1. Sensitivity of Wild Type (0) and mycophenolic acid resistant
neuroblastoma cells at passage 25 ( ) and passage 99 ( )
to growth inhibition by mycophenolic acid.

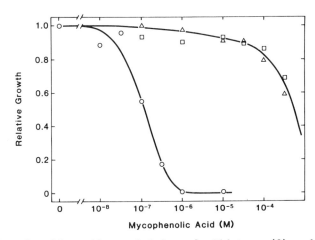

Figure 2. Mycophenolic acid sensitivity of wild type (0) and mycophenolic
acid resistant clones, Myco-0.2A ( ) and Myco-0.2B ( ).

The increased k_m for IMP and increased k_i for mycophenolic acid
suggest a structural mutation involving IMP dehydrogenase as opposed to
a mechanism of gene amplification accounting for the increased activity.
We therefore examined the stability of the mycophenolic acid resistant
phenotype by growing the cells in the absence of drug and monitoring the
IMP dehydrogenase activity.   The results of these studies show a progressive decline in IMP dehydrogenase activity to 25-35% of the initial
level by 17 to 18 days (Table 2).   The phenotype of the Myco-0.2A and
Myco-0.2B cells is therefore unstable with respect to IMP dehydrogenase
activity and shows a progressive normalization toward the wild type
level of activity.

The possibility that the guanine ribonucleotide pool might increase
upon removal of drug from the mycophenolic acid resistant clones was

Table 1. IMP dehydrogenase activity during selection of
mycophenolic acid resistant neuroblasts

	IMP Dehydrogenase Activity (nmole/min/mg protein)	Fold Increase
Wild Type	6.5 ± 0.6	1.0
Passage 93	22.9	3.5
Passage 97	26.6	4.1
Passage 102	56.3	8.7
Myco-0.2A	73.2	11.3
Myco-0.2B	68.5	10.5

IMP dehydrogenase activity was assayed spectrophotometrically
according to the method of Anderson and Sartorelli[7]. Myco-0.2A
and Myco-0.2B are two clones selected at passage 109 for
resistance to 200 μM mycophenolic acid.

Table 2. Instability of Mycophenolic Acid Resistant Phenotype

DAYS AFTER REMOVAL OF MYCOPHENOLIC ACID	IMP DEHYDROGENATE ACTIVITY Myco-0.2A Myco-0.2B (nmole/min/mg protein)	
0	73.2	68.5
4/5	52.5	38.1
10/11	29.2	27.8
17/18	24.6	17.8

investigated. The ratio of the guanine to adenine ribonucleotide pools
were the same for wild type cells in the absence of mycophenolic acid,
0.26 ± 0.03 and the resistant clones in the presence of 200 μM mycophenolic
acid, 0.25 ± 0.03. Removal of mycophenolic acid from the resistant clones
showed no change in this ratio at a variety of times. GTP is required
for CTP synthesis and we also examined the ratio of cytidine to uridine
ribonucleotide. This ratio was unchanged amongst wild type, 0.33 ± 0.09,
and the resistant clones in the presence, 0.38 ± 0.09 and absence 0.32 ±
0.05 of 200 μM mycophenolic acid.

Examination of the nucleotide pools did reveal that removal of
mycophenolic acid from the resistant clones caused a 4-fold increase in
the XMP concentration (Table 3). The increase in XMP persisted for days
which is consistent with the longevity of the increased IMP dehydrogenase
activity and suggests that guanylate synthetase may be rate limiting for
GMP synthesis from IMP under these conditions. The initial concentration
of XMP in resistant clones exposed to mycophenolic acid was marginally
less than that of wild type cells.

These studies have described the selection and characterization of
an unstable phenotype in mouse neuroblastoma cells resistant to 200 μM
mycophenolic acid, having 10-fold increased IMP dehydrogenase activity
and apparently reduced affinity for both IMP and mycophenolic acid.

Acknowledgments: This work was supported by the Alberta Cancer
Board Grant H-105.

Table 3. Accumulation of Xanthosine-5'-monophosphate upon
withdrawal of Mycophenolic acid

DAYS AFTER REMOVAL OF MYCOPHENOLIC ACID		XMP (pmole/10^6 cell)	FOLD INCREASE
Myco-0.2A	0	51 $\pm$ 28	
	1	187	3.7
	2	201	3.9
	6	199	3.9
Myco-0.2B	0	55 $\pm$ 5	
	0.25	228	4.1
	1	162	3.0
	7	137	2.5
Wild Type		84 $\pm$ 27	

Nucleotide concentrations were determined by liquid chromatography
using a partisil-1025-SAX column and eluting with sodium acetate and
potassium phosphate buffers.

REFERENCES

1. T.J. Franklin and J.M. Cook, The inhibition of nucleic acid synthesis by mycophenolic acid, Biochem. J., 113: 515 (1969).
2. M.J. Sweeney, D.H. Hoffman, and M.A. Esterman, Metabolism and biochemistry of mycophenolic acid, Cancer Res., 32: 1803 (1972).
3. R.C. Willis and J.E. Seegmiller, Increases in purine excretion and rate of synthesis by drugs inhibiting IMP dehydrogenase or adenylosuccinate synthetase activities, in: "Purine Metabolism in Man-III, Biochemical, Immunological and Cancer Research", A. Rapado, R.W.E. Watts and C.H.M.M. de Bruijn, ed., Plenum, New York, 122B: 237 (1980).
4. F.F. Snyder, R.J. Trafzer, M.S. Hershfield and J.E. Seegmiller, Elucidation of aberrant purine metabolism. Application to hypoxanthine-guanine phosphoribosyltransferase- and adenosine kinase-deficient mutants and IMP dehydrogenase and adenosine deaminase inhibited human lymphoblasts. Biochim. Biophys. Acta, 609: 492 (1980).
5. B. Ullman, Characterization of mutant murine lymphoma cells with altered inosinate dehydrogenase activities, J. Biol. Chem., 258: 523 (1983).
6. J.K. Lowe, L. Brox and J.F. Henderson, Consequences of inhibition of guanine nucleotide synthesis by mycophenolic acid and virazole, Cancer Res., 37: 736 (1977).
7. E. Huberman, C.K. McKeown, and J. Friedman, Mutagen-induced resistance to mycophenolic acid in hamster cells can be associated with increased inosine 5'-monophosphate dehydrogenase activity, Proc. Natl. Acad. Sci. U.S.A., 78: 3151 (1981).
8. F.F. Snyder, M.K. Cruikshank, and J.E. Seegmiller, A comparison of purine metabolism and nucleotide pools in normal and hypoxanthine-guanine phosphoribosyltransferase-deficient neuroblastoma cells. Biochim. Biophys. Acta, 543: 556 (1978).
9. J.H. Anderson and A.C. Sartorelli, Inosinic acid dehydrogenase of Sarcoma 180 cells, J. Biol. Chem., 243: 4762 (1968).

MURINE LYMPHOCYTES AND LYMPHOCYTE CELL LINES SECRETE ADENOSINE DEAMINASE

Phyllis R. Strauss

Department of Biology
Northeastern University
Boston, MA 02115

INTRODUCTION

Adenosine deaminase (E.C. 3. 5. 4. 4, ADA), first enzyme in the degradative pathway of adenosine (Ado) and 2'-deoxyadenosine (dAdo), converts Ado to inosine (Ino) and dAdo to dIno. Purine nucleoside phosphorylase (E.C. 2. 4. 2. 1, PNP), the second enzyme in the pathway converts Ino to hypoxanthine (HX). Cells from patients lacking ADA or PNP tend to accumulate dATP or dGTP respectively. The patients usually present with the clinical picture of acute immunodeficiency possibly due to a block in lymphocyte differentiation caused by the effects of the specific enzyme deficiency (1,2). It has been proposed that increased levels of purine deoxynucleoside triphosphates allosterically inhibit ribonucleotide reductase so that in dividing cells other deoxynucleotide levels are altered and ultimately DNA synthesis is impaired (3-5). Evidence has also been presented that accumulation of dATP may act at a point before the onset of DNA synthesis (6,7) and, indeed, dAdo has been shown to induce strand breaks in the DNA of resting cells and to deplete cellular NAD and ATP.

Lymphocytes are extremely sensitive to the presence of extra-cellular nucleosides. For example, some lymphocyte lines such as the murine T lymphoma S49 are so sensitive to exogenous thymidine that there is no concentration which blocks cell division that does not also result in cell death (8). It is particularly striking that plasma levels of certain nucleosides such as Ado are extremely low (<1 μM) in healthy individuals (9). Presumably, low extracellular levels are required for the survival of lymphocytes during transit through the vascular system. The origin of plasma ADA has never been determined. In this paper we demonstrate that ADA is selectively secreted by murine spleen cells and by murine T lymphoma cells (S49). The appearance of activity is time and cell concentration dependent and cannot be accounted for by dead or leaky cells. Between 7-10% of the initial intracellular activity is secreted during an 8 hr period.

* Supported by NSF PCM 77-25434, ONR 14-82-K-0283 and funds from Northeastern University; P.R.S. is the recipient of Career Development Award CA 00460 awarded by the National Cancer Institute, DHEW.

METHODS AND RESULTS
Conditioned medium contains ADA
Lymphocytes were obtained from a variety of sources including
wild type S49. 1 thymoma cells, the transport deficient AE - 1 variant,
and an adenosine kinase deficient variant, as well as splenocytes
from concanavalin A stimulated and unstimulated (outbred CD - 1 and
inbred AKR mice). Cells (100 µl containing 10^6 cells in the presence
or absence of deoxycoformycin (DCF), the tight binding inhibitor of
ADA (10, 11), were incubated at 37° C for 15 minutes in phosphate
buffered saline containing 5 mM glucose and 0.1% BSA and separated from
the conditioned incubation medium (CM) by the oil microfuge method (12).
CM was tested for its ability to metabolize ^{3}H-Ado (Table 1). Note that
^{3}H-Ado was added to CM <u>after</u> cells had been removed from the medium. In
this series of experiments the reaction was stopped 1 hr after addition
of isotope in order to maximize product formation. Note also that the
concentration of Ado at which ADA was measured was adjusted initially to
1 µM which is well below the Km level, since plasma Ado levels under
normal physiological conditions are <1 µM. Control experiments where ADA
activity was determined under standard conditions (0. 1 mM Ado range)
demonstrated that the Km of the measured activity was similar to
published values (13). CM from all the cells examined showed ability to
metabolize ^{3}H-Ado, although CM from cultured S49 cells (wild type and
nucleoside transport deficient) contained the greatest activity.

The products of the metabolized ^{3}H-Ado were determined by means of
ascending paper chromatography in 3 different systems (14). About 50%️ of
the radiolabeled material was recovered as Ino and 30% was recovered as
HX. CM from cells pre-incubated for 15 minutes with 6 µM DCF was also
tested for the presence of Ado metabolizing activity and the nature of
the tritiated products determined. DCF prevented the metabolism of Ado
by 90%. Because of the nature of the products and because of the DCF
inhibition, we inferred that CM contained ADA and to a lesser degree PNP.
That DCF did not entirely inhibit ADA activity is in line with reports by
Tedde et al. (15) of some DCF insensitive ADA in mouse tissues.

Secretion of ADA is time dependent
The appearance of Ado metabolizing activity in CM as a function of
time was examined by preparing CM from wild type S49 cells (1 x 10^6/ml)
incubated for periods up to 31 hrs. CM was then assayed for ADA
activity. Figures 1a and 1b represent the results of two typical
experiments among the eight experiments performed. For this and
subsequent experiments the reaction was stopped 15 min after Ado was
added to CM. This interval was chosen on the basis of time dependence
studies. The time of incubation of CM prepared from 1 x 10^7 CD-1 cells
in 1.0 ml with ^{3}H-Ado was examined in detail between 0.5 and 60 min. The
percentage remaining Ado decreased exponentially with time, $t_{1/2}$ being 5
min and t_{90} being 30 min. Therefore, unless stated otherwise, Ado
metabolizing activity was assayed over a 15 min interval. ^{3}H-Ado
metabolizing activity continued to appear in CM for up to 24 hrs. On
several occasions, cells which had been incubated for 20-23 hrs were
washed 3X and resuspended at 1 x 10^6/ml. CM was collected at various
times thereafter and tested for the ability to metabolize ^{3}H-Ado. The
appearance of ^{3}H-Ado metabolizing activity was again a function of time
at a rate similar to that of the original culture.

Extracellular ADA activity cannot be accounted for by dying cells
To investigate whether Ado metabolizing activity was released by
dying or leaky cells, the LDH and ADA content of S49 cells was determined
over a broad range of cell concentrations and an LDH:ADA ratio computed
(Figure 2). Cells were resuspended at the desired concentration in 1 ml

Table I. Metabolism of ^{3}H-Adenosine by Conditioned Medium
(percentage of total measured radioactivity)

Source of Cells	Ado		Ino		HX+Ad		Ad		Phosphorylated Compounds	
	-DCF	+DCF	-DCF	+DCF	-DCF	+DCF	-DCF	+DCF	-DCF	+DCF
CD-1, unstim	25	82	51	2	34	7	ND	ND	4	4
CDC-1, stim	21	83	46	12	23	5	<1	<1	2	5
AKR, unstim	11	86	55	30	2	1	2	2	2	
S49.WT	2	78	35	2	51	4	5	4	2	9
S49 AE-1	3	89	56	2	36	3	2	3	1	2
No Cells	92	ND	ND	ND	ND	ND	ND	ND	ND	ND

Cells (10^6 in 100 µl phosphate buffered saline containing 0.1% BSA and 5 mM glucose) from the indicated source were incubated at 37°C for 15 min. Wild type S49.1 thymoma cells were obtained from the Salk Collection. The nucleoside transport deficient mutant AE-1 was the gift of Dr. Buddy Ullman (21,22). The cell lines were cultured as described (22,23). Nonadherent spleen cells from outbred CD-1 mice (Charles River Laboratories, Wilmington, MA) or from AKR mice (Jackson Laboratories, Bar Harbor, ME) were prepared as described previously (12). Cells from concanavalin A stimulated mice were obtained in a similar fashion 24-48 hours after IV injection with 250 µg Con A/animal (12,16,17). These were the same conditions used in earlier transport studies (12,16,17). Cells were mixed with 200 µl of the same medium containing 300 pmoles Ado. This concentration of Ado was chosen because plasma Ado levels are <1 µM. After 4 secs, 200 µl was withdrawn and spun through a layer of silicone oil. The medium remaining above the oil was removed, mixed with 2 µl ^{3}H-Ado (1 µCi/ml final concentration; repurified as described in refs 12,14) and incubated for 60 min at RT. Ten or 20 µl aliquotes were analyzed by ascending paper chromatography in 3 solvent systems. Solvent System I consisted of t-butanol:methyl ethyl ketone:NH_4OH:H_2O, 4:3:1:2 (14). Solvent System II consisted of n-butanol:glacial acetic acid:H_2O, 2:1:1, and Solvent System III consisted of n-butanol:formic acid:H_2O, 77:10:13. When the effect of DCF was determined, the initial incubation containing cells included the drug at 6 µM. Standards of Ado, Ino, HX, AMP, ADP and ATP were run on the same paper. One cm wide strips in the lane of the unkown were cut out and the amount of radioactive material determined by scintillation spectrometry. The percentages are averages of duplicate experiments.

hypotonic phosphate buffer (10 mM PO_4, pH 7.4) and lysed by freezing and thawing three times in dry ice-ethanol. LDH and ADA activities were determined as described in the Legend to Figures 1a and 2.

At each cell concentration between 8 x 10^4 cells/ml and 8 x 10^5 cells, it was possible to compare ADA and LDH activities directly. Therefore, in other experiments, if the LDH in CM was taken as an indication of cell leakage or lysis, one could predict the amount of ADA that should have been released by dead cells into CM.

Figures 1a and 1b present both the ADA activity and the LDH values found in CM as a function of time in 2 separate experiments. The ^{3}H-Ado metabolizing activity predicted on the basis of LDH values is also given. Figure 1b illustrates an experiment where LDH levels remained low, while Figure 1a illustrates an experiment where LDH levels increased slightly.

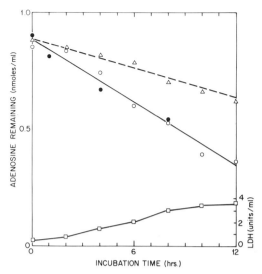

Figure 1a. Secretion of adenosine deaminase by wild type S49.1 cells. Cells were washed 3X and resuspended at 10^6 cells/ml in fresh medium. At the indicated time interval, cells were separated from conditioned medium (CM) which was then tested for ADA and LDH (25) activities. All incubations and washes were carried out in the absence of serum. Short incubations (up to 2 hrs) were carried out in phosphate buffered saline containing 5 mM glucose and 0.1% bovine serum albumin (BSA) in a Gilson shaking water bath at $37^\circ C$. Longer incubations (1–24 hrs) were performed in the CO_2 incubator by distributing 2 ml cell suspension in 35 mm Costar plastic dishes at $37^\circ C$ in DMEM supplemented with glutamine, antibiotics, and 0.1% BSA. In initial studies, cells were separated from CM by the oil-microfuge technique (12). Later studies separated cells from CM by pelleting in the microfuge without an intervening layer of silicone oil.

ADA was measured by observing the disappearance of ^{3}H–Ado or the appearance of HX+Ino as determined by ascending paper chromatography in System I (see Table Legend). Ado was added to conditioned medium (CM) to a final concentration of 1 μM, and ^{3}H–Ado (Schwartz–Mann Inc., Div. Becton Dickinson, Orangeburg, NY) was added to a final concentration of 1 μCi/ml. The reaction was terminated after 15 min. The amount of ^{3}H–Ado metabolized was calculated from the percentage of radiolabel appearing as Ado metabolites (HX + Ino + Ad) or the percentage of labeled ^{3}H–Ado which remained as the starting compound and from the fact that all incubations were carried out in 1 ml initially containing 1 nmole Ado. Control incubation in the absence of cells was carried out with each experiment. Routinely 88–92% of the control radioactivity was recovered as Ado. DCF was the kind gift of Dr. Ram Agarwal.

In this experiment S49.1 cells were incubated for intervals up to 31 hrs. During the first 12 hrs the incubation medium was removed at the specified intervals and tested for ADA (0) and LDH (□) activities. The amount of Ado remaining which would be expected if the source of ADA were lysed cells is shown as open triangles (△) (see Figure 2 and text). "Adenosine remaining" is a function of the ability of ADA in the medium to metabolize Ado (See text). Thus, the less Ado remaining at the end of the assay period, the more ADA present. In the same experiment the amount of Ado metabolites formed (HX + Ino) was also measured. In general, the amount of Ado metabolites (nmoles) = 0.92 – (Ado remaining, nmoles). After incubation for 24 hrs, the cells were washed 3X and resuspended. Secretion of ADA was followed for an additional 8 hrs (●).

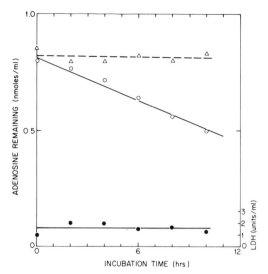

Figure 1b. Secretion of adenosine deaminase. An experiment similar to the one presented in Figure 1a is shown. In this experiment LDH values (●) were low and remained constant during the incubation period. Nevertheless, more ADA (0) is seen in the medium than can be accounted for on the basis of dead or leaky cells (△).

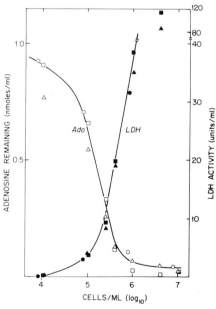

Figure 2. Adenosine deaminase and lactic dehydrogenase in lysates of S49.1 cells. One ml of cells at the indicated concentration was lysed by freezing and thawing in hypotonic phosphate and tested for ADA as described above and LDH (25). LDH-closed symbols; ADA-open symbols. Different shaped symbols refer to three independent experiments.

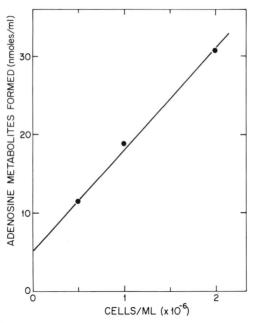

Figure 3. Adenosine deaminase secretion depends on cell number. Two ml cell suspension at 0.5×10^6/ml, 1.0×10^6/ml or 2.0×10^6/ml was plated in 35 mm Costar dishes and incubated for 6 hrs. Since S49 cells tend to adhere weakly to a plastic surface in the absence of serum, CM was removed gently by Pasteur pipet and spun in the microfuge to pellet cells. One ml H_2O was added to the cell pellet and then transferred to the original dish to lyse adherent cells. Lysates and CM were tested at the start and finish of the experiment for protein (BioRad Laboratories, Richmond, CA), with BSA as standard, and analyzed for LDH and ADA activity.

In both cases, progressively more ADA activity was observed than could be accounted for by dead cells. Therefore we conclude that ADA was secreted under these conditions.

Extracellular ADA activity depends on cell concentration

Cells (0.5×10^6/ml, 1.0×10^6/ml, and 2.0×10^6/ml) were incubated for 6 hrs in serum-free DME containing 0.1% BSA and separated from CM, which was assayed for ADA (Figure 3) and LDH as described above. The amount of metabolites formed by CM was clearly a function of cell number. In a second experiment, 7.2 nmoles, 17.0 nmoles, and 27.6 nmoles Ado metabolites were formed by 0.5×10^6 cells/ml, 1.0×10^6 cells/ml, and 2.0×10^6 cells/ml respectively. Under these conditions S49 cells continue to divide. Nevertheless the intracellular ADA activity per 100 µg protein remained constant or increased slightly, implying that ADA was continuously made.

DISCUSSION

In this work we have measured ADA at non saturating concentrations of Ado. Our rationale involved the fact that plasma levels of Ado are <1 µM and that intracellular concentrations of Ado are also extremely low (9). Moreover, the high affinity transport system for Ado in murine lymphocytes

and S49 cells is well below 10 μM (16-18, unpublished data). That the
activity was attributable to ADA was confirmed by the inhibition by DCF
and also the distribution of products in the presence and absence of DCF.
Since the appearance of HX was also observed when cells were incubated in
phosphate buffered saline, purine nucleoside phosphorylase (PNP) was
probably also secreted. It should be noted that when incubation was
carried out in DMEM, no HX was observed.

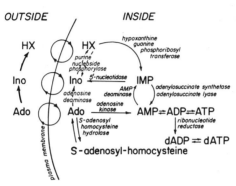

Figure 4. Major metabolic pathways of adenosine metabolism and their
location. Outside-extracellular events; Inside-intracellular events.

The presence of ADA in extracellular medium cannot be accounted for
by lysed cells for several reasons. First, LDH levels are low and do not
change very much during the course of the experiment. Second, the LDH/ADA
ratio found in cell lysates allows one to calculate the degradation of
^{3}H-Ado which would be expected on the basis of lysed cells in the
incubation medium. The observed Ado degradation is greater than the ADA
activity calculated on the basis of LDH levels. From the secretory data
and the data describing intracellular levels of ADA in 2 x 10^{6} cells
(Figure 2), it is possible to estimate the percentage of initial
intracellular ADA which is secreted as about 7-10% after 8 hrs.

Several authors (3,19-20) note the presence of ADA in the
extracellular fluid of different cell types. Since they did not examine
whether the source was dead or leaky cells, it is difficult to interpret
their results. Figure 4 summarizes the data presented here. Namely, ADA
can be added to the list of lymphocyte secretory products and an external
function for ADA may be sought to explain its role in immunodeficiency
disease.

REFERENCES

1.   Giblett, E., Chen, S.-H., and Osborne, W.R. In: Contemp. Hematol. 2.
     Ed R. Silber, A.S. Gordon, J. LoBue and F.M. Muggia, Plenum Med.
     Book Co., N.Y. (1981).
2.   Martin, D.W. Jr. and Gelfand, E.W. Ann. Rev. Biochem. 50:845 (1981).

3. Seegmiller, J.E., Watanabe, T., and Schreier, M.H. In: <u>Purine and Pyrimidine Metabolism</u>, Ciba Foundation Symposium 48 (new series), Excerpta Medica, Amsterdam (1977).

4. Hunting, D. and Henderson, J.F. CRC Crit. Rev. Biochem. 13:325 (1982).

5. Cohen, A., Hirschhorn, R., Horowitz, S.D., Rubinstein, A., Polmar, S.H., Hong, R., and Martin, D.W., Jr. Proc. Natl. Acad. Sci. 75:472 (1978).

6. Uberti, J., Lightbody, J.J., and Johnson, R.M. J. Immunol. 123:189 (1979).

7. Bagnara, A.S. and Hershfield, M.S. Proc. Natl. Acad. Sci. (USA) 79:2673 (1982).

8. Dorney, E. and Strauss, P.R. J. Cell Biol. 83:13a (1979).

9. Arch, J.R.S. and Newsholme, E.A. Essays Biochem. 14:82 (1978).

10. Agarwal, R.P., Spector, T., and Parks, R.E., Jr. Biochem. Pharmacol. 26:359 (1977).

11. Suhadolnik, R.J. <u>Nucleosides as Biological Probes</u>. John Wiley & Sons, Inc., N.Y. (1979).

12. Strauss, P.R., Sheehan, J.M., and Kashket, E.R. J. Exp. Med. 144:1009 (1976).

13. Peters, G.J., Oosterhof, A. and Veerkamp, J.H. Int. J. Biochem. 13:445 (1981).

14. Fink, K., Cline, R.E., and Fink, R.M. Analyt. Chem. 35:389 (1963).

15. Tedde, A., Bates, M.E., Schonberg, R. and Trotta, P.P. Cancer Res. 39:3044 (1979).

16. Strauss, P.R., Sheehan, J.M., and Kashket, E.R. J. Immunol. 118:1328 (1977).

17. Strauss, P.R., Sheehan, J.M., and Taylor, J. Can. J. Biochem. 58:1405 (1980).

18. Paterson, A.R.P., Harley, E.R. and Cass, C.E. In: <u>Methods in Pharmacol.</u> 6. Plenum Press, N.Y. (1985).

19. Snyder, F.F., Mendelsohn, J., and Seegmiller, J.E. J. Clin. Invest. 58:654 (1976).

20. Baer, H.P., and Drummond, G.I. Proc. Soc. Exp. Biol. Med. 127:33 (1968).

21. Cohen, A., Ullman, B. and Martin, D.W., Jr. J. Biol. Chem. 254:112 (1979).

22. Cass, C.E., Kolassa, N. Uehava, Y., Dahlig-Harley, E.R. and Paterson, A.R.P. Biochim. Biophys. Acta 649:769 (1981).

23. Kredich, N.M., and Martin, D.W., Jr. Cell 12:931 (1977).

24. Steinberg, R.A., Wetters, T. van D., and Coffino, P. Cell 15:1351 (1978).

25. Wroblewski, F. and LaDue, J.S. Proc. Soc. Exp. Biol. Med. 90:210 (1955).

IDENTIFICATION OF A PURINE 5'-NUCLEOTIDASE IN HUMAN ERYTHROCYTES

F. Bontemps, G. Van den Berghe and H.G. Hers

Laboratory of Physiological Chemistry, International Institute of
Cellular and Molecular Pathology and University of Louvain
Avenue Hippocrate 75, B-1200 Brussels, Belgium

INTRODUCTION

In 1975, Paglia and Valentine (1) demonstrated the existence in human
erythrocytes of a specific pyrimidine 5'-nucleotidase, which they found
inactive on purine nucleotides. A deficiency of this enzyme is known which
provokes an hemolytic anemia (reviewed in 2). In accordance with the reported enzymic specificity, the erythrocytes of these patients have markedly
elevated concentrations of cytidine and uridine nucleotides but not of
purine nucleotides. Although it is evident that dephosphorylation of AMP
and IMP should occur in erythrocytes (see Bontemps et al., this volume),
the enzyme(s) catalysing this process have hitherto not been identified.
This report describes the partial purification and the kinetic properties
of a purine-specific 5'-nucleotidase present in human erythrocytes.

MATERIALS AND METHODS

Preparation of hemolysates

Fresh blood, taken from a cubital vein of healthy human volunteers and
collected on heparin, was centrifuged to remove plasma and buffy coat and
freed of leucocytes and platelets as described in (3). After 3 washes in
0.9 % NaCl, the erythrocytes were hemolyzed by the addition of 4 volumes of
10 mM Tris-Cl buffer pH 7.2, containing 1 mM EDTA and 1 mM dithiothreitol
(buffer A). Where indicated, hemolysates were dialysed overnight against
500 vol. of buffer A; this first dialysis was followed by a second one of 2
hours duration.

Enzyme assays

5'-nucleotidase activity was measured by a radiochemical assay.
Appropriate amounts of hemolysate or enzyme were incubated at 37°C with
$[8-^{14}C]AMP$, $[8-^{14}C]IMP$, $[U-^{14}C]GMP$ or $[U-^{14}C]CMP$ at various
concentrations, 10 mM $MgCl_2$ and 50 mM Tris-Cl pH 7.2 or 50 mM MES pH 6.3,
in a final volume of 50 µl. The reaction was stopped by spotting 10 µl of
the incubation medium on TLC cellulose plates on which carrier solutions of
nucleotides and of the corresponding nucleosides had been applied. The plates were developed in butan-1-ol/methanol/water/conc. NH_3 (60:20:20:1,
by vol.) (4). When crude preparations were used, the dephosphorylation of

AMP into adenosine was measured in the presence of 1 µM deoxycoformycin in order to inhibit adenosine deaminase. In the same preparations the rate of dephosphorylation of IMP was calculated from the sum of the radioactivity appearing in inosine and hypoxanthine.

The activity of acid phosphatase was determined by measuring the production of Pi from 10 mM α-glycerophosphate in 50 mM MES buffer pH 5.

RESULTS

Dephosphorylation of nucleoside monophosphates in dialysed hemolysates

When 20 µl of a dialysed hemolysate were incubated at 37°C and pH 7.2 with either 0.25 mM [^{14}C]IMP or [^{14}C]AMP in a total volume of 50 µl, a production of respectively [^{14}C] inosine or [^{14}C] adenosine was clearly detectable (not illustrated). Under these conditions, the rates of dephosphorylation of AMP and IMP were linear during at least 2 hours and averaged respectively 0.34 ± 0.08 and 1.25 ± 0.12 µmol/h per g of Hb (mean ± SEM for n=3). In the same hemolysates, the activity of pyrimidine 5'-nucleotidase, measured with 0.25 mM [^{14}C]CMP, was 6.24 ± 1.64 µmol/h per g of Hb, which is 18-fold the rate of dephosphorylation of AMP and 5-fold the rate of dephosphorylation of IMP. All these activities were not modified by destromatization of the hemolysates by high speed centrifugation.

Purification procedures

All manipulations were carried out at 0°-6°C.

(1) DEAE-Trisacryl chromatography   Seven hundred and fifty ml of a 1/5 hemolysate, prepared from 1 unit of fresh ACD blood, were mixed during 1 hour with 150 ml of DEAE-Trisacryl diluted 1/5 and equilibrated with buffer A. Under this condition, more than 90 % of the various 5'-nucleotidase activities were bound to the resin. It was then washed abundantly with buffer A to remove hemoglobin and poured in a column (2.6 x 40 cm). The 5'-nucleotidase activities were eluted with a linear gradient from 250 ml of buffer A, to 250 ml of the same buffer, containing 0.4 M NaCl, and fractions of 6 ml were collected (Fig. 1). With IMP as substrate, two peaks of activity were distinguished : the first one (I) was eluted with the bulk of the proteins at 50 mM NaCl; the second one (II), the most active one, was eluted at 240 mM NaCl. The activities of acid phosphatase and of the pyrimidine 5'-nucleotidase were also measured throughout the gradient; these enzymes were also eluted with the protein peak at 50 mM NaCl. The pH curves of the IMPase activity of peaks I and II revealed (results not illustrated) that the activity of peak I had its optimum at pH 5, whereas the activity of peak II had an optimum pH between 6 and 7, which corresponds to the optimum pH of cytosolic 5'-nucleotidase from other tissues (5,6,7).

(2) Chromatography on blue sepharose   The 5'-nucleotidase activity of peak II was found to be strongly adsorbed on blue sepharose, whereas more than 90 % of the proteins of this peak were not retained. About 40 mU (measured with 0.25 mM IMP as substrate) of 5'-nucleotidase could be adsorbed on 0.2 ml of blue sepharose. After washing of the resin with buffer A containing 0.5 M NaCl to remove some additional proteins, the 5'-nucleotidase was eluted with 2 M NaCl. This eluate was dialysed during 1 hour against 500 vol. of buffer A to eliminate NaCl. After this step, the purine 5'-nucleotidase activity was purified at least 4000-fold and found free of nucleoside phosphorylase, adenosine deaminase, adenosine kinase, AMP deaminase, adenylate kinase, ATPase and acid phosphatase.

Properties of the purine 5'-nucleotidase

(1) Effect of pH and Mg^{++} The pH curve of the purified enzyme was determined with buffers containing equimolar (50 mM) amounts of Tris and MES. With 1 mM IMP as substrate, the activity of the 5'-nucleotidase dis-

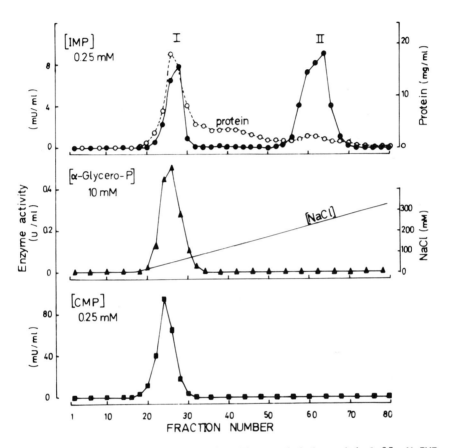

Fig. 1 DEAE-Trisacryl chromatography The activities with 0.25 mM IMP or 0.25 mM CMP as substrate were measured at pH 7.2.

played a fairly sharp optimum at pH 6.3. At pH 7.2, the activity was reduced by approx. 50 %. The 5'-nucleotidase was inactive in the absence of MgCl$_2$. With 1 mM IMP as substrate, a half maximal stimulatory effect was obtained at 1.7 mM MgCl$_2$ (not illustrated).

(2) Effect of the concentration of substrates As shown in Fig. 2, the substrate saturation curves of the 5'-nucleotidase for GMP and IMP were

hyperbolic. Km values were 0.4 mM for IMP and 0.8 mM for GMP. The activity measured with AMP as substrate was very low, remaining at the limit of detectability. Nevertheless, a Km or $S_{0.5}$ of about 1 mM and a Vmax of about one tenth of that reached with IMP and GMP could be estimated.

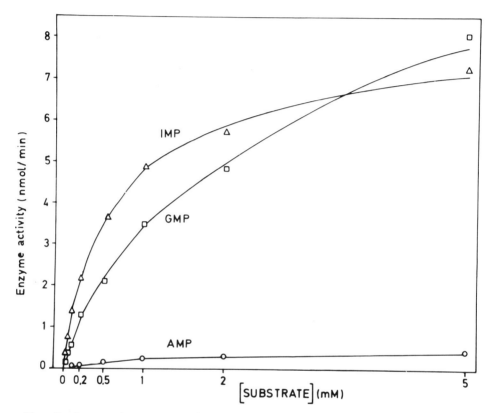

Fig. 2 Saturation curves of purine 5'-nucleotidase for IMP, GMP and AMP Activities were measured in the presence of 50 mM MES pH 6.3.

(3) Effect of metabolites   Table 1 shows that the activity of the purified 5'-nucleotidase, measured in the presence of 0.2 mM IMP or AMP, was strongly stimulated by all the nucleoside triphosphates tested, the most potent ones being ATP and dATP. ADP was also stimulatory, although to a lesser extent than ATP. As expected for a cytoplasmic 5'-nucleotidase, Pi was inhibitory. 2,3-bisphosphoglycerate (2,3-DPG), the most abundant phosphate ester in erythrocytes, was found to have a strong stimulatory effect on the activity of the 5'-nucleotidase, being equally potent as ATP at 1 mM concentration.

Table 1   Effect of various metabolites on the purine 5'-nucleotidase Activities were measured at pH 6.3.

5'-NUCLEOTIDASE ACTIVITY

(nmol/min)

Additions [1 mM]	with IMP [0.2 mM]	with AMP [0.2 mM]
none	1.9	0.08
ATP	19.3	0.69
dATP	20.3	1.16
GTP	10.0	0.14
ITP	7.8	0.10
CTP	5.0	0.11
UTP	3.4	0.07
ADP	5.9	0.13
GDP	1.9	0.03
Pi	0.3	0.01
2,3-DPG	20.4	1.46
2-PG	3.8	0.10
3-PG	2.6	0.08

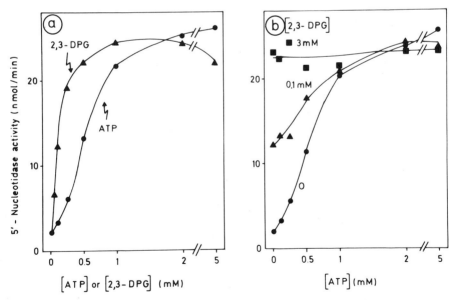

Fig. 3   Effect of ATP and 2,3-DPG on the purine 5'-nucleotidase The activity of 5'-nucleotidase was measured at pH 6.3 with 0.2 mM IMP as substrate, in the presence of increasing concentrations of 2,3-DPG (▲) or of ATP (●) (panel a). The effect of ATP was also investigated in the presence of 0.1 mM (▲) or 3 mM (■) 2,3-DPG (panel b).

(4) Influence of the concentrations of ATP and of 2,3-DPG As shown in Fig. 3a, the maximal stimulatory effects were the same with ATP and with 2,3-DPG. The affinity of the enzyme for the latter effector was, however, higher than that for ATP, since the half-maximal effect was obtained with 0.1 mM 2,3-DPG as compared with 0.5 mM for ATP. A study of the influence of ATP in the presence of 2,3-DPG (Fig. 3b) indicated that the stimulation by ATP was observed only at non-saturating concentrations of 2,3-DPG. In the presence of 3 mM 2,3-DPG, which is supposedly the concentration of free 2,3-DPG inside the erythrocyte, ATP did not any more influence the 5'-nucleotidase activity.

(5) Effect of the concentration of Pi Fig. 4a shows that the half-maximal inhibitory effect of Pi was obtained at the concentration of 0.25 mM. The ATP-stimulated enzyme (Fig. 4b) was also easily inhibited by Pi (Ki=0.4 mM). In the absence as well as in the presence of 1 mM ATP, the enzyme was

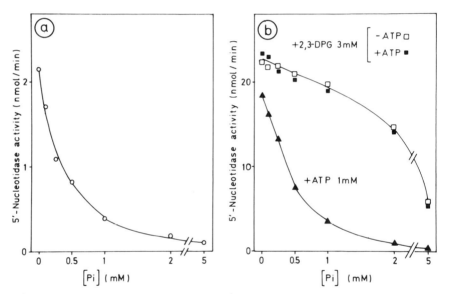

Fig. 4 Effect of Pi on the purine 5'-nucleotidase The enzymic activity was measured at pH 6.3 with 0.2 mM IMP as substrate. The effect of Pi was studied in the absence of stimulators (panel a) or in the presence of 1 mM ATP (▲), 3 mM 2,3-DPG (□) or both (■) (panel b).

thus approx. 80 % inhibited by a physiological concentration of Pi (1 mM). On the other hand, the enzyme stimulated by 2,3-DPG was clearly less sensitive to Pi, since it was only 10 % inhibited by a physiological concentration of Pi (Fig. 4b). Addition of 1 mM ATP in the presence of saturating concentrations of 2,3-DPG did not modify the inhibitory effect of Pi.

DISCUSSION

Our studies provide the first demonstration that human erythrocytes contain a purine 5'-nucleotidase, which is distinct from the pyrimidine 5'-nucleotidase described by Paglia and Valentine (1) and, as also apparent from the accompanying paper (see Bontemps et al., this volume) is located in the cytosol. Similarly to cytosolic 5'-nucleotidase from other sources (5,6,7,8), the enzyme has a pH optimum between 6 and 7, and displays a higher affinity for IMP and GMP than for AMP. It is also stimulated by nucleoside triphosphates, mainly ATP, dATP and GTP, and inhibited by Pi. Most noteworthy, it is stimulated by 2,3-DPG which, in addition, appears the most potent regulator of the enzyme. Indeed, at the concentration of free 2,3-DPG prevailing in human erythrocytes under physiological conditions (approx. 3 mM) the enzyme is not only maximally stimulated (Fig. 3a), but also insensitive to variations of the concentration of its other stimulator, ATP (Fig. 3b), and nearly not influenced by concentrations of its inhibitor in the physiological range (Fig. 4b). These kinetic characteristics most likely explain why the dephosphorylation of AMP and IMP occurs at relatively high rates in erythrocytes in which ATP catabolism has been induced. Indeed, if this catabolism is triggered by glucose deprivation, 2,3-DPG is lowered by not more than 25 % although intracellular Pi is increased 3-fold, whereas if it is induced by alkalinization, 2,3-DPG is not modified and Pi decreased by 50 % (see Bontemps et al., this volume).

It is also worth mentioning that red blood cell purine 5'-nucleotidase is the only enzyme which has hitherto been reported to be stimulated by 2,3-DPG. Indeed, this phosphate ester is known to influence several erythrocytic enzymes (reviewed in 9), among them AMP deaminase (10) but always in an inhibitory fashion.

The stimulatory effect of dATP on the erythrocytic purine 5'-nucleotidase may play a role in the pathogenesis of the depletion of red blood cell ATP, which has been reported in patients with adenosine deaminase deficiency (11) and following therapeutic trials with the adenosine deaminase inhibitor deoxycoformycin (12). Studies are in progress to verify this point.

ACKNOWLEDGEMENTS

The authors wish to express their appreciation to Mrs. Anne Delacauw for expert technical assistance.

This work was supported by grant nr 3.5463.82 of the "Fonds National de la Recherche Médicale". G. Van den Berghe is "Maître de Recherches" of the Belgian "Fonds National de la Recherche Scientifique".

REFERENCES

1. D.E. Paglia & W.N. Valentine, Characteristics of a pyrimidine-specific 5'-nucleotidase in human erythrocytes, J. Biol. Chem. 250:7973 (1975).
2. D.E. Paglia & W.N. Valentine, Hereditary and acquired defects in the pyrimidine nucleotidase of human erythrocytes, Curr. Top. Hematol. 3:75 (1980).
3. E. Beutler & C. West, The removal of leukocytes and platelets from whole blood, J. Lab. Clin. Med. 88:328 (1976).
4. G.W. Crabtree & J.F. Henderson, Rate-limiting steps in the interconversion of purine ribonucleotides in Ehrlich ascites tumor cells in vitro, Cancer. Res. 31:985 (1971).

5. R. Itoh, A. Mitsui & K. Tsushima, Properties of 5'-nucleotidase from hepatic tissue of higher animals, J. Biochem. Tokyo 63:165 (1968).

6. G. Van den Berghe, C. Van Pottelsberghe & H.G. Hers, A kinetic study of the soluble 5'-nucleotidase of rat liver, Biochem. J. 162:611 (1977).

7. D.A. Carson & D.B. Wasson, Characterization of an adenosine 5'-triphosphate and deoxyadenosine 5'-triphosphate activated nucleotidase from human malignant lymphocytes, Cancer Res. 42:4321 (1982).

8. R. Itoh, C. Usami, T. Nishino & K. Tsushima, Kinetic properties of cytosol 5'-nucleotidase from chicken liver, Biochim. Biophys. Acta 526:154 (1978).

9. H. Chiba & R. Sasaki, Functions of 2,3-bisphosphoglycerate and its metabolism, Curr. Top. Cell Reg. 14:75 (1978).

10. A. Askari & S.N. Rao, Regulation of AMP deaminase by 2,3-diphosphoglyceric acid : a possible mechanism for the control of adenine nucleotide metabolism in human erythrocytes, Biochim. Biophys. Acta 151:198 (1968).

11. M.S. Coleman, J. Donofrio, J.J. Hutton, L. Hahn, A. Daoud, B. Lampkin & J. Dyminski, Identification and quantitation of adenine deoxynucleotides in erythrocytes of a patient with adenosine deaminase deficiency and severe combined immunodeficiency, J. Biol. Chem. 253:1619 (1978).

12. M.F.E. Siaw, B.S. Mitchell, C.A. Koller, M.S. Coleman & J.J. Hutton, ATP depletion as a consequence of adenosine deaminase inhibition in man, Proc. Natl. Acad. Sci. USA 77:6157 (1980).

HUMAN PLACENTAL CYTOPLASMIC 5'-NUCLEOTIDASE:
KINETICS AND MOLECULAR PROPERTIES

V. Madrid-Marina, J. Kaminska, and I.H. Fox

Departments of Internal Medicine and Biological Chemistry

Ann Arbor, Michigan 48109

INTRODUCTION

Dephosphorylation of nucleoside 5'-monophosphates is the first and irreversible reaction of purine nucleotide catabolism.[1] This reaction is catalyzed by 5'-nucleotidase 5'-ribonucleotide phosphohydrolase, EC 3.1.3.5. Most of this activity is present on the external surface of the cell. This activity has no clear role in intracellular nucleotide catabolism. However, recent studies indicate that cytoplasmic 5'-nucleotidase may have an important role in intracellular nucleotide degradation.[2]

Cytoplasmic 5'-nucleotidase has been described in many animal tissues.[3-11] There is only limited information concerning human cytoplasmic 5'-nucleotidase in studies of extracts and partially purified enzymes from malignant cells or virus transformed cells.[2,12,13] To elucidate the regulation of intracellular nucleotide degradation at the level of dephosphorylation in normal human tissue, we examined the properties of a highly purified cytoplasmic 5'-nucleotidase from human placenta.

MATERIALS AND METHODS

Materials and methods are described elsewhere in detail.[14] The cytoplasmic 5'-nucleotidase used in our kinetics studies was purified about 3000-fold with a specific activity of 12 umol/min/mg. The purified enzyme has no nonspecific phosphatases, ATPase, or ADPase activities.

The enzyme activity was routinely assayed at physiological conditions, measured by release of either inorganic phosphate or nucleoside. To measure the release of nucleosides, we used $[8-^{14}C]$ AMP or $[8-^{14}C]$ IMP, the radioactive products were separated by high voltage electrophoresis and counted in a scintillation counter.

To determine the Michaelis constants (Km), Lineweaver-Burk plots of initial velocity versus substrate concentration were used. Secondary replots were used to determine Km or Ki values, using least squares analysis for linear replots, according to Cleland.[15]

RESULTS

We purified cytoplasmic 5'-nucleotidase from human placenta 8075-fold to a specific activity of 58.85 umol/min/mg. These final preparations show a single band in sodium dodecyl sulphate-polyacrylamide gel electrophoresis, corresponding to a molecular weight of 76,000. The human placental enzyme had a native molecular weight of 143,000 (Biogel P-300, gel filtration).[14] Therefore, the native enzyme appeared to be a dimeric protein, with two identical subunits. The cytoplasmic 5'-nucleotidase was activated by magnesium with an apparent Km value of 6 mM. Other divalent cations including zinc produce only a moderate activation. The pH studies showed that human placental enzyme had an optimal activity at pH from 7.4 to 9.0 using AMP, CMP or IMP as substrates. All properties of cytoplasmic 5'-nucleotidases are summarized in Table I. Initial velocity studies with AMP

Table I

Properties of Human Placenta Cytoplasmic 5'-Nucleotidase

1- 8070-fold purification, to 59 umol/min/mg of specific activity.

2- Molecular weight of 143,000.

3- Subunit molecular weight: 76,000 of two identical subunits.

4- Substrate specificity for pyrimidine and purine nucleoside 5'-monophosphates.

5- Absolute requirement for Mg^{2+}.

6- pH optimum from 7.4 to 9.0.

7- The enzyme was stable in concentrated form at $-70^{\circ}C$

8- AMP dephosphorylation was inhibited by purine, pyrimidine and pyridine nucleotides.

---

and IMP as substrates showed hyperbolic kinetics (Figure 1). This was observed with all substrates studied. Double reciprocal plots of initial velocity data at varying concentrations of nucleoside 5'-monophosphates are linear with Km values of 18 and 30 uM for AMP and IMP respectively.

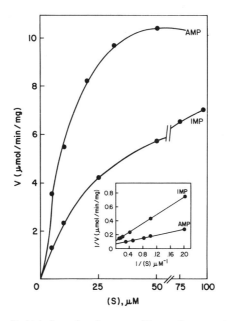

Figure 1. Initial velocity studies of cytoplasmic 5'-nucleotidase with variable AMP and IMP concentrations. Inset, double reciprocal plots of initial velocity data and different concentrations of substrates. The Km value for AMP is 18 uM and 30 uM for IMP.

The Km value and Vmax for other nucleoside 5'-monophosphates were performed.[14] The data show that cytoplasmic 5'-nucleotidase purified from human placenta has the lowest Km and the highest Vmax values for CMP and UMP (Table II). This indicates that these enzyme have a high preference for pyrimidine nucleotides.

Table II

Kinetics Parameters of Human Placenta Cytoplasmic 5'-Nucleotidase

Compound	Vmax (umol/min/mg)	Km (uM)	Ki (uM)
AMP	11.8	18.0	---
IMP	8.2	30.0	22
GMP	16.0	17.3	---
CMP	33.0	2.2	---
UMP	31.8	14.8	---
ADP	---	---	15
ATP	---	---	100
Pi	---	---	16 and 42 mM[a]
NAD	---	---	56
NADH	---	---	33
AMPCP[b]	---	---	5

a. Ki values for slope and intercept, resectively.
b. alpha, beta methylene adenosine diphosphate.

A large variety of purine, pyrimidine and pyridine nucleotides exert an inhibitory effect on AMP dephosphorylating activity. Nucleoside 5'-monophosphates were more potent inhibitors than diphosphates and these more than nucleoside triphosphates.[14] IMP inhibition studies at different concentrations and fixed concentrations of AMP, showed competitive inhibition with an apparent Ki value of 22 uM, which is similar to the Km value (Table II). Ribonucleoside 5'-monophosphates were more potent inhibitors than the respective deoxyribonucleotides.[14] When human placenta cytoplasmic 5'-nucleotidase was assayed in the presence of different concentrations of AMP and fixed concentrations of ADP and ATP, produce competitive inhibition, with Ki values of 15 and 100 uM, respectively (Table II). Similar results were found when IMP was used as substrate. Ribonucleotides were more potent inhibitors than the respective deoxyribonucleotides.[14] Inorganic phosphate was a noncompetitive inhibitor, with apparent Ki (slope) of 16 mM and Ki (intercept) of 42 mM. NAD, NADH, and AMPCP were competitive inhibitors with respect to AMP with Ki values of 56, 33, and 5 uM, respectively (Table II). Finally, we have studied whether there is any interaction between Pi and, ADP and ATP inhibition. The inhibitory properties of these compounds appear to be additive, at physiological concentrations.[14]

DISCUSSION

The enzyme cytoplasmic 5'-nucleotidase purified, from human placenta
had a native apparent molecular weight of 143,000 with a subunit molecular
weight of 76,000, suggesting that the enzyme has two identical subunits.
These data are in the same range of molecular weight to other cytoplasmic
5'-nucleotidase from a variety of tissues.[4,8] In contrast with the enzyme
purified from chicken liver which has a molecular weight of 205,000.[3] There
is an absolute requirement for magnesium in human placental cytoplasmic 5'-
nucleotidase as well as in other cytoplasmic preparations from a large
variety of tissues.[2-4,6,7,12] The pH curve for placental enzyme resembles
the cytoplasmic 5'-nucleotidases from bovine brain and liver,[9,11] as well as
many preparations of plasma membrane enzyme.[2,17] However, this in contrast
with other cytoplasmic 5'-nucleotidases which have peaks of activity from pH
6.3 to 6.5.[2-7,12]

The Km values for AMP and IMP for human placental cytoplasmic 5'-
nucleotidase are in contrast with the Km value reported for chicken and rat
liver, and human lymphoblasts,[5-7,12] which are one or two orders of
magnitude higher. They are in accord with values from bovine brain and
liver, and values for membrane bound 5'-nucleotidase.[8,11,17] The low Km and
high Vmax values for CMP, make this compound the favored substrate and
appear to be a unique property of this enzyme and the bovine liver
enzyme.[11] Preferential hydrolysis of IMP and GMP as compared to AMP was not
observed in contrast to previous reports.[5-10] In contrast with other
cytoplasmic 5'-nucleotidases, human placental enzyme activity was inhibited
for ATP, ADP and inorganic phosphate.[14] These compounds, ATP and ADP,
produce activation on other cytoplasmic 5'-nucleotidases.[9,11] Heart, liver
and lymphoblast cytoplasmic 5'-nucleotidases are activated by 3 to 10 mM
ATP and 3 to 5 mM ADP, and are inhibited by 5 to 20 mM of inorganic
phosphate.[5-7,10] Cooperative kinetics are apparent in this regulation;
however, there was no evidence of cooperative kinetics in our studies. The
Ki values for human placental cytoplasmic enzyme resemble the kinetic
constants derived for membrane 5'-nucleotidase.[17] Inorganic phosphate did
not modify the regulatory properties of ATP and ADP in our studies in
contrast to its properties with the liver enzymes.[6,7] Our results indicate
that the enyzme is regulated by magnesium, nucleoside mono, di- and
triphosphates. These suggest that the enzyme is regulated by a mechanism
of heterogenous metabolic pool.[16]

The role of cytoplasmic 5'-nucleotidase in human intracellular purine
and pyrimidine 5'-monophosphate dephosphorylation has been motive of several
years of investigation. The regulatory properties of human placental enzyme

imply that the activation of the purine nucleotide degradation pathway may occur with the depletion of intracellular purine nucleoside triphosphates and an accumulation of IMP and AMP. Such activation may explain the drop of nucleotide degradation which is observed in fructose-induced hyperuricemia, strenous muscular exercise, hypoxia and ethanol ingestion.[1,18-20]

According to previous reports, and based on results of our studies, there are at least two cytoplasmic 5'-nucleotidase enzymes. Both enzymes have properties totally different. The exact roles of these two types of cytoplasmic 5'-nucleotidases in the regulation of intracellular nucleoside 5'-monophosphate degradation remain under investigation.

REFERENCES

1. I.H. Fox. Metabolic basis for disorders of purine nucleotide degradation. Metab 30:616-634 (1981).

2. N.L. Edwards, D. Recker, J. Manfredi, R. Rembecki, and I.H. Fox. Regulation of purine metabolism by plasma membrane and cytoplasmic 5'-nucleotidases. Am J Physiol 243:C270-277 (1972).

3. Y. Naito and K. Tsushima. Cytosol 5'-nucleotidase from chicken liver. Purification and properties. Biochem Biophys Acta 438:159-168 (1976).

4. P. Fritzson. Regulation of nucleotidase activity in animal tissues. Adv Enz Reg 16:43-61 (1978).

5. G. Van den Berghe, C. Van Pottelsberghe and H-G Hers. A kinetic study of soluble 5'-nucleotidase of rat liver. J Biochem 162:611-616 (1977).

6. R. Itoh, C. Usami, T. Nishino and K. Tsushima. Kinetic properties of cytosol 5'-nucleotidase from chicken liver. Biochem Biophys Acta 526:154-162 (1978).

7. R. Itoh. Purification and some properties of cytosol 5'-nucleotidase from rat liver. Biochem Biophys Acta 657:402-410 (1980).

8. J.M. Montero and J.B. Fes. Purification and characterization of bovine brain 5'-nucleotidase. J. Neurochem 39:982-989 (1982).

9. J. Mallol and J. Bozal. Modification of 5'-nucleotidase activity by cations and nucleotides. Neurochem 40:1205-1211 (1983).

10. J.M. Lowenstein, M. Yu and Y. Naito. in "Regulatory Function of Adenosine", R.M. Berne, T.W. Rall, and R. Rubio, eds., Martinus Nijhoff Publishers, Boston (1983).

11. R. Itoh and K. Tsushima. 5'-Nucleotidase of bovine liver. Int J Biochem 2:651-656 (1971).

12. D.A. Carson, J. Day and D.B. Wasson. The potential importance of soluble deoxyribonucleotidase activity in mediating deoxyadenosine toxicity in human T-lymphoblasts. J Immunol 126:348-352 (1981).

13. A.S. Bagnara and M.S. Hershfield. Mechanism of deoxyadenosine-induced catabolism of adenine nucleotides in adenosine deaminase inhibited T-lymphoblastoid cells. Proc Natl Acad Sci, USA 79:2673-22677 (1982).

14. V. Madrid-Marina, I.H. Fox. Human placental cytoplasmic 5'-nucleotidase: Kinetic properties and inhibition. J Biol Chem, 1985 (Submitted).

15. W.W. Cleland. in "Enzymes", Vol. 2, P.D. Boyer, ed., Academic Press, New York, pp. 1-66 (1970).

16. Stadtman, E.R. in "Enzymes", Vol. 2, P.D. Boyer, ed., Academic Press, New York, pp. 397-459 (1970).

17. I.H. Fox. in "Uric Acid", W.N. Kelley and I.M. Werner, eds., Spring-Verlag, Berlin, pp. 93-124 (1978).

18. I.H. Fox and W.N. Kelley. Studies on the mechanism of fructose-induced hyperuricemia. Metab Clin Exp 21:713-721 (1972).

19. J.O. Woolliscroft, H. Colfer and I.H. Fox. Hyperuricemia in acute illness: A poor diagnostic sign. Am J Med 72:58-62 (1982).

20. J. Faller and I.H. Fox. Ethanol-induced hyperuricemia. N Eng J Med 307:1598-1602 (1982).

ACKNOWLEDGEMENTS

The authors wish to thank Paul Nowak for excellent technical assistance and Holly Gibson for typing the manuscript. This work was supported by grants from the United States Public Health Service 2-R01-AM-19674, 5M01 RR00042, and F05 TW03371 (Fogarty award to Dr. V. Madrid-Marina).

REGULATION OF THE CYTOSOL 5'-NUCLEOTIDASE OF THE HEART BY ADENYLATE ENERGY

CHARGE

Roichi Itoh, Jun Oka and Hisashi Ozasa

The National Institute of Nutrition
Toyama, Shinjuku-ku, Tokyo
Japan

INTRODUCTION

Adenosine formed by the heart has been proposed to participate in the metabolic regulation of coronary blood flow [1]. Adenosine can be formed from AMP by action of 5'-nucleotidase or alkaline or acid phosphatase, or from S-adenosylhomocysteine by means of S-adenosyl-homocysteine hydrolase. A 5'-Nucleotidase in the heart was shown by histochemical studies to be associated with myocardial cell membrane [2], and functional studies indicate the enzyme to be an ectoenzyme [3].

Schütz et al. [4] suggested that in hypoxic heart, a) the ecto-5'-nucleotidase most likely is not involved in the formation of adenosine, b) release of adenosine from heart requires adenosine to be transported across the membrane, and c) adenosine is predominantly formed intra-cellularly by a process involving dephosphrylation of AMP and/or hydrolysis of S-adenosylhomocysteine.

Presence of another 5'-nucleotidase in the cytosolic fraction of rat heart has been suggested by Lowenstein et al [5]. Recently a 5'-nucleo-tidase has also been purified from soluble fraction of chicken heart [6], and kinetic properties of the enzyme were reported to be very similar to those of the cytosol 5'-nucleotidase, whose presence in the liver of chicken and rat has previously been confirmed [7, 8].

In this report, it is described that the AMP-hydrolysing activity of the cytosol 5'-nucleotidase highly purified from chicken and rat heart increases sharply with a decrease in intracellular energy level reflecting on a decrease in adenylate energy charge [9]. The results of the work suggest a possibility of participation of the cytosol 5'-nucleotidase in production of adenosine in hypoxic heart

MATERIALS AND METHODS

Assay of 5'-nucleotidase

5'-Nucleotidase was assayed by determination of [8-^{14}C]inosine from [8-^{14}C]IMP or [8-^{14}C]adenosine from [8-^{14}C]AMP as described previously [8]. The reaction mixture contained 100 mM imidazole-HCl (pH 6.5), 50 mM MgCl$_2$, 500 mM NaCl, 0.1% bovine serum albumin, an appropriate amount of enzyme preparation and various amount of a substrate and an effector.

Estimation of kinetic parameters

V and Km values for various substrates were determined by the direct linear plot described by Eisenthal and Cornish-Bowden [10]. When the substrate-velocity plot of the reaction was sigmoidal, v, $s_{0.5}$ and h were estimated by the method of Atkinson [11].

Purification of 5'-nucleotidase from chicken and heart

Chicken heart cytosol 5'-nucleotidase was prepared by the method described previously [6]. The procedure is composed of the five steps as follows: extraction, first phosphocellulose column chromatography, second phosphocellulose column chromatography, ammonium sulfate fractionation and precipitation at pH 6.5.

The rat heart enzyme was prepared by essentially the same procedure used for the purification of the enzyme from chicken heart.

Preparation of reaction mixtures of various adenylate energy charge

Reaction mixture of various energy charges, (ATP + $\frac{1}{2}$ADP)/(ATP + ADP + AMP), [9] were prepared according to the method of Chapman and Atkinson [12]. The total Mg^{2+} concentration corresponding to 1mM free Mg^{2+} at pH 6.5 was calculated by the method described by Chapman et al. [13].

Chemicals

AMP, ADP, ATP and IMP were products of Yamasa Shoyu Co. Concentrations of all nucleotide solutions were determined spectrophotometrically. [8-^{14}C] IMP (61 Ci/mol) and [8-^{14}C]AMP (59 Ci/mol) were purchased from Amersham UK.

All other chemicals were of reagent grade or of the highest quality available.

RESULTS

Purity of the enzyme preparations

Chicken heart enzyme was purified about 7000-fold from crude extract with 1% of recovery and specific activity of the final preparation was 31.5 μmol inosine formed from IMP per min per mg pritein. Rat heart enzyme was purified about 3000-fold with 3% recovery and specific activity of the final preparation was 28.5 μmol inosine formed from IMP per min per mg protein.

On SDS-polyacrylamide gel electrophoresis, the purified 5'-nucleotidase preparations from chicken and rat heart gave one Coomasie blue staining band, respectively. On polyacrylamide gel electrophoresis without SDS at pH 8.9, each preparation gave one Coomasie blue staining band, which corresponded to the position of 5'-nucleotidase activity detected with IMP as substrate.

Effect of ATP, ADP and Pi on the enzyme activity

We studied effects of adenine nucleotides, which compose intracellular adenylate energy storage system, and Pi on the activity of the cytosol 5'-nucleotidase of the heart. Changes in some kinetic parameters of both enzymes induced by various effectors are summarized in Table 1.

When IMP was used as a substrate with the cytosol 5'-nucleotidase from chicken heart or rat heart, the substrate-velocity plot was hyperbolic. ATP or ADP activated both enzymes by decreasing Km's and increasing V. Pi inhibited chicken heart enzyme competitively and rat heart enzyme by changing the substrate velocity plot sigmoidal and increasing $s_{0.5}$.

Table 1. Effect of ATP, ADP and Pi on the kinetic parameters of the cytosol
5'-nucleotidase of the heart

Substrates	Effectors	Chicken heart enzyme			Rat heart enzyme		
		$V^{a)}$	$s_{0.5}^{b)}$	$h^{c)}$	$V^{a)}$	$s_{0.5}^{b)}$	$h^{c)}$
IMP	None	1.0	1.0	1.0	1.0	0.9	1.0
	5 mM ATP	3.3	0.3	1.0	2.8	0.1	1.0
	5 mM ADP	2.3	0.1	1.0	3.4	0.3	1.0
	1 mM Pi	1.0	3.5	1.0	1.0	3.2	1.3
AMP	None	2.8	40	1.8	3.4	25	1.8
	5 mM ATP	2.8	8.0	1.0	3.4	4.5	1.0
	5 mM ADP	2.8	12	1.0	3.4	9.8	1.0
	1 mM Pi	2.8	59	2.3	3.4	40	2.5

a) V for IMP without effectors was taken as the standard.
b) Substrate concentration at half saturation of enzyme (mM).
c) Hill coefficient.

When AMP was used as a substrate, substrate-velocity plot was sigmoidal.
ATP or ADP activated the enzyme by changing the curve to hyperbolic and Pi
inhibited the enzyme by increasing the sigmoidicity of the curve.

Response of the cytosol 5'-nucleotidase activity to variations in adenylate energy charge

The response of AMP-hydrolysing activity of both chicken heart and
rat heart cytosol 5'-nucleotidase activity to variations in adenylate energy

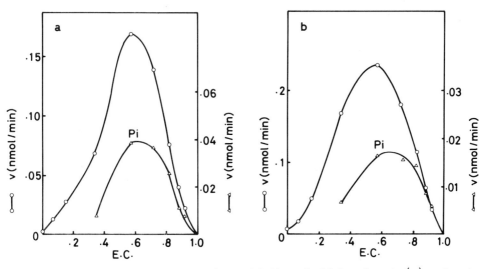

Fig. 1. Response of AMP-hydrolysing activity of chicken heart (a) and rat
heart (b) 5'-nucleotidase to variation in the adenylate energy
charge in the absence (O) and in the presence of 5 mM Pi (△).

charge was studied. The total adenine nucleotide concentration and free Mg^{2+} concentration were held at 4 mM and 1 mM, respectively. AMP-hydrolysing activity of both enzymes was low at normal physiological energy charge value (0.9) and it was greatly activated when the energy charge decreased within the range of surviving cells (0.7-0.9). Maximum activity was obtained at energy charge value of about 0.6 for both enzymes (Fig.1., a and b).

AMP-hydrolysing activity of both enzymes were highly suppressed in the presence of 5 mM Pi, but the response to the change in adenylate energy charge was still maintained.

DISCUSSION

Jennings et al. [14] and Swain et al. [15] reported that a decrease in ATP during ischemia is associated with a marked increase in myocardium content of adenosine, a vasodilator. These results suggest that the pathway of degradation of adenine nucleotides in the heart during ischemic condition is, at least in part, through dephosphorylation of AMP. Based on the observation of the experiments with isolated guinea pig hearts, Shütz et al. [3] proposed that adenosine is formed intracellularly by a process involving dephosphrylation of AMP by cytosol 5'-nucleotidase.

Liang and lowenstein observed that adenosine content of myocardium, hypoxanthine concentration in coronary sinus plasma, and coronary blood flow in the anesthetized dog increased under the condition of AMP accumulation induced by infusion of acetate [16], suggesting that intra-cellular AMP concentration may be the primary determinant of adenosine production in the heart.

To characterize the enzyme which participates in adenosine formation in cardiac tissue, we purified a 5'-nucleotidase from soluble fraction of chicken and rat heart. As described in this report, AMP-hydrolysing activity of the 5'-nucleotidase purified from the heart increased sharply with decreasing adenylate energy charge within the physiological range, as has been reported for rat liver cytosol 5'-nucleotidase [17]. The response was observed in the presence of physiological concentrations of total adenine nucleotides, free Mg^{2+} and Pi. This implies that change in intracellular AMP concentration is the primary determinant of the rate of hydrolysis of AMP by the cytosol 5'-nucleotidase, and is consistent with the observation by Liang and Lowenstein [16] which has been reffered to above.

A similar response of AMP deaminase of mammalian liver to changes in adenylate energy charge has already been reported [12, 18]. This implies that two enzymes that participate in the first steps of two pathways of AMP catabolism in the liver share common regulatory properties. In contrast to the liver enzyme, the cardiac muscle AMP deaminase has been reported to show virtually no variation in activity with the energy charge between zero and 0.9 [18].

It is highly possible that AMP catabolism in the heart through dephosphorylation catalysed by the cytosol 5'-nucleotidase is effectively regulated by changes in energy charge in the cells, and is one of the important pathways involved in production of adenosine for the purpose of increasing coronary blood flow during anoxic condition.

ACKNOWLEDGEMENT

This study was supported in part by the Research Grant for the Intractable Diseases from the Ministry of Health and Welfare of Japan.

REFERENCES

1. J. Schrader, Site of action and production of adenosine in the heart, in: "Purinergic Receptor," G. Burnstock, ed., Chapman and Hall, London (1981)
2. R. Rubio, R. M. Berne, and J. G. Dobson, Jr., Site of adenosine production in cardiac and skeletal muscle, Am. J. Physiol. 225:938 (1973)
3. H. P. Bear, and G. I. Drummond, Catabolism of adenine nucleotides by the isolated perfused rat heart, Proc. Soc. Exp. Biol. Med. 127: 33 (1968)
4. W. Shütz, J. Schrader, and E. Gerlach, Different sites of adenosine formation in the heart, Am. J. Physiol. 240: H963 (1981)
5. J. M. Lowenstein, M.-K. Yu, and Y. Naito, Regulation of adenosine metabolism by 5'-nucleotidase, in: "Regulatory Function of Adenosine," R. M. Berne, T. W. Rall, and R. Rubio, ed., Martinus Nijhoff, Boston (1983)
6. R. Itoh, and J. Oka, Evidence for existence of a cytosol 5'-nucleotidase in chicken heart, Comp. Biochem. Physiol. (in press)
7. Y. Naito, and K. Tsushima, Cytosol 5'-nucleotidase from chicken liver, Biochim. Biophys. Acta 438: 159 (1976)
8. R. Itoh, Purification and some properties of cytosol 5'-nucleotidase from rat liver, Biochim. Biophys. Acta 651:402 (1981)
9. D. E. Atkinson, The energy charge of adenylate pool as a regulatory parameter, Biochemistry 7: 4030 (1968)
10. R. Eisenthal, and A. Cornish-Bowden, The direct linear plot, Biochem. J. 139: 715 (1974)
11. D. E. Atkinson, "Cellular Energy Metabolism and its Regulation," Academic Press, London (1977)
12. A. G. Chapman, and D. E. Atkinson, Stabilization of adenylate energy charge by the adenylate deaminase reaction, J. Biol. Chem. 248: 8309 (1973)
13. A. G. Chapman, A. L. Millar, and D. E. Atkinson, Role of adenylate deaminase reaction in regulation of adenine nucleotide metabolism in Ehrlich ascites tumor cells, Cancer. Res. 36: 1144 (1976)
14. R. B. Jennings, K. A. Reimer, M. L. Hill, and S. E. Mayer, Total ischemia in dog heart in vitro, Circ. Res. 49: 892 (1981)
15. J. L. Swain, R. L. Sabina, P. A. McHale, J. C. Greenfield, and H. W. Holmes, Prolonged myocardial nucleotide depletion after brief ischemia in open-chest dog, Am. J. Physiol. 242: H818 (1982)
16. C-s Liang, and J. M. Lowenstein, Metabolic control of the circulation, J. Clin. Invest. 62: 1029 (1978)
17. R. Itoh, Regulation of cytosol 5'-nucleotidase by adenylate energy charge, Biochim. Biophys. Acta 659: 31 (1981)
18. C. Solano, and C. J. Coffee, Differential response of AMP deaminase isozymes to changes in the adenylate energy charge, Biochem. Biophys. Res. Commun. 85: 564 (1978)

CHARACTERIZATION OF HYDROXYUREA (HYU) RESISTANT S49 T LYMPHOMA CELLS

Daniel A. Albert and Lorraine J. Gudas

Department of Medicine, University of Chicago, Chicago
IL, 60637 and Pharmacology Department, Dana Farber Cancer
Institute, Harvard Medical School, Boston, MA 02115

Ribonucleoside diphosphate reductase (EC 1.17.4.1) catalyzes the reduction of ribonucleoside diphosphates to their deoxyribonucleotide counterparts for DNA synthesis (1). The mammalian enzyme is composed of two subunits: the M1 protein has binding sites that are involved in the complex allosteric feedback control of the enzyme (2,3); whereas, the M2 subunit contains the free radical binding site of hydroxyurea (4,5).

Previous studies have suggested that _in vivo_ ribonucleotide reductase activity and consequent deoxyribonucleoside triphosphate production is rate-limited by the availability of M2 activity (6). We tested these hypotheses by selecting and characterizing cell lines with variable resistance to hydroxyurea and comparing them with wild type S49 T-lymphoma cells.

METHODS

Cell Growth. Cells were cultured in Dulbecco's modification of Eagle's Medium containing heat inactivated 10% horse serum at 37° C in a 10% CO_2 in air atmosphere. Toxicity curves were performed in Falcon 24 well tissue culture plates.

Determination of Deoxyribonucleotide Pools. 2.5 to 5×10^7 cells were washed in PBS, centrifuged in a microfuge, resuspended in 200 ul ice cold 0.1 M potassium phosphate buffer (pH 7.2) and an aliquot was counted on a Coulter model ZBI. 1.5 N perchloric acid (40 ul) was added, followed 60 seconds later by the addition of 80 ul 1.2 N KOH (final pH = 7.0). Samples remained on ice for 5 to 10 minutes, followed by centrifugation at 4° C in a microfuge. These extracts were analyzed for ribonucleotides. For deoxyribonucleotide pool measurements, periodation of the extract with 0.2 M sodium periodate for 90 minutes at 37° C was performed by the method of Garrett and Santi (7). The reaction was terminated by the addition of 10 ul of 0.3 M rhamnose. Extracts were analyzed on a Beckman Altex high performance liquid chromatograph using an SAX Partisil column with a 0.38 M ammonium phosphate 1.5% acetonitrile pH 3.45, elution buffer. Peaks were measured at both 280 and 254 nM and concentration was proportional to peak height above baseline.

Purification of M1. The M1 subunit of ribonucleotide reductase was purified according to the method of Gudas (8). Briefly, this involved harvesting and washing 2-3 x 10^9 cells in ice cold PBS, followed by sonication in 50 mM Tris, 2 mM DTT (pH 7.4). RNA and DNA were removed by streptomycin sulfate precipitation (.65%) and the supernatant was precipitated with 40% ammonium sulfate. This precipitate was dialyzed or desalted in Tris buffer, and loaded onto a 3 ml column of Dextran blue sepharose. The flow through contained M2. The column was washed with Tris buffer and M1 was eluted at a salt concentration of between 50 and 500 mM NaCl.

Ribonucleotide Reductase Assay. Ribonucleotide reductase assay was measured by the conversion of CDP to deoxyCDP by permeabilized cells or partially purified cell extracts. 1.2 x 10^7 cells were permeabilized by exposure to 1% Tween 80 by the method of Lewis (9). The assay mixture included final concentrations of 39 mM HEPES, 6.6 mM ATP, 8 mM MgCl$_2$, 22 mM DTT, 50 uM CDP, and 0.42 uCi ^{14}C-CDP in final volume of 300 ul after addition of 60 ul of a five-fold concentrated cocktail. To assay the partially purified enzyme preparation, 5 mM NaF was added to the other chemicals to diminish phosphatase activity. The reaction was terminated by boiling the samples for 4 minutes. 6.0 mg of Crotalus atrox venom per assay was then added, and samples were incubated for 4 hours at 37° C. Samples were diluted with 500 ul H$_2$O, loaded onto 1 ml borate Dowex-1 columns, and [^{14}C] deoxycytidine product was eluted from the column with 2-3 ml water. Samples were then counted by liquid scintillation.

Cell-cycle Analysis. Experimental cell suspensions (5 ml) containing 5-10 x 10^5 cells/ml were centrifuged and resuspended in 1-3 ml of hypotonic solution containing 0.05 mg/ml propidium iodide, 0.1% sodium citrate, and 0.1% Triton X-100. After staining, cell fluorescence was determined on a FACS analyzer.

RESULTS

Selection of Hydroxyurea Cell Lines. Wild type S49 T-lymphoma cells are sensitive to hydroxyurea with an EC$_{50}$ (50% survival at 72 hours) of 50 uM (Table 1). Wild type cells were exposed to progressively higher concentrations (an increase of 50 uM at two week intervals) for a one year period. After one year, increments in HYU-resistant concentration were increased progressively to 100 uM, 200 uM, then 500 uM. At 500 uM, 1 mM, 2 mM, 3 mM and 4 mM concentrations subcultures were grown in the absence of hydroxyurea. Cell characteristics remained relatively stable out of hydroxyurea for up to one year. HYU-resistant cells grew at the same rate as wild type cells (doubling time 14 to 18 hours) but appeared slightly larger and lobulated under phase contrast microscopy.

Hydroxyurea Resistance. S49 cells grown in increasing concentrations of hydroxyurea are resistant to the growth inhibitory effects of hydroxyurea. There is a slight decrement in hydroxyurea resistance in cells grown in the absence of hydroxyurea but phenotypic reversion to normal did not occur.

HYU Resistant Cells Have Increased Ribonucleotide Reductase Activity. Increasing resistance to hydroxyurea is associated with increasing CDP reductase activity as shown in Table 1. Similarly, HYU-resistant cells grown in the absence of HYU retain the increased CDP reductase activity. However, this increased CDP reductase activity is not directly proportional to the degree of hydroxyurea resistance.

M2 But Not M1 Activity Is Increased In HYU R Cells. Crude extracts of hydroxyurea-resistant cells have increased CDP reductase activity which is proportional to the degree of resistance to hydroxyurea. Using affinity chromatography to purify M1 away from M2 activity then recombining fractions from wild type and hydroxyurea resistant cells, M1 activity is not increased in hydroxyurea resistant cells, however M2 activity is (Table 1).

Deoxyribonucleoside Triphosphate Pools In Hydroxyurea-Resistant Cells. Deoxyribonucleoside triphosphate pools are increased in hydroxyurea-resistant cells (Table 1). In general, the pool size varies more closely with the rise in CDP reductase activity than the degree of resistance to hydroxyurea. Cells washed out of hydroxyurea increase their dNTP pools within 24 hours and thereafter pools remain stable.

Deoxyadenosine Resistance In Hydroxyurea-Resistant Cells. Hydroxyurea-resistant cells are resistant to the growth inhibitory effects of deoxyadenosine under conditions where adenosine deaminase is inhibited by deoxycoformycin. Deoxyadenosine resistance varies more closely with ribonucleotide reductase activity and dNTP pools than with degree of hydroxyurea resistance (Table 1).

Table 1. Comparison of Wild Type and HYU R Cells

	EC_{50} (HYU)	EC_{50} (AdR)	CDP Reductase Activity (CPM/ 10^6 Cells)	dNTP Pools dATP	dTTP	dCTP	dGTP (picomoles/10^6 cells)	Relative M2 Activity (cpm/ug)
Wild Type	50 uM	2.3 uM	25,050	33.8	38.7	39.6	17.6	8.1
HYU R 2.0	1.9 mM	6.1 uM	149,800	70.4	75.3	58.5	28.2	39.4

EC_{50} is concentration of drug that 50% of cells survive after 72 h exposure. HYU R 2.0 cells are selected for resistance to 2.0 mM HYU.

Hydroxyurea-Resistant Cells Do Not Arrest In G1 When Exposed to Cylic AMP. Cyclic AMP reversibly arrests wild type S49 cells in G1 (Fig.1) phase of the cell cycle. However, HYU R cells are resistant to cyclic AMP at concentrations up to 10 mM. Similarly, they are resistant to the activator of adenyl/cyclase, Forskolin. At low concentrations (30 uM), HYU R cells are resistant to the phosphodiesterase inhibitor RO 1274 (WT cells arrest in G1 at this concentration) but at higher concentrations (1 mM) they partially arrest.

DISCUSSION

We selected and characterized a series of cell lines resistant to increasing concentrations of hydroxyurea. Hydroxyurea reversibly binds to the tyrosine radical of the M2 subunit of ribonucleotide reductase (10) that is required for catalytic activity (11). This interaction is similar to that in E. coli except that hydroxyurea irreversibly inactivates the tyrosine radical in those bacteria.

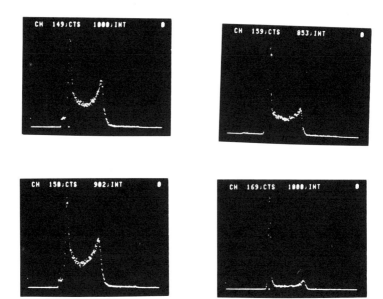

Fig.1   Log phase HYU R cells (top left) and WT cells (top right)
have a normal cell cycle distribution of 30% in G1 phase
(tall peak on left), 40% in S phase (broad plateau in the
middle), and 30% in G2/M phase (small peak on right). After
treatment with 1 mM dibutyryl cyclic AMP, HYU R cell cycle
distribution is unchanged (bottom left) whereas WT cells are
arrested in G1 (bottom right).

Hydroxyurea resistant L cells have been studied previously (12) and
documented to have increased CDP reductase activity. More recently, 3T3
fibroblasts resistant to hydroxyurea have been selected by stepwise
increased hydroxyurea concentrations. These cells have been studied
extensively. Similar to our findings, these cells have more resistance
to hydroxyurea than the increased CDP reductase activity suggests e.g.
100 fold increase in hydroxyurea resistance was associated with a 3 to
15 fold increase in CDP reductase activity (13). These cells have been
used in subsequent studies to characterize the free radical of the
enzyme (14). More recently, these cells have been used to purify the M2
subunit of ribonucleotide reductase to apparent homogeneity (15) which
is in increased amounts in hydroxyurea-resistant cells. However,
deoxyribonucleoside triphosphate metabolism has not been studied in
these fibroblasts.

We have documented elevated deoxyribonucleoside triphosphates in
hydroxyurea-resistant cells. In general, these pools are increased in
proportion to the CDP reductase activity, not the degree of resistance
to hydroxyurea. This suggests that M2 may be over-produced as a
response to hydroxyurea but that some other aspect of deoxynucleotide
metabolism limits CDP reductase activity and dNTP pools. One
possibility is that instead of M2 being rate limiting for CDP reductase
activity (16), M1 is rate limiting in M2 overproducing cells.
Alternatively, some other mechanism in addition to M2 overproduction
might protect against hydroxyurea toxicity.

The CDP reductase activity and dNTP pool elevation is paralleled by an increase in deoxyadenosine resistance. Deoxyadenosine is toxic to adenosine deaminase inhibited S49 cells (17). This is thought to be due to elevated pools of deoxyadenosine triphosphate which feedback inhibits ribonucleotide reductase, causing depletion of the dCTP and dTTP pools. Indeed, mutant S49 cell lines which have altered feedback inhibition probably due to structural alterations in the allosteric binding sites on M1 subunit (18) are resistant to deoxyadenosine toxicity. We reasoned that increased CDP reductase activity unrelated to allosteric effector mechanisms should make cells resistant to deoxyadenosine because the larger dCTP pools would be more resistant to depletion. This appeared to be the case.

We observed that HYU R cells are not arrested in G1 by cyclic AMP or agents that increase endogenous cyclic AMP concentration. The explanation of this is unclear and may be due to a separate effect of hydroxyurea on cyclic AMP metabolism. Conversely, the elevation in M2 activity in HYU R cells may prevent G1 arrest by cyclic AMP by an unknown mechanism.

In summary, high levels of hydroxyurea resistance are associated with moderate increases in CDP reductase activity, dNTP pools, and deoxyadenosine resistance which appears to be due to increased production of the M2 subunit of ribonucleotide reductase.

REFERENCES

1. L. Thelander and P. Reichard, Ann. Rev. Biochem. 48:133 (1979).
2. E. C. Moore and R. B. Hurlbert, J. Biol. Chem. 241:4802 (1966).
3. E. C. Moore, Adv. Enzyme Reg. 15:101 (1977).
4. L. Akerblom, A. Ehrenberg, A. Graslund, H. Lankinen, P. Reichard, and L. Thelander, Proc. Natl. Acad. Sci. USA 78:2159 (1981).
5. P. Reichard and A. Ehrenberg, Science 221:514 (1983).
6. S. Eriksson and D. W. Martin, Jr., J. Biol. Chem. 255:9436 (1981).
7. C. Garrett and D. Santi, Anal. Biochem. 99:268 (1979).
8. L. Gudas, S. Eriksson, B. Ullman, and D. W. Martin, Jr., Adv. Enzyme Regul. 19:129 (1981).
9. W. H. Lewis, B. A. Kuzik, and J. A. Wright J. Cell Physiol. 94:287 (1978).
10. P. Reichard and A. Ehrenberg, Science 221:514 (1983).
11. I. W. Caras, T. Jones, S. Eriksson, and D. W. Martin, Jr. J. Biol. Chem. 258:3064 (1983).
12. B. A. Kuzik and J. A. Wright, Biochem. Genetics 18:311 (1980).
13. L. Akerblom, A. Ehrenberg, A. Graslund, H. Lankinen, P. Reichard, and L. Thelander, Proc. Natl. Acad. Sci. USA 78:2159 (1981).
14. A. Graslund, A. Ehrenberg, and L. Thelander, J. Biol. Chem. 257:5711 (1982).
15. A. Graslund, A. Ehrenberg, and L. Thelander, J. Biol. Chem. 260:2737 (1985).
16. S. Eriksson and D. W. Martin, Jr. J. Biol. Chem. 156:9436 (1981).
17. B. Ullman, L. J. Gudas, A. Cohen, and D. W. Martin, Jr. Cell 14:365 (1978).
18. B. Ullman, S. M. Clift, L. J. Gudas, B. B. Levenson, M.A. Wormsted, and D. W. Martin, Jr., J. Biol. Chem. 255:8308 (1980).

PURIFICATION AND PROPERTIES OF HUMAN DEOXYCYTIDINE KINASE

Christina Bohman and Staffan Eriksson

Department of Biochemistry I, Medical Nobel
Institute, Karolinska Institute, Stockholm Sweden

INTRODUCTION

In tissues with active deoxyribonucleoside salvage pathways, such as the thymus and the spleen, it is likely that
the salvage of extracellular deoxyribonucleosides is a major
source of precursors for DNA - synthesis. Direct reduction
of ribonucleotides to the corresponding deoxyribonucleotides
via ribonucleotide reductase is the only other possible route
for the synthesis of deoxynucleosidetriphosphates. There is
evidence that in S-phase thymocytes, ribonucleotide reduction
is restricted mainly to the synthesis of purine dNTPs, while
pyrimidine dNTPs are synthesized primarily by extracellular
deoxyribonucleoside salvage (1). We report, here, the purification procedure and some properties of pure human deoxycytidine kinase, one of the principal enzyme responsible for
deoxyribonucleoside phosphorylation in human tissues (2-4).

EXPERIMENTAL PROCEDURE

Deoxycytidine kinase activity was assayed with 11 uM
5-^{3}H deoxycytidin, 4,7 mM ATP, 4,7 mM MgCl$_2$, 0,6 mM deoxyuridine, 58 mM potassium phosphate PH 7,6, 19 mM dithiothreitol, 14 mM sodium flouride and 0,5 mM cytidine. The reaction
was run for 30 and 60 minutes at 37C, and produced 5-^{3}H dCMP
was determined as in ref. (2). One unit is defined as 1 nmol
5-^{3}H dCMP formed per minute. Specific activity was defined
as units/mg protein, determined by Bradford.

Human leucemic spleen (1000 g) stored at -70C was homogenized in 1000 mls of 50 mM potassium phosphate buffer
(PH. 7,6), contained 5 mM dithiothreitol, 0,25 M sucros,
5 mM benzamidine and 0,5 mM phenylmethylsulfonylflyoride.
The homogenate was centrifugated at 9000 rpm for 40 minutes.
Deoxycytidin kinase was purified from the supernatant by
streptomycinsulphateprecipitation, ammoniumsulphate fractionation, 0 - 20% was discarded and 20 - 60% was used, chromatography on Tris-acryl DEAE,chromatography on hydroxylapatite
and affinity chromatography on dTTP-sepharose.

Table 1. Purification of deoxycytidine kinase

Fraction	Total volume	Total activity	Total protein	Specific activity
	ml	units	mg	
crude extract	365	260	14600	0,018
$(NH_4)_2SO_4$ (20 - 60%)	172	208	4990	0,042
DEAE Tris-acryl	690	76	445	0,17
Hydroxylapatite	200	51	49	1,03
dTTP-sepharose	4	30	0,330	107

RESULTS AND DISCUSSION

    The purification of deoxycytidin kinase is summarized in
table 1. The specific activity of the final preparation is
about 110 nmol/min/mg protein and represents a more than 6000
fold purification compared to the crude extract. The overall
yield is about 10%. Deoxycytidin kinase activity was not
separated from deoxyguanosine kinase activity during DEAE
chromatography. Deoxycytidine/deoxyguanosine kinase activity
eluated in two distinct peaks. A minor activity peak at
2 m si (0,07 M KCl) and the main activity peak was eluated at
around 5 m si (0,2 M KCl). When the final preparation was
tested for purity by SDS-gel electrophoresis, one major band
was observed. The band showed an estimated molecular weight
of 28 K. dalton. The correlation of the kinase activity and
the 28 K. dalton band during a dTTP gradient used to elute
the dTTP-sepharose (Fig. 1) suggested that this peptide
corresponds to the enzyme. From the results of Ultrogel
ACA-44 a native molecular weight of 56,000 was estimated and
the deoxycytidin kinase activity was directly correlated to
the 28 K. dalton band (Fig. 2-3). These results strongly
indicated that the deoxycytidin kinase has a molecular weight
of 56,000 and consists of two identical subunits.

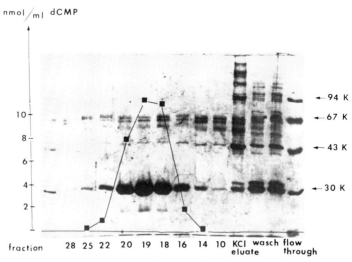

Figure 1.

Affinity chromatography on dTTP-sepharos, followed by SDS-
gelelectrophoresis. The pool from a hydroxylapatitechromato-
graphy containing 49 mg protein in a total volume of 200 ml,
were applied to a 10 ml column of dTTP-sepharose.

Deoxycytidine kinase was eluated with 28 ml of a 0 to 1,5 mM linear dTTP-gradient. The vertical axis in figure 1 indicates the enzyme activity per one ml fraction ■——■. 0,125 ml of a eluate fraction was loaded on a 10% SDS-gel electrophoresis.

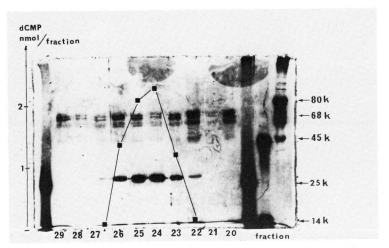

Figure 2.
Size fractional chromatography of deoxycytidine kinase on Ultro-gel ACA-44, followed by a 10% SDS-gelelectrophoresis. 5 ml of 1 mM dTTP eluate fraction containing deoxycytidine kinase activity was ultradialysed and applied to 18 ml of Ultro-gel ACA-44 in a column (45 x 0,75 cm). The column was eluted with 18 ml of 50 mM Tris PH 7,6, Mg Cl$_2$ 5 mM, DTT 5 mM glycerol 20% and KCl 0,1 M. The vertical axis indicates the enzyme activity per fraction ■——■ . 0,1 ml of a eluate fraction was loaded on a 10% SDS-gel.

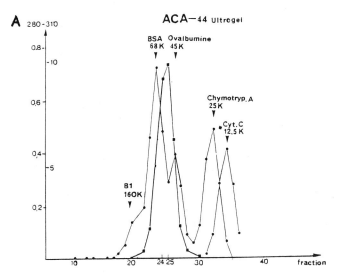

Figure 3.
Localisation of proteins with known molecular weight and native deoxycytidine kinase on a Ultro-gel-ACA-44 ●——● The vertical axis indicates the enzyme activity per fraction ■——■ .

The final preparation catalyzed the phosphorylation of deoxycytidine, deoxyguanosine and deoxyadenosine. The Km value
for deoxycytidine was 3,5 uM and the Km value for deoxyguanosine was 33 uM and finally the Km value for deoxyadenosine was 38 uM.

SUMMARY

The purification of deoxycytidine kinase from human
leucemic spleen reported here result in a pure protein of
molecular weight 28K. The enzyme eluates during gel filtration as a dimer and the same enzyme phosphorylates both
deoxycytidine, deoxyguanosine and deoxyadnosine, but with
different Km and Vmax values. Our results are in agreement
with earlier studies with partially purified calf thymus
deoxycytidine (5), but clearly different from some studies
on human deoxycytidine kinase (6).

REFERENCES

1.   A. Cohen, J. Barankiewicz, H. M. Lederman, E. W. Gelfand
      J. Biol. Chem. 258, 12334 - 12340, (1983).
2.   M. B. Meyers, W. Kreis, Arch. Biochem. Biophys. 177,
      10 - 15, (1976).
3.   M. C. Hurley, T. D. Palella, I. Fox, J. Biol. Chem. 288,
      15021 - 15027, (1983).
4.   I. P. Durham, D.H. Ives, J. Biol. Chem. 245, 2276 - 2284
      (1970).
5.   D. H. Ives, S. M. Wang, Methods enzymol. vol. LI, 337 -
      345.
6.   Y-C. Cheng, B. Domin, L-S. Lee, Biochem. Biophys. Acta
      487, 481 - 492, (1977)

PURIFICATION OF MAMMALIAN GLYCINAMIDE

RIBONUCLEOTIDE (GAR) SYNTHETASE

R.G. Hards, S.L. Graw, and D. Patterson

Eleanor Roosevelt Institute for Cancer Research
4200 East Ninth Avenue, B-129
Denver, CO 80262

Glycinamide ribonucleotide (GAR) synthetase is the second enzyme of the purine biosynthetic pathway (1). The gene coding for the enzyme has been regionally assigned to human chromosme 21, specifically the q22 band (2,3). This region of chromosme 21 when present in the trisomy state has been shown to lead to the Down Syndrome phenotype (3). It may be that an overproduction of this enzyme plays a role in the development of this disease as hyperuricemia has been associated with Down Syndrome (4).

Although GAR synthetase has been purified from bacteria (5,6,7) and yeast, (8) progress with the mammalian enzyme has been slower. This communication reports the purification of the enzyme rat liver.

METHODS

Enzyme activity was determined using a modification of the method described by Oates and Patterson (9). Assays were run for 10 minutes in 100 mM TRIS pH9, 8 mM $MgCl_2$, 1.5 mm ATP, 1.5 mM ribose-5'-phosphate, 250 mm NH_4Cl and 43 mM ^{14}C-glycine (2.55×10^5 cpm/nmol) in a total volume of 200 µl. A five minute pre-incubation in the absence of radioactive substrate was used to generate the second substrate required for the generation of GAR, 5'-phosphoribosyl-1-amine. The reaction was stoped by the addition of 50 µl of 30 percent trichloroacetic acid. The precipitated protein was chilled on ice for 10 minutes and pelleted by centrifugation at 13,000 xg in a Fisher microcentrifuge. Two hundred microliters of the supernatant was applied to a 0.5 x 3 cm column of AG50W-X8 and the GAR eluted with 1.6 ml of 50 mM ammonium formate pH 3.3.

Protein determinations were based on the Lowry modification of the Folin phenol method (10) using bovine serum albumin as a standard.

Antiserum to the purified rat liver GAR synthetase was raised in New Zealand white rabbits using the following protocol. On day zero, 131.6 µg of purified protein in a mixture of antigen, PBS and Freud's complete adjuvant (one half of volume adjuvant) was split into four 0.5 ml aliquots which were subcutaneously injected into each shoulder and hauch. On day 21, the same protocol was used except that only 13.2 µg of protein was used. Boosts of 19.8 µg of antigen were given subcutaneously without adjuvant on days 28 and 29, first in the shoulders and then in the hauches. A final boost of 98.8 µg was administered interperontoneally on day 30. Antiserum was obtained by bleeding the rabbits from the ear on day 37.

Proteins run on various polyacrylamide gels were transfered to nitrocellulose using the electrophoretic blotting system of American Bionuclear. Proteins were transferred from the gradient gels in 25 mM TRIS pH 8.3, 192 mM glycine at 60 volts for 4 hours. Transfer from SDS gels was effected in 25 mM TRIS pH 8.3, 192 mmglycine /20% (V/v) methanol at 60 voles for 4 hours. Proteins from the isoelectric focusing gels were transferred in 0.7% acetic acid at 100 volts for 3.5 hours.

After transfer to nitrocellulose, GAR synthetase was detected using the antibody raised against the enzyme in conjunction with the BioRad Immuno-Blot Assay Kit using the manufacturers suggested methodology.

RESULTS AND DISCUSSION

Using the protocol presented in Table I GAR synthetase from rat liver was purified 1800 fold with a final yield of 2.5 percent.

Table 1

Purification of Glycinamide Ribonucleotide Synthtase From Rat Liver

Fraction	Vol. (ml)	Protein (mg/ml)	Protein (mg)	Spec. Act. (cpm/mg/10^7)	Total Act. (cpm/10^7)	% Yield	Purification Factor
Homogenate	962	37.31	35396	1.68×10^4	6.02×10^8	100.0	1.00
15000 x G	818	19.90	16276	3.43×10^4	5.58×10^8	92.7	2.04
100000 x G	766	18.85	14143	4.01×10^4	5.79×10^8	96.2	2.39
0.2% protamine sulfate	700	9.53	6671	2.86×10^4*	1.91×10^8*	31.7*	1.70*
20-70% $(NH_4)_2SO_4$	164	25.41	4167	3.47×10^4*	1.44×10^8*	23.9*	2.07*
DEAE Cellulose	16.5	27.02	445.8	1.94×10^5	0.86×10^8	14.3	11.5
Glycine Sepharose	3.2	84.12	269.2	2.75×10^5	0.74×10^8	12.2	16.4
ATP(C_8) Agarose	45	5.22	234..9	6.60×10^6	0.78×10^8	13.0	39.3
ATP(OH) Agarose	1.8	1.61	2.9	1.87×10^7	0.40×10^8	6.6	815.5
ATP(C_8) Agarose	1.3	2.56	3.2	3.03×10^7	0.15×10^8	2.5	1803.6

*assays run in presence of protamine or ammonium sulfate

Examination of the final preparation (ATP(C8)) on a 5 to 20%
polyacrylamide gel revealed the presence of several closely spaced
protein bands. Subsequent enzymatic analysis of the bands showed that
each possessed GAR synthetase activity. As the structure of these
bands changed during storage at 4°C with the higher molecular weight
bands disappearing and the new smaller proteins emerging, it would
seem the enzyme was subject to a slow degredation. This problem made
the accurate estimation of protein molecular weights and subunit composition difficult if not impossible.

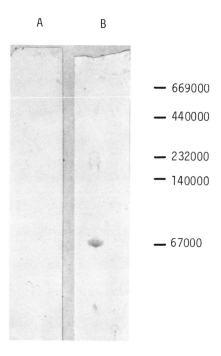

A   B

— 669000

— 440000

— 232000

— 140000

— 67000

Fig. 1. Western blot analysis of rat liver extract run on a 5
to 20% acrylamide gel and probed with non-immune (A) and anti-GAR
synthetase (B) sera.

To overcome this difficulty, fresh rat livers were extracted
using several techniques designed to reduce the degradation effect,
run on various polyacrylamide gels and probed with the anti-GAR
synthetase serum using the Western blot methodology. As shown in
figure 1, native gradient gels detected a single protein band of
70,000 ± 1500 daltons.

Similar Western blot analysis of SDS and isoelectric focusing gels also revealed single protein bands. In the case of the SDS denaturing gels, monomer molecular weight was determined to be 65,500 ± 3000 daltons. The isoelectric focusing gels indicated rat liver GAR synthetase to have a pI value of 4.8.

Taken together, the results from the three types of gels strongly indicate that GAR synthetase was purified to homogeneity. Moreover, the gradient and SDS gels indicate the native enzyme is a monomer of 70,000 daltons.

Various reports exist which suggest that GAR synthetase may be part of a multifunctional enzyme. A single protein containing both GAR synthetase and 5'-aminoimidazole ribonucleotide (AIR) synthetase activities has been purified from Schizosaccharomyces pombe (8). Complementation studies with ade 5 and ade 7 mutants of Saccharomyces cerevisiae (11) are also suggestive of the GAR and AIR synthetase activities residing on the same polypeptide. Biochemical analysis of mammalian auxotrophs also suggest GAR synthetase and AIR synthetase are found on a single polypeptide (12) and both activities have been mapped to human chromosome 21 (13). Very recently it has been shown that a single polypeptide from Drosophila contains GAR synthetase, AIR synthetase and GAR transformylase (unpublished results).

To investigate this possibility in the system under study here, the antibody directed against GAR synthetase was used to attempt precipitation of AIR synthetase and GAR transformylase as well as GAR synthesis. As shown in Table II, the polyclonal antibody was able to precipitate GAR synthetase and GAR transformylase. Immunoprecipitation of AIR synthetase was more difficult to determine as the sera seemed to have an adverse effect on measurement of the activity. In a preliminary experiment using partially purified preparations of the immune and nonimmune sera, the antibody raised against GAR synthetase did seem to precipitate AIR synthetase activity. Thus like the Drosophila situation, the rat liver protein would seem to be trifunctional.

Table II - Immunoprecipitation of GAR Synthetase
and GAR Transformylase

| Activity | Enzyme Levels (nmol/mg/min) in Supernatants Treated With | |
	Immune Serum	Nonimmune Serum
GAR Synthetase	0.000	0.078
GAR Transformylase	0.000	0.105

A modified procedure based on the protocol developed for purification of GAR synthetase from rat liver has been used to partially purify the activity from CHO cells. To date, the most pure preparation has five protein bands. In the very near future we expect to purify CHO and human GAR synthetase to homogeneity.

REFERENCES

1) D. Patterson Som. Cell Genet. 1:91-100 (1975)

2) D. R. Cox, H. Teshima,   S. Vora and C. J. Epstein Cytogenet. Cell Genet. 31:441-442 (1985).

3) B. Chadefaux, D. Allard, M. O. Rethore, O. Raoul, M. Poissonier, S. Gelgenkrantz, C. Cheury and H. Jerome Human Genet 66:190-192.

4) S. S. Pant, H. W. Moser and S. M. Krane J. Clin. Endocrinol. Metabol. 28:412-478 (1968).

5) D. P. Nierlich and B. Magasanik J. Biol. Chem. 240:366-374 (1965).

6) D. P. Nierlich Met. Enzymol. 51:179-185 (1978).

7) N. R. Gandhi and C. A. Westby Enzymologia 42:185-200 (1972).

8) R. Fluri, A. Coddington and V. Flury Molec. Gen. Genet. 147:211-282 (1976).

9) D. C. Oates and D. Patterson Som. Cell Genet. 3:561-577 (1977).

10) O. H. Lowry, N. J. Rosebrough, A. L. Farr and R. J. Randall J. Biol. Chem. 193:205- (1951).

11) T. S. Gross and R. A. Woods Biochem. Biophys. Acta. 247:13-21 (1971).

12) D. Patterson, S. Graw and C. Jones Proc. Natl. Acad. Sci. USA 78:405-409 (1981).

13) D. Patterson, C. Jones, C. Socggin, Y.E. Miller and S. Graw Ann. N.Y. Acad. Sci. 396:69-81 (1982).

STEADY-STATE KINETICS

OF THE REACTION CATALYZED BY GMP REDUCTASE

A.Spadaro, A.Giacomello, and C.Salerno

Institutes of Rheumatology and Biological Chemistry
University of Rome, and C.N.R. Centre for Molecular
Biology, Rome, Italy

INTRODUCTION

The continuous replenishment of the IMP pool through the concerted
action of GMP reductase, AMP deaminase, and phosphoribosyltransferases most
likely provides a flexible mechanism for adjusting the intracellular level
of the individual purine nucleotides to both the varying supply of exoge-
neous purines and the momentary metabolic needs of the cell (1). Such a
mechanism can be of vital importance in tissues which possess a limited
capacity for synthesizing its own nucleotides and depend to a large extent
on the supply of preformed purines.

The task of maintaining constant and properly balanced pool of purine
nucleotides poses particular problems in human erythrocytes which have lost
completely the ability to produce purines de novo and to synthesize AMP from
IMP (2). In these cells the catabolism of purine nucleotides appears to be
mainly directed via IMP and inosine by hypoxanthine (1,3). The irreversible
reductive deamination of GMP to IMP is catalyzed by GMP reductase (EC 1.6.
6.8) which utilizes NADPH as hydrogen donor. This enzyme is markedly inhibited by XMP and activated by GTP (4). In the present paper, we report
kinetic studies on GMP reductase purified from human erythrocytes.

EXPERIMENTAL PROCEDURES

GMP reductase was purified from human erythrocytes by the procedure
described in reference (4). IMP from Boehringer was purified by ion-exchange
chromatography (5). Other reagents were high purity commercial samples
from Boehringer, Merck, and Sigma. The reductive deamination of GMP was
followed spectrophotometrically at 340 nm. All kinetic experiments were
performed in 1 mM ethylendiaminetetraacetate, 10 mM 2-mercapto ethanol,
and 0.1 M Tris(hydroxymethyl)aminomethane buffer pH 7.4 at 37°C.

RESULTS AND DISCUSSION

Preliminary kinetic studies on human GMP reductase were made by Spector
et al. (4). The initial velocity analysis and the inhibition patterns by
substrate analogs (XMP, 2'-dXMP, and 8-azaXMP) were consistent with a sequential mechanism for the reaction, either ordered with GMP binding first or
random.

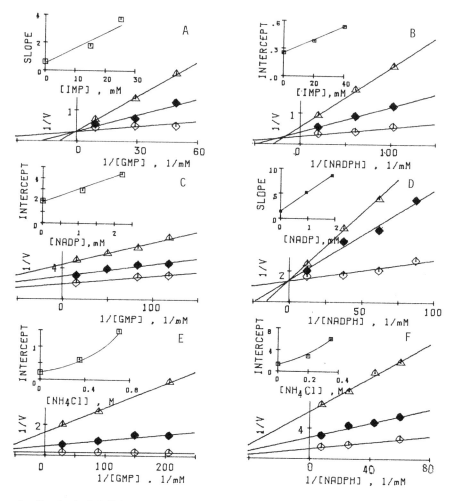

Fig. 1. Product inhibitions of the reaction catalyzed by GMP reductase by
IMP (A,B), NADP (C,D), and NH_4Cl (E,F). Left: NADPH concentration
was held constant (A, 80 uM; C, 24 uM; E, 100 uM). Right: GMP concentration was held constant (B, 60 uM; D, 4.8 uM; F, 68 uM). Inhibitor concentrations were as shown in the insets. All other conditions were as described under "Experimental procedures".

Further informations about the reaction mechanism can be obtained by
product inhibition studies. As shown in fig. 1, IMP inhibition is competitive with respect to GMP and noncompetitive with respect to NADPH. NADP
is competitive with respect to NADPH and noncompetitive with respect to GMP.
Ammonium ions are noncompetitive with both substrates. Thus, it is likely
to assume that (i) substrates bind to the reductase in an ordered sequence,
first GMP and then NADPH; (ii) NADP is released from the enzyme-substrate
complex by a Theorell-Chance mechanism; (iii) ammonia and IMP are sequentially liberated in this order. Dead-end complexes have to be included in
the model. In fact, as shown in figs. 1E and 1F, ammonia affects both the
1/v-axis intercept and the slope of the double reciprocal plots obtained by
varying substrate concentrations in the absence of other reaction products.
This observation suggests, according to the general Cleland's rules for
predicting inhibition patterns (6), that ammonia binds also to the free enzyme. Another dead-end complex forms most likely between ammonia and the
enzyme-IMP-ammonia ternary complex since replots of 1/v-axis intercept

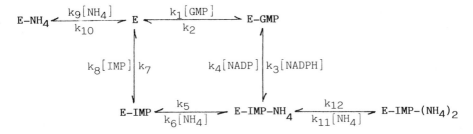

Fig. 3. Basic King-Altman interconversion figure of the reaction mechanism
proposed for GMP reductase.

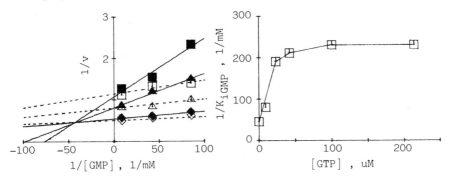

Fig. 4. (left) Activation of the reaction catalyzed by GMP reductase by GTP.
NADPH concentration was: 18 uM (◇◆), 30 uM (△▲), and 90 uM (▯■).
GTP concentration was: 0.21 mM (---), none (—). Other conditions
were as described under "Experimental procedures".

Fig. 5. (right) Abscissa value (= $-1/K_{iGMP}$) of the interception point
of families of lines obtained as shown in fig. 4 as a function of
GTP concentration.

versus ammonia concentration give rise to parabolic curves (figs. 1E and
1F). The scheme of the reaction mechanism is given in fig. 3.

    GTP does not appreciably affect the inhibition patterns by IMP, NADP,
and ammonia. When in Lineawever-Burk plots GMP is varied in the absence of
products, the more evident effect exerted by GTP is a decrease of the abscissa value of the interception point of the lines obtained at a series of
fixed NADPH concentrations (figs. 4,5). Both Vmax and Km for NADPH seem
to be independent of GTP concentration. When NADPH is the varied substrate,
the interception point of the lines obtained at a series of fixed concentrations of GMP does not appear to be affected by GTP. Since the lines con-
verge very near to 1/v axis, it is not possible to accurately study the
effect of GTP on the Km for GMP.

    If we assume that the model reported in fig. 3 still holds in the case
of the GTP-activated enzymatic reaction, the effect exerted by the activator
could be attributed, at least in part, to an increase of the binding constant
($1/K_{iGMP}$) of the enzyme-GMP complex. On the other hand, the dissociation
first-order rate constant of the enzyme-GMP complex (= $Vmax/Km_{NADPH}[E]$) does
not appear to be affected by GTP. An explanation for these seemingly conflicting results could be that GTP increases the apparent second-order rate
constant for GMP binding. This effect could result from protein conformational changes associated with a decrease of the activation energy for substrate binding. If this is true, the reductase is an example of an enzyme
modulated by the activator in such way that the affinity for only one of its

323

substrates increases while the turnover number is unchanged (6).

ACKNOWLEDGEMENTS

This work was supported in part by a grant from Ministero della Pubblica Istruzione (Italy).

REFERENCES

1. A.W. Murray, The biological significance of purine salvage, Ann. Rev. Biochem. 40:773 (1971).
2. B.A. Lowy and B. Dorfman, Adenosylsuccinase activity in human and rabbit erythrocyte lysate, J. Biol. Chem. 245:3043 (1970).
3. C. Salerno and A. Giacomello, Hypoxanthine-guanine exchange by intact human erythrocytes, Biochemistry 24:1306 (1985).
4. T. Spector, T.E. Jones, and R.L. Miller, Reaction mechanism and specificity of human GMP reductase, J. Biol. Chem. 254:2308 (1979).
5. I.C. Caldwell, Ion-exchange chromatography of tissue nucleotides, J. Chromatog. 44:331 (1969).
6. I.H. Segel, "Enzyme Kinetics: Behavior and Analysis of Rapid Equilibrium and Steady-State Enzyme Systems", John Wiley and Sons, New York (1975).

PURINE DE NOVO SYNTHESIS AND SALVAGE DURING TESTICULAR DEVELOPMENT

IN THE RAT

Jennifer Allsop and Richard W.E. Watts

Division of Inherited Metabolic Diseases
MRC Clinical Research Centre, Watford Road
Harrow, Middlesex HA1 3UJ, U.K.

We have encountered several patients with the Lesch-Nyhan syndrome in whom, although the testes were clinically normal in early childhood, they failed to develop at puberty and in one they could not be found at autopsy when he was 18 years old. This patient showed no evidence of pre-pubertal development although the plasma testosterone and gonadotrophin levels were consistent with approaching puberty and with his bone age of 13.2 years[1]. The testes of another clinically prepubertal Lesch-Nyhan syndrome patient who died at age 14 years showed tubules lined by simple germinal epithelium with no evidence of spermatogenesis. The interstitial tissue contained numerous fibroblastic cells but no interstitial cells (Leydig cells) were seen.

We have previously shown that rat testicular tissue has hypoxanthine phosphoribosyltransferase (HPRT: EC 2.4.2.8), adenine phosphoribosyltrans-ferase (APRT: EC 2.4.2.7) and amidophosphoribosyltransferase (PRPP-At; EC 2.4.2.14) activities and therefore, by inference, both the salvage and purine de novo synthesis metabolic pathways[2].

The present study was designed to determine if there is a critical stage in testicular development which might be specifically susceptible to HPRT deficiency.

METHODS

We have compared the time course of testicular development in the Sprague-Dawley rat as judged histologically and endocrinologically with changes in the activity of the purine de novo synthesis pathway and purine salvage pathways. Purine de novo synthesis activity was measured by the incorporation of [^{14}C]formate into the testicular tissue total purines. Approximately 50 mg of tissue were used for each assay under the conditions described previously for liver[3,4]. 25 µg and 50 µg tissue per assay were used to measure HPRT and APRT activities respectively[5]. Plasma testosterone was measured by the DPC Coat-A-Count Assay (Diagnostic Products, Wallingford Oxfordshire). Protein was measured by the method of Lowry et al.[6]. The histology of the testes was assessed on haematoxylin and eosin stained and on Feulgen stained paraffin embedded sections. The biochemical and histo-logical observations were made at 3, 7, 14, 17, 28, 35, 42, 56 and 70 days of age.

RESULTS

The results of the enzyme and plasma testosterone assays are shown in Fig. 1. In the histological preparations, meioses were prominent in spermatids and spermatogonia at age 17 days but not before, and active spermatogenesis was only seen in the specimens from age 35 days onwards.

DISCUSSION

Both the HPRT and APRT activities are two to three orders of magnitude greater than the purine de novo synthesis activity. The peak of purine de novo synthesis activity at 17 days coincides with the sample in which meioses were prominent in spermatids and spermatogonia indicating that the germinal epithelium had begun to differentiate. At this time, the ratio of [HPRT activity]:[purine de novo synthesis activity] is about 300:1. The HPRT activity climbs more slowly to reach its maximum in the first specimen showing complete spermatogenesis. The ratio [HPRT activity]: [purine de novo synthesis activity] is about 800:1 at 3 days and after spermatogenesis is established. We interpret this as indicating that purine de novo synthesis is relatively more important later during the period before there is complete spermatogenesis than later on. However, at all stages of development, purine salvage catalysed by HPRT is a very much more active and important pathway for purine nucleotide production. This adequately explains the failure of testicular maturation in the patients and shows that there is no stage during postnatal organogenesis of the testis at which HPRT deficiency is likely to act selectively. Testosterone was not detectable until after the peak purine de novo synthesis rate had been achieved. This suggests that the progressive rise in the activity of this pathway is initiated by a testosterone-independent signal, possibly FSH. The slower rise in HPRT activity (Fig. 1) also begins before testosterone is detectable although its steepest rise is after the Leydig cells become active as judged by testosterone production. The changes in the activities of the alternative pathways of purine nucleotide production are complete well before the animals are sexually mature at 56 days of age.

These studies were performed on whole testicular tissue so that biochemical changes in the tubule cells and the interstitial Leydig cells cannot be distinguished from one another. It seems most probable that most of what is observed represents the metabolic function of the tubule epithelium and that HPRT-deficiency acting there is sufficient to explain the lack of structural development in the Lesch-Nyhan syndrome. Thus, HPRT-deficiency can be viewed as inducing a state of end-organ resistance to hormonal influences on the testicular tubule. However, it could also impair the function of the hypothalamic-pituitary-testicular axis by blunting the modulation of the gonadotrophin releasing factor and the gonadotrophins. It could also impair Leydig cell development and function, either stopping or delaying the onset of androgen production by these cells. There is increasing evidence for the view that HPRT deficiency produces its neurological concomitants by impairing neurotransmitter production or release[7], and similar considerations may well apply to the hypothalamic neuroendocrine system. However, it seems contradictory that developing Leydig cells were not identified in the testis of the 14 year old boy with the Lesch-Nyhan syndrome although in the 18 year old patient testosterone was being produced and the gonadotrophin levels were normal.

Patients with the sex linked recessive gout syndrome who have a partial HPRT deficiency have fathered children[8] but there are no data either on their fertility or on the degree of HPRT deficiency which is still compatible with testicular development.

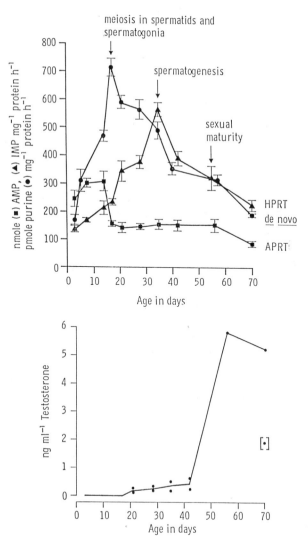

Fig. 1. The activities of the purine de novo synthesis pathway
and the purine phosphoribosyltransferases in rat
testicular tissue at different ages, together with
the plasma testosterone concentrations. The time
points at which meioses of the spermatids and spermatogonia and complete spermatogenesis were first seen
are shown together with the age at which the animals
become sexually mature.

REFERENCES

1. R.W.E. Watts, E. Spellacy, D.A. Gibbs, J. Allsop, R.O. McKeran and
    G.E. Slavin, Clinical, post-mortem, biochemical and therapeutic
    observations on the Lesch-Nyhan syndrome with particular reference
    to the neurological manifestations, Quart. J. Med., N.S., 51:43
    (1982).

2. J. Allsop and R.W.E. Watts, Activities of amidophosphoribosyltransferase and purine phosphoribosyltransferases in developing rat brain,
   Adv. exp. Biol. Med., 122A:361 (1980).

3. J. Allsop and R.W.E. Watts, Purine synthesis and salvage in brain and
   liver, Adv. exp. Biol. Med., 165B:21 (1984).

4. J. Allsop and R.W.E. Watts, Purine de novo synthesis in liver and
   developing rat brain, and the effect of some inhibitors of purine
   nucleotide interconversion, Enzyme, 30:172 (1983)

5. J. Allsop and R.W.E. Watts, Activities of amidophosphoribosyltransferase (EC 2.4.2.14) and the purine phosphoribosyltransferases
   (EC 2.4.2.7 and 2.4.2.8) and the phosphoribosylpyrophosphate content
   of rat central nervous system at different stages of development,
   J. Neurol. Sci., 46:221 (1980).

6. O.H. Lowry, N.J. Rosebrough, A.L. Farr and R.J. Randall, Protein
   measurement with the Folin phenol reagent, J. biol. Chem., 193:265
   (1951).

7. R.W.E. Watts, Advances in Enzyme Regulation, Ed. G. Weber, 23 (1984)
   pp. 25-58.

8. W.N. Kelley, M.L. Greene, F.M. Rosenbloom, J.F. Henderson and
   J.E. Seegmiller, Hypoxanthine-guanine phosphoribosyltransferase
   deficiency in gout, Ann. Int. Med., 70:155 (1969).

PATHWAYS OF ADENINE NUCLEOTIDE CATABOLISM IN HUMAN ERYTHROCYTES

F. Bontemps, G. Van den Berghe and H.G. Hers

Laboratoire de Chimie Physiologique, Université de Louvain and
International Institute of Cellular and Molecular Pathology
avenue Hippocrate 75, B-1200 Brussels, Belgium

INTRODUCTION

The catabolism of the adenine nucleotides proceeds from AMP and its initial steps can theoretically occur via two possible enzyme sequences : either a prior deamination by AMP deaminase, followed by dephosphorylation by 5'-nucleotidase, or a prior dephosphorylation by the latter enzyme, followed by deamination by adenosine deaminase (scheme 1). Both pathways lead to the formation of hypoxanthine, which in erythrocytes constitutes the terminal product of adenine nucleotide catabolism. The second one can be interrupted at the level of adenosine which can be reconverted to AMP by adenosine kinase, resulting in a substrate cycle which has been demonstrated in the liver (1). In the present study, the inhibitors of adenosine metabolism indicated on the scheme were used to determine, under various experimental conditions, the respective participation of both enzyme sequences in the formation of hypoxanthine and the occurence of recycling of adenosine in human erythrocytes.

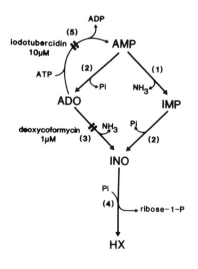

Scheme 1 Pathway of degradation of adenine nucleotides in human erythrocytes ADO, adenosine; HX, hypoxanthine; INO, inosine; (1) AMP deaminase, (2) 5'-nucleotidase, (3) adenosine deaminase, (4) nucleoside phosphorylase, (5) adenosine kinase.

METHODS

Fresh blood, collected on heparin, was centrifuged to remove plasma and buffy coat. The red cells were washed three times with Krebs-Ringer-bicarbonate (KRB) buffer pH 7.4, gassed with 95 % O_2 - 5 % CO_2, containing 5 mM glucose. They were resuspended, unless otherwise noted, in the same medium at an haematocrit of 20 % and their adenine nucleotides were labelled by a 60 min preincubation at 37° with 1 µM [U-^{14}C] adenine. This was followed by two washes and resuspension in KRB buffer or modifications thereof, with or without inhibitors, as described under results. Incubations were performed in carefully regassed and stoppered vials to avoid modifications of pH by CO_2 loss.

For analytical determinations on the cell suspensions as a whole, 0.5 ml aliquots were transfered into 0.25 ml of ice-cold 10 % $HClO_4$ and neutralized as described in (2). For separate analysis of the medium and red cell content, 0.5 ml of cell suspension were centrifuged through a 0.5 ml layer of silicone as given in (3).

Nucleotide concentrations were measured by HPLC and their radioactivity, as well as that in nucleosides and bases, by TLC, as described previously (2). The concentration of the latter compounds was calculated from the specific radioactivity of the adenine nucleotides, which remained constant over the duration of the experiment.

A concentration of deoxycoformycin (dCF) of 1 µM was found to inhibit by 95-98 % the deamination of 10-100 µM [8-^{14}C]-adenosine by intact erythrocytes. The incorporation of 10 µM [8-^{14}C] adenosine into the erythrocytic adenine nucleotides was 98 % inhibited by 10 µM of the adenosine kinase inhibitor iodotubercidin (ITu). The adenosine transport inhibitor R-51469 (Mioflazine, a gift from Dr. H. Van Belle, Janssen Pharmaceutica, Beerse, Belgium) was used at the concentration of 10 µM, which inhibited the influx as well as the efflux of 100 µM [8-^{14}C]-adenosine from erythrocytes by 99 %.

RESULTS AND DISCUSSION

1) Catabolism of the erythrocytic adenine nucleotides under physiological conditions

Erythrocytes incubated in KRB containing 5 mM glucose and 1.2 mM Pi maintained constant adenine nucleotide levels for several hours and produced only a small amount of hypoxanthine (4.2 ± 0.7 nmol/h per ml of packed cells; mean ± SEM of 8 experiments). No adenosine, inosine or adenine could be detected in the cell suspension. Addition of dCF did not decrease the production of hypoxanthine. This indicates that under physiological conditions the degradation of AMP proceeds by way of AMP deaminase, followed by dephosphorylation of IMP. No recycling of adenosine could be evidenced since inhibition of adenosine kinase by ITu did not increase the production of hypoxanthine and addition of both ITu and dCF did not provoke an accumulation of adenosine (results not shown).

2) Degradation of the adenine nucleotides induced by glucose deprivation

Incubation of the erythrocytes in KRB buffer devoid of glucose resulted in a fall of the concentration of ATP to 20 % of its initial value after 4 h (Fig. 1a), in a 2- to 3-fold increase in ADP (Fig. 1b) and in a 30-fold elevation of AMP (Fig. 1c). IMP was barely modified, but there was a progressive loss of the adenine nucleotide pool, one third of which was lost after 4 h (Fig. 1d) and quantitatively recovered under the form of hypoxanthine (Fig. 1e). The rate of formation of the purine base increased approx. 20-fold, to 79.5 ± 5.6 nmol/h per ml of packed cells (mean ± SEM of 7 experiments).

The arrest of glycolysis, provoked by lack of glucose, results in the suppression of the regeneration of ATP, which explains the elevation of AMP. This elevation, in turn, causes an increase in the rate of adenine nucleotide catabolism. The observation that inhibition of adenosine deaminase by 1 uM dCF decreased the rate of production of hypoxanthine by 75 % (Fig. 1e) indicates that during glucose deprivation 3/4 of the catabolism of AMP proceeds by way of the sequence 5'-nucleotidase/adenosine deaminase and 1/4 through the sequence AMP deaminase/5'-nucleotidase. The absence of accumulation of adenosine during the first hour of incubation in the presence of dCF (Fig. 1f), and the noticeable protection of the adenine nucleotide pool (Fig. 1d), reflected in a decrease in the production of the sum of purine catabolites (Fig. 1h), indicate that adenosine was in great part rephosphorylated by adenosine kinase during this time interval. This is confirmed by the observation that in the presence of both dCF and ITu, adenosine accumulated from the beginning of the experiment (Fig. 1f) and together with adenine, quantitatively accounted for the decrease in the production of hypoxanthine (Fig. 1h). The recycling of adenosine is also evidenced by the effect of ITu alone to increase the rate of production of hypoxanthine (Fig. 1e). In the experiment shown, this rate was elevated from 100 to 120 nmol/h per ml of packed cells, indicating that recycling proceeded at the rate of 20 nmol/h per ml of packed cells. The formation of adenine, which was only recorded when adenosine accumulated (Fig. 1g), is most likely explained by a phosphorolysis of adenosine favoured by the

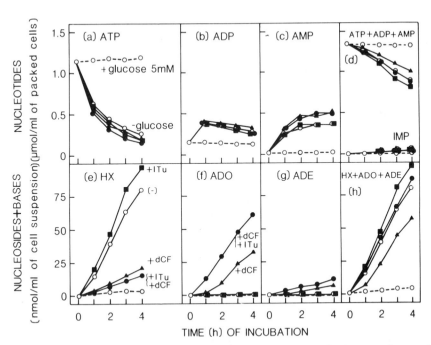

Fig. 1 Effect of lack of glucose on the catabolism of the erythrocytic adenine nucleotides After two washes performed at 20°C in glucose-free medium, the cells were resuspended in KRB in the presence (dotted lines) or in the absence (full lines) of glucose, without inhibitors (o), with 10 uM ITu (■), 1 μM dCF (▲) or both (●) and incubated immediately at 37°C.

elevation of Pi (4). Indeed, deprivation of glucose for 2 h resulted in an increase of the concentration of Pi inside the erythrocytes from 0.88 ± 0.05 to 2.42 ± 0.15 µmol/ml of packed cells (mean ± SEM of 4 experiments).

Experiments with the adenosine transport inhibitor R-51469 demonstrated that adenosine accumulated intracellularly, indicating that it was formed by a cytosolic 5'-nucleotidase (not illustrated).

3) Degradation of the adenine nucleotides provoked by alkalinization of the medium

Incubation of the erythrocytes in KRB containing 5 mM glucose and 50 mM NaHCO$_3$ (instead of the physiological concentration of 25 mM), resulting in a pH of about 7.7, provoked a loss of ATP, which was less marked than with glucose deprivation and came to an arrest after 2 h (Fig. 2a). The concentration of ADP increased approx. 2-fold (Fig. 2b) and that of AMP about 5-fold (Fig. 2c) after 1 h. Both concentrations decreased progressively after 2 h. These modifications are believed to be caused by an imbalance between the first half of glycolysis, which consumes ATP and is accelerated by alkalinization, and the second half, which produces ATP. This imbalance is evidenced by the more than 50-fold increase in fructose 1,6-bisphosphate and triose phosphates recorded during alkalinization

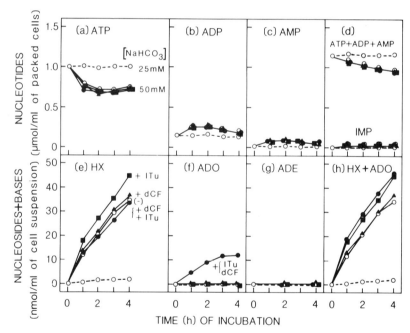

Fig. 2 Effect of alkalinization of the incubation medium on the catabolism of the adenine nucleotides The cells were resuspended in the presence of glucose in the usual KRB buffer in which the concentration of NaHCO$_3$ is 25 mM (dotted lines) or in a modified KRB in which the concentration of NaHCO$_3$ was increased to 50 mM and the concentration of NaCl commensurately decreased (full lines). They were incubated without inhibitors (o) or with 10 µM ITu (■), 1 µM dCF (▲) or both (●).

(5-8). Like in glucose deprivation, the rise in AMP was accompanied by a loss of the total adenine nucleotide pool (Fig. 2d), which was quantitatively recovered under the form of hypoxanthine (Fig. 2e), the rate of formation of which was increased 15-fold, to 63.6 ± 5.8 nmol/h per ml of packed cells (mean $\pm$ SEM of 8 experiments).

In marked contrast with glucose deprivation, dCF did not influence the production of hypoxanthine (Fig. 2e), indicating that it was formed by way of the sequence AMP-deaminase/5'-nucleotidase. However, the observation that ITu alone increased the production of hypoxanthine from 54 to 75 nmol/h per ml of packed cells (Fig. 2e) indicated that adenosine was formed from AMP by 5'-nucleotidase but completely rephosphorylated by adenosine kinase, none of it being converted to hypoxanthine. This recycling was confirmed by the observation, upon addition of dCF and ITu together, of an accumulation of adenosine (Fig. 2f). Also in contrast with glucose deprivation, alkalinization lowered the intraerythrocytic concentration of Pi from 0.95 ± 0.02 to 0.43 ± 0.02 (mean $\pm$ SEM for n = 4) μmol per ml of packed cells. This lowering is caused by the trapping of Pi under the form of fructose 1,6-bisphosphate and triose-phosphates and explains the absence of formation of adenine when adenosine accumulated (Fig. 2g).

4) Influence of the concentration of Pi in the incubation medium

The observation that the intraerythrocytic concentration of Pi increased during glucose deprivation and decreased after alkalinization prompted a study of the influence of variations of the concentration of Pi in the incubation medium on the production of hypoxanthine. To distinguish between the effects of Pi on 5'-nucleotidase and on AMP deaminase, experiments were also performed in the presence of dCF and ITu together. Under this condition, the production of adenosine reflects the activity of 5'-nucleotidase and the residual formation of hypoxanthine, that of AMP deaminase.

Fig. 3a shows that at physiological pH and in the presence of 5 mM glucose, suppression of Pi in the incubation medium resulted in an approx. 3-fold enhancement of the rate of production of hypoxanthine. Elevation of Pi to 10 mM, abolished the production of the purine base. Identical results were recorded in the presence of both dCF and ITu, indicating a pronounced effect of the concentration of Pi on the production of hypoxanthine proceeding via AMP deaminase and, by the same token, a great sensitivity of this enzyme to variations of the concentration of Pi. As evidenced by the absence of accumulation of adenosine in the presence of both dCF and ITu (Fig. 3d), no dephosphorylation of AMP occured at 1.2 and 10 mM Pi, but the small accumulation of the nucleoside recorded in the absence of extracellular Pi indicated that the cytoplasmic 5'-nucleotidase became slightly active under this condition. In accordance with the very low rates of catabolism at physiological pH and glucose concentration, no significant changes in erythrocytic ATP were recorded over the duration of the experiment (not illustrated). An approx. 2-fold increase in AMP in the absence of Pi, and a 50 % decrease in the presence of 10 mM Pi (not shown), were nevertheless observed and most likely also contribute to the modifications of the rate of catabolism recorded. Intracellular Pi decreased by 60 % in the absence of Pi and increased progressively in the presence of 10 mM Pi (Fig. 3g).

As depicted in Fig. 3b, the high rates of hypoxanthine formation induced by the suppression of glucose were markedly less influenced by modifications of the concentration of Pi in the incubation medium. The residual rate of hypoxanthine production recorded in the presence of both dCF and ITu, which reflects the activity of AMP deaminase was, nevertheless, approx. doubled in the absence of Pi and reduced by about 75 % in the presence of 10 mM Pi. This confirms the Pi-sensitivity of AMP deaminase.

The small effect of the modifications of Pi on the production of adenosine recorded in the presence of dCF and ITu (Fig. 3e), indicates that the dephosphorylation of AMP is much less sensitive to Pi than its deamination, and mainly responsive to variations in substrate concentration.

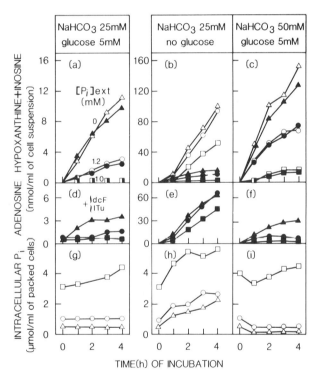

Fig. 3 Effect of the concentration of Pi on the catabolism of the adenine nucleotides Preincubation of the erythrocytes with [^{14}C] adenine, and incubations were performed in the presence of three concentrations of Pi : 0 mM (△,▲), 1.2 mM (o,●) or 10 mM (□,■). The concentration of Ca^{++} in the KRB buffer was also reduced from 2.5 mM to 1.25 mM. The incubation medium contained 5 mM glucose (left and right pannels) or not (central pannels) or was alkalinized (right pannels). Erythrocytes were incubated without (open symbols) or with (closed symbols) 10 μM ITu and 1 μM dCF. Inosine was only detected at alkaline pH and in the absence of extracellular Pi and amounted then to 60 % of the sum of hypoxanthine and inosine.

At alkaline pH, like under physiological conditions, modifications of the extracellular concentration of Pi markedly influenced the catabolism of the erythrocytic adenine nucleotides (Fig. 3, right hand pannels). In the absence of Pi, the rate of production of purine catabolites was nearly doubled (Fig. 3c). Due to the very low intracellular Pi concentrations recorded in the Pi-free medium (Fig. 3i), resulting in a limitation of the activity of nucleoside phosphorylase, the formation of hypoxanthine was accompanied by a nearly equivalent production of inosine. In the presence of 10 mM Pi, the production of hypoxanthine decreased by more than 80 %. These effects were observed in the absence as well as in the presence of dCF and ITu, demonstrating again the Pi-sensitivity of AMP deaminase. The limited accumulation of adenosine recorded in the presence of the inhibitors was also enhanced by the suppression of Pi and reduced by its elevation to 10 mM (Fig. 3f). In the presence of 0, 1.2 and 10 mM Pi, erythrocytic AMP increased respectively 12-, 6- and 3-fold over 90 min (results not shown).

CONCLUDING REMARKS

The use of specific inhibitors of adenosine metabolism allows us to conclude that the very slow physiological catabolism of the erythrocytic adenine nucleotides and its marked acceleration induced by alkalinization of the medium occurs by deamination of AMP, followed by dephosphorylation of IMP. In contrast, 75 % of the rapid degradation of the adenine nucleotides resulting from glucose deprivation proceeds by a prior dephosphorylation of AMP, followed by deamination of adenosine. Recycling of adenosine could not be evidenced under physiological conditions, but became apparent both in the absence of glucose and at alkaline pH. These results differ markedly from those obtained with isolated rat hepatocytes, in which the physiological catabolism of the adenine nucleotides and that induced by fructose (2) or by anoxia (9) proceed exclusively by way of AMP deaminase, and in which an AMP-adenosine cycle operates at a velocity reaching about 2-fold the rate of production of allantoin (1), the terminal purine catabolite in these cells. This work also demonstrates the existence of a cytoplasmic 5'-nucleotidase in human erythrocytes, which catalyzes the dephosphorylation of purine nucleotides. This enzyme has been purified and is now under investigation (see Bontemps et al. this volume).

ACKNOWLEDGEMENTS

The authors wish to express their appreciation to Mrs. Anne Delacauw for expert technical assistance.
This work was supported by grant nr 3.5463.82 of the "Fonds National de la Recherche Médicale". G. Van den Berghe is "Maître de Recherches" of the Belgian "Fonds National de la Recherche Scientifique".

REFERENCES

1. F. Bontemps, G. Van den Berghe & H.G. Hers, Evidence for a substrate cycle between AMP and adenosine in isolated hepatocytes, Proc. Natl. Acad. Sci. USA. 80:2829 (1983).
2. G. Van den Berghe, F. Bontemps & H.G. Hers, Purine catabolism in isolated rat hepatocytes. Influence of coformycin, Biochem. J. 188:913 (1980).
3. J.M. Whelan & A.S. Bagnara, Factors affecting the rate of purine ribonucleotide dephosphorylation in human erythrocytes, Biochim. Biophys. Acta 563:466 (1979).

4. G.C. Mills, R.M. Goldblum & F.C. Schmalstieg, Catabolism of adenine nucleotides in adenosine deaminase deficient erythrocytes, Life Sci. 29:1811 (1981).
5. G.C. Mills, Effects of pH on erythrocyte metabolism. Arch. Biochem. Biophys. 117:487 (1966).
6. S. Minakami, S. Takayasu, C. Suzuki & H. Yoshikawa, The hydrogen ion concentrations and erythrocyte glycolysis, Biochem. Biophys. Res. Commun. 17:748 (1964).
7. V. Albrecht, H. Roigas, M. Schultze, G. Jacobasch & S. Rapoport, The influence of pH and methylene blue on the pathways of glucose utilisation and lactate formation in erythrocytes of man, Eur. J. Biochem. 20:44 (1971).
8. I. Rapoport, S. Rapoport & R. Elsner, Accumulation of phosphate esters and decline of ATP in red cells incubated in vitro is caused by lack of pyruvate, Acta Biol. Med. Germ. 40:115 (1981).
9. M.F. Vincent, G. Van den Berghe & H.G. Hers, The pathway of adenine nucleotide catabolism and its control in isolated hepatocytes subjected to anoxia, Biochem. J. 202:117 (1982).

MECHANISM OF INCREASED RATE OF DE NOVO PURINE BIOSYNTHESIS IN RAT LIVER AFTER BILATERAL ADRENALECTOMY

Mitsuo Itakura, Noriaki Maeda and Kamejiro Yamashita

Division of Endocrinology and Metabolism
Institute of Clinical Medicine
The University of Tsukuba, Ibaraki 305, Japan

ABSTRACT

The incorporation of [^{14}C]glycine to hepatic purines increased proving the increased rate of de novo purine biosynthesis in rat liver after bilateral adrenalectomy in comparison to sham-operated controls. In the liver of adrenalectomized animals 24 hours after adrenalectomy when the rate of de novo purine biosynthesis is increased above control by 70% there was a 200% increase of 5-phosphoribosyl 1-pyrophosphate (PRPP) concentration. The concentrations of purine ribonucleotides showed a 33 and a 24% decrease of ATP and GTP, and 245 and 38% increase of AMP and ADP respectively associated with the unchanged total adenine and guanine nucleotide concentration and a 18% decrease of adenylate energy charge. The specific activity of amidophosphoribosyltransferase (ATase) was not changed. The replacement with corticosterone acetate to adrenalectomized animals for 24 hours partially restored the rate of de novo purine biosynthesis and the concentrations of purine ribonucleotides.

These results suggest that the increased rate of de novo purine biosynthesis in adrenalectomized rat liver is compensatory against the increased catabolism of purine ribonucleotides as a result of the increased AMP concentration and that it is mediated by the increased concentration of PRPP. Our study has demonstrated the importance of the physiological amount of adrenocortical hormone to sustain the normal concentrations and the metabolism of purine ribonucleotides in liver.

INTRODUCTION

The incidence of hyperuricemia and gout have been reported by H.D. Itskovitz and A.M. Sellers to be as high as 67 and 6.4% respectively in the population who underwent therapeutic bilateral adrenalectomy for their malignant hypertension[1]. It is however not clear whether the underexcretion of uric acid from kidney or the overproduction of uric acid by the increased catabolism is the mechanism of hyperuricemia in these population. The possibility of regulation of the renal excretion of uric acid by ACTH or cortisone has been reported by Ingbar et al.[2]. On the other hand it has been reported by P. Feigelson and M. Feigelson[3] or by A.A. Letter et al.[4] respectively that the intraperitoneal administration of 1.0 mg/200 g B.W. of cortisone acetate to adrenalectomized

rats or 2 mg of hydrocortisone sodium succinate per a mouse, the
relatively large dose of adrenocortical hormone increases the incorporation of intraperitonealy administered [^{14}C]glycine into acid-soluble
adenine and RNA fraction of liver.

To study the role of the physiological amount of adrenocortical
hormone in the regulation of purine biosynthesis in liver we have
directly studied in rat liver after bilateral adrenalectomy or sham-
operation, 1) the rate of de novo purine biosynthesis, 2) the hepatic
concentration of PRPP and purine ribonucleotides, 3) the specific
activity of ATase and 4) the effect of the corticosterone acetate
replacement on the rate of de novo purine biosynthesis and concentrations
of purine ribonucleotides.

MATERIALS AND METHODS

Materials

Experimental animals. 7 week old male rats of Wistar strain were
purchased from Shizuoka Agricultural Cooperative Association for
Laboratory Animals and kept for one week prior to the experiment on 12
hours of light from 0700 to 1900 hour and 12 hours of dark period alternatively with the laboratory feed and water accessible at libitum. The
average body weight of the animals was 208 g with the range from 170 to
220 g.

Reagent. [^{14}C]glycine (110-113 μCi/μmol), [^{14}C]adenine (49.0-55.6
μCi/μmol) were purchased from New England Nuclear. PRPP Mg salt, AMP,
ADP, ATP, GMP, and GTP of their salts were purchased from Sigma. DE 52
was purchased from Whatman, Dowex AG 50 W H$^+$, x 200-400 was purchased
from Bio Rad Laboratory. Polyethylene tubes of PE 10 and 50 were
purchased from Cray Adams. Polyoxyethylene dialkyl ether phosphate salt
(HCO 60), tris n-octylamine, 1,1,2-trichloro 1,2,2-trifluoroethane
(freon) and scintisol EX-H were purchased from Wako Pure Chemical
Industries, Ltd., Tokyo, Japan. Other chemical reagents were of the
highest quality available.

Methods

Adrenalectomy and sham-operation. Under light open ether anesthesia
dorsal skin and muscle layers were incised along the paravertebral
direction at the distance of 1.5 cm from vertebrate as long as 2 cm.
Bilateral adrenal glands were removed by clamping with the forceps.
After bilateral adrenalectomy 0.9% saline was supplied as the drinking
water. The sham-operation was performed exactly in the same way but the
skin was ligated without adrenalectomy. Surgical operations were
performed at 10:00 to 11:00 AM.

Sampling of liver specimens. Under light open ether anesthesia the
abdominal cavity was opened and liver samples were freeze-clamped by the
tongs precooled in liquid nitrogen. The liver samples thus obtained were
kept under liquid nitrogen until they were used for the assay of the rate
of de novo purine biosynthesis, PRPP, purine ribonucleotides or ATase
activity.

Assay of [^{14}C]glycine incorporation into hepatic purine and proteins.
Liver specimen was obtained at exactly 30 min after the iv administration
of 5 μCi/44-45 nmol [^{14}C]glycine into the jugular vein of the etherized
animals. According to the published methods[5,6], purine ribonucleotides
extracted from liver specimen by ice cold 2 N perchloric acid were

converted to purine bases by heat at 100°C for 1 hour. Purine bases were precipitated by silver nitrate and the redissolved sample by acid was applied to Dowex W H$^+$ column (0.8 x 3 cm). The specific activity of the purine fraction which is eluted between 1 and 6 N HCl was calculated by dividing the radioactivity by the amount of purine bases assayed by the optic density at 260 nm. The rate of de novo purine biosynthesis was expressed by this specific activity as dpm/1 O.D. at 260 nm. The specific activity of protein was calculated by dividing the radioactivity of the redissolved protein fraction by the amount of protein assayed by Lowry's methods[7]. The rate of protein biosynthesis was expressed by this specific activity as dpm/mg protein.

Assay of glycine concentration in liver tissue or plasma. Under light open ether anesthesia liver tissue was obtained by freeze-clamping and 0.5 ml of blood was drawn from jugular vein into a heparinized syringe to obtain plasma sample from the separate groups of animals. Acid soluble free glycine was extracted from plasma by adding 10% volume of 50% sulfosalycilic acid and from liver tissue by 12% trichloroacetic acid. This acid extract was applied to Durrum Amino Acid Analyzer Model d-500 after adjusting pH to 2.0-2.2 by NaOH.

Assay of PRPP concentration in liver. The concentration of PRPP in liver was assayed by the methods reported by Hisata et al.[9] with the following modifications. PRPP was extracted by 12% trichloroacetic acid at 4°C within 90 seconds. At exactly 90 seconds of extraction after homogenization by polytrone homogenizer, extract was neutralized with the charcoal containing neutralizing buffer which contains 6 N KOH and 1 N Tris pH 7.4 at the ratio of 1 to 6.5 and the activated charcoal at the final concentration of 100 mg/ml. 60 μl of the filtrate through 0.45 μm filter was introduced to the assay mixture which contains 20 μl of adeninephosphoribosyltransferase (APRT) preparation and 20 μl of 2 mM [^{14}C]adenine with the specific activity of 9.0 μCi/μmol. APRT was partially purified by the reported methods[10] from human red blood cells. After incubating the assay mixture at 37°C for 60 min the formed radioactive AMP was separated from adenine by the high voltage paper electrophoresis and the [^{14}C]AMP spot was cut and counted in the liquid scintillation counter. The concentration of PRPP was expressed as nmol/g protein. The study on the degradation rate of PRPP in 12% trichloro-acetic acid at 4°C during the extraction prodedure showed that 82% of the initial amount is present at the end of 90 seconds exposure to acid. Accordingly the concentration of PRPP was calculated by dividing the assayed amount by 0.82.

Assay of purine ribonucleotide concentration. According to the published methods[6,11] purine ribonucleotides were extracted from a piece of frozen liver weighing 10 to 15 mg by 1 ml of ice cold 12% trichloro-acetic acid by a glass homogenizer. At 30 min of extraction on ice the supernatant was neutralized by 0.5 N tris n-octylamine in freon on ice for 30 min. 20 μl were applied to HPLC system containing Whatman Partisil SAX 10 anion exchange column and eluted by the linear gradient starting at 5 mM NH$_4$PO$_4$ pH 2.8 and ending at 750 mM NH$_4$PO$_4$ pH 3.9 at the flow rate of 2 ml/min. The concentrations of purine ribonucleotide were calculated by comparing the peaks manually with the appropriate standards and were expressed as μmol/g wet liver weight.

Corticosterone acetate replacement to adrenalectomized animals. Corticosterone acetate was replaced in adrenalectomized animals at the dose of 0.1 mg/24 hours by the continuous infusion via iv cannula. Two polyethylene tubes of PE 10 were cannulated into the jugular vein under light ether anesthesia in the previous day. Two tubes were guided subcutaneously to the nuchal area and further guided through the coiled

wire to the infusion pump. 7% HCO 60 in saline was used as a vehicle.
Infusion was performed through one tube at the speed of 0.59 ml/hour.
[^{14}C]glycine was injected through another tube. The first group of
animals had the bilateral adrenalectomy with vehicle infusion, the second
group had the bilateral adrenalectomy with replacement and the third
group had the sham-operation with vehicle infusion.

Assay of the specific activity of ATase. Liver samples pulverized
in liquid nitrogen were introduced to 1 ml of ice-cold 25 mM pottasium
phosphate buffer pH 7.4 containing 60 mM beta-mercaptoethanol. After
extraction of the enzyme on ice for 30 min, 60 μl of the supernatant of
1,000 g for 30 min was introduced to the assay mixture containing 5 mM
PRPP, 5 mM MgCl$_2$, 50 mM pottasium phosphate pH 7.4, 1 mM dithiothreitol
and 5 mM [^{14}C]glutamine with the specific activity of 0.15 μCi/μmol.
[^{14}C]glutamate formed was separated from [^{14}C]glutamine by the high
voltage paper electophoresis[12]. The difference of [^{14}C]glutamate formed
in the assay mixture with or without PRPP was regarded as the activity of
ATase. The pellet was treated with 0.5 N perchloric acid and the amount
of DNA was assayed by Schmidt-Thaunhuser-Schneider's methods[13]. The
specific activity of ATase was assayed in 4 animals each with
adrenalectomy, sham-operation, adrenalectomy and replacement and it was
expressed as nmol glutamine deaminated/hr/μg DNA.

Statistical analysis. Statistical significance was tested by
Student's unpaired t test and the marks of * or ** were used to show the
P less than 0.05 or 0.01.

RESULTS

Incorporation of [^{14}C]glycine into hepatic purines and proteins

The time course of [^{14}C]glycine incorporation to hepatic purines and
proteins are shown in Fig. 1. [^{14}C]glycine incorporation to hepatic
purines in the liver of nonoperated animals was 95 ± 7 dpm/1 O.D. at 260
nm (n=4). This rate is 195 ± 10 dpm/1 O.D. at 260 nm (n=4) 24 hours
after adrenalectomy which is significantly higher than that of 117 ± 10
in the corresponding sham-operated control. It remained increased 48 and
72 hours after the surgery as high as 1.7 to 2.2 folds. [^{14}C]glycine
incorporation to hepatic proteins was not significantly different between
the two groups. The ratio of the specific activity of purine/protein
showed the significant increases in adrenalectomized rat liver in
comparison to sham-operated control.

Glycine concentration in liver and plasma

The time course of the change of glycine concentration in liver and
plasma is shown in Fig. 2. Liver and plasma glycine concentration in
nonoperated animals was 1.16 ± 0.13 (n=6) and 0.16 ± 0.02 (n=6)
respectively. No significant differences of tissue or plasma glycine
concentration was observed between the two groups at the time points of
6, 24, 48 and 72 hours after the surgery.

PRPP concentration

The time course of the change of PRPP concentration in liver is
shown in Fig. 3. The PRPP concentration in liver of nonoperated animals
was 33.1 ± 2.2 nmol/g protein (n=30) and it started to increase 6 hours
after adrenalectomy and reached the peak value of 92.8 ± 18.2 nmol/g
protein (n=5) 24 hours after the surgery which is significantly higher
than control value of 33.6 ± 3.4 nmol/g protein (n=6). It remained

significantly increased 72 hours after the surgery. The pattern of the
increased concentration of PRPP beared a parallel to the increase of the
rate of de novo purine biosynthesis in Fig. 1.

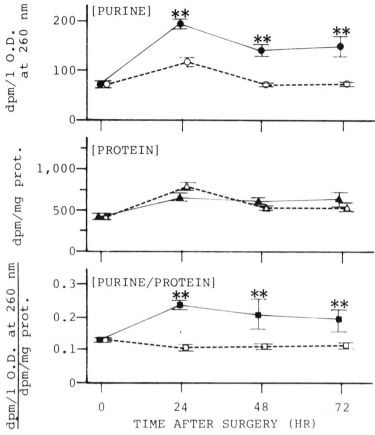

Fig. 1.   Time course of [^{14}C]glycine incorporation: Time in
hours after adrenalectomy or sham-operation is plotted
on the abscissa.   The upper panel shows the specific
activity of hepatic purines as dpm/1 O.D. at 260 nm,
the middle panel that of hepatic proteins as dpm/mg
protein and the lower panel their ratio of purine/
protein.   Data after adrenalectomy are shown by the
solid marks with solid lines and those after sham-
operation are shown by the open marks with dotted
lines.   Each point is a mean ± S.E.M. of 4 animals.

Effect of replacement on purine ribonucleotide concentration

The concentrations of purine ribonucleotides in liver 24 hours after
surgery with or without replacement are summarized in Table 1.   In liver
of adrenalectomized animals, ATP and GTP concentration of 1.52 and 0.22
μmol/g liver wt respectively are significantly lower than those of 2.28
and 0.29 in control.   The concentration of AMP and ADP of 0.38 and 0.99
respectively in adrenalectomized animals are significantly higher than
0.11 and 0.72 in control.   The replacement significantly restored the
concentrations of purine ribonucleotides toward those in control.

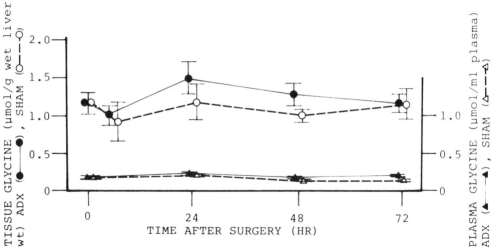

Fig. 2. Time course of the free glycine concentration in liver tissue
and plasma: Time in hours after adrenalectomy or sham-
operation is plotted on the abscissa. The tissue concentrations of free glycine in liver after adrenalectomy or sham-
operation are respectively shown by the solid circles with
solid lines and by the open circles with dotted lines. The
plasma concentration of free glycine after adrenalectomy or
sham-operation are respectively shown by the solid triangles
with solid lines and by the open triangles with dotted lines.
Each point is a mean ± S.E.M. of 5 animals.

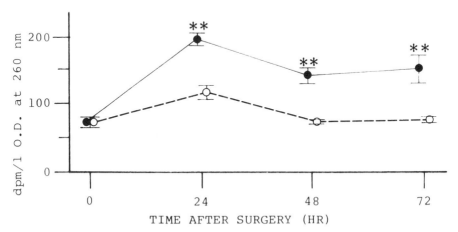

Fig. 3. Time course of PRPP concentration in liver: Time in hours
after adrenalectomy or sham-operation is shown on the
abscissa. The concentration of PRPP in liver as nmol/g
protein is plotted on the ordinate. The solid triangles
with solid lines represent data after adrenalectomy and
the open triangles with dotted lines represent those after
sham-operation. Each point is the mean ± S.E.M. of 4 animals.

Table 1. Concentration of purine ribonucleotides in liver 24 hours after adrenalectomy, adrenalectomy plus replacement or sham-operation

	ADRENALECTOMY (ADX)	ADRENALECTOMY PLUS REPLACEMENT (REP.)	SHAM-OPERATION (SHAM)
AMP	0.38 ± 0.09 *	0.27 ± 0.07	0.11 ± 0.03
ADP	0.99 ± 0.05 *	0.89 ± 0.09	0.72 ± 0.11
ATP	1.52 ± 0.17 *	1.96 ± 0.41	2.28 ± 0.22
GMP	0.02 ± 0.01	0.02 ± 0.01	0.01 ± 0.01
GDP	* 0.19 ± 0.02	0.13 ± 0.01	0.14 ± 0.01
GTP	0.22 ± 0.02	* 0.28 ± 0.06	0.29 ± 0.01
T.A.N.	2.89 ± 0.13	3.12 ± 0.42	3.11 ± 0.20
T.G.N.	0.43 ± 0.03	0.42 ± 0.07	0.45 ± 0.01
T.A.G.N.	3.31 ± 0.16	3.54 ± 0.48	3.56 ± 0.21
A.E.C.	0.70 ± 0.04 *	0.76 ± 0.04	0.85 ± 0.03
(N)	(5)	(4)	(4)
[S.S.]	[ADX vs. REP.]	[ADX vs. SHAM]	[REP. vs. SHAM]

(T.A.N., T.G.N., T.A.G.N., A.E.C., [S.S.] and (N) stand for total adenine nucleotides, total guanine nucleotides, total adenine and guanine nucleotides, adenylate energycharge, statistical significance and the number of animals respectively. Each number is a mean ± S.E.M.)

Effect of replacement on [^{14}C]glycine incorporation to hepatic purines

The effect of replacement on [^{14}C]glycine incorporation to hepatic purines is summarized in Table 2. The value of 350 dpm/1 O.D. at 260 nm in adrenalectomized rats is significantly higher than 118 in control animals. The increased [^{14}C]glycine incorporation to hepatic purines was partially reversed by the replacement reaching the level of 188 dpm/1 O.D. at 260 nm.

Table 2. Rate of de novo purine biosynthesis in liver 24 hours after adrenalectomy, adrenalectomy plus replacement and sham-operation

	ADRENALECTOMY (ADX)	ADRENALECTOMY PLUS REPLACEMENT (REP.)	SHAM-OPERATION (SHAM)
	* 319 ± 32	** 187 ± 32	138 ± 15
(N)	(7)	(5)	(12)
[S.S.]	[ADX vs. REP.]	[ADX vs. SHAM]	[REP. vs. SHAM]

([S.S.] and (N) stand for statistical significance and the number of animals respectively. Each number is a mean ± S.E.M.)

The specific activity of ATase

The specific activity of ATase in rat liver after adrenalectomy plus vehicle infusion, sham-operation plus vehicle infusion and adrenalectomy plus the replacement infusion with corticosterone acetate were respectively 0.93 ± 0.03 (n=4), 1.00 ± 0.04 (n=4) and 0.94 ± 0.02 nmol glutamine deaminated/hr/μg DNA (n=4). There was no statistical differences from each other.

DISCUSSION

The increased rate of de novo purine biosynthesis in the liver of adrenalectomized animals was demonstrated in this study. This conclusion is drawn from 1) the increased [^{14}C]glycine incorporation to hepatic purine 24 through 72 hours after adrenalectomy, 2) the specific activity of protein in the assay of [^{14}C]glycine incorporation which is also subject to the glycine pool size is not changed and 3) the free glycine concentration in liver and plasma are not significantly changed. The results of replacement on [^{14}C]glycine incorporation and the concentration of purine ribonucleotide clearly demonstrated that the increased rate of de novo purine biosynthesis and the changes in the concentration of purine ribonucleotides are causally related to the presence or the absence of the physiological amount of adrenocortical hormone. Although our results disagree with the reported stimulation by corticosterone reported by Feigelson et al.[3] and Letter et al.[4], the difference can be explained by the difference of the dose of corticosterone acetate, in other words the dose of 0.1 mg/200 g BW/24 hours used in our study is far less thn those of 1 mg/200 g BW of rats or 2 mg of hydrocortisone sodium succinate per a mouse as a bolus used in their studies.

The mechanism of the increased rate of de novo purine biosynthesis in adrenalectomized rat liver is firstly due to the increased concentration of hepatic PRPP concentration. The possibility of the decreased feedback inhibition is unlikely since the ATP concentration is decreased and the AMP concentration is increased resulting in the decreased adenylate energy charge with the conservation of total adenine and guanine nucleotide concentration. The decreased concentration of ATP in adrenalectomized rat liver with its recovery by corticosterone acetate is compatible with the previous report[14]. The increased rate of de novo purine biosynthesis is regarded as the compensatory mechanism against the increased catabolism of purine ribonucleotides as a result of the increased concentration of AMP.

The replacement with corticosterone acetate partially restored not only the rate of de novo purine biosynthesis but also the concentration of purine ribonucleotides. This observation suggests the importance of the physiological amount of adrenocortical hormone to maintain the normal concentration and the metabolism of purine ribonucleotides.

REFERENCES

1. H. D. Itskovitz, and A. M. Sellers, Gout and hyperuricemia after adrenalectomy for hypertension. N. Eng. J. Med. 268:1,105 (1963).
2. S. H. Ingbar, E. H. Kass, C. H. Burnett, A. S. Relman, B. A. Burrows, and J. H. Sisson, The effects of ACTH and cortisone on the renal tubular transport of uric acid, phosphorus and electrolytes in patients with normal renal and adrenal function. J. Lab. Clin. Med. 38:533 (1951).

3. P. Feigelson, and M. Feigelson, Studies on the mechanism of regulation of the metabolism of liver purine and ribonucleic acid. J. Biol. Chem. 238:1073 (1963).

4. A. A. Letter, G. Zombor, and J. F. Henderson, Tryptophan as a source of one-carbon units for purine biosynthesis de novo. Can. J. Biochem. 51:486 (1973).

5. M. P. Edmonds, and G. A. LePage, The incorporation of glycine-2-C^{14} into acid soluble nucleotide purines. Cancer Res. 15:93 (1955).

6. M. Itakura, R. L. Sabina, P. W. Heald, and E. W. Holmes, Basis for the control of purine biosynthesis by purine ribonucleotides. J. Clin. Invest. 67:994 (1981).

7. O. H. Lowry, N. J. Rosenbrough, A. L. Farr, and R. J. Randall, Protein measurement with Folin phenol reagent. J. Biol. Chem. 193:265 (1951).

8. T. Hisata, An accurate method for estimating 5-phosphoribosyl 1-pyrophosphate in animal tissues with the use of acid extraction. Analyt. Biochem. 68:448 (1975).

9. C. B. Thomas, W. J. Arnold, and W. N. Kelley, Human adenine phosphoribosyltransferase: Purification and properties. J. Biol. Chem. 248:2529 (1973).

10. L. M. Rose, and R. W. Brockman, Analysis by HPLC of 9-beta-D arabinofuranosyladenine-5'-triphosphate level in murine leukemia cells. J. Chromatogr. 133:335 (1977).

11. M. Itakura, and E. W. Holmes, Human amidophosphoribosyltransferase. An oxygen-sensitive iron-sulfur protein. J. Biol. Chem. 333:999 (1879).

12. W. C. Schneider, Phosphorus compounds in animal tissues. III. A comparison of methods for the estimation of nucleic acids. J. Biol. Chem. 164:747 (1946).

13. P. Ove, S. Takai, T. Umeda, and I. Lieberman, Adenosine triphosphate in liver after partial hepatectomy and acute stress. J. Biol. Chem. 242:4963 (1967).

INCREASE OF AMIDOPHOSPHORIBOSYLTRANSFERASE ACTIVITY AND
PHOSPHORIBOSYLPYROPHOSPHATE CONCENTRATION AS THE BASIS FOR
INCREASED DE NOVO PURINE BIOSYNTHESIS IN THE REGENERATING RAT LIVER

Mitsuo Itakura, Masami Tsuchiya and Kamejiro Yamashita

Division of Endocrinology and Metabolism
Institute of Clinical Medicine
The University of Tsukuba, Ibaraki 305, Japan

SUMMARY

The labeling of hepatic purines with [^{14}C]glycine increased in the
regenerating rat liver 12 hours after a 70% hepatectomy and it reached
the 2.4 folds higher peak 12 hours after the surgical operation in
comparison to sham-operated control. These observations suggest that the
rate of de novo biosynthesis increases in the regenerating rat liver to
supply purine ribonucleotides to the cells. In the regenerating rat
liver the activity of amidophosphoribosyltransferase (ATase) significantly
increased and reached the 1.8 folds higher peak than control 18 hours
after surgery. Hepatic 5-phosphoribosyl 1-pyrophosphate (PRPP) concentration significantly increased and reached the 3.0 folds higher peak 12
hours after surgery. Although the mean hepatic concentration of ATP and
GTP showed a 13 and 15% decrease respectively in the regenerating rat
liver 12 hours after surgery, they were counterbalanced by the increased
concentrations of AMP and GMP by 13 and 44% respectively. These results
suggest that the increased rate of de novo purine biosynthesis in the
regenerating rat liver is mediated by the increased enzymatic activity of
ATase and by the increased concentration of PRPP.

INTRODUCTION

Regenerating rat liver after a 70% hepatectomy has been widely used
as the experimental model for the study of the rapid growth of liver[1,2].
Although the time course of the increase of DNA synthesis[1] and DNA
polymerase activity[3] have been reported, the mechanism of the activations
of multiple metabolic pathways leading to the increase in DNA synthesis
and the rapid growth of the remnant liver has not been well understood.
The de novo purine biosynthetic pathway, one of the typical anabolic
pathways, supplies cells the purine ribonucleotides which are required
for cell growth as the essential components of RNA, DNA, ATP etc. The
rate of this pathway in liver has been supposed to be regulated at the
level of ATase (EC 2. 1. 2. 14) by the relative concentration of PRPP, one
of the substrate and purine ribonucleotides, the feedback inhibitors[4-6]
and also possibly by the amount of this enzyme protein itself[7]. This
pathway is supposed to be activated in the regenerating rat liver unless
the supply of purine ribonucleotides is attained by the salvage pathway.
Since the regulation of this pathway is relatively well understood, the

time course and the mechanism of the activation of de novo purine biosynthetic pathway in this experimental system was studied to understand the basis of activations of multiple metabolic pathways in the regeneration process of liver.

MATERIALS AND METHODS

Materials

Seven-week-old male Wistar rats were purchased from Shizuoka Agricultural Cooperative Association for Laboratory Animals, Shizuoka, Japan, and kept on standard laboratory feed and on 12 hour light and dark cycle for one week prior to the experiments. Their body weights ranged from 180 to 230 g. Standard laboratory feed was purchased from Oriental Yeast Co. Ltd., Tokyo, Japan. Radioactive [^{14}C]glycine (110-113 μCi/μmol), [^{14}C]adenine (49-55.6 μCi/μmol) and [^{14}C]glutamine (258-292 μCi/μmol) were purchased from New England Nuclear. Polyethylene tubes were purchased from Cray Adams. Dowex 50W x 8, 200-400 mesh H$^+$ form was purchased from BioRad Laboratory. HPLC system was purchased from Waters Associate containing 710 autosampler equipped with the sample cooling unit, 730 data module and Whatman SAX 10 anion exchange column. Other reagents are of the highest quality available.

Methods

Hepatectomy and sham-operation. The methods for a 70% hepatectomy were those previously described[8]. It was performed under light open ether anesthesia between 9:00 and 11:00 AM. Sham-operation was performed in a similar way by exposing liver lobes out of the abdominal cavity but they were returned to the abdominal cavity without ligation.

Intravenous administration of [^{14}C]glycine and liver tissue sampling. In the pulse labeling experiment, 5 μCi [^{14}C]glycine was administered intravenously via jugular vein after its exposure under light open ether anesthesia. Liver tissue was sampled in 30 minutes under light open ether anesthesia by freeze-clamping the liver with metal tongs precooled with liquid nitrogen. The same sampling technique was used for the assay of ATase activity, PRPP and purine ribonucleotides concentration. This procedure was performed within 45 seconds from skin incision to liver sampling. Liver tissue samples thus obtained were stored under liquid nitrogen.

Assay of [^{14}C]glycine incorporation to hepatic purines. The methods for the assay of [^{14}C]glycine incorporation to hepatic purines were those previously described[9] except for the following modifications: Firstly 5 μCi/44-45 nmol of [^{14}C]glycine were administered intravenously into the external jugular vein under light open ether anesthesia in stead of intraperitoneally. Secondly liver tissue was sampled by the freeze-clamping method exactly 30 minutes after the injection. Following these steps the liver sample was pulverized under liquid nitrogen and liver powder containing about 50 mg of protein was put into 4 ml of ice cold 2 N perchloric acid. The methods to degrade purine ribonucleotides to purine bases by heat at 100°C for 1 hour, precipitation of purine bases by silver nitrate and partial purification of purine bases by Dowex 50 W H$^+$ column were those of the reported methods[6]. The specific activity of this purine base fraction was expressed as dpm/1 O.D. at 260 nm. Data are expressed as a mean ± S.E.M. of 5 animals. The efficiency of counts per minute to disintegration per minute was 0.95.

Assay of PRPP concentration. The methods for the assay of PRPP were those previously described[10] except for the following modifications. Liver tissues specimen was pulverized in liquid nitrogen and liver powder containing about 15 mg of protein was put into 1.0 ml of ice cold 12% trichloroacetc acid at time 0. The mixture was mixed with a polytron homogenizer for exactly 20 seconds. The mixture was then centrifuged in a microfuge between time points 45 and 75 seconds. 400 μl supernatant were neutralized with 400 μl of charcoal containing neutralizing buffer at exactly 90 seconds composed of 100 mg activated charcoal in 1 ml of buffer containing 6 N KOH and 1 N Tris pH 7.4 at a ratio of 1 to 6.5. 60 μl of the filtrate through a 0.45 μm filter were mixed with 20 μl of 2 mM radioactive adenine with the specific activity of 9.0 μCi/μmol and 20 μl of dialyzed adenine phosphoribosyltransferase (APRT) preparation. The methods for the partial purification of APRT from human red blood cells by DE 52 were those previously described[11]. The enzyme preparation was dialyzed against 1,000 volumes of 5 mM $MgCl_2$ and 20 mM Tris pH 7.4 for 3 and a half hours before the assay at 4°C in the cold room. The mixture was incubated at 37°C for 1 hour and 20 μl were spotted on Whatmann 3MM chromatography paper. The methods for separation of AMP from adenine were those previously described[11].

Degradation rate of PRPP in 12% trichloroacetic acid at 4°C in 90 seconds was 82% residual activity which was calculated by plotting the residual PRPP concentration in semilographythmic scale against the time interval between 15 and 120 seconds. Protein in the pellet was assayed by the Lowry's methods[12]. Assays were performed in duplicate. Results were expressed as a mean ± S.E.M. of 3 to 4 animals in nmol/g of liver protein.

Assay of purine ribonucleotides concentrations. The methods for the assay of purine ribonucleotides in liver were those previously described[6,13]. In brief, purine ribonucleotides were extracted from weighed frozen liver sample in ice-cold 12% trichloroacetic acid and the 3,000 g for 2 min supernatant was neutralized with the equal volume of 0.5 N tris n-octylamine in freon for 30 min on ice. 20 μl of the neutralized sample was introduced to HPLC system containing Whatman SAX 10 column eluted with the linear gradient starting at 5 mM NH_4PO_4 pH 2.8 and ending at 750 mM NH_4PO_4 pH 3.9 at the rate of 2 ml/min. The concentrations of six purine ribonucleotides, i.e. AMP, ADP, ATP, GMP, GDP and GTP were determined in triplicate in 9 regenerating rat liver, 11 sham-operated rat liver 12 hours after the surgical operation and 7 control rat liver at time 0. Data were expressed as a mean ± S.E.M. in μmol/g of well liver weight.

Assay of ATase. To assay the activity of ATase liver samples pulverized in liquid nitrogen were introduced to 1 ml of ice-cold 25 mM potassium phosphate buffer pH 7.4 containing 60 mM betamercaptoethanol. 60 μl of the 3,000 g for 30 min supernatant was introduced to the assay mixture containing 4 mM PRPP, 5 mM $MgCl_2$, 50 mM potassium phosphate pH 7.4, 1 mM dithiothreitol and 5 mM [^{14}C]glutamine with the specific activity of 0.15 μCi/μmol. [^{14}C]glutamate was separated from [^{14}C]glutamine by high voltage paper electrophoresis[14]. The difference of [^{14}C]glutamate formed in the assay mixture with or without PRPP was regarded as the activity of ATase. The pellet was treated with 0.5 N perchloric acid and the amount of DNA was assayed by Schmidt-Thaunhauser-Schneider's methods[16] and that of protein by Lowry's method[12]. Data were expressed as a mean ± S.E.M. in nmol glutamine deaminated/hr/μg DNA or nmol glutamine deaminated/hr/mg protein.

Statistical analysis. Statistical analysis was performed by Student's t test for unpaired samples between hepatectomized and sham-

operated control animals. The marks of * or ** were used to show
respectively the P less than 0.05 or 0.01.

RESULTS

Time course of the specific activity of hepatic purines in [^{14}C]glycine incorporation assay

The changes of the specific activity of hepatic purine in
[^{14}C]glycine incorporation assay after hepatectomy of sham-operation are
shown in Fig. 1. The specific activity of hepatic purines in nonoperated
rat liver was 74 ± 8 dpm/1 O.D. at 260 nm (n=6), which increased sharply
for the first 4 hours after either hepatectomy or sham-operation. The
rate reached a maximal value of 254 ± 8 dpm/1 O.D. at 260 nm 12 hours
after 70% hepatectomy which is 2.4 folds higher than sham-operated
control. After this peak there was a gradual decline and the increase in
hepatectomized animals in comparison to sham-operated animals were
significant 6 through 48 hours.

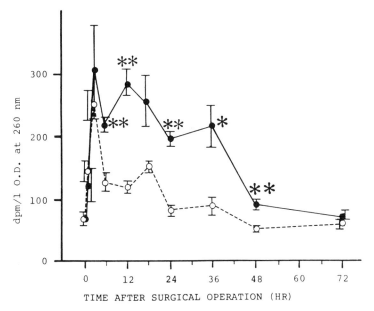

Fig. 1.   The changes in the specific activity of hepatic purines
in [^{14}C]glycine incorporation assay in the regenerating
and sham-operated rat liver. Time in hours after the
surgical operation is plotted on the abscissa. The
specific activities of hepatic purines as dpm/1 O.D. at
260 nm are plotted on the oridinate. Data for the
regenerating rat liver are shown by the solid lines and
those for the sham-operated control rat liver by the
dotted lines. This type of delineation was used in the
following figures. Each point is the mean ± S.E.M. of
5 animals.

PRPP concentration

The changes in PRPP concentration in liver after the hepatectomy or sham-operation are shown in Fig. 2. The PRPP concentration in non-operated rat liver at time 0 was 25.7 ± 1.2 nmol/g liver protein (n=30). After the hepatectomy PRPP concentration gradually increased and reached a maximal value of 79.2 ± 6.5 nmol/g liver protein (n=3) 12 hours after the surgical operation. The PRPP concentration in liver tissue after the sham-operation did not show significant changes after the surgical operation until 72 hours with the range of means between 21.2 and 34.5 nmol/g liver protein. The differences between hepatectomized and sham-operated rat liver were significant at 6, 12 and 18 hours. Although statistically insignificant, the PRPP concentration in the regenerating rat liver showed the higher tendency 24, 48 and 72 hours when compared to sham-operated control rat liver.

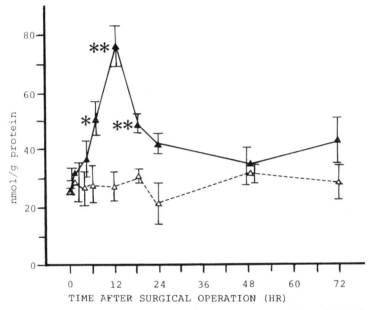

Fig. 2. The changes in the hepatic concentration of PRPP in the regenerating and sham-operated rat liver. Time in hours after the surgical operation is plotted on the abscissa. The concentrations of PRPP in nmol/g protein are plotted on the ordinate. Each point is the mean ± S.E.M. of 5 animals.

Purine ribonucleotide concentrations

The concentrations of purine ribonucleotides in rat liver 12 hours after the surgical operations are summarized in Table 1. Although there were not any statistically significant changes of purine ribonucleotides concentration, the mean concentration of each purine ribonucleotide showed the following results. The concentration of ATP and GTP respectively of 3.021 and 0.382 μmol/g wet liver weight in the regenerating rat liver are lower than those in sham-operated rat liver of 3.466 and 0.451 respectively by 13 and 15%. The concentration of AMP and GMP respectively of 0.123 and 0.013 μmol/g wet liver weight are higher than those of 0.109 and 0.009 respectively in sham-operated control by 13 and 44%. These changes resulted in a 10% decrease in the total adenine and guanine nucleotides concentration of 4.660 in comparison to 5.196 μmol/g wet liver weight and the decreased adenylate energy charge of 0.843 in comparison to 0.680 respectively in regenerating rat liver and in sham-operated control rat liver. The concentrations of purine ribonucleotides in control rat liver 12 hours after the surgical operation are comparable to those in the control rat liver at 0 hour but the concentration of GMP and GDP respectively of 0.009 and 0.075 μmol/g wet liver weight in sham-operated rat liver 12 hours after the surgical operation is lower than those of 0 hour control of 0.016 and 0.136 μmol/g wet liver weight respectively.

Table 1. Concentration of Purine Ribonucleotides in Regenerating or Control Rat Liver Respectively 12 Hours after 70 Percent Hepatectomy or Sham-operation

(μmol/g wet liver weight, Mean±S.E.M.)

	Time after surgical operation (hour)	
	0	12
Hepatectomy		
AMP		0.123 ± 0.013
ADP		1.048 ± 0.046
ATP		3.021 ± 0.184
GMP		0.013 ± 0.001
GDP		0.074 ± 0.004
GTP		0.382 ± 0.030
T.A.N.		4.192 ± 0.188
T.G.N.		0.469 ± 0.031
T.A.G.N.		4.660 ± 0.218
A.E.C.		0.843 ± 0.009
(N)		(9)
Sham-operation		
AMP	0.131 ± 0.011	0.109 ± 0.009
ADP	1.135 ± 0.067	1.086 ± 0.072
ATP	3.549 ± 0.181	3.466 ± 0.147
GMP	0.016 ± 0.003	0.009 ± 0.001
GDP	0.136 ± 0.054	0.075 ± 0.004
GTP	0.452 ± 0.034	0.451 ± 0.022
T.A.N.	4.838 ± 0.218	4.661 ± 0.145
T.G.N.	0.604 ± 0.070	0.534 ± 0.024
T.A.G.N.	5.443 ± 0.261	5.196 ± 0.166
A.E.C.	0.853 ± 0.006	0.860 ± 0.004
(N)	(7)	(11)

(T.A.N., T.G.N., T.A.G.N., A.E.C. and (N) stand for total adenine nucleotides, total guanine nucleotides, total adenine and guanine nucleotides, adenylate energy charge and the number of animals respectively.)

ATase activity

The changes of the specific activity of ATase per unit amount of DNA or protein in the regenerating rat liver or sham-operated rat liver are shown in Fig. 3. The specific activity of ATase per unit amount of DNA at time 0 was 0.93 ± 0.10 nmol glutamine deaminated/hr/μg DNA which increased steeply from 6 hours and reached the peak value of 1.68 ± 0.06 18 hours and decreased but remained significantly increased thereafter until 72 hours after the surgical operation. The magnitude of the increase in comparison to sham-operated control at peak 18 hours after the surgical operation is 1.93 folds. In the sham-operated rat liver the mean specific activity remained relatively stable within the range from 0.63 to 1.03 as a mean until 72 hours after the surgical operation.

The specific activity of ATase per mg liver protein at time 0 was 10.3 ± 1.0 nmol glutamine deaminated/hr/mg protein which started to increase 3 hours and reached the significantly increased level 12 through 72 hours with the peak value of 19.8 ± 1.0 18 hours after the surgical operation. The magnitude of the increase in comparison to sham-operated control at peak 18 hours after the surgical operation is 1.91 folds. In the sham-operated rat liver the mean specific activity remained relatively stable within the range from 7.2 to 11.6 until 72 hours after the surgical operation.

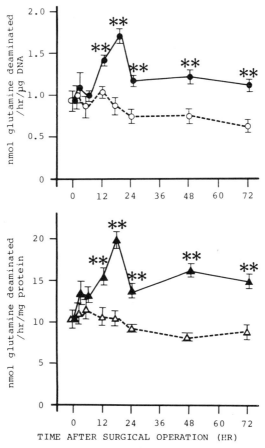

Fig. 3. The changes in the specific activity of ATase in the regenerating and sham-operated rat liver. Time in hours after the surgical operation is plotted on the abscissa. The specific activities per unit amount of DNA as nmol glutamine deaminated/hr/μg DNA are plotted on the ordinate in the upper panel and those per unit amount of protein as nmol glutamine deaminated/hr/mg protein are plotted on the ordinate in the lower panel. Each point is the mean ± S.E.M. of 5 animals.

DISCUSSION

This study has convincingly shown that the rate of de novo purine biosynthesis increases in the regenerating rat liver based on the increased specific activity of hepatic purines in [^{14}C]glycine incorporation assay 6 through 48 hours after the surgery.

Although there are two theoretically possible pathways to supply purine ribonucleotides to the cells, i.e. de novo and salvage pathway when cells are rapidly growing, this study has demonstrated that at least the de novo synthetic pathway is activated in the regenerating rat liver. The increased rate of de novo purine biosynthesis is apparently playing a role to answer the demands for these substances for cell growth.

As the mechanism of the increased rate of de novo purine biosynthesis, there are several possibilities including 1) increased supply of the substrate, PRPP, 2) decreased feedback inhibition by purine ribonucleotides and 3) increased enzyme activity of ATase. This study has firstly shown that the increased concentration of PRPP is one of the mechanism of the increased rate of de novo purine biosynthesis. Although the mechanism by which the concentration of PRPP is increased in the regenerating rat liver is not clear, it is speculated that either the increased supply of ribose 5-phosphate, a substrate for PRPP synthetase and/or the increased enzymatic activity of PRPP synthetase is the mechanism. Although the increased concentration of PRPP is analogous to the increased rate of de novo purine biosynthesis observed in fructose infused mouse liver[6], which is interpreted as the compensatory mechanism against the catabolism of purine ribonucleotides, the increased activity of ATase in the regenerating rat liver shown in our study and in the report by Prajda et al.[7] is regarded as the second mechanism to increase the rate of de novo purine biosynthesis which is unique to the rapidly growing cells. Although the magnitude of the increase of the specific activity of ATase 24, 48 and 72 hours after the surgery is comparable in our study and the report by Prajda et al. our study has in addition shown that the specific activity of ATase is increased sharply with the peak 18 hours after the surgery. The difference in timing of the peak of PRPP concentration 12 hours after the surgery and that of ATase 18 hours after the surgery suggests that there are several different mechanisms to increase the rate of de novo purine biosynthesis in the regenerating rat liver.

The role of the decreased feedback inhibition on ATase by purine ribonucleotides due to their increased consumption were unlikely since purine nucleoside triphosphates decreased and monophosphates were contrarily increased in the regenerating rat liver with the relative conservation of total adenine and guanine nucleotide concentration. The presence of the slightly lower adenylate energy charges, although not statistically significant, in the regenerating rat liver suggests that the increased adenylate energy charge is not the essential factor for the cell growth.

In conclusion the rate of de novo purine biosynthesis increases in the regenerating rat liver as one of the prerequisite for the cell growth and that it peaks 6 thorough 18 hours after the surgical operation preceeding the reported peak of DNA synthesis. This increase is mediated by the increased specific activity of ATase and the increased concentration of PRPP, and not by the decreased feedback inhibition by purine ribonucleotides due to their increased consumption.

REFERENCES

1. E. Bresnick, Chapter VIII. Regenerating liver: An experimental model for the study of growth. Methods in Cancer Res. 6:347 (1970).
2. N. L. R. Bucher, and R. A. Malt, Liver regeneration. in: "Regeneration of Liver and Kidney," pp.17-178, Little Brown and Company, Boston (1971).
3. K. M. Hwang, S. A. Nurphree, C. W. Shansky, and A. C. Sartorelli, Sequential biochemical events related to DNA replication in the regenerating rat liver. Biochim. Biophys. Acta 366:143 (1974).
4. J. B. Wyngaarden, and W. N. Kelley, Gout. in: "The Metabolic Basis of Inherited Disease," J. B. Stanbury, J. B. Wyngaarden, and D. S. Fredrickson, eds., pp.1043-1114, McGraw-Hill, Inc., New York (1983).
5. E. W. Holmes, J. B. Wyngaarden, and W. N. Kelley, Human glutamine phosphoribosyltransferase, Two molecular forms interconvertible by purine ribonucleotides and phosphoribosylpyrophosphate. J. Biol. Chem. 248:6035 (1973).
6. M. Itakura, R. L. Sabina, P. W. Heald, and E. W. Holmes, Basis for the control of purine biosynthesis by purine ribonucleotides. J. Clin. Invest. 67:994 (1981).
7. N. Prajda, N. Katunuma, H. P. Morris, and G. Weber, Inbalance of purine metabolism in hepatomas of different growth rates as expressed in behavior of glutamine PRPP amidotransferase. Cancer Res. 35:3061 (1975).
8. G. M. Higgins, and R. M. Anderson, Experimental pathology of liver; restoration of liver of white rat following partial surgical removal. Arch. Pathol. 12:186 (1931).
9. M. P. Edmonds, and G. A. LePage, The incorporation of glycine-2-C^{14} into acid soluble nucleotide purines. Cancer Res. 15:93 (1955).
10. T. Hisata, An accurate method for estimating 5-phosphoribosyl 1-pyrophosphate in animal tissues with the use of acid extraction. Analyt. Biochem. 68:448 (1975).
11. C. B. Thomas, W. J. Arnold, and W. N. Kelley, Human adenine phosphoribosyltransferase: Purification and properties. J. Biol. Chem. 248:2529 (1973).
12. O. H. Lowry, N. J. Rosebrough, A. L. Farr, and R. J. Randall, Protein measurement with the Folin phenol reagent. J. Biol. Chem. 193:265 (1951).
13. L. M. Rose, and R. W. Brockman, Analysis by HPLC of 9-beta-D arabinofuranosyladenine-5'-triphosphate level in murine leukemia cells. J. Chromatogr. 133:335 (1977).
14. M. Itakura, and E. W. Holmes, Human amidophosphoribosyltransferase. An oxygen-sensitive iron-sulfur protein. J. Biol. Chem. 333:999 (1979).
15. W. C. Schneider, Phosphorus compounds in animal tissues. III. A comparison of methods for the estimation of nucleic acids. J. Biol. Chem. 164:747 (1946).

PURINE METABOLISM IN HUMAN NEUROBLASTOMA CELL LINES

Chaim Kaplinsky, Jerzy Barankiewicz, Herman Yeger*, and Amos Cohen

Division of Immunology/Rheumatology and The Department of
Pathology*, Research Institute, The Hospital for Sick
Children, Toronto, Ontario Canada M5G 1X8

Neuroblastoma is the second most common solid tumor of childhood,
originating from the embryonic neural crest (1). The clinical picture
is polymorphic due to varied histological patterns, different
localization and different degrees of malignancy. Despite aggressive
treatment, the overall prognosis has not changed in the last 2
decades. There are only few reports on clinical trials with drugs
affecting nucleotide metabolism, such as methotrexate (2) and
trifluoromethyl-2-deoxyuridine (3). In the present study we describe
in detail the metabolism of purine nucleotide as a possible target for
chemotherapy in neuroblastoma. Two different human neuroblastoma cell
lines established from surgically resected tumor were used in this
study, NUB-6, from a patient with a localized disease and EW-2 from a
patient with recurrent metastatic disease. The cells were grown in
alpha-medium supplemented with 20% inactivated fetal calf serum. EW-2
cells formed monolayers whereas NUB-6 cells formed spheroid-like
aggregates with an average size of 400 μm. All assays were
performed on a single cell suspension achieved by gentle shaking in
citrate-saline and mild trypsinization for two minutes. Viability of
over 95% was kept through all experiments. The availability of two
biologically different cell lines may allow us to relate changes in
nucleotide metabolism to these phenotypic differences. For the
assessment of nucleotide metabolism, labelled adenine, hypoxanthine,
adenosine, guanosine, formic acid and glycine were used, and the
incorporation of radioactivity into the respective purine
ribonucleotides, ribonucleosides and bases was assayed as previously
described (4). The cytotoxic effect of a number of drugs was
evaluated in a clonogenic assay (6).

Metabolism of purine bases

Radioactive adenine and hypoxanthine were efficiently salvaged to
AMP and IMP respectively by both cell lines (Tables I and II). EW-2
cell line expressed 2 to 3 times higher activity of the salvage
pathway in comparison to NUB-6. Adenine salvage moderately exceeded
that of hypoxanthine in NUB-6, but not in EW-2. In both cell lines
AMP formed from adenine was mainly phosphorylated to ATP via ADP. It
was only slightly deaminated to IMP or dephosphorylated to adenosine.
IMP formed from hypoxanthine was interconverted to AMP and GMP at a

Table I: Adenine metabolism in human neuroblastoma.

Reaction	NUB- 6		EW- 2	
	CPM	%	CPM	%
APRTase	29982	100	55198	100
AMP Kinase	22235	74.1	36655	66.4
AMP Deaminase	6286	20.9	15107	27.3
IMP Dehyrogenase	3233	10.7	8795	15.8
GMP Synthetase	2214	7.3	7295	13.2
IMP Dephosphorylase	2932	9.7	6167	11.1
AMP Dephosphorylase	1204	4.0	42	0.07

Human neuroblastoma cell lines NUB-6 and EW-2 (2 x 10^6 cells) were incubated with 1 µCi of [8-^{3}H] adenine for 2 hrs. After incubation cells were collected by centrifugation and the radioactivity in nucleotides, nucleosides and bases were determined according to Henderson et al (4). Cell pellet was extracted with 50 µl of 0.4 M PCA for 10 minutes on ice, neutralized with Alamine-Freon mixture (5) and the radioactivity of the respective metabolites was determined using TLC chromatography. Counts are expressed per 1x10^6 cells and as a percentage of total amount of radioactivity passed by the studied reaction.

similar rate in NUB-6 but in EW-2 interconversion to AMP significantly exceeded the interconversion to GMP (Table II). Relatively high activity of IMP dephosphorylation was observed in both cell lines (Table II).

Purine nucleoside metabolism

Adenosine and guanosine were efficiently incorporated by both NUB-6 and EW-2 cells. Adenosine was mainly deaminated to inosine in NUB-6 cell and only one third was phosphorylated to AMP (Table III).

Table II: Hypoxanthine metabolism in human neuroblastoma.

Reaction	NUB- 6		EW- 2	
	CPM	%	CPM	%
HGPRTase	14945	100	59355	100
AMP Synthetase	5310	35.5	38594	65
AMP Kinase	3701	24.7	3166	53.5
IMP Dehydrogenase	4296	28.7	10767	18.1
AMP Dephosphorylase	1520	10.1	3639	6.1
IMP Dephosphorylase	4500	30.1	7285	12.2
GMP Dephosphorylase	3344	22.3	1079	1.8

Human neuroblastoma cell lines NUB-6 and EW-2 (2 x 10^6) were incubated with 2 µCi of [8-^{3}H] hypoxanthine. Incorporation of radioactivity into purine nucleotides, nucleosides and purine bases was determined as described in table 1.

Table III: Adenosine metabolism in human neuroblastoma.

Reaction	NUB-6		EW-2	
	CPM	%	CPM	%
AR Kinase	7140	17.8	7285	37.5
AMP Kinase	6755	16.8	6867	66
ADP Kinase	6195	15.4	6069	35.24
AR Deaminase	32969	82.19	12200	62.6
AR Phosphorylase	11841	29.52	11314	58.0
Xanthine Oxidase	2456	6.12	6390	32.79

Human neuroblastoma cell lines NUB-6 and EW-2 ($2x10^6$) were incubated with 1 µCi of adenosine. Incorporation of radioactivity into purine nucleotides, nucleosides and bases was determined according to Henderson et al (4). For details see table 1.

In EW-2 almost the same amount of adenosine was deaminated and phosphorylated. Guanosine was markedly phosphorylized to guanine which was subsequently salvaged to GTP by both cell lines. The latter was equally phosphorylated to GMP and deaminated to IMP with similar rate in both cell lines, though NUB-6 phosphorylation slightly exceeded deamination and in EW-2 deamination exceeded phosphorylation (Table IV).

Purine nucleotide biosynthesis de novo

Incorporation of radioactive formate or glycine into purine nucleotides was minimal in both cell lines suggesting that de novo nucleotide biosynthesis in human neuroblastoma contributes minimally to the synthesis of purine nucleotides in cultured neuroblastoma cells (results not shown).

Purine nucleotide catabolism

Hypoxanthine was the major product of noninduced ATP catabolism in both neuroblastoma cell lines (Figure 1). EW-2 cells excreted more hypoxanthine than NUB-6 but it should be noted that they also

Table IV: Guanosine metabolism in human neuroblastoma.

Reaction	NUB-6		EW-2	
	CPM	%	CPM	%
GR Phosphorylase	41228	100	41643	100
GPRTase	11528	27.9	15177	36.4
GMP Kinase	3414	8.2	1333	32.6
GMP Reductase	2114	5.1	1504	9.9

Neuroblastoma cell lines NUB-6 and EW-2 ($2x10^6$) were incubated with 1 µCi guanosine. Incorporation of radioactivity into purine nucleotides, nucleosides and bases was measured according to Henderson et al (4). For details see table 1.

synthesized more ATP (Table I). Only minor amounts of inosine, adenosine and uric acid were excreted to the medium. Preincubation with 1 to 40 μM deoxycoformycin did not have any effect on the amount of the catabolic products (data not shown). Tubercidin (100 μM) markedly induced ATP catabolism in both neuroblastoma cell lines, but did not change the proportion of excreted catabolites, where hypoxanthine remained the major product (Fig. 1).

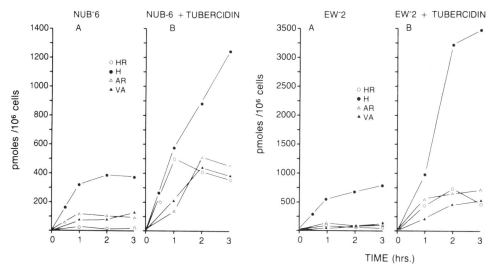

Fig. 1. ATP catabolism in neuroblastoma cell lines NUB-6 and EW-2

Cells (10x10⁶) were incubated with 2 μCi of radioactive adenine for 1 hr. Unincorporated radioactive adenine was washed out and cells were incubated in presence or absence of 100 μM tubercidin for 3 hrs. Excretion of radioactive products of ATP catabolism was measured. HR-inosine, H-hypoxanthine, AR-adenosine, UA-uric acid.

Cytotoxic effect of purine and pyrimidine analogs on colony formation

We tested the cytotoxic effect of some drugs in a clonogenic assay (5). 5-fluorouracil and 6-mercaptopurine did not affect colony formation at concentration ranging from 10^{-8} to 10^{-6} M. However, ara-C inhibited 50% of colony formation at 10^{-7} M, and 100% at 10^{-6} M. Methotrexate was quite toxic to NUB-6, and less to EW-2 at concentrations considered toxic to many tissues, in conventional protocols. Tubercidin, an adenosine analog, was toxic to neuroblastoma cells at a wide range of concentrations, and inhibited 100% of colony formation at 10^{-7} M (Table V).

This study established for the first time the overall purine metabolism in human neuroblastoma cell lines. Low activity of purine de novo synthesis and a highly active salvage pathway suggest that the latter is crucial for nucleotide synthesis in neuroblastoma cells. Although the overall purine nucleotide metabolism in neuroblastoma cells seems to be similar to other tissues, a few differences appear to be unique to neuroblastoma: a) interconversion of synthesized IMP with similar rate to AMP and GTP; b) a relatively high rate of GMP deamination. No significant differences in purine metabolic pathways were found between non-metastatic (NUB-6) and metatastic (EW-2) neuroblastoma cell lines, but increased activity of overall nucleotide metabolism was observed in the latter. On the basis of these data, we would like to suggest that drugs affecting de novo synthesis, such as methotreoxate, will be ineffective in the therapy of neuroblastoma, at concentrations that are non-toxic to the host. However, drugs that augment catabolic pathways, such as tubercidin, may be an effective adjuvant chemotherapeutic agent.

Table V: Cytotoxic effect of nucleoside analogs on colony formation in human neuroblastoma[a].

	ARA-C		6-MP		MTX		TUBERCIDIN		5-FU	
	EW-2	NUB-6	EW-2	NUB-6	EW-2	NUB-6	EW-2	NUB-6	EW-2	NUB-6
10^{-10} M	100	100	100	100	100	100	33	45	100	100
10^{-9} M	100	100	100	100	65	38	0	0	100	100
10^{-8} M	90	92	100	100	ND	ND	0	0	100	100
10^{-7} M	50	56	92	87	42	22	0	0	100	100
10^{-6} M	0	0	90	82	30	18	0	0	59	39

[a] The clonogenic assay reported in a modification of the basic two stage system using low gelling temperature agarose (6). Cultures were fixed with 1% glutaraldehyde and 4% formaldehyde after 1-3 wks, and colonies containing a minimum of 30 cells were enumerated. Number of colonies is expressed as a percentage of control (non-treated) cultures.
Abbreviations: Ara C - arabinosylcytosine; 6-MP - 6-mercaptopurine; MTX - methotrexate, 5-FU - 5 fluorouracil.

REFERENCES

1. P.A. Voûte, Neuroblastoma, in: "Clinical Pediatric Oncology", 3rd edition, W.W. Sutow, D.J. Ferenbach, T.J. Vieti, eds., The C.V. Mosby Company (1984).
2. A.R. Albin, W.A. Blyer, J.Z. Finklestein, et al., Failure of moderate dose prolonged-infusion methotrexate and citrovorum factor rescue in patients with previously treated metastatic neuroblastoma: A phase II study, Cancer Treat. Rep., 62(2):1097 (1978).
3. R. Nitschke, A. Cangir, W. Christ, and D.H. Berry, Intensive chemotherapy for metastatic neuroblastoma: a southwest oncology groups study, Med. Pediatr. Oncol., 8(3):281 (1980).
4. J.F. Henderson, J.H. Fraser, and E.E. McCoy, Methods for the study of purine metabolism in human cells in vitro, Clin. Biochem., 7:339-358 (1974).
5. J.X. Khym, An analytical system for rapid separation of tissue nucleotides at low pressure on conventional anion exchange. Clin. Chem. 21:1245-1251, 1975.
6. M. Rahman, et al., Characterization of human epidymona cell line, Submitted for publication (1985).

ACKNOWLEDGMENT: This research was supported by grants from the Medical Research Council of Canada and the National Cancer Institute of Canada. Amos Cohen is a recipient of NCI Scholarship. Chaim Kaplinsky is a recipient of Terry Fox Scholarship.

PURINE BIOSYNTHESIS IN CHINESE HAMSTER CELL MUTANTS AND HUMAN

FIBROBLASTS PARTIALLY DEFICIENT IN ADENYLOSUCCINATE LYASE

Paul K. Laikind, Harry E. Gruber, Inga Jansen, Laurie
Miller, Michael Hoffer, J. Edwin Seegmiller, and Randall C.
Willis. University of California San Diego, Department of
Medicine. La Jolla, California USA 92093

Jaak Jaeken, and Georges Van den Berghe. University of
Leuven and International Institute of Cellular and Molecular
Pathology. Brussels, Belgium

INTRODUCTION

Adenylosuccinate lyase (ASMP lyase, EC4.3.2.2) is the terminal enzyme
in de novo adenylate synthesis and as such acts to convert
adenylosuccinate (AS) monophosphate to adenosine monophosphate. This
enzyme also participates in the de novo synthesis of IMP converting 5'-
phosphoribosyl-4-(N-succinocarboxamide)-5-aminoimidazole (SAICAR) to 5'-
phosphoribosyl-5-amino-4-imidazolecarboxamide (AICAR). Recently Jaeken,
et al. described three patients with neurologic dysfunctions including
infantile autism syndrome (1) who had elevated levels of AS and a compound
tentatively identified as SAICA riboside in urine, blood, and cerebral
spinal fluid. These patients have a markedly decreased but varying
activity of ASMP lyase in various tissues studied. Deficiency of ASMP
lyase has previously been described in mutagenized and nutrient-selected
Chinese Hamster Ovary fibroblasts (CHO-Ade I cells) (2,3). These cells
excrete excess ASMP and SAICAR into the culture medium.
    In the present study of purine nucleotide metabolism of CHO-Ade I
cells and fibroblasts from a AS and SAICA riboside over-producing patient
we examined the effect of ASMP lyase deficiency on branchpoint enzyme
activities and on de novo synthesis by comparison to the control cell
lines, wild type parent cells (CHO-K1) and fibroblasts from age-matched
normal individuals, respectively.

EXPERIMENTAL PROCEDURES

Cells and Medium. The CHO-K1 cell line, the ASMP lyase deficient (Ade
I) derivative of this line, and culture of these cells have been described
(3). The stocks were maintained in -MEM containing 10% FBS (fetal bovine
serum). In addition media for CHO-Ade I contained 10^4 M guanine. Human
cells lines were maintained in -MEM containing 10% FBS. Fibroblasts of
the patient were obtained by a skin biopsy. The patient has been
previously described by Jaaken, et al. (1, patient B). Fibroblasts from a
normal individual (14-year-old female) were also obtained by skin biopsy.
    Cell Harvest. Cells at approximately 80% confluency were briefly
trypsinized, washed twice with minimal Hanks, and resuspended in buffer
containing .3 M sucrose, 10mM Tris-HCl (pH 7.4), 0.5 mM NaEDTA, 0.1 mM

dithiothreoitol. Cell-free lysates were prepared by three 20-second sonications, centrifugation at 10^5 x g for 60 min at 4°C, and dialysis against 100 volumes of the lysis buffer at 4°C with three buffer changes over a 16-hour period. Aliquots were taken for ASMP lyase assay and the remainder was frozen at -20°C for branchpoint assays. Protein contents of the cell-free lysates were determined by the Coomossie Blue method.

Enzyme Assays. Branchpoint assays (ASMP synthetase, IMP dehydrogenase, and 5'-nucleotidase-ATP stimulated) were done as described by Gruber, et al. (4). ASMP lyase was done by a modification of the assay for ASMP synthetase. Briefly, a 26 ul premix containing 1 mM GTP, 5 mM aspartate, 10 mM magnesium acetate, 0.05 M Tris-Hepes (pH 6.9), 105 uM [8-^{14}C] IMP, 4.2 mM dithiothreitol and 1.0 mg/ml CHO-Ade I extract (to increase the level of ASMP synthetase) was incubated for 10 minutes before addition of 10 ul of the appropriate protein at 0.5 mg/ml. The reaction was stopped by pipetting 9 ul of the mixture into 9 ul of 8 M formic acid. Nine ul of this ;mix was applied to PEI cellulose TLC plates and separated with 1 M acetic acid/0.2 M lithium chloride (1:1 v/v). The substrate and products were located by radioautography, cut out and radioactivity determined by scintillation counting.

Measurement of de novo Purine Synthesis and Excretion. The rate of newly-formed purine synthesis and excretion was measured as previously described by determining the amount of ^{14}C formate incorporated into adenine (Ade), hypoxanthine (Hyp), and guanaine (Gua) moieties (5). Cells from stock cultures were trypsinized, transferred to either 35 mm diameter culture dishes at 0.4-.5 x 10^6 cells/well or to 100 mm diameter culture dishes at 1 x 10^6 cells per dish and cultured 16-20 hours in appropriate media prior to start of assay. The nucleotide, nucleoside and base content of cells and media were analyzed, where indicated, by high-pressure liquid chromatography.

RESULTS

ASMP Lyase Enzyme Activity. ASMP lyase catalyzes the conversion of SAICAR to AICAR during de novo synthesis of IMP and in a similar reaction acts at the second step in the synthesis pathway from IMP to AMP converting ASMP to AMP. The activity of this enzyme for the later of these reactions was measured, in vitro (Figure 1) in CHO-Ade I cells, fibroblasts from a patient with ASMP and SAICAR overproduction, and fibroblasts from a normal individual. All assays were conducted on proliferating cell lines. As expected ASMP lyase is markedly deficient in CHO-Ade I cells and fibroblasts from the patients with purine autism when compared to normal fibroblasts.

IMP Branchpoint Enzyme Activities. Three enzymes compete for IMP. They are ASMP synthetase, which initiates guanylate synthesis, IMP dehydrogenase, which initiates guanylate synthesis, and ATP-stimulated 5'-nucleotidase which degrades IMP to inosine. The activity of the three enzymes in CHO-K1 and CHO-Ade I are shown in Table I. The specific activities in CHO-K1 are not markedly different from replicating human cell lines (data from studies of human B-lymphoblasts are also given in Table I for comparison) except in the case of the ATP stimulated, 5'-nucleotidase which is higher in CHO-K1 cells. The CHO-Ade I cells display a marked (2.5-fold) elevation of ASMP synthetase activity when compared to the level in the parent CHO-K1 cells. The difference in IMP dehydrogenase activity between the two lines is correlated with the difference in growth rates observed between the lines; CHO-K1 growth rate being approximately 2-fold faster than CHO-Ade I.

Purine Nucleotide Synthesis. The rates of purine synthesis de novo and the distribution of this purine analogue among adenine, hypoxanthine and guanine components of the cell and media fractions of CHO-K1 and CHO-Ade I cultures are compared in Table II. The rate of de novo purine synthesis is increased almost 2-fold in the ASMP lyase-deficient cell

364

Table I
Branchpoint Enzyme Assays[1]

ENZYME	CELL LINE	
	CHO-K1	CHO-Ade I[2]
ASMP synthetase	1.7 (1.8 0.4)[3]	4.2
IMP dehydrogenase	0.6 (0.7 0.2)	0.4
5'-nucleotidase, ATP stimulated	3.1 (2.1 0.3)	2.5

[1] Enzyme activities are in units of nmol/min per mg. Values shown are the average of three seperate determinations. Standard deviations were less than 20% of the mean values.

[2] CHO-Ade I, ASMP lyase deficient.

[3] The average activity values determined for 22 normal human B-lymphoblast lines are shown in parenthesis.

Table II
Rates of Purine Synthesis and Excretion[1]

Distribution of ^{14}C-formate		CELL LINE[2]	
Fraction	Base	CHO-K1	CHO-Ade I[2]
Cell, acid soluble	Ade	16397 (71)	776 (2)
	Hyp	462 (2)	7362 (19)
	Gua	6236 (27)	30619 (79)
	Total	23094	38757
, nucleic acid	Ade	5582 (26)	1586 (5)
	Gua	15888 (74)	30119 (95)
	Total	21470	31705
, Total		44564	70462
Medium	Ade	<100 (<3)	<100 (<3)
	Hyp	3294 (86)	15305 (92)
	Gua	427 (11)	1167 (7)
	Total	3821	16472
Total Purine Synthesized		48385	86934
% Excreted of Total		8	19

[1] Cells were plated at 0.4 to 0.5 x 10^6 per 35 mm well and cultured approximately 24 hours. Culture media were removed, cells were washed once with 0.1% FBS media and 0.6 ml of 0.1% FBS media were added to each well. The cells were cultured for 30 min. prior to addition of 5 uCi ^{14}C-formate (59 uCi/umole) to each well. The labeling period was 90 min.

[2] Cell number per well at the time of the experiments were: CHO-K1, 1.1 x 10^6; CHO-Ade I, 0.7 x 10^6. Results are in cpm/hr per 10^6 cells and are the average value from 2 wells. Values in perenthesis are the distribution of adenine, hypoxanthine, and guanine in each fraction of the culture normalized to 100.

[3] ASMP lyase deficient.

ASMP Lyase Activity

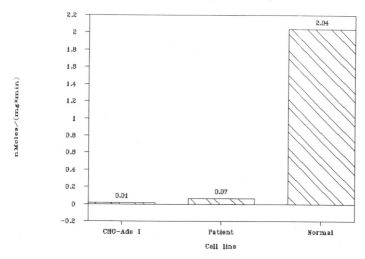

FIGURE 1. Comparison of ASMP lyase activity in mutagenized and nutrient selected CHO-cells (CHO-Ade I) with the activity in fibroblasts from a AS and SAICA riboside over-producing patient and a normal individual.

line. The largest increase (4-fold increase over CHO-Kl) is in the media (excreted) fraction, mainly in the form of hypoxanthine. The sum of the rates of adenylate (Ade) and guanylate (Gua) synthesis are nearly equal in CHO-Kl, i.e., Ade:Gua in the cell fraction is 49:51. The Ade:Gua ratio is altered in CHO-Ade I (5:88) reflecting the genetic deficiency of ASMP lyase activity. The alteration of the ratio results from both an apparent stimulation of rate of guanylate synthesis and accumulation of hypoxanthine-containing compounds in the cell and media (excreted) fractions.

The rate of de novo purine synthesis in ASMP lyase deficient fibroblasts from the patient compared to fibroblasts of a normal control is shown graphically in Fig. 2. These results are consistent with CHO cells both in the increase in total synthesis and in the increase in the fraction of total purine excreted to the media.

DISCUSSION

The data presented here confirms previous results demonstrating a reduced level of ASMP lyase activity in the adenine requiring mutant CHO-Ade I (2,3) and in fibroblasts of a patient with neurologic dysfunctions, including infantile autism syndrome, who overproduces the dephosphorylated products of the adenylate precursor ASMP as well as the IMP precursor SAICAR (6).

Although it was previously reported that CHO-Ade I fibroblasts contain no ASMP lyase activity (2,3) both the enzyme activity measurements (Fig. 1) and the demonstration of de novo adenylate synthesis (Table II) suggest that the enzyme activity is reduced but not absent in these cells. In contrast, the percent activity in the autistic patients´ fibroblasts

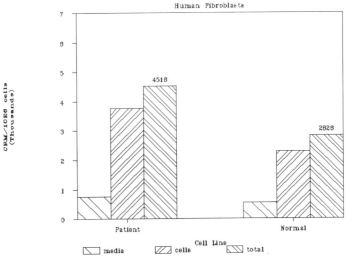

FIGURE 2. Comparison of de novo purine synthesis and excretion in fibroblasts from a AS and SAICA riboside over-producing patient and a normal individual. See Table II for details.

compared to normal (4%) is significantly lower than previously reported (63%) (6). A different method of assay was used in the previous study than used here but, probably of more significance, the preparation of the crude cell extract in the two studies was markedly different. In the previous study enzyme extracts were used within three hours of preparation by tissue homogenation in water whereas in the present study cell extracts were dialyzed against neutral buffer for 16 hours. As the activity in the control cells in the present study remained active during the time of preparation (and had equivalent activity to that reported by Jaaken, et al.) the results suggest an increased lability of the ASMP lyase from the autistic patient compared to the normal enzyme.

The CHO-Ade I (ASMP lyase-deficient) cells display a marked (2.5-fold) elevation of the ASMP synthetase activity suggesting that the ASMP synthetase expression may be regulated by levels of either ASMP or the ultimate product adenylates. ASMP synthetase is the apparent rate-limiting step in the two-step conversion of IMP to AMP. The increase in activity of ASMP synthetase as the ASMP lyase decreases apparently reflects an up regulation of the first step in the adenylate synthesis pathway to overcome the deficiency in the second step.

This study provides further evidence that adenine nucleotides are involved in the regulation of de novo purine synthesis. In both cell lines a reduction in the synthesis of adenylates due to a loss of activity of ASMP lyase causes an increase in total de novo synthesis. These results are consistent with previous studies using alanosine as an inhibitor of ASMP synthetase. Most of the increase observed in the CHO-Ade I cells is due to an increase in guanylate synthesis suggesting either spillover of the increased IMP into the guanylate bcranch during an attempt to elevate adenylates or direct regulation of a rate-limiting step in the guanylate synthesis pathway by adenylates or adenylate precursors.

In conclusion, these and other studies have shown that changes in the adenylate synthesis pathway can lead to alterations in other parts of the de novo purine synthesis pathway. In addition, evidence is presented suggesting that the mutagenically-induced deficiency of ASMP lyase in CHO-Ade I fibroblasts behaves biochemically similar to an inherited deficiency of ASMP lyase in patients with neurologic dysfunctions including infantile autism syndrome who have elevated levels of AS and SAICA riboside in urine, blood, and cerebral spinal fluid.

REFERENCES

1. Jaeken, J., and Van den Berghe, G. (1984) Lancet 2: 1058-1061.
2. Tu, A. S., and Patterson, D. (1977) Biochem. Genet. 15: 195-210.
3. Patterson, D. (1976) Somat. Cell. Genet. 2: 189-203.
4. Gruber, H. E., Jansen, I., Willis, R. C., and Seegmiller, J. E. (1985) Biochim. Biophys. Acta. (in press).
5. Herschfield, M. S., and Seegmiller, J.E. (1977) J. Biol. Chem. 252: 6002-6010.
6. Van den Berghe, G., and Jaeken, J. (1985) Ped. Res. 19: 780, Abs.

PURINE SYNTHETIC CAPACITIES OF DE NOVO AND SALVAGE PATHWAYS IN RAT

HEPATOMA 3924A CELLS

Yutaka Natsumeda, Tadashi Ikegami and George Weber

Indiana University School of Medicine, Laboratory for

Experimental Oncology, Indianapolis, IN, U.S.A.

INTRODUCTION

The relative contribution of de novo and salvage pathways for
nucleotide synthesis in neoplastic tissues is one of the important factors
in cancer chemotherapy. Although almost all antipurine and antipyrimidine
metabolites aim at de novo pathways as target sites, our previous
investigations have suggested the possible role of salvage pathways in
circumventing the action of the inhibitors of de novo synthesis (1-4). For
purine nucleotide synthesis the de novo and the salvage synthetic enzymes,
amidophosphoribosyltransferease (EC 2.4.2.14), adenine phosphoribosyltransferase (APRT; EC 2.4.2.7) and hypoxanthine-guanine phosphoribosyltransferase (HGPRT; EC 2.4.2.8), share the common substrate, PRPP, and are
inhibited by the end products, purine nucleotides (5-7). These facts
suggest that the inhibition of purine de novo synthesis might be compen-
sated by purine salvage synthesis in chemotherapy. The purpose of this
study was to compare the potential synthetic capactities of purine de novo
and salvage pathways in neoplastic cells as reflected in enzyme specific
activities and incorporation rates of [^{14}C]formate, [^{14}C]adenine, [^{14}C]
hypoxanthine and [^{14}C]guanine into purine nucleotides and nucleic acids.

EXPERIMENTAL PROCEDURES

Animals and Tumors

The maintenance of tumor-bearing animals in individual cages under
standardized conditions, and the procedures for killing animals and
excising tumors were outlined (1). Human colon adenocarcinomas, kidney
tumors, and hepatocellular carcinomas were processed immediately after
surgical resection as reported (3,4). Acute myelogenic leukemia cells were
obtained by bone marrow aspiration from the patient.

Cell Culture

Rat hepatoma cells were grown in monolayer cultures as previously
described (8) in McCoy's 5A medium supplemented with 10% dialyzed fetal
calf serum, penicillin, 100 units/ml and streptomycin, 100 μg/ml. Growth
properties, cytogenetics, karyotype and several biochemical parameters were
reported (8-10).

Table 1. Specific Activities of Purine Synthetic Enzymes in Various Human, Rat and Mouse Tumors

Tumors	No. of samples	Amido-transferase	APRT	HPRT	GPRT
Human					
Colon adenocarcinoma	15	83.3 ± 11.1[a] (100)[b]	516 ± 85 (619)[c]	255 ± 41 (306)[c]	655 ± 93 (786)[c]
Hepatocellular carcinoma	12	31.2 ± 3.8 (100)	323 ± 64 (1035)[c]	134 ± 30 (429)[c]	310 ± 67 (994)[c]
Kidney tumor	1	35.0 (100)	672 (1920)	109 (311)	309 (833)
Acute myelogenic leukemia	1	5.3 (100)	34 (642)	17 (321)	58 (1094)
Rat					
Hepatoma 16	4	101 ± 6 (100)	532 ± 3 (527)[c]	996 ± 30 (986)[c]	3240 ± 77 (3208)[c]
Hepatoma 3924A	10	155 ± 5 (100)	851 ± 23 (549)[c]	325 ± 11 (210)[c]	1010 ± 32 (652)[c]
Mouse					
Lewis lung carcinoma	4	43.0 ± 0.1 (100)	952 ± 20 (2214)[c]	267 ± 6 (621)[c]	802 ± 6 (1865)[c]

[a] Means ± S.E. expressed in nmol/hr/mg protein.

[b] Values in parentheses, % of amidotransferase activities.

[c] Significantly different from amidotransferase activities ($p < 0.05$). In all tumors examined, the salvage enzyme activities were higher than that of amidotransferase.

Enzyme Assays

All enzyme activities were measured in crude 100,000 x g supernatants from tissues or cells. Assays for APRT and HGPRT were conducted (3) and amidotransferase activity was measured as reported (5).

Assays for Purine Synthesis in Cells

Prior to the assays the culture medium was replaced by the serum free medium. Five to ten x 10^5 cells in each 25 cm^2 flask were pulsed at 37 °C for various time periods (0, 10, and 20 min in the standard assay) with either [14C]formate (5.8 to 59 mCi/mmol), [8-^{14}C]adenine (25 mCi/mmol), [8-^{14}C]hypoxanthine (25 mCi/mmol) or [8-^{14}C]guanine (25 mCi/mmol). To terminate the pulse, medium was suctioned, and 5 ml of ice-cold phosphate buffered saline (PBS) was added to the flask. PBS was discarded immediately; 0.5 ml of 0.2 N NaOH was added to dissolve the cells, and 0.5 ml of 1N $HClO_4$ was added.

For measuring salvage synthesis from [^{14}C]adenine, [^{14}C]hypoxanthine, or [^{14}C]guanine the contents of flasks were filtered onto a glass fiber filter (Grade 934AH). Acid soluble filtrates were neutralized with 1/10 volume of 4.42 N KOH and paper-chromatographed in butanol: methanol: water: ammonia (60:20:20:1) to separate [^{14}C]nucleotides from [^{14}C]base. The glass fiber filters with acid insoluble materials were washed repeatedly with 0.4 N $HClO_4$ and 95% ethanol, dried, and counted. Assays for purine de novo synthesis from [^{14}C]formate were carried out as described (11) after the acid extraction.

RESULTS AND DISCUSSION

Enzymic Activities of Purine de Novo and Salvage Pathways

The specific activities of the purine de novo rate-limiting enzyme, amidotransferase, and the purine salvage enzymes, APRT and HGPRT, were compared in 7 different human, rat and mouse neoplasias (Table 1). In all tumors examined the activity of APRT was 5.3- to 22-fold, HPRT was 2.1- to 9.7-fold, and GPRT was 6.5- to 19-fold higher than that of amidotransferase. Table 2 compares K_m's for PRPP of the purine synthetic enzymes in rat liver. The results indicated that the affinities for PRPP of the salvage enzymes were 2 orders of magnitude higher than that of amidotransferase. The higher activities and affinities for PRPP would confer an advantage on the salvage enzymes for nucleotide synthesis over the de novo enzyme in neoplastic tissues.

Table 2. K_m's for PRPP of Purine Synthetic Enzymes in Rat Liver

Enzymes	$K_m = \mu M$
Amidotransferase	600
APRT	2
HGPRT	4

The kinetic properties of hepatoma enzymes are similar to those of liver enzymes (3).

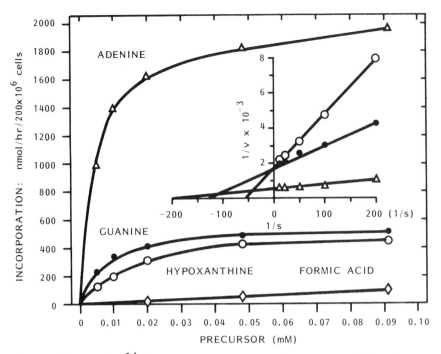

Fig. 1   Effects of [14C]precursor concentration on overall purine
          nucleotide synthesis in hepatoma 3924A cells. The initial
          incorporation rates of the precursors into the sum of purines in
          acid-soluble nucleotides and in acid–insoluble macromolecules
          were measured as described in "EXPERIMENTAL PROCEDURES."

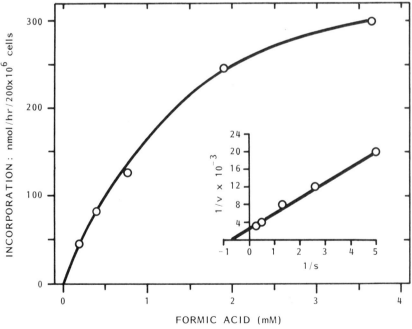

Fig. 2   Effects of [14C]formate concentration on overall purine
          nucleotide synthesis in hepatoma 3924A cells. The initial rates
          were measured as described in "EXPERIMENTAL PROCEDURES."

Initial Rate Kinetics of Purine de Novo and Salvage Synthesis in Hepatoma 3924A Cells

Under standard pulse-labeling conditions with $[^{14}C]$formate, $[8-^{14}C]$adenine, $[8-^{14}C]$hypoxanthine or $[8-^{14}C]$guanine described above, the total labeling for purine in acid-soluble nucleotides and in acid-insoluble nucleic acids was linear with incubation time, up to 20 min, and with numbers of cells, within the range tested.

Figure 1 shows the effect of concentration of the precursors on total purine nucleotide synthesis in hepatoma 3924A cells in log phase. The salvage synthesis from adenine, hypoxanthine and guanine followed Michaelis-Menten kinetics and the double-reciprocal plots yielded apparent K_m's of 5, 16 and 8 μM, respectively. The formate concentration curve for purine de novo synthesis was also hyperbolic (Fig. 2) and the apparent K_m for formate was 1,500 μM.

Table 3 summarizes the enzymic activities and overall metabolic fluxes of purine de novo and salvage synthetic pathways in hepatoma 3924A cells in log phase. The maximal rates of the synthetic fluxes were calculated from the double reciprocal plots. Since 2 mol of $[^{14}C]$ formate could label 1 mol of purine ring at the maximal incorporation rate, the de novo synthetic rate in Table 3 was tentatively given as 1 mol incorporation of formate = 1/2 mol formation of purine. In Table 3 the potential capacities of salvage pathways were higher than that of the purine de novo pathway in hepatoma 3924A cells in log phase. In resting phase the de novo synthetic rate was reduced to 14% of that in log phase in parallel with the decrease of PRPP concentration in cells, whereas salvage capacities were nearly constant during the culture growth (data not shown). Thus, both in stationary and logarithmically growing phases the potential synthetic capacities of the salvage pathway were higher than that of the de novo pathway.

The high enzymic activities, the high affinities for PRPP, and the great potential synthetic capacities of salvage pathways in neoplastic cells may explain, in part at least, the apparent failure of inhibitors of purine de novo synthesis to produce lasting chemotherapeutic results. The evidence in this study provides further support for a rational basis for combination chemotherapy of de novo synthesis inhibitors with blockers of the salvage pathway (1,2).

Table 3. Purine Synthetic Capacities of Hepatoma 3924A Cells in Log Phase

Enzymes	Activity	Precursors	Maximal rate of flux	% of enzyme activity
Amidotransferase	11,800[a]	Formate	213[b]	1.8
APRT	61,600	Adenine	2,200	3.6
HPRT	31,200	Hypoxanthine	530	1.7
GPRT	70,800	Guanine	590	0.8

Values are expressed in nmol/hr/200 x 10^6 cells and % of enzyme activity.
[a]Assayed in 100,000 x g supernatant fluid.
[b]Calculated from the double reciprocal plots of initial rates.

ACKNOWLEDGEMENTS

This work was supported by Grants CA 13526 and CA 05034 from the National Cancer Institute, NIH and by the American Cancer Society Institutional Grant.

REFERENCES

1. G. Weber, Biochemical strategy of cancer cells and the design of chemotherapy: G.H.A. Clowes Memorial Lecture. Cancer Res. 43: 3466 (1983).
2. G. Weber, M. S. Lui, Y. Natsumeda and M. A. Faderan, Salvage capacity of hepatoma 3924A and action of dipyridamole. Advan. Enzyme Regul. 21: 53 (1983).
3. Y. Natsumeda, N. Prajda, J. P. Donohue, J. L. Glover and G. Weber, Enzymic capacities of purine de novo and salvage pathways for nucleotide synthesis in normal and neoplastic tissues. Cancer Res. 44: 2475 (1984).
4. Y. Natsumeda, M. S. Lui, J. Emrani, M. A. Faderan, M. A. Reardon, J. N. Eble, J. L. Glover and G. Weber, Purine enzymology of human colon carcinomas. Cancer Res. 45: 2556 (1985).
5. E. W. Holmes, J. A. McDonald, J. M. McCord, J. B. Wyngaarden and W. N. Kelley, Human glutamine phosphoribosylpyrophosphate amidotransferase. J. Biol. Chem. 248: 144 (1973).
6. A. W. Murray, Inhibition of purine phosphoribosyltransferases from Ehrlich ascites tumour cells by purine nucleotides. Biochem. J. 100: 671 (1966).
7. Y. Natsumeda, Purification and properties of hypoxanthine-guanine phosphoribosyltransferase from rat liver. Yokohama Med. Bull. 26: 107 (1975).
8. E. Olah, M. S. Lui, D. Y. Tzeng and G. Weber, Phase and cell cycle specificity of pyrazofurin action. Cancer Res. 40: 2869 (1980).
9. E. Olah and G. Weber, Giemsa-banding karyotype of rat hepatomas of different growth rates. Cancer Res. 39: 1708 (1979).
10. G. Weber, E. Olah, J. E. Denton, M. S. Lui, E. Takeda, D. Y. Tzeng and J. Ban, Dynamics of modulaton of biochemical programs in cancer cells. Advan. Enzyme Regul. 19: 87 (1981).
11. M. S. Hershfield and J. E. Seegmiller, Regulation of de novo purine biosyhthesis in human lymphoblasts. J. Biol. Chem. 251: 7348 (1976).

METABOLISM OF PURINE NUCLEOTIDE IN THE LIVER AND KIDNEY OF CASTRATED RATS

Maria Pizzichini, Anna Di Stefano, Giacomo Matteucci, and
Enrico Marinello

Univ. of Siena, Department of Biological Chemistry, Italy

INTRODUCTION

The positive influence of testosterone on kidney RNA polymerase[1] and
its effect on RNA metabolism in the sexual organs[2] is well known : however
the action of the hormone on the free nucleotide metabolism and of RNA in
different organs has not yet been sufficiently clarified. Avdalovic et al.[3]
have reported that testosterone enhances the biosynthesis of kidney RNA,
but has no effect on its catabolism : the same Authors gave no conclusive
evidence concerning the behavior of RNA and the free nucleotides in the
liver of castrated rats, following testosterone administration[4]. We have
compared the metabolism of free purine nucleotides and of RNA in the liver
and kidney of adult male Wistar rats before and after castration.

MATERIAL AND METHODS

Male albino Wistar (250 g b.w.) were castrated under ether; they
were then allowed water and normal diet ad libitum. 15 days later, the
animals were fasted for 20 hours before decapitation. We followed the incorporation of ^{14}C-formate into the acid-soluble bases and into RNA, a
method which is still widely used as an index of the overall rate of purine nucleotide synthesis. We studied the liver and kidney of normal and
castrated rats. ^{14}C-formate (54.8 mCi/mmole) was injected intraperitoneally (10 μCi/100 g b.w.) and the animals killed 60 minutes later.
Immediately after killing, the organs were rapidly homogenized with 0.5 N
perchloric acid (1 g + 5 ml). The acid-soluble purines were purified by
acid hydrolysis and separated by ion exchange chromatography[5]. RNA was
determined by method described by Munro and Fleck[6].

RESULTS

The total base content does not significantly change after castration
the specific activity shows no variation in the kidney, while a not significant increase is observed in the liver. The weight falls in both organs (table 1).

TABLE 1 - Organ weight, content (total mg), total dpm and specific activity (SA=dpm/μmole) of acid-soluble purines in liver and kidney of normal (N) and castrated (C) rats.

	WEIGHT	CONTENT	DPM	SA
N liver (8)	8.97 + 0.36	5.39 + 0.15	185824 + 17703	4481 + 400
C liver (11)	°7.53 + 0.24	4.90 + 0.17	194341 + 19472	5394 + 563
N kidney (12)	2.03 + 0.07	0.87 + 0.03	110001 + 10973	16828 + 1544
C kidney (12)	°1.78 + 0.04	0.84 + 0.03	98050 + 12102	15875 + 2130

° $p < 0.01$. Results are the means + SE. No. cases indicated between brackets

Table 2 shows the content in adenine, guanine and hypoxanthine, in the liver and kidney, before and after castration (the values refer to the total weight of organs); their specific activity is slightly enhanced in both organs and especially in the liver, indicating an accelerated turn over. The G/A labeling ratio (counts per min in guanine / counts per min in adenine), which is an index of the " IMP branch point regulation ", is moderately augmented in both tissues.

TABLE 2 - Content (total mg) and specific activity (SA= dpm/μmole) of acid soluble purines in liver and kidney of normal (N) and castrated (C) rats.

	ADENINE content	SA	GUANINE content	SA	HYPOXANTHINE content	SA	G/A
N liver (6)	3.49 + 0.11	2303 + 233	0.69 + 0.04	1385 + 76	0.51 + 0.04	5653 + 607	0.10
C liver (7)	3.26 + 0.11	°°4428 + 754	0.59 + 0.04	°3025 + 446	°°0.37 + 0.03	°°9570 + 1482	0.11
N kidney (10)	0.53 + 0.02	14301 + 1836	0.15 + 0.01	4038 + 274	0.14 + 0.01	15274 + 1498	0.071
C kidney (9)	0.53 + 0.03	17510 + 1964	0.15 + 0.01	5614 + 749	°°0.10 + 0.01	20334 + 2700	0.081

° $p < 0.01$, °° $p < 0.05$. Results are the means + SE. No. cases indicated between brackets.

TABLE 3 – Content (total mg), total dpm and specific activity (dpm/γ) of RNA in liver and kidney of normal (N) and castrated (C) rats.

	TOTAL CONTENT	TOTAL DPM	SPECIFIC ACTIVITY
N liver	70.16 + 1.43	945672 + 92002	13.58 + 1.38
C liver	71.51 + 6.98	°639263 + 31472	9.18 + 1.28
N kidney	8.74 + 0.11	88923 + 4091	10.16 + 0.48
C kidney	°6.30 + 0.15	°45447 + 3842	°°7.20 + 0.56

° $p<0.01$, °° $p<0.005$. The values are the means + SE. of 3 cases.

Table 3 shows the total content in RNA in both organs, the total dpm and the specific activity, before and after castration. The content is unchanged in the liver and significantly decreased in the kidney; the ^{14}C-formate incorporation decreases in both organs, but only in the kidney is the decrease, in specific activity, statistically significant.

DISCUSSION

Our results show: 1) no significant variation in the base content either in the liver or the kidney following castration 2) the specific activity of the bases increases after castration, in the liver and kidney, but the increase is significant only in the liver 3) a decreased content in RNA only in the kidney. This would indicate an accelerated turnover of the free nucleotides and a delayed turnover of RNA due to castration. The increased turnover, which occurs in the kidney and the liver after castration, is difficult to interpret. It is very likely that, in absence of testosterone, the turnover of purine nucleotide is affected by other hormones and specially by glucocorticoid, all of which are known to increase purine nucleotide turnover in the liver[7], and by aldosterone which plays the same role in the kidney[8]. The decreased RNA turnover which we observed, may be due to an increased catabolism of the newly formed free nucleotides and to the decreased RNA polymerase activity after castration[1]. Our results would suggest that testosterone plays a regulatory role in purine nucleotide synthesis, in along with other hormones and specially the glucocorticoids.

REFERENCES

1) N. Avdalovic and C.D. Kochakian – Androgen regulation of RNA polymerase activity in isolated mouse kidney nuclei. Biochim.Biophys.Acta, 182, 382, 1969.

2) S.J. Higgins and J.M. Burchell - Effect of testosterone on RNAm and protein synthesis in rat seminal vesicle. Biochem.J., 174, 543, 1978.

3) N. Avdalovic and M. Bates - The influence of testosterone on the synthesis and degradation rate of various RNA species in the mouse kidney Biochim.Biophys.Acta, 407, 299, 1975.

4) E. Zoref-Shani, A. Shainberg, O. Sperling - Characterization of purine nucleotide metabolism in primary rat muscle cultures. Biochim.Biophys. Acta, 716, 324, 1982.

5) M.M. Welch and F.B. Rudolph - Regulation of purine biosynthesis and interconversion in the chick. J.Biol.Chem., 257, 13259, 1982.

6) H.N. Munro and A. Fleck - Recent developments in the measurement of nucleic acid in biological material. Analyst. 91, 78, 1965.

7) M. Pizzichini, A Di Stefano, G. Bruni, R. Leoncini, E. Marinello - The hormonal regulation of purine biosynthesis. Adv.Exp.Med.Biol. 165 A, 433, 1984.

8) R.K. Mishra, J.F. Weldrake, L.A.W. Feltham - Adrenocorticoids and RNA metabolism in kidney. Biochim.Biophys.Acta, 287, 106, 1972.

THE HORMONAL REGULATION OF PURINE BIOSYNTHESIS:

CONTROL OF THE INOSINIC ACID BRANCH POINT

Maria Pizzichini[°], Anna Di Stefano [°], Giuseppe Pompucci [°°],
and Enrico Marinello[°]

Univ. of Siena, Department of Biological Chemistry, Italy [°]
Univ. of Sassari, Dept. of Physiol. and Biochem., Italy [°°]

INTRODUCTION

We previously studied the behavior of " de novo " purine biosynthesis
and of RNA metabolism in the levator ani muscle (LAM) of castrated rats[1],
demostrating that 1) " de novo " purine biosynthesis decreases in the LAM
of the rat after castration, as   shown by a diminishing incorporation of
^{14}C-formate in all bases; 2) the ratio G/A (= counts per min in guanine /
counts per min in adenine), an index of IMP branched point regulation[2],
is unchanged under such conditions; 3) since only the RNA content decreases, but not its specific activity, we concluded that the formation of new
molecules of RNA remains the same, while the RNA breakdown is accelerated
during the atrophy of the LAM which follows castration. Now we have studied the behavior of purine biosynthesis " de novo " in the LAM of adult
male Wistar rats, before, after castration and after testosterone administration.

MATERIALS AND METHODS

Male albino Wistar rats (250 g b.w.) were castrated under ether;
they were then allowed water and normal diet ad libitum. All the animals
were fasted for 20 hours before decapitation, 15 days later. We followed
the incorporation of ^{14}C-formate into the acid-soluble bases, a method
widely used, as an index of the overall rate of purine nucleotide synthesis. One group of rats, 15 days after castration, was treated subcutaneously with testosterone propionate, dissolved in arachis oil, (2.5 mg /
100 g b.w.), for 24 h, or daily for 10 days. Immediately after killing,
the LAM was homogenized with 0.5 N perchloric acid. The acid-soluble purines were purified by acid hydrolysis and separated by ion exchange chromatography[3].

RESULTS

Table 1 shows the organ weight, content, dpm in total bases and their

TABLE 1 – Effect of castration and testosterone propionate (TP) on the weight of LAM and on the incorporation rate of ^{14}C-formate into acid-soluble purines.

	weight g	total content mg	total radioactivity dpm	specific activity dpm/μmole
Normal (N) (8)	0.88 + 0.04	0.56 + 0.02	6673 + 445	1586 + 108
Castrated (C) (7)	°0.39 + 0.03	°0.24 + 0.02	°1782 + 209	°°1015 + 128
C + TP$_{24 h}$ (4)	°0.44 + 0.03	°0.25 + 0.03	8924 + 1073	°4716 + 848
C + TP$_{10 d}$ (5)	0.93 + 0.03	0.55 + 0.05	9660 + 1079	2341 + 225

° p<0.01, °° p<0.05. Results are the means + SE. No. of cases indicated between brackets.

specific activity in the LAM, before, after castration, and at different times after testosterone treatment.

The LAM weight, the content in total bases and their specific activity decrease after castration : no change in the content of the bases is observed when the values refer to 1 g of tissue and not to the absolute weight; therefore, the purine acid-soluble pool decreases at same rate as the weight.

24 h after testosterone propionate administration, the specific activity is remarkably higher then the baseline; 10 days after treatment, a normal incorporation of ^{14}C-formate can be observed and the weight is also normal.

Table 2 shows the content in adenine and guanine of the LAM before and after castration, the specific activity of the various bases, and G/A ratio.

The content in the single bases and the specific activity decrease after castration, while the G/A ratio is unchanged.

24 h after testosterone administration, both the guanine and the G/A ratio increase.

The administration of testosterone restores the content and the specific activity of the bases only after at least 10 days.

TABLE 2 - Effect of castration and testosterone propionate (TP) on total
content and on specific activity (SA = dpm/μmole) of acid-
soluble adenine and guanine.

	ADENINE		GUANINE		G/A
	total mg	SA	total mg	SA	
Normal (N) (6)	0,45 + 0,02	1391 + 60	0,019 + 0,004	4184 + 600	0.11
Castrated (C) (4)	°0,20 + 0,04	°769 + 56	0,010 + 0,002	°°2189 + 302	0.11
C + TP$_{24 h}$ (4)	°0,20 + 0,01	°3573 + 313	0,014 + 0,003	°20274 + 5889	°0.32
C + TP$_{10 d}$ (4)	0,50 + 0,06	1953 + 154	0,022 + 0,001	4646 + 1325	0.10

° $p<0.01$, °° $p<0.005$. Results are the means + SE. No. of cases indicated
between brackets.

DISCUSSION

We have confirmed our previous observations: castration reduces the
content, the specific activity of total bases and of the single bases in
the LAM, indicating an inferior turnover.

The normal turnover is rapidly restored by testosterone: the hormone
seems to increase the guanine content and its specific activity more than
the content and specific activity of adenine.

The G/A ratio remains the same after castration, increases in the
course of brief treatment and after 10 days returns to the basal values.

In interpreting the results we assumed that testosterone plays a specific role in the control of the " inosinic acid branch point ". There is
a strict metabolic regulation of this step, as reported by Crabtree[4]; it
is very likely that the " inosinic acid branch point " also undergoes
hormonal regulation and that testosterone will stimulate the turnover of
guanylic acid.

The increased turnover of this nucleotide - which is always present
although not as much as adenylic acid - will favour, in the sexual organs, the synthesis of RNA, which is known to be stimulated by testosterone, through the activation of RNA polymerase[5].

REFERENCES

1) M. Pizzichini, A. Di Stefano, E. Marinello - Effect of castration on
the incorporation of ^{14}C-formate into the ribonucleic acid and acid-
soluble purines in rat perineal muscle. J.Nucl.Med.All.Sci., 29, n°1-2
137, 1985.

2) M.S. Hersfield and J.E. Seegmiller - Regulation of " de novo " purine biosynthesis in human lymphoblasts. J.Biol.Chem. 251, 7348, 1976.

3) M.Pizzichini, A. Di Stefano, L. Ciccoli, E. Marinello - Purine nucleotide biosynthesis in rat muscles after castration. Boll.Soc.It.Biol. Sper. LX, 721, 1984.

4) G.W. Crabtree and J.F. Henderson - Rate limiting steps in the interconversion of purine ribonucleotides inEhrlich Ascites Tumor Cells in vitro. Cancer Res. 31, 985, 1971.

5) W.I.P. Mainwaring, F.R. Mangan, B.M. Peterken - Studies on the solubilized Ribonucleic acid polymerase from rat ventral prostate gland. Biochem J. 123, 619, 1971.

PRODUCTION AND PROPERTIES OF MONOCLONAL ANTIBODIES AGAINST

HUMAN ECTO-5´-NUCLEOTIDASE

Wolf Gutensohn, Udo Kummer and Roswitha Forstpointner

Institut für Anthropologie und Humangenetik der
Universität
Institut für Hämatologie der GSF
D 8000 München 2, Fed.Rep.Germany

INTRODUCTION

The membrane enzyme 5´-nucleotidase (5´-N) has been characterized as an ectoenzyme in many different tissues of different species. The ecto-property means that a major portion
of this molecule including the active center and the carbohydrate side chains is located in the extracellular space. Thus,
the enzyme represents an interesting surface marker. This is
especially true in lymphoid cells, where we observe charcteri-
stic differences in expression of this marker in various subpopulations, in cell lines and in different forms of leukemias
and lymphomas. The latter observation can even be used diagno-
stically in selected cases[1,2]. Therefore, a method of detection
of the molecule independent of the overall measurement of enzyme
activity, for example by monoclonal antibodies (MABs) seemed
desirable. In addition specific immunological reagents should
be useful tools for further biochemical studies. In the following a number of monoclonal antibodies against the human enzyme shall be described.

RESULTS

Mice were immunized with a partially purified preparation
of 5´-N from human placenta. After the classical fusion procedure with myeloma cells hybridoma supernatants were screened
for the ability to inhibit 5´-N-activity. Out of 400 about 40
showed significant inhibition[3]. Selected from these, 5 hybrids
were established by cloning and recloning. For most of the following studies the MABs were used in the form of ascites, since
titers there were higher than in culture supernatants.

Characterization. The inhibitory potency of the MABs was
observed whether solubilized and partially purified 5´-N or
enzyme in the membrane bound form (placental microsomes) were
used as antigen (Fig. 1). Likewise there was no tissue specificity, the MABs also blocked the enzyme on lymphoid cells.
The antigen-antibody reaction is on a quantitative basis. Enzyme and antibodies can be titrated against one another.

The MABs of some clones (IFH-5N1, 5N4, and 5N5) show a strong temperature dependence of their inhibitory action, whereas others (IFH-5N2) do not (Fig. 2). The binding sites for the antibodies most probably do not include determinants of the carbohydrate chains, since neuraminidase treatment of the enzyme

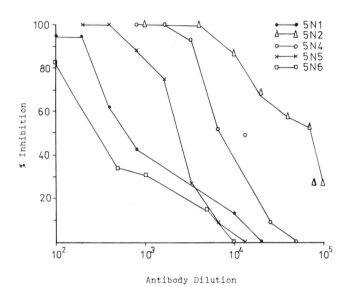

Fig. 1   Immunotitration with MABs from 5 different clones.
Source of 5´-N is human placental microsomes.

sample does not change the inhibitory effect of the MABs (Fig. 3 Antibodies 5N1, 5N4, 5N5 and 5N6 belong to the IgG 1 subclass, whereas 5N2 is IgG 2b . It may be a coincidence, but this latter antibody exhibits the highest titers in inhibition experiments and is the least temperature dependent in its action (Fig. 2).

Immunofluorescence. Immunofluorescence studies mainly on lymphoid cells were all performed by the indirect method using goat anti-mouse-IgG-FITC as second antibody. Some general observations can be made: As with polyclonal antibodies fluorescence on lymphoid cells is rather weak. Titers for optimal fluorescence are much lower than those found in inhibition experiments. On normal peripheral blood lymphocytes a somewhat lower percentage of cells (5 - 15%) are found positive for 5´-N with the MABs than in tests with polyclonal antibodies or with histochemical stains. With lymphoblastoid cell lines and leukemic cells with extremely high or absent 5´-N fluorescence with the MABs shows the expected pattern. Human fibroblasts with a high 5´-N-activity show intense fluorescence.

In cytofluorometric studies (kindly performed by Prof.G. Valet, Martinsried) fluorescence was related to cell size.

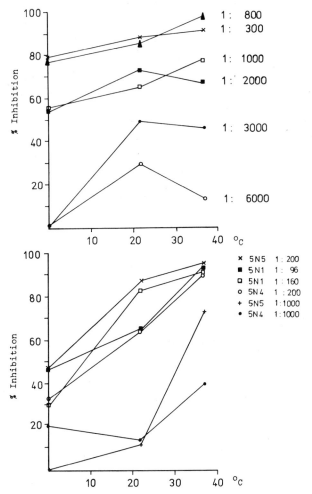

Fig. 2   Temperature effect of enzyme inhibition by MABs.
Upper: Clone IFH-5N2, various dilutions.
Lower: 3 other clones.
Enzyme source: Human placental microsomes

Surprisingly, in peripheral blood lymphocytes - using MAB IFH-5N2 and a polyclonal serum - cytofluorometry could only discern a very small percentage (2%) of positive cells grouped predominantly into a class with large cell volumes. This fluorescence is not abolished by monocyte depletion leading to the

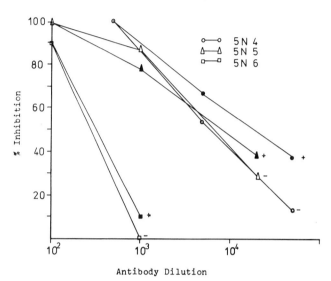

Fig. 3   Immunotitration of human placental microsomes with MABs from 3 different clones. Microsomes were either untreated (-) or preincubated with neuraminidase (+).

conclusion that possibly large granular lymphocytes could be the labeled cells. This result is clearly in contrast to the fluorescence data mentioned above, obtained by visual inspection of the samples, and earlier histochemical data, where subpopulations of small lymphocytes (T- and B-cells) are labeled as 5´-N-positive. The reason for this discrepancy may be that fluorescence on the small lymphocytes is too weak and stippled to be detectable by the cytofluorometer.

Immunoprecipitation. The most elegant scheme here would be immunoblotting after SDS-gel electrophoresis and electrophoretic transfer (Western blot) of proteins on nitrocellulose. With our MABs this was not successful. This is not too surprising, since it is known that SDS destroys 5´-N-activity indicating that the active center and epitopes in its neighborhood are drastically changed.

Another method tried was immunoadsorption to the MABs covalently coupled to diazocellulose. Out of detergent extracts of membrane material 5´-N-activity could be adsorbed to this carrier. However, electrophoretic analysis of the adsorbed material revealed that the amount and variety of proteins unspecifically adsorbed was so high that a clearcut identification of the antigen (5´-N) was not possible.

Somewhat more successful was immunoadsorption to the MABs coated on plastic surfaces. The tests were performed in PVC-microtiterplates. Again 5´-N-activity out of microsomal extracts was specifically adsorbed to the MABs on the plastic surface and this was titer dependent. When a ^{125}I-labeled microsomal extract was used and the material adsorbed to the MABs analysed by SDS-gelelectrophoresis one faint band with a molecular weight of 72 000 D was seen. This band did not coincide with a major labeled band of the total extract and the molecular weight is similar to that given for 5´-N in other systems. This band was not adsorbed when the plastic surface was precoated either with unspecific proteins (fetal calf serum), with nonimmune mouse ascites (NS-1) or a monoclonal antibody directed against a different antigen. However, the yields of adsorbed material are still very low and the method clearly requires further improvements.

DISCUSSION

The method of selecting monoclonal antibodies against an enzyme according to their ability to block the catalytic activity has advantages and drawbacks. One definite advantage is the relative ease of the screening procedure which allows the testing of many samples in a reasonable time. Moreover, in the case of an ectoenzyme the procedure will automatically pick up those antibodies reacting with the extracellular portion of the molecule.

On the other hand the immediate vicinity of the active center of an enzyme may not represent the most stable epitopes available in the molecule. In this respect antibodies selected for tight binding would be preferable. This is clearly demonstrated by the MABs described here. Wherever binding is the criterium - in immunofluorescence and immunoadsorption - our results, though encouraging, are not yet completely satisfactory.

ACKNOWLEDGEMENTS

This work was supported by the Deutsche Forschungsgemeinschaft. Grant No. Gu 123/13-6

REFERENCES

1. W.Gutensohn and E.Thiel, High levels of 5´-nucleotidase activity in blastic chronic myelogenous leukemia with common ALL-antigen, Leukemia Res. 5:505 - 510 (1981)

2. W.Gutensohn, E.Thiel and B.Emmerich, Evaluation of 5´-nucleotidase as biochemical marker in leukemias and lymphomas, Klin.Wochenschr. 61:57 - 62, (1983)

3. U.Kummer, J.Mysliwietz, W.Gutensohn, S.Buschette, H.Jahn, D.Neuser and R.Munker, Development and properties of a monoclonal antibody specific for human ecto-5´-nucleotidase, Immunobiol. 166:203 - 211 (1984)

AMP-DEAMINASE AND CYTOSOLIC 5'NUCLEOTIDASE

IN HUMAN AND MURINE LYMPHOCYTE SUBPOPULATIONS

Jacques Dornand, Jean Favero and Jean-Claude Mani

Laboratoire de Biochimie des Membranes, ER CNRS 228
ENSCM, 8 rue de l'Ecole Normale
F-34075 Montpellier (France)

INTRODUCTION

The lymphotoxicity occurring in the pathogenesis of immunodeficient states associated with inherited adenosine deaminase (ADA) deficiency and in the treatment of leukemia by deoxycoformycin, an ADA inhibitor, results from an intracellular accumulation of deoxy-ATP (dATP) (1-4). It was recently reported that this accumulation depends not only on intracellular dATP synthesis but also on the capacity of the cells to degrade accumulated dATP (5). AMP-deaminase (AMP-DA) which catalyzes the irreversible hydrolytic deamination of AMP and dAMP and cytosolic 5'nucleotidase (c5'N) which hydrolyzes intracellular mononucleotides are involved in nucleotide catabolism and might participate to the regulation of intracellular dATP levels.

AMP-DA and c5'N have been scarcely studied in lymphoid cells. It was only reported (6) that T lymphoblasts display AMP-DA activities 5-10-fold lower than those of B lymphoblasts and that different isoenzymes could be expressed in these cells which differently accumulate dATP when incubated with deoxyadenosine in the presence of an ADA inhibitor. It was also shown that AMP-DA activity is greatly altered during the *in vitro* differentiation of muscle cells (7). These data and the fact that the sensitivity of thymocyte subpopulations to ADA inhibitors changes with their differentiation stage (8) prompted us to study AMP-DA activity in different lymphocyte subpopulations. We showed that, unlike c5'N, AMP-DA displayed activities dependent on the differentiation level; these differences in activity seemed to result from the presence of different AMP-DA isoenzymes. AMP-DA activities appeared inversely related to dATP accumulation which coincided with the inhibition of RNA synthesis and cell death.

MATERIALS AND METHODS

Cells

Mouse cells were obtained from 4- to 7-week-old Swiss male mice. Separation of cortical and medullary mouse thymocytes was performed by agglutination with peanut agglutinin (PNA); T and B splenocytes were separated by agglutination with soybean agglutinin (SBA) or wheat germ agglutinin (WGA). The purification of normal peripheral blood lymphocytes (PBL) and tonsil lymphocytes, and the origin of the different cell lines have been reported elsewhere (9,10). The cells were packed, washed twice with normal saline, and disrupted by two successive freezings;after two centri- fugations (100,000g, 20 min, 4°C), the supernatants were dialyzed against 10 mM Tris-HCl, pH 7.4, and assayed for their enzymatic activities. Pro- tein concentrations were determined by Lowry's method.

Enzymatic activities

AMP-DA was assayed by following the conversion of 5'-[³H]AMP or 5'-d [³H]AMP to 5'-[³H]IMP or 5'-d[³H]IMP, at 37°C. The standard assay contain- ed in 200 µl 50 mM sodium cacodylate, pH 6.7, 150 mM KCl, 0.2 mM EDTA, 20 mM α,β-methylene ADP (AOPC), 0.5 mM 5'-[³H]-AMP or 1 mM 5'-d[³H]AMP, and 5 to 300 µg supernatant. AOPC, a specific inhibitor of ecto-5'nucleotidase (ecto-5'N) and EDTA which inhibits c5'N and adenosine kinase, do not af- fect AMP-DA activity. After 30 min incubation at 37°C the reaction was stopped at 100°C and 5 mM 5'AMP and 5'IMP or 5 mM 5'dAMP and 5'dIMP were added to the reaction mixture. 5'IMP or 5'dIMP were separated from 5'AMP or 5'dAMP by thin layer chromatography on silica gel plates (11) and coun- ted.

c5'N was determined, at 37°C, with 5'-[³²P]AMP as substrate. The re- action mixture contained in 200 µl, 50 mM imidazole, pH 6.5, 3 mM MgCl$_2$, 50 µM AOPC (to inhibit ecto-5'N), 0.25 mM substrate and 5 to 25 µg super- natant. After 30 min incubation at 37°C the reaction was stopped with 0.5 ml 1% TCA. ³²Pi produced was determined as reported by Dornand *et al* (12).

Purine nucleoside kinase activities of cell cytosols were measured under the experimental conditions described by Lukey and Snyder (13): for adenosine kinase, 100 mM Tris-maleate, pH 6.0, 5 mM ATP, 1 mM MgCl$_2$, 5 µM erythro-9-[3-(2-hydroxynonyl)]-adenosine (EHNA), 45 µM [³H]adenosine, 100 µg protein cytosol; for deoxyadenosine kinase, 100 mM Tris-HCl, pH 7.6, 1 mM ATP, 5 mM MgCl$_2$, 5 µM EHNA, 450 µM d[³H]adenosine, 300 µg protein cy- tosol. After 30 min incubation, at 37°C, the reaction was stopped at 100°C the nucleoside and the nucleotide formed separated as described (11) and their relative concentration determined.

All enzymatic reactions were studied under such conditions that the product formation was linear with time and with the cytosol concentration. Measurements were performed in triplicates.

Cell incubation and deoxyadenosine incorporation

5 µM EHNA was added at 37°C to 10 ml RMPI 1640 containing 10⁷ cells/ ml. After 15 min the cultures were supplemented with 10 µM deoxyadenosine, then incubated for various times at 37°C under a CO$_2$ wet atmosphere. The cells were sedimented (5 min, 500g at 4°C) and the supernatant was discar- ded. The cell pellets were extracted with 0.5 ml 0.4 M perchloric acid, neutralized and their dATP content determined as described (14).

RESULTS

AMP-DA activity in lymphocyte subsets

Measurements of AMP-DA activities in mouse lymphoid cells reveal that thymocytes express an activity 10-20-fold lower than that of splenocytes, independently of the substrate used. Lymph node lymphocytes and bone mar- row cells display intermediary values (Table 1).

We showed that the differences in AMP-DA levels do not result from different percentages of T and B cells in different organs. WGA- or SBA- agglutinated splenocytes, which represent a B-cell enriched population,

Table 1. Specific activities (nmole/hr/mg) of AMP-DA in mouse and human lymphocyte subsets and in human lymphoblastoid cell lines.

Cell populations	AMP	dAMP
Mouse lymph node lymphocytes	1800–2500	200–220
Mouse bone marrow lymphocytes	600–800	n.d.
Mouse unseparated thymocytes	150–200	20–25
Mouse cortical (PNA$^+$) thymocytes	100–150	0–15
Mouse medullary (PNA$^-$) thymocytes	850–900	70–80
Mouse hydrocrotisone-resistant thymocytes	800–1000	80–120
Mouse unseparated splenocytes	2400–2900	250–300
Mouse T-enriched (SBA$^-$) splenocytes	2250–2700	190–250
Mouse B-enriched (SBA$^+$) splenocytes	1700–2400	200–230
Human blood unseparated lymphocytes	1200–1500	120–220
Human blood T lymphocytes	1000–1200	150–190
Human blood B lymphocytes	1100–1400	180–190
Human tonsil unseparated lymphocytes	1200–1700	130–140
Human unseparated thymocytes	150–200	0–15
Human cortical thymocytes	100–130	0
Human medullary thymocytes	400–600	40–50
MOLT-4 (Human T-ALL^a)	40–60	0
MOLT-3 (Human T-ALL)	20–50	0
CEM (Human T-ALL)	30–100	0
H-SB2 (Human T-ALL)	20–40	0
1301 (Human T-ALL)	60–70	0
RAJI (Human Burkitt lymphoma)	70–100	0
CCRF-SB (Human B-cells^b)	450–600	30–50
EBV lymphoblasts (Human B cells^c) (4)	500–630	45–65
K-562 (Human Null CML^d)	450–600	50–70

a   ALL: acute lymphoblastic leukemia.
b   CCRF-SB appear as normal B cells from a T-ALL patient (10).
c   Four different cell lines were established after infection of normal PBL with Epstein-Barr virus.
d   Non-T non-B chronic myelocytic leukemia.

display activities similar to those of non-agglutinated splenocytes (T-cell enriched fractions) or unseparated populations.

The separation of mouse thymocytes into two broad subpopulations, cortical and medullary cells, was performed by a method based on the availability of binding sites for PNA. Agglutinated cortical thymocytes (PNA$^+$) which are the majority of the thymic cells (80–90%) are functionally immature, while the non-agglutinated medullary population (PNA$^-$) exhibits an immunocompetence similar to that of mature cells. PNA$^-$ negative displayed AMP-DA activities 5-6-fold higher than those of PNA$^+$ cells; in two experiments over six, AMP-DA was found undetectable in PNA$^+$ cells when dAMP was the substrate used. We also studied medullary cells by depleting the cortical population by hydrocortisone treatment; two days after an intraperitoneal injection of hydrocortisone acetate, resistant thymocytes pooled from 10 mice (15×10^6 cells per thymus) displayed AMP-DA activities identical to those of PNA$^-$ thymocytes.

The relative AMP-DA activity of lymphocyte populations does not arise from the presence of an AMP-DA inhibitor in cortical cells or that of an activator in the other cells: no inhibition or activation of AMP-DA activities was observed when PNA[-] thymocytes or splenocytes were incubated with cytosols from PNA[+] thymocytes (data not shown). The low AMP-DA activity of the thymocytes appears related to the high percentage of cortical cells, the PNA[-] cells displaying higher activities which however remain 2-3-fold lower than those of T splenocytes.

AMP-DA and dAMP-DA activities were also measured in human lymphocytes and lymphoblastoid cell lines. PBL and tonsil lymphocytes displayed activities more than 8-fold higher than those of thymocytes. The separation of cortical and medullary cells showed again that the low activity of thymocytes was due to the immature population. Most lymphoblastoid cell lines had abnormally low AMP-DA levels, specially T-ALL lymphoblasts with activities 10-fold lower than those of normal PBL and close to those of cortical thymocytes. Under our experimental conditions, undetectable dAMP-DA levels were found in these cell lines. AMP-DA activities of the B cell lines tested were higher than those of T-ALL lymphoblasts (with the exception of RAJI cell line which had similar activity), but remained lower than those of normal lymphocytes. The null cell line K-562 displayed a relatively high AMP-DA level. The B and null cell lines studied, which derive from normal cells, are considered as mature cells with low terminal deoxy-nucleotidyl transferase (TdT) activity (9), while T-ALL lymphoblasts display high TdT levels which reflect early stages of maturation. Our results agree with those of Fishbein (6) and suggest that AMP-DA, like ecto-5'N and unlike ADA, increase during cell differentiation.

Kinetic parameters of AMP-DA or dAMP-DA activities in cortical thymocytes and T splenocytes

The differences of AMP-DA or dAMP-DA activities observed among lymphoid cells at various stages of differentiation could arise from either different amounts of the same enzyme or different forms of the enzyme. AMP was found a better substrate than dAMP both in T splenocytes and cortical

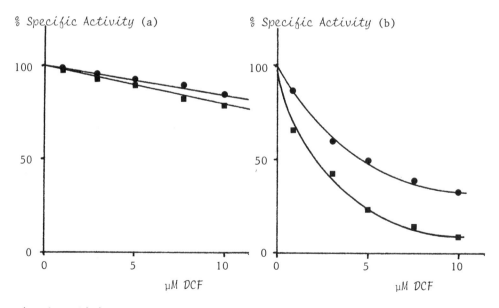

Fig. 1. Inhibitory effect of deoxycoformycin (DCF) on AMP-DA activities of mouse splenocytes (1-a) or thymocytes (1-b), determined with either AMP ($\bullet$-$\bullet$) or dAMP ($\blacksquare$-$\blacksquare$) as substrate.

thymocytes, but differences in the relative substrate specificity were observed: Km for AMP was similar for the two cell types (0.4-0.5 mM) while that for dAMP was higher for cortical thymocytes (4 mM) than for T splenocytes (1.5 mM). The maximal velocity of AMP-DA was 13-fold higher (2900 versus 220 nmole/hr/mg) in T splenocytes, while that of dAMP-DA was 25-fold higher (380 versus 15 nmole/hr/mg) in the same cells.

EHNA (0.5 μM), an inhibitor of ADA, had no effect on AMP-DA of both cell types, with either AMP or dAMP as substrate. Deoxycoformycin which is a better ADA inhibitor than EHNA, inhibited dAMP-DA and to a lesser extent AMP-DA in cortical cells, but it was less efficient for mature T splenocytes (Fig. 1). These data are consistent with the presence of different forms of AMP-DA in the two cell populations. The enzyme form(s) present in immature thymocytes poorly deaminates dAMP and is highly inhibited by deoxycoformycin, unlike that present in mature T cells. Such a difference in deoxycoformycin sensitivity has been reported for the enzymes expressed in T-ALL and B or null lymphoblastoid cell lines (6).

Cytosolic 5'nucleotidase, adenosine and deoxyadenosine kinase, adenosine deaminase activities in lymphocyte subsets

The study of dATP accumulation in lymphocyte subpopulations led us to compare some other enzymes involved in dATP metabolism, like c5'N, adenosine and deoxyadenosine kinases, ADA. We found only small variations for c5'N and the kinases among the different subpopulations studied, and these differences have probably no biological implication, unlike the large changes in AMP-DA or ADA activities (Table 2).

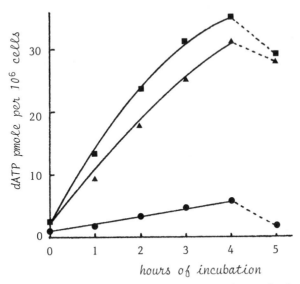

Fig. 2. Accumulation of dATP in T splenocytes and cortical thymocytes. T splenocytes (10^7 cells/ml) in RPMI 1640 (10 ml) were incubated at 37°C in the presence of 10 μM deoxyadenosine plus 5 μM EHNA or DCF (●-●). The same experiment was performed with cortical thymocytes (EHNA ▲-▲, DCF ■-■). dATP content was determined after 0, 1, 2, 3, 4 hr incubation. The cells were then washed and resuspended in RPMI 1640 for 1 hr more and their dATP content measured.

Table 2. Cytosolic 5'nucleotidase, nucleoside kinase and adenosine deaminase activities (nmole/hr/mg) of mouse lymphocyte subsets.

| Cells | c5'N | | Kinases | | ADA |
	AMP	dAMP	Ado	dAdo	
Unseparated thymocytes	34–42	90–104	180–240	40–60	18400–24000
Cortical thymocytes	30–36	105–130	180–200	47–50	22000–24000
Medullary thymocytes	20–26	87–95	165–200	27–30	6000–7000
Unseparated splenocytes	42–47	110–125	120–160	22–35	2400–3300
T splenocytes	48–56	65–75	130–150	30–34	2200–2600
B splenocytes	40–44	130–154	130–140	28–32	2900–3400

dATP accumulation in cortical thymocytes and T splenocytes

When incubated for 4 hr with 10 µM deoxyadenosine and EHNA, both cell populations accumulated dATP. The dATP content of cortical thymocytes increased 10 times, was linear during one hour and significantly higher when 10 µM deoxycoformycin substituted for EHNA. T splenocytes accumulated 6-7-fold less dATP and deoxycoformycin did not improve this accumulation. After the 4 hr incubation with deoxycoformycin and deoxyadenosine, the cells were washed and incubated one hour more in RPMI 1640 alone. In T splenocytes the dATP content decreased to its initial level, while in cortical thymocytes it remained very high (Fig. 2). In medullary thymocytes or B splenocytes, dATP accumulation was found inversely related to the AMP-DA activity, when the cells were incubated under the same conditions.

CONCLUSION

Little attention has been paid to the role of AMP-DA in the mechanism of deoxyadenosine cytotoxicity. Fishbein reported that this enzyme involved in nucleotide metabolism was not uniformly distributed among human lymphoblastoid cell lines (6). We found that AMP-DA activity is widely different among the various human and mouse lymphocyte populations. Its distribution is similar to that of ecto-5'N (9,10): immature lymphocytes display 10-fold lower activity than mature cells, independently of their T or B character; AMP-DA level of medullary thymocytes is 2-3-fold lower than that of mature lymphocytes but 4-fold higher than that of immature cortical thymocytes.

In human lymphoblastoid cell lines, our results (Table 1) agree with those of Fishbein (6), but we think that AMP-DA activity correlates with the cell maturation stage rather than with their T or B character: high AMP-DA levels were found in B or null lymphoblastoid cell lines with mature characters (high ecto-5'N/ADA ratio (10) and low TdT) while immature T-ALL lymphoblasts, with low ecto-5'N/ADA ratio and high TdT, displayed low, and sometimes undetectable, AMP-DA activities.

Although the molecular characterization and purification of AMP-DA have not yet been performed, the presence of different isoenzymes in cortical thymocytes and T splenocytes is suggested by kinetic studies and by the specificity of deoxycoformycin inhibition.

As the other enzymes involved in nucleotide metabolism, c5'N, adenosine and deoxyadenosine kinases (Table 2), S-adenosylhomocysteine-hydrolase (15), have similar values in cortical thymocytes and T splenocytes, the differences in AMP-DA could account for the differences in dATP accumulation between the two cell types, when ADA is inhibited. These data suggest the hypothesis of an important role of AMP-DA in dATP detoxification. This hypothesis is consistent with the good correlation between the level of dATP accumulation and the AMP-DA activity and could explain why, unlike in T splenocytes, the dATP content of cortical thymocytes does not significantly drop after deoxyadenosine and EHNA have been washed off. In resting lymphocytes this dATP accumulation correlates with the inhibition of RNA synthesis and cell death 24 hr later (16,17). On the view of our results it appears difficult to compare the sensitivity of cells to deoxyadenosine toxicity without considering AMP-DA activity.

REFERENCES

1. E.R. Giblett, J.E. Anderson, F. Cohen, B. Pollara, and H.J. Meuwissen, Adenosine deaminase deficiency in two patients with severely impaired cellular immunity, Lancet 2:1067 (1972).
2. C.A. Koller, B.S. Mitchell, M.R. Grever, E. Mejias, L. Malspeis, and E.N. Metz, Treatment of acute lymphoblastic leukemia with 2'-deoxycoformycin. Clinical and biomedical consequences of adenosine deaminase inhibition, Cancer Treat. Rep. 63:1949 (1979).
3. A.L. Yu, F.H. Kung, B. Bakay, and W.L. Nyhan, In vitro and in vivo effect of deoxycoformycin in human T cell leukemia, Adv. Exp. Med. Biol. 122:373 (1980.
4. D.A. Carson, J. Kaye, and J.E. Seegmiller, Differential sensitivity of human leukemic T cell lines and B cell lines to growth inhibition by deoxyadenosine, J. Immunol. 121:1726 (1978).
5. T. Sylwestrowicz, A. Piga, P. Murphy, F. Ganeshaguru, N.H. Russel, H.G. Prentice, and V.A. Hoffbrand, The effects of deoxycoformycin and deoxyadenosine on deoxyribonucleotide concentration in leukemic cells, Brit. J. Haematol. 51:623 (1982).
6. W.N. Fishbein, J.I. Davis, J.W. Winkert, and D.M. Strong, Levels of adenosine deaminase, AMP-deaminase, and adenylate kinase in cultured human lymphoblast lines: exquisite sensitivity of AMP-deaminase to adenosine deaminase inhibitors, Biochem. Med. 26:377 (1981).
7. E. Al Zoref-Shani, A. Shainberg, and O. Sperling, Alteration in purine nucleotide metabolism during muscle differentiation in vitro, Biochem. Biophys. Res. Commun. 116:507 (1984).
8. M. Ballow, and A.G. Pantschenko, In vitro effects of adenosine deaminase inhibitors on lymphocyte mitogen responsiveness in the mouse, Cell. Immunol. 64:29 (1981).
9. J. Dornand, J.C. Bonnafous, J. Favero, and J.C. Mani, 5'Nucleotidase activity of two populations of mouse thymocytes separated by PNA agglutination, FEBS Lett. 118:225 (1980).
10. J. Dornand, J.C. Bonnafous, J. Favero, and J.C. Mani, Ecto-5'nucleotidase and adenosine deaminase activities of lymphoid cells, Biochem. Med. 28:144 (1982).
11. N. Kolassa, H. Ross, and K. Pfleger, A separation of purine derivatives by thin-layer chromatography on silica gel plates suitable for metabolic studies, J. Chromatogr. 66:175 (1972).
12. J. Dornand, C. Reminiac, and J.C. Mani, Activité 5'AMPase des membranes plasmiques de lymphocytes. Effet de la concanavalin A, Biochimie 86:425 (1977).
13. T. Lukey, and F.F. Snyder, Purine ribonucleoside and deoxyribonucleoside kinase activities in thymocytes. Specificity and optimal conditions for phosphorylation, Canad. J. Biochem. 58:677 (1980).
14. G.J. Peters, R.A. De Abreu, A. Oosterhof, and J.H. Verkamp, Concentration of nucleotide and deoxynucleotide in peripheral and phytohemagglutinin stimulated mammalian lymphocytes. Effects of adenosine and deoxyadenosine, Biochim. Biophys. Acta 759:7 (1983)

15. J.C. Bonnafous, J. Dornand, J. Favero, and J.C. Mani, Role of adenosine in lymphocyte maturation: Possible involvement of adenosine receptors coupled to adenylate cyclase, *in*: Intercellular Communication in Leucocyte Function, J.W. Parker, and R.L. O'Brien,eds., Wiley, New York, p. 437 (1983).

16. J. Dornand, G. Clofent, J.C. Bonnafous, J. Favero, and J.C. Mani, Purine metabolizing enzymes of lymphocyte cell populations: Correlation between AMP-deaminase activity and dATP accumulation in murine lymphocytes, *Proc. Soc. Exp. Biol. Med.* 179 (1985), in press.

17. S.S. Matsumoto, J. Yu, and A.L. Yu, Inhibition of RNA synthesis by deoxyadenosine plus deoxycoformycin in resting lymphocytes, *J. Immunol.* 131:2762 (1983).

AN EXPLANATION FOR THE HETEROGENEITY IN B LYMPHOCYTE ECTO-5'-NUCLEOTIDASE

ACTIVITY IN PATIENTS WITH HYPOGAMMAGLOBULINEMIA

Linda F. Thompson, Julie M. Ruedi, Richard D. O'Connor, and
John F. Bastian

Department of Immunology, Scripps Clinic and Research Foundation, La Jolla, California; Division of Pediatric Immunology
and Allergy, University of California, San Diego; and
Children's Hospital and Health Center, San Diego, California

A deficiency of lymphocyte ecto-5'-nucleotidase (ecto-5'-NT)[1] activity
in patients with common variable immunodeficiency (CVI), congenital X-linked
agammaglobulinemia (CAG), severe combined immunodeficiency (SCID), and a
variety of other immunodeficiency diseases is well described (1-6). Adult
peripheral B cells have three- to five-fold higher ecto-5'-NT activity than
peripheral T cells (3). Thus, any disease which causes a loss of surface
immunoglobulin-positive (sIg$^+$) B cells from the peripheral blood, such as
CAG, will cause a decrease in total lymphocyte ecto-5'-NT activity (3).
Ecto-5'-NT deficiency may also be caused by decreased enzyme activity in T
cells, B cells, or both (1,3,4). We have previously shown that decreased
ecto-5'-NT activity in T cells of patients with CVI and CAG is due to
decreased numbers of ecto-5'-NT$^+$ cells from both the helper (OKT4$^-$) and
suppressor (OKT8$^+$) subpopulations (4). Furthermore, increased percentages
of ecto-5'-NT$^-$ T cells in these patients appears to be correlated with an
increased capacity for the suppression of pokeweed mitogen (PWM) driven
immunoglobulin synthesis in vitro. Ecto-5'-NT$^-$ maternally engrafted T
cells in a patient with SCID and maternal graft versus host disease also
suppressed PWM-driven immunoglobulin synthesis (6). Thus, T cell ecto-5'-
NT deficiency does not necessarily indicate T cell immaturity.

The significance of B cell ecto-5'-NT deficiency in patients with
immunodeficiency diseases has been less thoroughly investigated. Rowe et
al. described B cell ecto-5'-NT activity in a series of patients with CVI
and CAG and found a wide range of activities from normal to severely depressed (1). Since cord blood B cells have low ecto-5'-NT activity compared
to adult peripheral B cells (1,7), the low B cell ecto-5'-NT activity in
these patients was presumed to be due to a block in B cell maturation.
However, Rowe's patients were not characterized for the cellular defect
responsible for their hypogammaglobulinemia; i.e., it was not determined
whether they had primary B cell defects, a lack of T cell help, or excess

---

[1]Abbreviations: Ecto-5'-NT, ecto-5'-nucleotidase; CVI, common variable
immunodeficiency; CAG, congenital agammaglobulinemia; SCID, severe combined
imunodeficiency; sIg, surface immunoglobulin; PWM, pokeweed mitogen; E$_N$,
neuraminidase-treated sheep erythrocytes; ME$_N$, neuraminidase-treated mouse
erythrocytes; E$_O$-anti-kappa, fresh ox erythrocytes coated with anti-kappa;
E$_O$-anti-lambda, fresh ox erythrocytes coated with anti-lambda.

T cell suppression. Therefore, no direct correlation between B cell ecto-5'-NT activity and B cell function could be made. Thus, we decided to measure ecto-5'-NT activity in B cells at various stages of differentiation and in a series of patients with hypogammaglobulinemia in order to determine whether ecto-5'-NT activity is a useful biochemical marker for defining sites of blocks in B cell differentiation. Purified B cell preparations from all of our patients with CVI were also studied for the ability to synthesize IgM and IgG when co-cultured with irradiated normal T cells and PWM so that the relationship between B cell ecto-5'-NT activity and B cell function could be determined.

METHODS

Patient population. The diagnosis of CVI was based upon increased incidence of infections beginning in late adolescence or early adulthood and depressed concentrations of serum immunoglobulins. Control subjects were healthy laboratory volunteers between the ages of 24 and 36 years.

Lymphocyte isolation. For experiments to evaluate the percentage of B cells which are ecto-5'-NT+ by histochemical stain, peripheral blood lymphocytes were isolated from freshly drawn heparinized venous blood by dextran sedimentation, carbonyl iron treatment to remove monocytes, and Ficoll-Hypaque density gradient centrifugation (4). T lymphocytes were removed by one cycle of rosetting with neuraminidase-treated sheep erythrocytes, E_N, for 15 min at 37°C and 1 h at 4°C followed by Ficoll-Hypaque density gradient centrifugation (3). These preparations contained 63 ± 19% B cells, <5% T cells, <2% monocytes as determined by esterase staining, and variable percentages of third population lymphocytes. In some experiments, B cells were further purified by rosetting 10 x 10^6 non-T cells with 1 ml 1% fresh ox erythrocytes coated with anti-kappa and anti-lambda (E_O-anti-kappa + E_O-anti-lambda, a gift from Dr. Hans Spiegelberg) for 1 hr at 4°C. After Ficoll-Hypaque density gradient centrifugation, the ox erythrocytes were lysed with 0.83% NH_4Cl. In other experiments, mouse erythrocyte rosetting cells were isolated from non-T cell preparations. Twenty million non-T cells were mixed with 1.0 ml of 10% neuraminidase-treated mouse erythrocytes, ME_N, sedimented at slow speed, and incubated overnight at 4°C. The next day, the pellet was gently resuspended and the rosetting cells separated from non-rosetting cells by Ficoll-Hypaque density gradient centrifugation. The mouse erythrocytes were removed by lysing with 0.83% NH_4Cl. The purity of all B cell preparations was determined by rosetting with E_O-anti-kappa + E_O-anti-lambda.

For experiments to evaluate PWM-driven immunoglobulin biosynthesis in vitro, peripheral blood mononuclear cells were isolated from freshly drawn heparinized venous blood by Ficoll-Hypaque density gradient centrifugation. These cells were then rosetted with E_N as above to yield rosetting (T) cells and non-rosetting (B cell-enriched) cells.

5'-NT assays. Ecto-5'-NT assays with ^{14}C-IMP as substrate and the histochemical stain for 5'-NT activity were performed as previously described (3).

PWM-driven immunoglobulin biosynthesis in vitro. One-half million patient non-T cells (B-cell enriched cells) were cultured with an equal number of irradiated (2000 R) control T cells in 1 ml of RPMI 1640 + 10% fetal bovine serum + 2 mM glutamine + streptomycin (100 µg/ml), penicillin (100 U/ml) and amphotericin B (2.5 µg/ml) at 37°C in an humidified atmosphere of 5% CO_2 in air for 7 days in the presence of 10 µl of PWM (Grand Island Biological Co., Grand Island, NY, ∿5 mg/ml stock solution). At the end of the culture period, the cells were spun down and the quantities of IgM and IgG secreted into the supernatant tissue culture medium were determined by radioimmunosorbent tests.

RESULTS

Ecto-5'-NT activity during human B cell development. Ecto-5'-NT activity was measured on B cell-enriched fractions isolated from fetal spleen, cord blood, adult peripheral blood, adult spleen, and adult lymph node (Table 1). Ecto-5'-NT activity was five- to six-fold higher in adult B cell preparations from all three sources than in B cell preparations from fetal spleen or cord blood (27.9 ± 12, 29.2, and 33.8 nmol/hr/10^6 cells vs. 5.08 and 5.59 ± 2.8 nmol/hr/10^6 cells, respectively). The percentages of B cells in all preparations were comparable.

Table 1. Ecto-5'-Nucleotidase Activity
During Human B Cell Development

Cell source	Ecto-5'-NT activity[*] (nmol/hr/10^6 cells)
Fetal spleen, 22-23 wks. gestation	5.08
Cord blood	5.59 ± 2.8
Adult peripheral blood	27.9 ± 12
Adult spleen	29.2
Adult lymph node	33.8

[*]Ecto-5'-NT activity was measured on B cell-enriched preparations isolated by Ficoll-Hypaque density gradient centrifugation, monocyte depletion by adherence to plastic, and one cycle of rosetting with neuraminidase-treated sheep erythrocytes to remove T cells.

Ecto-5'-NT activity in purified B cell fractions. Total ecto-5'-NT activity in nmol/hr/10^6 cells and the percentages of 5'-NT$^+$ cells were determined in purified B cell fractions rigorously depleted of monocytes with carbonyl iron (Table 2). Mouse erythrocyte rosetting B cells were prepared by rosetting non-T cells with ME$_N$. Since non-T, non-B cells lack ecto-5'-NT activity (3), the percentage of B cells which are 5'-NT$^+$ was calculated by the following formula:

$$\frac{\% \ 5'\text{-NT}^+ \ \text{cells}}{\% \ \text{sIg}^+ \ \text{cells}} \ \times \ 100$$

The data show that the 4.5-fold higher ecto-5'-NT activity in adult peripheral blood B cells as compared to cord blood B cells (51.5 ± 8.6 vs. 11.0 ± 6.9 nmol/hr/10^6 cells) is parallelled by a 2.2-fold higher percentage of 5'-NT$^+$ cells (71 ± 5 vs. 32 ± 17%). Mouse erythrocyte rosetting B cells had similar total ecto-5'-NT activity and percentages of 5'-NT$^+$ cells as total B cell preparations (58.2 ± 25 vs. 51.5 ± 8.6 nmol/hr/10^6 cells and 61 ± 10 vs. 71 ± 5% 5'-NT$^+$ cells, respectively).

Table 2.   Ecto-5'-Nucleotidase Activity in Purified B Cell Fractions

Cell fraction	Ecto-5'-NT activity nmol/hr/10^6 cells	% sIg$^+$ cells	% 5'-NT$^+$ B cells*
Total cord blood B cells (n=4)	11.0 ± 6.9	59 ± 23	32 ± 17
Total adult B cells (n=4)[†]	51.5 ± 8.6	69 ± 17	71 ± 5
Mouse erythrocyte-rosetting adult B cells (n=3)	58.2 ± 25	86 ± 3	62 ± 10

*Determined with a histochemical stain and calculated as described in the Results.

[†]Isolated by rosetting monocyte-depleted non-T cells with E_0-anti-kappa + E_0-anti-lamba.

PWM-driven immunoglobulin biosynthesis by patient B cells. B cells from eight patients with CVI were co-cultured with irradiated control T cells in the presence of PWM. The quantities of IgM and IgG secreted into the supernatant tissue culture medium after seven days are shown in Table 3. The ecto-5'-NT activity in non-T cells from these patients is included in this table. The five patients with the lowest B cell ecto-5'-NT activities made very low quantities of both IgM and IgG, quantities well below the range synthesized by our control donors. A sixth patient (#4) made low (beneath our normal range), but significant quantities of IgM, but not of IgG. His B cell ecto-5'-NT activity is approximately 1 S.D. below our mean normal value, and would probably be considered to be subnormal due to the high B cell percentage in the preparation assayed. Only one patient (#2) made normal quantities of IgM; the amount of IgG made by this patient was significant but still beneath the range for our control donors. When co-cultured with control B cells, his T cells could support IgM and IgG synthesis in response to PWM only after irradiation (data not shown). Thus, this patient appears not to have a pure B cell defect, but to have excess suppressor T cell activity. This patient's B cell ecto-5'-NT activity was within the normal range. The results for patient #5 are particularly interesting in that he had high B cell ecto-5'-NT activity, but made essentially no IgM or IgG in response to PWM. Thus, B cells from patients with CVI which are incapable of synthesizing IgM or IgG when co-cultured with normal irradiated T cells in the presence of PWM may still have normal ecto-5'-NT activity.

DISCUSSION

Ecto-5'-NT activity increases during B cell development. Enzyme activity in B cell preparations isolated from adult peripheral blood, spleen, or lymph node is five- to six-fold higher than that in B cell preparations isolated from fetal spleen or cord blood. The increased activity in B cell preparations isolated from adults cannot be explained by increased percentages of B cells, since the B cell percentages were similar in all preparations assayed. Instead, the increased ecto-5'-NT activity in adult peripheral B cells vs. cord blood B cells is due to both an increased proportion of ecto-5'-NT$^+$ cells as demonstrated by histochemical stain and to increased ecto-5'-NT activity per ecto-5'-NT$^+$ cell (see Table 2).

Table 3. In Vitro Immunoglobulin Biosynthesis and B Cell
Ecto-5'-Nucleotidase Activity in Patients with
Common Variable Immunodeficiency

Patient	B cell ecto-5'-NT activity[*] nmol/hr/10^6 cells	ng Ig synthesized by patient B-enriched cells[†]	
		IgM	IgG
#1	6.35 (15)[‡]	746	14
#2	34.7 (72)	9422	3054
#3	2.83 (37)	<5	<5
#4	15.4 (61)	3404	626
#5	58.2 (86)	31	<5
#6	5.33 (58)	714	17
#7	12.6 (38)	46	802
#8	6.00 (19)	<5	<5
Controls (mean) (± 1 S.D. range)	27.9 (41) 15.9–39.9 24–58	14120 7940–25100	10720 6610–17400

[*]Measured on non-T cell fractions (not monocyte-depleted).

[†]5×10^5 patient non-T cells were co-cultured with 5×10^5
irradiated control T cells for 7 days in the presence of PWM
as described in Methods.

[‡]Percentage of sIg^+ B cells in the preparation studied.

During B cell development, ecto-5'-NT activity increases prior to the
time that B cells acquire the ability to synthesize IgM and IgG in vitro in
response to PWM. This conclusion was reached by measuring ecto-5'-NT activity in purified B cell preparations separated into subsets according to
their ability to form rosettes with mouse erythrocytes. The mouse erythrocyte receptor is an antigen present on the majority of adult peripheral
blood B cells; it defines B cells which are less mature and cannot yet synthesize immunoglobulin in response to PWM (8). That is, those peripheral
blood B cells which synthesize immunoglobulin in response to PWM have lost
the mouse erythrocyte receptor and fail to form rosettes with mouse erythrocytes. Total ecto-5'-NT activity and the percentage 5'-NT[+] cells by histochemical stain were found to be comparable in total B cell preparations and
in the subset isolated by rosetting with ME_N. Thus, ecto-5'-NT activity
appears to be acquired prior to the time that B cells acquire the ability
to synthesize immunoglobulins in vitro in response to PWM. This finding is
consistent with our previous report that ecto-5'-NT activity increases in
the peripheral blood B cells of normal infants during the first six months
of life, prior to the time that they acquire the ability to synthesize IgG
in vitro in response to Epstein Barr virus (7).

The fact that ecto-5'-NT activity is acquired prior to the time that B cells gain the ability to synthesize immunoglobulins in vitro, and presumably in vivo, provides an explanation for the wide range of B cell ecto-5'-NT activities found in patients with CVI. If patients with CVI have blocks in the B cell differentiative pathway, then one might expect to find patients with both low and normal B cell ecto-5'-NT activity depending upon the site of their block. That is, it should be possible to have acquired normal B cell ecto-5'-NT activity, but still not have acquired the ability to synthesize immunoglobulins. Six of the eight patients studied had low B cell ecto-5'-NT activity and a greatly reduced capacity to synthesize immunoglobulins in vitro in response to PWM even when co-cultured with normal irradiated T cells. One patient with normal B cell ecto-5'-NT activity appeared not to have a primary B cell defect, but perhaps to be hypogammaglobulinemic due to excess suppressor T cell activity. B cells from the eighth patient, however, had high normal ecto-5'-NT activity and were still unable to synthesize immunoglobulins in vitro when co-cultured with either autologous or control irradiated T cells. The measurement of B cell ecto-5'-NT activity, therefore, makes it possible to classify patients with CVI who would otherwise be indistinguishable by conventional functional assays. Thus, measurement of this enzyme activity may aid in the definition of discrete stages in normal B cell differentiation and in understanding the defects in patients with hypogammaglobulinemia.

ACKNOWLEDGMENTS

The authors thank Dr. Richard Block at Kaiser Permanente in San Diego for providing cord blood samples and Drs. David Mathison and Stephen Wasserman for allowing us to study their patients. The secretarial expertise of Margaret Stone is gratefully acknowledged. This work was supported by grant AI 18220 and the Biomedical Research Support Program grant RRO 5514 from the National Institutes of Health. This is publication no. 4022 IMM from the Research Institute of Scripps Clinic, La Jolla, California 92037.

REFERENCES

1. M. Rowe, G.C. DeGast, T.A.E. Platts-Mills, G.L. Asherson, A.D.B. Webster, and S.M. Johnson, Lymphocyte 5'-nucleotidase in primary hypogammaglobulinemia and cord blood, Clin. Exp. Immunol. 39:337 (1980).
2. N.L. Edwards, D.B. Magilavy, J.T. Cassidy, and I.H. Fox, Lymphocyte ecto-5'-nucleotidase deficiency in agammaglobulinemia, Science 201:628 (1978).
3. L.F. Thompson, G.R. Boss, H.L. Spiegelberg, I.V. Jansen, R.D. O'Connor, T.A. Waldmann, R.N. Hamburger, and J.E. Seegmiller, Ecto-5'-nucleotidase activity in T and B lymphocytes from normal subjects and patients with congenital X-linked agammaglobulinemia, J. Immunol. 123-2475 (1979).
4. L.F. Thompson, A. Saxon, R.D. O'Connor, and R.I. Fox, Ecto-5'-nucleotidase activity in human T cell subsets: decreased numbers of ecto-5'-nucleotidase positive cells from both OKT4[+] and OKT8[+] cells in patients with hypogammaglobulinemia, J. Clin. Invest. 71:892 (1983).
5. A. Cohen, A. Mansour, H.-M. Dosch, and E.W. Gelfand, Association of a lymphocyte purine enzyme deficiency (5'-nucleotidase) with combined immunodeficiency, Clin. Immunol. Immunopathol. 15:245 (1980).
6. L.F. Thompson, R.D. O'Connor, and J.F. Bastian, Phenotype and function of engrafted maternal T cells in patients with severe combined immunodeficiency, J. Immunol. 133:2513 (1984).
7. J.F. Bastian, J.M. Ruedi, G.A. MacPherson, H.E. Golembesky, R.D. O'Connor, and L.F. Thompson, Lymphocyte ecto-5'-nucleotidase activity in infancy: increasing activity in peripheral blood B cells precedes their ability to synthesize IgG in vitro, J. Immunol. 132:1767 (1984).

8. G. Lucivero, A.R. Lawton, and M.D. Cooper, Rosette formation with mouse erythrocytes defines a population of human B lymphocytes unresponsive to pokeweed mitogen, Clin. Exp. Immunol. 45:185 (1981).

CELL SURFACE ADENOSINE DEAMINASE (ADA) AND ITS COMPLEXING PROTEIN (ADCP)

IN HUMAN T-LYMPHOID CELLS

Winand N.M. Dinjens,[1,2,3] Jeroen van der Boon,[3] Joop ten Kate,[2,3] Wim P. Zeijlemaker,[1] Chris H.M.M. De Bruijn,[4] Fré T. Bosman[2] and P. Meera Khan[3]

[1]Central Laboratory of the Netherlands Red Cross Blood Transfusion Service, incorporating the Laboratory for Experimental and Clinical Immunology, University of Amsterdam, Amsterdam, The Netherlands; [2]Department of Pathology, Biomedical Center, University of Limburg, Maastricht, The Netherlands; [3]Department of Human Genetics, State University of Leiden, Leiden, The Netherlands; [4]Project Group Biomedical and Health Care Technology, University of Technology, Eindhoven, The Netherlands

INTRODUCTION

Adenosine deaminase (ADA, EC no. 3.5.4.4), the enzyme which converts adenosine and deoxyadenosine to inosine and deoxyinosine respectively, plays an important role in the development of the immune system. The enzyme is probably involved in T-lymphocyte differentiation since a deficiency of the enzyme is associated with severe combined immunodeficiency disease.[1]

Two major molecular forms of ADA are known to occur in man and they are designated as ADA-S and ADA-L. ADA-S is a monomer with a molecular weight of about 35,000 D. ADA-L, with a molecular weight of about 280,000 D, is composed of one dimeric molecule of ADA complexing protein (ADCP) and two molecules of ADA-S.[2] ADA-S can be converted to ADA-L by complexing with ADCP.

ADCP, initially described as a soluble protein,[2,3] was found to be predominantly concentrated in membranes.[4-8] ADCP has been reported to be markedly decreased or completely absent in in vitro transformed and certain cancer derived human cell lines[9] and was suggested as a potential cancer marker.[3] In some studies[4,5,9,10] a fraction of the cellular ADA was reported to be attached also to the cell membrane.

Our research is concerned with the role of adenosine and ADA in T-lymphocyte differentiation. In this respect we are interested in the occurrence of ADA and ADCP in the membranes of T-lymphoid cells. In the present study ADA and ADCP were immunohistochemically detected in membranes of T-lymphoid cells.

Intact human peripheral blood lymphocytes (PBL) or human thymocytes were first incubated with rabbit anti-ADA or rabbit anti-ADCP antiserum followed by incubation with fluorescein isothiocyanate-labeled goat anti-

rabbit IgG (G/R-FITC). Since only cell surface ADA and ADCP would be exposed to the antisera fluorescence indicates respectively the ADA and ADCP
expression on the outer cell membrane. In double label experiments with T-
lymphocyte specific phycoerythrin (PE) conjugated monoclonal antibodies,
we investigated the distribution of membrane ADA and ADCP positive cells
among the T-lymphocyte subsets using a fluorescence activated cell sorter
(FACS).

MATERIALS AND METHODS

Preparation of Cell Suspensions

    Human peripheral blood mononuclear cells (MNC) were isolated from
blood of normal healthy donors by centrifugation over Percoll (d= 1.076 g/
cm^3; Pharmacia, Uppsala, Sweden). Contaminating platelets were removed by
two low-speed washing steps. Human peripheral blood T lymphocytes (PBL-T)
were obtained by nylon wool filtration of the MNC.[11] The final cell suspension contained more than 90% E-rosette-forming cells. Human thymocytes were
prepared as described by Astaldi et al.[12] Thymus tissues were obtained from
children undergoing cardiac surgery. After removal of adipose tissue, connective tissue and blood vessels, they were cut into small pieces and gently pressed through a metal sieve. The thymocytes were washed in phosphate
buffered saline (PBS).

Antisera and Monoclonal Antibodies

    Anti-human ADA-S and anti-human ADCP specific antisera were raised in
New Zealand and Chinchilla rabbits, respectively.[13,14] The anti-ADA-S
antiserum detects free ADA-S as well as ADA-S bound to ADCP. The anti-ADCP
antiserum detects free ADCP as well as ADCP complexed with ADA-S, but does
not recognize ADA-S.[14] G/R-FITC was from Nordic Immunological Laboratories,
Tilburg, The Netherlands. PE conjugated monoclonal antibodies anti-Leu-1,
anti-Leu-2a and anti-Leu-3a were purchased from Becton Dickinson, Sunny-
vale, CA, USA.

Flow Cytofluorometry

    At least 10^6 cells were incubated with normal rabbit (NRS) anti-ADA or
anti-ADCP serum. The cells were washed once and incubated with a combination of G/R-FITC and anti-Leu-1-PE, anti-Leu-2a-PE or anti-Leu-3a-PE. The
cells were washed twice and resuspended in 300 µl PBS containing 0.5% (w/v)
bovine serum albumin and 0.01% (w/v) sodium azide. In some experiments, the
cells were incubated with a lysate of human erythrocytes before the incubation with antiserum.

    Fluorescence intensity of viable lymphocytes was measured with a fluo-
rescence activated cell sorter (FACS IV, Becton Dickinson). Monocytes, erythrocytes, non-viable lymphocytes and cell debris were gated out electroni-
cally. The percentage of positive cells was calculated from the fluorescence
histogram (immunofluorescence profile) as follows. For the histograms obtained from incubation with anti-Leu-1-PE, anti-Leu-2a-PE or anti-Leu-3a-PE
a clear distinction between positive and negative cells was possible and
the threshold was set at the minimum between the two populations. The cells
with fluorescence exceeding the threshold were regarded as positive. For
the calculation of the percentage anti-ADA and anti-ADCP positive cells the
histogram obtained with NRS was used to determine the threshold. Double
fluorescence (FITC and PE) was measured by gating out the PE-negative cells
and determining the FITC fluorescence intensity among the PE-positive cells.

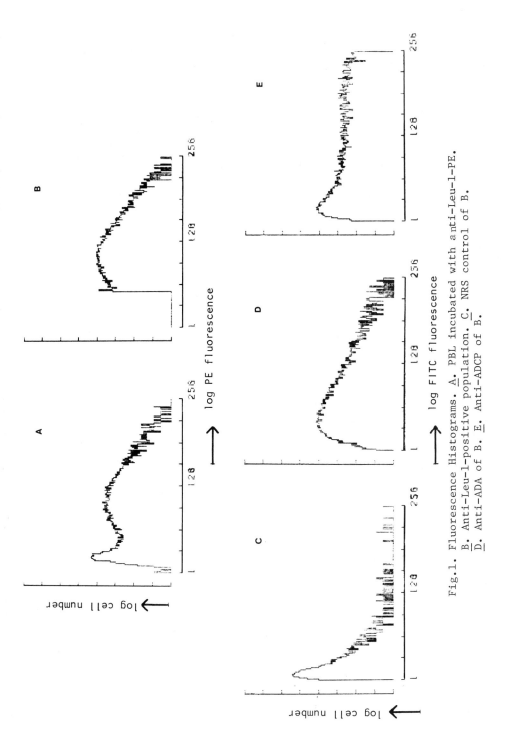

Fig.1. Fluorescence Histograms. A. PBL incubated with anti-Leu-1-PE. B. Anti-Leu-1-positive population. C. NRS control of B. D. Anti-ADA of B. E. Anti-ADCP of B.

409

RESULTS

The results of a representative experiment are shown in Figures 1-3. In Figure 1A the fluorescence histogram of PBL treated with anti-Leu-1-PE is shown. Seventy-five per cent of the cells is anti-Leu-1 positive. From the anti-Leu-1-positive cells (Fig. 1B) the histogram obtained from incubation with NRS, anti-ADA or anti-ADCP serum are shown in Figures 1C, 1D and 1E, respectively. From the anti-Leu-1-positive cells 15% are anti-ADA positive and 48% are anti-ADCP positive. The PBL population contains 28% anti-Leu-2a positive cells (Fig. 2A). From this positive population (Fig. 2B) 12% are anti-ADA positive (Fig. 2D) and 29% are anti-ADCP positive (Fig. 2E) with respect to the NRS control (Fig. 2C). From the PBL, 48% are anti-Leu-3a positive (Fig. 3A and 3B). Incubation of the cells with the monoclonal antibody anti-Leu-3a-PE in combination with NRS, anti-ADA or anti-ADCP serum resulted in 14% anti-ADA-positive (Fig. 3D) and 54% anti-ADCP-positive (Fig. 3E) cells.

A population of viable human thymocytes contained about 25% anti-ADA- and 10% anti-ADCP-positive cells (histograms not shown).

When isolated PBL-T were incubated with a lysate of human erythrocytes the percentage of anti-ADA-positive PBL-T increased from 10 to 50% (Fig. 4), whereas no difference in ADCP expression was found (not shown).

DISCUSSION

Among human PBL-T, detected by the pan T-cell specific monoclonal antibody anti-Leu-1, about 15% of the cells contained membrane bound ADA and 48% membrane bound ADCP (Fig. 1). About 25% of human thymocytes contained membrane bound ADA and only 10% showed membrane ADCP expression. We could demonstrate that within the PBL-T population the percentage membrane ADA-positive T-helper/inducer cells (defined by the anti-Leu-3a monoclonal antibody; Fig. 3) was comparable to the percentage of T-cytotoxic/inducer cells (defined by the anti-Leu-2a monoclonal antibody; Fig. 2). In contrast, T-helper/inducer cells contained about 54% membrane ADCP-positive cells (Fig. 3) whereas the T-cytotoxic/inducer cells contained only about 29% membrane-ADCP-positive cells (Fig. 2). Obviously, there is no quantitative correlation between the presence of ADA and ADCP in the membranes of PBL-T as well as in thymocytes. In a number of human and rabbit tissues also a lack of correlation between membrane bound ADCP and membrane bound ADA was found.[4,15]

Andy and Kornfeld[6] demonstrated that ADCP present in isolated membranes of fibroblasts has the capacity to bind ADA-S. Recent in vivo studies by Schrader et al.[15] showed that intravenously injected ADA-S in the rabbit was cleared from the blood by binding to ADCP present in the proximal convoluted tubules of the kidney. Intact PBL-T incubated with a lysate of human erythrocytes (as source of ADA-S) have also still the capacity to bind ADA-S (as shown in Fig. 4). Probably the ADA-S binds to membrane-associated ADCP. It was suggested that the membrane bound ADA functions to protect the cells from the toxic effects of adenosine and deoxyadenosine by extracellular metabolization.[5,10] The ADA-free ADCP, present in membranes, was suggested to serve as a receptor for the recovery or removal of free ADA-S.[4,16]

PBL-T and human thymocytes are known to possess adenosine receptors.[16] When this receptor is occupied with adenosine the intracellular concentration of cyclic AMP is increased. We measured the adenosine-induced cyclic AMP increase in PBL-T and thymocytes in the presence of anti-ADA and anti-ADCP serum. No effect of the antisera on the cyclic AMP level could be

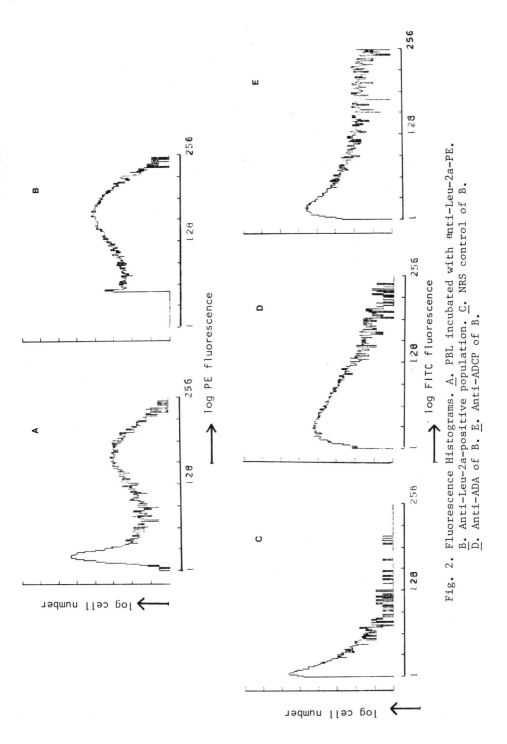

Fig. 2. Fluorescence Histograms. A. PBL incubated with anti-Leu-2a-PE. B. Anti-Leu-2a-positive population. C. NRS control of B. D. Anti-ADA of B. E. Anti-ADCP of B.

log PE fluorescence

log FITC fluorescence

log cell number

411

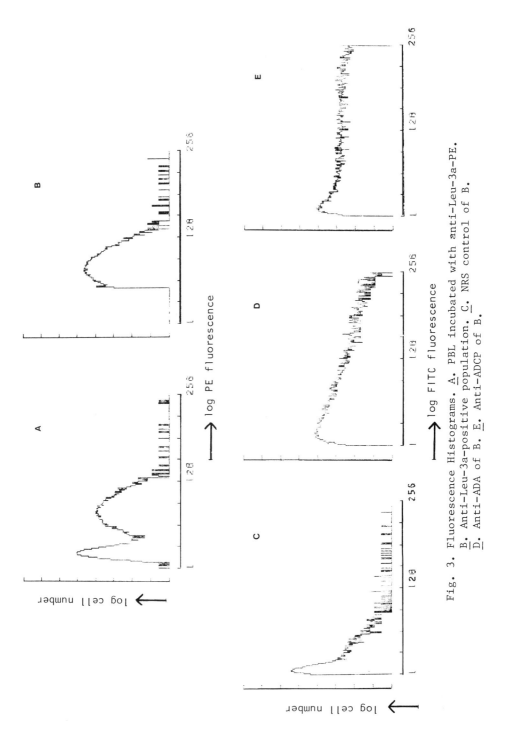

Fig. 3. Fluorescence Histograms. A. PBL incubated with anti-Leu-3a-PE.
B. Anti-Leu-3a-positive population. C. NRS control of B.
D. Anti-ADA of B. E. Anti-ADCP of B.

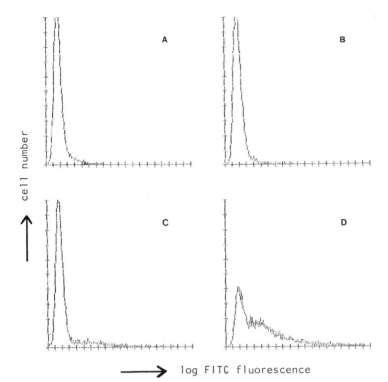

Fig. 4. Fluorescence histograms. <u>A</u>. PBL pre-incubated with PBS and incubated with NRS. <u>B</u>. PBL pre-incubated with a lysate of human erythrocytes and incubated with NRS. <u>C</u>. PBL pre-incubated with PBS and incubated with anti-ADA serum. <u>D</u>. PBL pre-incubated with a lysate of human erythrocytes and incubated with anti-ADA serum.

demonstrated (results not shown). Obviously, there is no relationship between the adenosine receptor and membrane bound ADA or ADCP.

A panel of cryopreserved lymphoblasts from 9 acute T-lymphoblastic leukemia (T-ALL) patients was investigated for membrane ADCP expression. Differences in membrane ADCP expression ranging from 0 to 70% were found (results not shown). This expression was found not to be correlated with the Leu-2 or Leu-3 antigen. Perhaps, membrane ADCP expression can be useful as an additional new marker in classifying T-ALL.

Studies concerning the relationship between functional biological characteristics and membrane ADA and ADCP expression in T-lymphoid cells will throw more light on the usefulness of these molecules as markers in T-lymphocyte maturation and differentiation studies. These studies together with those in progress will be published elsewhere.

ACKNOWLEDGEMENTS

We are grateful to Dr. Paul Oljans (University of Leiden) for the excellent technical contributions in the FACS analysis, and to Drs. Hans Tanke, Bert Schutte and Marlies Lagaaij (University of Leiden) for their kind gifts of PE-conjugated antibodies. This study was supported in part by grant no. 13-40-61 from the Foundation for Medical Research (FUNGO) which is subsidized by the Netherlands Organization for the Advancement of Pure Research (ZWO); and in part by the Queen Wilhelmina Cancer Foundation (KWF, projects A 81.98 and IKW 85.82). Participation in the ISHPPM was made possible by a grant from the foundation Simonsfund.

REFERENCES

1. E. R. Giblett, J. E. Anderson, F. Cohen, B. Pollara, and H. J. Meuwissen, Adenosine deaminase deficiency in two patients with severely impaired cellular immunity, Lancet ii:1067 (1972).
2. P. E. Daddona and W. N. Kelley, Analysis of normal and mutant forms of human adenosine deaminase. A review, Mol. Cell. Bio chem. 29:91 (1980).
3. E. Herbschleb-Voogt, J. ten Kate, and P. Meera Khan, Adenosine deaminase complexing protein: A transformation sensitive protein with potentials of a cancer marker, Anticancer Res. 3:95 (1983).
4. W. P. Schrader and P. Bryer, Characterization of an insoluble adenosine deaminase complexing protein from human kidney, Arch. Biochem. Biophys. 215:107 (1982).
5. P. P. Trotta, Identification of a membrane associated adenosine deaminase binding protein from human placenta, Biochemistry 21:4014 (1982).
6. R. J. Andy and R. Kornfeld, The adenosine deaminase binding protein of human skin fibroblasts is located on the cell surface, J. Biol. Chem. 257:7922 (1982).
7. J. ten Kate, J. Th. Wijnen, J. Boldewijn, P. Meera Khan, and F. T. Bosman, Immunohistochemical localization of adenosine deaminase complexing protein in intestinal mucosa and in colorectal adenocarcinomas as a marker for tumour cell heterogeneity, Histochem. J. 17:23 (1985).
8. W. P. Schrader and C. A. West, Adenosine deaminase complexing proteins are localized in exocrine glands of the rabbit, J. Histo chem. Cytochem. 33:508 (1985).
9. M. B. van der Weyden and W. N. Kelley, Human adenosine deaminase: distribution and properties, J. Biol. Chem. 251:5448 (1976).
10. B. E. Chechik, R. Baumal, and S. Sengupta, Localization and identity of adenosine-deaminase-positive cells in tissues of young rat and calf, Histochem. J. 15:373 (1983).
11. D. Roos and J. A. Loos, Changes in the carbohydrate metabolism of mitogenically stimulated human peripheral lymphocytes. I. Stimulation by phytohaemagglutinin, Biochim. Biophys. Acta 222:565 (1970).
12. G. C. B. Astaldi, A. Astaldi, M. Groenewoud, P. Wijermans, P. T. A. Schellekens, and V. P. Eijsvoogel, Effects of a human serum thymic factor on hydrocortisone-treated thymocytes, Eur. J. Immu nol. 7:836 (1977).
13. E. Herbschleb-Voogt, J. W. Scholten, and P. Meera Khan, Basic defect in the expression of adenosine deaminase in ADA⁻ SCID disease. II. Deficiency of ADA-CRM detected in heterozygote human-Chinese hamster cell hybrids, Hum. Genet. 63:121 (1983).
14. J. ten Kate, J. Th. Wijnen, R. G. M. van der Goes, R. Quadt, G. Griffioen, F. T. Bosman, and P. Meera Khan, Quantitative changes in adenosine deaminase isoenzymes in human colorectal adenocarcinomas, Cancer Res. 44:4688 (1984).
15. W. P. Schrader, C. M. Harder, D. K. Schrader, and C. A. West, Metabolism of several molecular forms of adenosine deaminase intravenously infused into the rabbit, Arch. Biochem. Biophys. 230: 158 (1984).
16. R. J. van de Griend, A. Astaldi, P. Wijermans, R. van Doorn, and D. Roos, Low β-adrenergic receptor concentration in human thymocytes, Clin. Exp. Immunol. 53:53 (1983).

ON THE ROLE OF CYTIDINE DEAMINASE IN CELLULAR METABOLISM

Per Nygaard
University Institute of Biological Chemistry B
Sølvgade 83
1307 Copenhagen K, Denmark

INTRODUCTION

Pyrimidine nucleotides are synthesized either de novo or via salvage pathways from preformed pyrimidine compounds. The de novo pathway of UTP and CTP synthesis appears to be almost identical in the various organisms studied, while there are major differences in the pathways by which pyrimidine deoxynucleotides are synthesized. A few organism have been found to be unable to synthesize UMP de novo (1). However pyrimidine salvage appears to occur in all organisms studied,the salvage pathways seem diverse in their nature and distribution, and the various organisms and tissues studied differ significantly in their ability to metabolize exogenously supplied pyrimidine compounds.

PYRIMIDINE SALVAGE ENZYMES

The pyrimidine salvage pathways appear to serve two different functions. One is to scavenge pyrimidine compounds for nucleotide synthesis, the other is degradative and leads to formation of compounds which may serve as carbon and nitrogen sources. This is accomplished by developing two types of pyrimidine salvage enzymes, the anabolic enzymes: Uracil phosphoribosyltransferase, pyrimidine nucleoside kinases and nucleoside deoxyribosyltransferases, and the catabolic enzymes: Nucleotidases, ribonucleotide glycosylases and nucleoside hydrolases. Some enzymes e.g. cytosine deaminase, cytidine deaminase and the various pyrimidine nucleoside phosphorylases

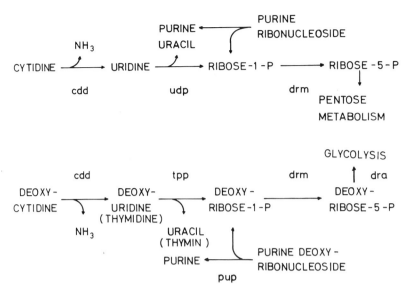

Figure 1. Schematic representation of enzyme reactions involved in the catabolism of cytidine and deoxycytidine in Salmonella and E.coli. cdd = cytidine deaminase; udp = uridine phosphorylase; tpp = thymidine phosphorylase; drm = phosphodeoxyribomutase; pup = purine nucleoside phosphorylase and dra = deoxyriboaldolase.

may functionally belong to boths groups listed above. The salvage pathways also define the sensitivity of a cell to the pyrimidine analogs which require conversion to nucleotides for toxicity.

In the present report the main emphasis will be devoted to the role of cytidine deaminase in the metabolism of cytidine and deoxycytidine. Other enzymes which react with cytidine are cytidine kinase and certain ribonucleoside hydrolases; deoxycytidine is also substrate for some deoxynucleoside kinases and for nucleoside deoxyribosyltransferases (1,2).

CYTIDINE DEAMINASE, PROPERTIES AND OCCURENCE

Cytidine deaminase (EC 3.5.4.5) catalyzes the hydrolytic deamination of cytidine and deoxycytidine to the corresponding uracil nucleosides. The enzyme also deaminates some nucleoside analogs, including antimetabolites such as cytosine arabinoside, 5-fluorodeoxycytidine and 5-azacytidine, but it does not

Table 1   Cytidine Deaminase and Uridine Phosphorylase
           Levels in Different Salmonella Mutants (11).

Relevant Genotype	Carbon Source	Inducer added (2mM)	Cytidine deaminase	Uridine phosphorylase
			nmoles/min/mg protein	
wt	glycerol	no	11	47
wt	-	cytidine	310	1333
wt	-	uridine	519	1798
wt	-	adenosine	39	83
add	-	adenosine	101	406
cdd,udk	-	no		449
cytR	-	no	1415	2305
cytR	glucose	no	376	681

react with free cytosine and with cytosine nucleotides. Potent
inhibitors of the enzyme are available (3,4). Enzyme activity
is determined spectrophotometrically (5) or by radiochemical
assays (6,7). The enzyme has been partly purified from human
liver and granulocytes, sheep liver, mouse kidney and spleen,
corn and yeast and purified to homogeneity from E.coli (3-6).
The enzyme most likely consists of a single polypeptide chain.
Cytidine deaminase from human granulocytes has three electrophoretic phenotypes representing the homozygous and heterozy-
gous expression of two common alleles (8).
     Cytidine deaminase activity is widely distributed in hu-
man tissues: The normal tissues, liver, spleen and marrow and
white blood cells contain high levels. Among tumor tissues
chondriocarcinoma has very high levels 254 µmoles/h/g (9).The
distribution of the enzyme in twenty mammalian cell lines of
different origin but not of hematopoietic origin showed variation in enzyme levels from 2 - 400 nmoles/h/mg protein, in 17
of the cell lines studied the enzyme levels were within 80-200
nmoles/h/mg protein (10). Cytidine deaminase activity has been
demonstrated in a few higher plants and in several microorga-

nisms. Some microorganisms belonging to Lactobacilli,Pseudomonas and the Moraxella group do not posses cytidine deaminase activity(1).

CYTOSINE NUCLEOSIDE METABOLISM IN ENTEROBACTERIACEAE

The most detailed study on cytidine deaminase has been performed in Escherichia coli and in Salmonella typhimurium. The enzyme, encoded for by the cdd gene, belongs to a group of enzymes (Figure 1) and transport proteins which are involved in the catabolism of nucleosides, regulated by a repressor encoded by the cytR gene (2,11).The low-molecular effectors are cytidine (and in Salmonella also uridine) and adenosine, adenosine functions most efficiently in an adenosine deaminase (add) deficient mutant, Table 1. Mutants of the cytR type contain high levels of cytidine deaminase and the other nucleoside catabolizing enzymes and transport proteins. The cytR regulated promotors are subjected to catabolite repression (Table 1) and are dependent on cyclic-AMP and the cyclic-AMP receptor protein. In adenylate cyclase and in cyclic-AMP receptor protein deficient mutants, only wild type levels of cytidine deaminase and uridine phosphorylase are expressed (12). The presence of cytidine deaminase activity ensures that E.coli and Salmonella can utilize the pentose part of both cytidine and deoxycytidine as sole carbon and nitrogen source. The uracil moiety will serve as a general pyrimidine source. By selecting for resistance to 5-fluorodeoxycytidine, mutants defective in the synthesis of cytidine deaminase, cdd mutants, can be obtained (1). In cdd mutants, deoxycytidine is no longer metabolized. E.coli and Salmonella do not possess deoxycytidine kinase activity, but cytidine can be phosphorylated to CMP (1). In cdd,udk double mutants defective in cytidine deaminase and cytidine kinase activities, increased levels of other nucleoside metabolizing enzymes is seen, Table 1, indicating an intracellular accumulation of cytidine , acting as an endogenous inducer (2,11). Thus in E.coli and in Salmonella cytidine deaminase activity is essential for their utilization of cytidine and deoxycytidine as pyrimidine, carbon and nitrogen sources.

Table 2 Cytidine Deaminase(CDD), Adenosine Deaminase (ADA) and Cytidine Kinase (CDK) Levels of Human Hematopoietic Cell Lines.

Cell Line	Origin	Differential stage (13)	CDD	ADD	CDK
			nmoles/h/mg protein		
CCRF-CEM	ALL	T-Blast I	29	7500	43
CCRF-HSB-2	ALL	T-Blast III	16	18000	22
BALL-1	ALL	B-Blast II	321	1600	44
SKW-4	BL	B-Blast I	<1	7600	31
HL-60	APL	Promyel.	5	4900	21
KG-1	AML	Pre-Mybl.	<1	3500	41

Nygaard and Sundstrøm unpubl. obs.

CYTIDINE DEAMINASE ACTIVITY IN HUMAN WHITE BLOOD CELLS AND HEMATOPOIETIC CELL LINES

Peripheral blood lymphocytes from normal individuals (N=16) had a mean cytidine deaminase activity of 279 (range 38-778) and granulocytes had a mean activty of 2443 (range 1513-4712) nmoles/h/ mg protein (7). The distribution of cytidine deaminase activity in some human hematopoietic cell lines is shown in Table 2. The profile of cytidine deaminase show levels that the activity is low and apparently absent in some cell lines. In contrast to this, these cell lines possess high levels of adenosine deaminase activity and almost identical levels of cytidine kinase activity (Table 2) Based on the comparison of the various cell lines studied (Table 2 and ref.10) it appears that cytidine deaminase activity is very low in immature cells and high in mature cells. In an earlier study of cytidine deaminase and adenosine deaminase in peripheral blood cells from patients with acute myeloblastic leukemia we found that the level of cytidine deaminase increased and that of adenosine deaminase de- creased at decreasing blast cell percentage (7,14).It seems   that cytidine deaminase may be

regarded as biochemical marker of cell immaturity in hematopoietic cells.

In human thymocytes it has been observed that salvage of
deoxycytidine inhibits the salvaging of deoxyadenosine and
deoxyguanosine (15). This may be of importance for the maintenance of a balanced DNA synthesis when the precursors of
DNA are synthesized both via ribonucleotide reduction and via
deoxynucleoside salvage , a situation which is likely to
occur in the immature lymphocytes (15). Thus, a low level of
cytidine deaminase activity may be of physiological significance for immature lymphocytes for the salvaging of deoxycytidine and for the maintenance of a proper balance between
the four DNA precursors.

REFERENCES

1. J. Neuhard.In: Metabolism of nucleotides,nucleosides
    and nucleobases in microorganisms, A.Munch-Petersen
    ed., Acad.Press N.Y. 3:95(1983)
2. K. Hammer-Jespersen.In. Ibid.5:203(1983)
3. B. A. Chabner, D. G. Johns, C. N. Coleman, J. C.
    Drake and W. H. Evans. J.Clin.Invest.53:922(1974)
4. G. W. Ashley and P. A. Bartlett. J.Biol.Chem.
    259:13615(1984)
5. D. F. Wentworth and R. Wolfenden.In. Meth.Enzymol.
    P.A.Hoffe and M.E.Jones ed.,Acad.Press N.Y.LI:401(1978)
6. K. Rothman, V. A. Malathi and R. Silber.In. Ibid.
    LI:408(1978)
7. J. Mejer and P. Nygaard. Leuk.Res.3:211(1979)
8. Y. S. Teng, J. E. Anderson and E. R. Giblett. Am.
    J.Hum.Genet.27:492(1975)
9. D. H. W. Ho. Canc.Res.33:2816(1973)
10. K. Ishii, H. Sakamoto, J. Furuyama, M. Hanaoka.
    Cell.Struct.Funct.9:117(1984)
11. P. Nygaard. Eur.J.Biochem.36:267(1973).
12. K. Hammer-Jespersen and P. Nygaard. Molec.Gen.Genet.
    148:49(1976).
13. B. I. S. Srivastava and J. Minowada. Leuk.Res.7:331(1983)
14. J. Mejer and P. Nygaard. Leuk.Res.3:211(1979)
15. A. Cohen, J. Barankiewicz, H. M. Lederman and E. W.
    Gelfand. J.Biol.Chem.258:12334(1983)

PURINE ENZYME ACTIVITIES AS MARKERS OF LYMPHOCYTIC
DIFFERENTIATION: STUDIES OF LYMPHOCYTES FROM HORSES
WITH SEVERE COMBINED IMMUNODEFICIENCY (SCID)

Nancy Magnuson, Lance Perryman, Carol Wyatt
and Patricia Mason

Department of Microbiology and Pathology
Washington State University
Pullman, WA   99164-7040   USA

INTRODUCTION

Severe combined immunodeficiency (SCID) occurs in horses of the Arabian
breed and is inherited as an autosomal recessive trait.[1] Affected foals are
severely lymphopenic with an absence of functional B and T lymphocytes.[2]
The few lymphocytes that can be found in affected foals resemble large
granular lymphocytes (LGL) and can be grown long-term with media containing
interleukin-2 (IL-2).[3] The disorder is restricted to the lymphoid system,
as other leukocytes are functionally and quantitatively normal.[4]

The importance of purine metabolism to the immune system has been
demonstrated by the finding of two inherited deficiency diseases in which
adenosine deaminase (ADA) and purine nucleoside phosphorylase (PNP) are
lacking.[5] The deficiencies of these enzymes have been causally associated
with selective impairment of lymphocyte differentiation and growth. The
discovery of these diseases focused attention on the relationship between
purine metabolism and normal lymphocyte maturation. The importance of
purine metabolism in lymphocyte differentiation appears to be related to the
regulation of certain nucleoside and nucleotide pools.[6] Studies of the
enzymes associated with purine salvage in man have shown that these enzymes
have restricted distribution in different lymphocyte subpopulations and,
therefore, may be indicators of discrete stages in lymphocyte maturation and
differentiation.[7-9]

The purpose of this study was to determine if the pattern of activities
for purine metabolic enzymes obtained for LGL from foals with SCID is a
marker for a lymphocyte subpopulation that survives the lethal defect or is
a result of a metabolic defect.

MATERIALS AND METHODS

Cells: Blood from foals with SCID and from nonaffected foals was
collected by venipuncture into a sterile bottle containing glass beads.
After constant agitation for 30 min, the blood was decanted from fibrin-
coated beads. Peripheral blood mononuclear cells (PBMC) were then isolated
as previously described by density centrifugation using Ficoll-Hypaque
(Ficoll; Sigma Chemical Co., St. Louis, MO; Hypaque; Winthrop Laboratories,
Division of Sterling Drug, Inc., New York).[3] It was also necessary to

separate the SCID LGL from contaminating monocytes and neutrophils. This was accomplished by Sephadex G-10 filtration (Pharmacia Fine Chemicals, Uppsala, Sweden). A nearly uniform population, as determined from morphologic characteristics of Wright-stained preparations, of between 0.5 to 2 x 10^6 lymphoid cells could be obtained from 100 ml of SCID blood. Bone marrow mononuclear cells and thymocytes were isolated by Ficoll-Hypaque density centrifugation as described for PBMC.

Long-term cultivation of lymphocytes: The cultivation medium consisted of human recombinant interleukin 2 (IL-2) (Biogen, Geneva, Switzerland) at a concentration of 100 U/ml in Dulbecco's modified Eagle's medium containing 2.5% fetal calf serum. Freshly prepared cells obtained from either SCID or nonaffected foals were established in long-term culture by initially supplementing the medium with 2 µg/ml concanavalin A. Cell density and viability were evaluated twice weekly at which time the medium was relaced with fresh IL-2-containing medium. Cultures were incubated at 37° C in a 5% CO_2 humidified atmosphere.

Surface marker analysis: Cell suspensions were incubated with monoclonal antibody at 4° C for 30 min, then washed in phosphate buffered saline containing 0.5% bovine serum albumin and 0.001 M sodium azide (PBS-BSA). The cells were resuspended in appropriately diluted fluoresceinated goat anti-mouse immunoglobulin (Cappel, Malvern, PA) and reincubated at 4° C for 30 min. Cells were washed and resuspended in PBS-BSA. Paraformaldehyde (1.5% in saline) was added and the cells were refrigerated until used. Analysis was performed by flow cytometry using a Becton-Dickinson FACS 420.

Assay for equine natural killer (NK) cells: A microassay was developed based on previously published procedures.[10] Briefly, cells from animals to be tested were adjusted to 5 x 10^5/ml medium and 0.1 ml/well plated out in V-bottomed microtiter plates. Each well also received either human recombinant IL-2 (1000 U/ml), recombinant interferon (A/D Bgl, a gift from NIH; 100 U/ml), or poly IC (Sigma Chemical Co., St. Louis, MO; 5 µg/ml). After 18 hrs incubation, 0.1 ml of target cells (5 x 10^3 YAC-1 cells labeled with ^{51}Cr) were added to achieve an effector to target cell ratio of 10:1. Plates were centrifuged (400 x g) for 1 min and then incubated at 37° C for 4 hr before recentrifugation. Supernatant aliquots (0.1 ml) were removed from each well to determine the amount of ^{51}Cr released. The results are calculated as the mean percent specific lysis.

Enzyme assays: The activities of 5'-nucleotidase (5'-N), adenosine deaminase (ADA), purine nucleoside phosphorylase (PNP), adenine phosphoribosyl transferase (APRT), hypoxanthine guanine phosphoribosyl transferase (HGPRT) and AMP deaminase were measured by a modified microassay originally described for human peripheral blood lymphocytes.[7] In brief, 3 µl of a lymphocyte suspension containing 7 x 10^7 cells/ml of 0.9% NaCl were pipetted into small incubation vessels prepared from parafilm. The cells were frozen at -20° C for 30 min and subsequently lyophilized. Depending on the enzyme being analyzed, the appropriate substrate was added to the lyophilized cells and allowed to incubate. The enzyme activities were quantitated by separating products and substrates by thin layer chromatography. All buffers, substrates, incubation times and chromatography conditions are given in detail elsewhere.[11] Enzyme activities were calculated and expressed as nmol of product formed/hr/10^6 cells. Enzyme activities were linear for up to twice the assay times used.

RESULTS

Surface markers of SCID LGL: Monoclonal antibodies have been produced for identification of equine lymphocyte subpopulations. The binding

characteristics of the four monoclonal antibodies used in this study for
PBMC, thymocytes and bone marrow are listed in Table 1. Complete
characterization of these monoclonal antibodies is to be published elsewhere
(Wyatt and Perryman, manuscript in preparation). In Table 2 the surface
phenotype of normal PBMC are compared with those of cells cultured in human
recombinant IL-2 for 30 days. The number of normal PBMC expressing the
macrophage marker increased after long-term culture. The normal PBMC also
lost marker for mature T lymphocytes and gained the immature T lymphoid
marker after long-term culture. The phenotype of freshly isolated SCID LGL
was also compared with that of SCID LGL maintained in long-term culture with
human recombinant IL-2 for 15 and 20 days. After 20 days in culture, the
proliferation rate of SCID LGL slows down and the viability of the culture
begins to decline. Between 15 and 20 days, therefore, was chosen as the
best time to analyze the phenotype of SCID LGL in long-term culture. In
contrast to normal PBMC, the percentage of SCID LGL expressing both markers
for T lymphocytes increased as did the number of cells expressing the
macrophage marker.

Effector function of SCID LGL: Speculation that the SCID LGL were
natural killer cells came after observing that their morphological
appearance, best described as large granular lymphocytes, was a feature of
NK cells described for man and rodents.[12] It had also been shown for some
children with SCID that cells with NK cell acitivity were present and that
these cells could be cultured long-term with exogenous IL-2.[13] In order to
evaluate the relatively small numbers of LGL from the SCID foals for NK cell
activity, a microassay for NK cells actiyity had to be established. By
adapting an assay that used as few as 10^3 target cells, we were able to
establish that LGL from SCID foals are NK cells[10]. It was found that LGL
from SCID foals did not kill ^{51}Cr-labeled YAC-1 until after the cells had
been incubated for 18 hr with human recombinant IL-2 (Table 3). The data
for three foals are shown, but a total of six SCID foals have been tested
with identical results. Recombinant interferon and poly IC were not as
effective as IL-2 in increasing NK cell activity. NK cell activity was not
readily detectable in PBMC from normal horses (1.4 - 8%) whether or not
these had been treated in vitro with IL-2 for 24 hr (data not shown).

Table 1. Reactivities of monoclonal antibodies

Antibody reactivity	Percent labeled		
	PBMC	Thymocytes	Bone marrow
Monocyte and B cell (EqM1)	26.6+12.1[a]	2.2+2.5	2.0+2.1
Macrophage (EqM2)	3.0+2.0	3.1+2.6	5.9+6.5
Mature T cell (EqT3)	49.4+2.0	51.3+7.4	3.9+2.0
Immature T cell (EqT6)	8.0+6.0	16.7+7.3	14.1+5.2

[a]The values represent the mean + standard deviation determined
from assays on tissue from four different animals.

Table 2. Phenotype of peripheral blood mononuclear cells

Antibody reactivity	Percent labeled			
	Normal[a]	SCID[b]	Normal-LT[c]	SCID-LT[d]
Monocyte and B cell	23.0	32.7+5.2	4.3	14.2; 32.7
Macrophage	3.0	7.7+6.5	19.2	52.3; 24.4
Mature T cell	62.0	14.9+11.7	1.9	25.6; 33.8
Immature T cell	7.8	18.8+13.0	34.1	55.3; 31.3

[a]The PBMC from a normal foal were separated over Sephadex G-10 with the same procedures used for SCID cells.
[b]The mean + standard deviation, n = eight different SCID foals.
[c]LT indicates normal PBMC were cultured long-term (30 days) with IL-2.
[d]LT indicates SCID LGL were cultured long-term with IL-2. First value from LGL of foal 1 cultured 15 days; second value from LGL of foal 2 cultured 20 days.

Table 3. Natural killer cell activity of SCID LGL

Cells	Addition	Percent cytotoxicity[a]
SCID		
1881	none	<5
1884	none	<5
1885	none	<5
SCID		
1881	IL-2	47, 48.7
1884	IL-2	40.6
1885	IL-2	20.7
SCID		
1881	interferon	22.7
SCID		
1881	poly IC	26.1
Normal		
1880	none	<5
1882	none	<5
Normal		
1880	IL-2	7.2
1882	IL-2	<5

[a]The effector:target ratio was 10:1 in all cases. All assays were run after an 18 hr incubation with or without IL-2, interferon or poly IC. Neither normal PBMC nor SCID LGL had any NK activity when analyzed immediately after isolation.

Enzyme activities in SCID LGL: Data from previous studies suggested that the purine enzyme activity pattern for SCID LGL was unique to this cell population. It was proposed that this pattern either reflected the genetic defect or represented the unique biochemistry of this subpopulation of cells.[14] In the previous study, high activities for ADA, 5'N, APRT and HGPRT were found compared to those for normal PBMC. In subsequent studies which evaluated other cellular elements of the blood, it was found that platelets have high activities for APRT and HGPRT relative to activities found in PBMC (Table 4). To be certain that high values originally found for SCID LGL were not due to platelets which are difficult to separate from the SCID LGL with the usual cell preparation procedures, a method using glass beads to remove platelets was adapted. By removing the platelets, the activities of APRT and HGPRT were reduced to ranges found for normal PBMC (Table 4). In comparison to normal PBMC, however, 5'N and ADA still remained 10- and 4-fold higher, respectively. These high activities were not due to neutrophil contamination as the SCID LGL preparations did not contain neutrophils. The resulting pattern is still unique to SCID LGL.

DISCUSSION

Our results confirmed and extended previous findings that purine enzymatic activity patterns may serve as additional markers for the identification of distinct lymphocyte subpopulations.[11,14] In particular we were interested in characterizing the LGL found in the blood of SCID foals. Using newly developed monoclonal antibodies, we determined that these cells had a surface phenotype that suggested a relationship to T lymphocytes or possibly NK cells. The finding of potent NK cell activity after IL-2 treatment confirmed that a significant number of functional NK cells exist in the peripheral blood of these animals.

The previous findings of altered purine metabolism in tissues of SCID foals [15,16] along with the finding of a distinct pattern of purine enzymatic

Table 4. Activities[a] of purine metabolic enzymes

Enzyme	SCID LGL	Normal PBMC	Normal PBMC-LT[b]	Neutrophil	Platelet
5'-N	2.0+1.6[c]	0.2+0.1[c]	1.6+0.4[c]	4.9+2.2[c]	<0.5[d]
ADA	15.0+1.7	3.8+0.6	16.9+2.0	16.9+2.0	<0.5
PNP	3.4+1.2	2.5+0.5	3.3+1.7	2.4+0.7	<0.5
APRT	2.5+0.3	2.8+0.2	10.0+0.5	4.7+1.4	0.8
HGPRT	2.0+0.3	2.2+0.8	6.3+1.0	1.5+0.5	1.5
AMP deaminase	69 + 20	71 + 18	139 + 18	129 + 23	3.5

[a]Activities are expressed as nmol product formed/hr/10^6cells or platelets.
[b]LT indicates PBMC were maintained in long-term culture for 30 days.
[c]The numbers represent the mean + standard deviation of values obtained from 5 individuals.
[d]The numbers represent the means of 2 values.

activities for SCID LGL[14] originally suggested that the altered enzyme activities might be a reflection of the metabolic defect. Our recent finding of NK activity for LGL suggests that SCID LGL may be normal and, thus, the pattern of enzymatic activity is a characteristic of this subpopulation of cells. Furthermore, LGL from SCID horses have a life time in culture with IL-2 of approximately 6 weeks.[3] This is consistent with the growth patterns of NK active cell lines obtained from children with SCID.[13] If this short-term growth pattern is not a result of the genetic defect, then our observations suggest to us that the NK cell in the SCID foals represents a normal subpopulation of NK cells.

With respect to the enzyme activities that were measured, only those enzymes were used that appeared to have high activites in lymphocytes.[11] It is interesting to note that PNP appears to vary the least among the cell populations. This indicates that PNP is not a useful marker. In contrast, the other five enzymatic activities examined appear to vary significantly among lymphocyte populations. This panel can serve as a useful marker in conjunction with surface antigens.

ACKNOWLEDGMENTS

This work was supported in part by the National Institutes Health Grant HD 08886 and the Morris Animal Foundation.

REFERENCES

1. L.E. Perryman and R.L. Torbeck, Combined immunodeficiency of Arabian horses: confirmation of autosomal recessive mode of inheritance, J. Am. Vet. Med. Assoc. 176:1250 (1980).
2. T.C. McGuire, K.L. Banks, and M.J. Poppie, Combined immunodeficiency in horses: characterization of the lymphocyte defect, Clin. Immunol. Immunopathol. 3:555 (1975).
3. N.S. Magnuson, L.E. Perryman, C.R. Wyatt, T. Ishizaka, P.H. Mason, A.E. Namen, K.L. Banks, and J.A. Magnuson, Continuous cultivation of equine lymphocytes: Evidence for occasional T cell-like maturation events in horses with hereditary severe combined immunodeficiency, J. Immunol. 133:2518 (1984).
4. K.L. Banks and T.C. McGuire, Surface receptors on neutrophils and monocytes from immunodeficient and normal horses. Immunology 28:581 (1975).
5. D.W. Martin, Jr. and E.W. Gelfand, Biochemistry of diseases of immunodevelopment, Ann. Rev. Biochem. 50:845 (1981).
6. D.A. Carson, J. Kaye, S. Matsumoto, J.E. Seegmiller, and L. Thompson, Biochemical basis for the enhanced toxicity of deoxyribonucleosides toward malignant human T cell lines, Proc. Natl. Acad. Sci. U.S.A. 76:2430 (1979).
7. J.P.R.M. Van Laarhoven, G.Th. Spierenburg, and C.H.M.M. de Bruyn, Enzymes of purine nucleotide metabolism in human lymphocytes, J. Immun. Meth. 39:47 (1980).
8. J.P.R.M. Van Laarhoven, G.Th. Spierenburg, C.H.M.M. de Bruyn, and E.D.A.M. Schritlen, Enzymes of purine interconversions in subfractions of lymphocytes. Adv. Exp. Med. Biol. 122B:283 (1980).
9. M. Massaia, D.D.F. Ma, T.A. Sylewtrowicz, N. Tidman, G. Price, G. Janossay, and A.V. Hoffbrand, Enzymes of purine metabolism in human peripheral lymphocyte subpopulations, Clin. Exp. Immun. 50:148 (1982).
10. L. Lefrancois, F.R. Klein, V. Paetkau, and M.J. Bevan, Antigen-independent activation of memory cytotoxic T cells by interleukin 2, J. Immunol. 133:1845 (1984).

11. N.S. Magnuson, L.E. Perryman, P.H. Mason and K.M. Marta, Distribution of enzymes of purine metabolism in lymphocytes of horse, Equus caballus, Comp. Biochem. Physiol. 81B:459 (1985).
12. J.R. Ortaldo and R.B. Herberman, Heterogeneity of natural killer cells, Ann. Rev. Immunol. 2:359 (1984).
13. N. Flomenberg, K. Welte, R. Mertelsmann, R. O'Reilly, and B. Dupont, Interleukin 2-dependent natural killer (NK) cell lines from patients with primary T cell immunodeficiencies, J. Immunol. 130:2635 (1983).
14. N.S. Magnuson and L.E. Perryman, Metabolic investigations of horses with severe combined immunodeficiency, Ann. N.Y. Acad. Sci. (in press).
15. N.S. Magnuson and L.E. Perryman, In vitro effects of adenosine on lymphocytes and erythrocytes from horses with severe combined immunodeficiency, J. Clin. Invest. 64:89 (1979).
16. N.S. Magnuson, D.M. Decker, and L.E. Perryman, Increased susceptibility of fibroblasts from horses with severe combined immunodeficiency to growth inhibition by 2'-deoxyadenosine. Clin. Immunol. Immunopathol. 29:391 (1983).

THE EFFECTS OF PNP INHIBITION ON RAT LYMPHOID CELL POPULATIONS

Randall W. Barton* and William R. A. Osborne#

*Pulmonary Division, University of Connecticut Health Center
Farmington, CT 06032; and #The Department of Pediatrics, University of Washington School of Medicine, Seattle, WA 98195

INTRODUCTION

In humans an inherited deficiency of the enzyme purine nucleoside phosphorylase (PNP; EC 2.4.2.1) is associated with a T cell deficiency and apparently normal B cells (1,2). Although a deficiency of PNP is present in all cells of deficient individuals, only the cells within the T lymphocyte lineage are affected. The immunodeficiency state associated with PNP deficiency is thought to arise from block(s) in T lymphocyte differentiation resulting from the accumulation of toxic metabolites of the enzyme's substrate(s) (3-5).

The growth inhibition of lymphoid cell lines by PNP substrates revealed that deoxyguanosine was the most potent and that T cell lines were markedly more sensitive than B cell lines (6,7). Within the T cell lineage thymocytes were more sensitive to growth inhibition by deoxyguanosine than peripheral lymphocytes (8). Growth inhibition of T cell lines by deoxyguanosine has been shown to be associated with increased dGTP (6,8-10) and dGTP has been shown to accumulate in the erythrocytes of some PNP-deficient children (11). This provides support for the thesis that the accumulated dGTP acts as an inhibitor of the enzyme ribonucleotide reductase (10,12-14).

The nucleoside analogue, 8-aminoguanosine, is a potent inhibitor of PNP (15) and has been reported to be selectively toxic towards T lymphoblasts (16). In the present study we have examined the effect of PNP inhibition on the in vitro proliferation of differing lymphoid populations in the rat and have compared the metabolic consequences of PNP inhibition among the lymphoid populations.

METHODS

Spontaneous ^{3}H-thymidine incorporation assay. Cells were dispensed into a 96 well tissue culture plate at a concentration of 5×10^5 cells per well. 2'-deoxyguanosine (Sigma Chemical Co.) was prepared at 5 mg/ml. Dilutions of GdR were made in RPMI 1640 medium and 20 µl added per well. 8-aminoguanosine (Sigma Chemical) was prepared in the same manner as GdR at 5 mg/ml. Dilutions were made in RPMI 1640 and 20 µl added per well. Cells were then pulsed with 1 µCi per well 6-^{3}H-thymidine (Amersham Corp.). Samples were incubated for one hour before harvesting.

Mitogenesis assay. Cells were dispensed into a 96 well tissue culture plate at a concentration of 5×10^5/well. The T cell mitogen, concanavalin A (ConA), was added at a final concentration of 5 µg/ml. Dextran sulfate (DxS) (both from Sigma Chemical) and Salmonella typhimurium mitogen (STM) from RIBI Immunochem Research, Inc. were added at final concentrations of 25 µg/ml and 30 mg/ml with STM and DxS used in combination for B cell stimulation (18). After 72 hours GdR, 8-aminoguanosine, and ^{3}H-TdR were added as described above. Cells were incubated at 37°C for one hour, harvested and counted.

Nucleoside uptake. Single cell suspensions from thymus and spleen were added to a 96 well tissue culture plate at 2×10^6 cells per well in RPMI media containing either 187 µM guanosine (Sigma Chemical) and 30 µCi/ml ^{3}H-guanosine (ICN Radiochem) or 187 µM deoxyguanosine (Sigma) and 30 µCi/ml ^{3}H-deoxyguanosine (ICN). Some wells also contained 8-aminoguanosine (400 µM). After incubating at 37°C in 5% CO_2 for 2 hours, replicate wells were pooled. Protein-free cell extracts were prepared using perchloric acid by the method of Garret and Santi (19). Guanosine nucleotides were quantitated by high pressure liquid chromatography using a gradient elution procedure (20) and the radioactivity counted as previously described (14).

Deoxynucleoside kinase activity and PNP assays. Deoxyguanosine and deoxycytidine kinase were assayed as previously described (20) with the exception that the former assay was performed in the presence of 25 µM 8-aminoguanine in order to inhibit PNP and prevent substrate loss and alternate product formation. PNP was measured spectrophotometrically (17).

RESULTS

Cell proliferation. The effects of varying concentrations of deoxyguaonsine with and without the PNP inhibitor 8-aminoguanosine on the spontaneous proliferation of lymphoid cells from thymus and spleen were compared. Cell suspensions from freshly removed tissues were cultured immediately in order to study the lymphoid cells proliferating in vivo at the time of sacrifice. As shown in Figure 1, the spontaneously proliferating thymocytes were exquisitely sensitive to deoxyguanosine; even at 1 µM the incorporation of ^{3}H-thymidine was only approximately 75% of control levels. In contrast, at 100 µM the proliferation of spleen lymphocytes was 50-60% of control values. The addition of 8-aminoguanosine, 400 µM, alone had a slight effect on spontaneous proliferation. However, when 8-aminoguanosine was added to the cultures containing deoxyguanosine, inhibition of ^{3}H-thymidine incorporation was markedly enhanced at the lower concentrations of deoxyguanosine.

The effects of an induced PNP deficient state on the proliferation of mitogen-induced lymphoblasts from thymus and spleen were also compared. As seen in Figure 2, thymocytes that were induced to proliferate by ConA were much less sensitive to deoxyguanosine than the thymocytes that were spontaneously proliferating at the time of sacrifice. The sensitivities of ConA-induced lymphoblasts from thymus and spleen to deoxyguanosine were similar, although the thymocytes were inhibited to a somewhat greater extent. The addition of 8-aminoguanosine to cultures containing deoxyguanosine enhanced the inhibition of thymidine incorporation.

Rat B lymphoblasts were slightly less sensitive to deoxyguanosine than rat T lymphoblasts. Incorporation of ^{3}H-thymidine was only slightly inhibited at deoxyguanosine concentrations as high as 100 µM. However, the addition of 8-aminoguanosine plus deoxyguanosine did partially inhibit thymidine incorporation.

Nucleoside uptake. The concentrations of purine nucleotides and the percentage distribution of radioactive purine from incubation of thymocytes and spleen lymphocytes with deoxyguanosine or guanosine are shown in Table 1. Both thymocytes and spleen lymphocytes accumulated comparable amounts of guanosine nucleotides from exogenous guanosine. When 8-aminoguanosine was added to the cultures the amount of guanosine nucleotides was reduced.

Unlike the guanosine nucleotides neither spleen lymphocytes nor thymocytes contained detectable deoxyguanosine nucleotides when cultured in media alone. However, thymocytes incubated with exogenous deoxyguanosine formed significant amounts of deoxyguanosine nucleotides that contained greater than 90% of the incorporated radiolabel. With spleen lymphocytes there was a virtually total metabolism of exogenous deoxyguanosine to ribonucleotides. In spleen lymphocytes the total radioactivity incorporated into nucleotides from both guanosine and deoxyguanosine was similar. Thymocytes incorporated radiolabel into nucleotides from deoxyguanosine at a nearly three-fold greater level than with guanosine.

The addition of 8-aminoguanosine with deoxyguanosine increased the amount of deoxyguanosine nucleotides synthesized in thymocytes and decreased the formation of guanosine nucleotides in spleen lymphocytes without promoting deoxyribonucleotide synthesis.

<u>Deoxyguanosine kinase and PNP activities</u>. The kinase assays showed thymocytes to have 5-fold higher deoxyguanosine kinase levels than spleen lymphocytes (149.7 nmol/hr/mg vs. 27.2 nmol/hr/mg). The reverse relationship was observed with PNP activities; spleen lymphocytes having 5-fold higher levels than thymocytes (5.3 vs. 1.1 nmol/30 min/10^8 cells).

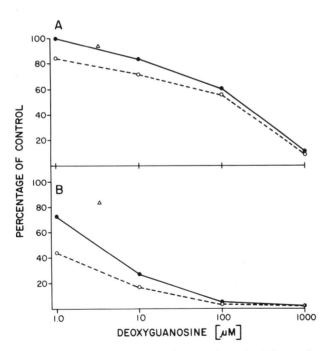

Figure 1. The effects of deoxyguanosine with and without 8-aminoguanosine on the spontaneous proliferation of spleen lymphocytes (A) and thymocytes (B). Deoxyguanosine alone (●————●); plus 400 μM 8-aminoguanosine (o----------------o).

DISCUSSION

The results of the present study also showed that thymocytes were more sensitive to an induced PNP deficiency than peripheral, splenic lymphocytes. The thymocytes that were spontaneously proliferating at the time of removal were profoundly more sensitive than thymocytes induced to proliferate by ConA. Interestingly, mitogen-stimulated spleen B lymphocytes were the least sensitive. This is consistent with the fact that a deficiency of PNP causes a T cell but not a B cell deficiency (1,2).

Exogenous deoxyguanosine produced significant increases of phosphory-lated product in spontaneously proliferating thymocytes but not in spleen lymphocytes, and this was correlated with the 5-fold higher deoxyguanosine

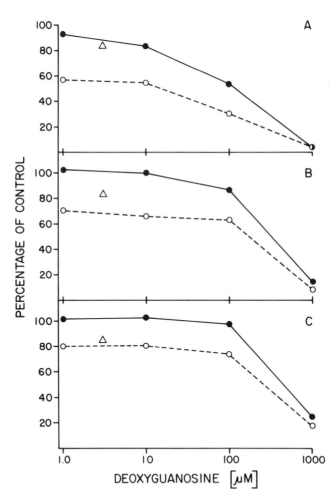

Figure 2. The effects of deoxyguanosine with and without 8-aminoguanosine on mitogen-induced thymocytes and spleen T and B lymphocytes. (A) ConA-induced thymocytes; (B) ConA-induced spleen lymphocytes; (C) STM/DxS-in-duced spleen lymphocytes; deoxyguanosine alone (● ──────── ●); plus 8-aminoguanosine (○ ------------- ○); 8-aminoguanosine alone (△).

Table I

Effects of 8-Aminoguanosine on the Incorporation of

^{3}H-Guanosine and ^{3}H-Deoxyguanosine into Guanine Nucleotides

	GTP + GDP		dGDP + dGTP	
	nmoles[a]	% radioactivity	nmoles[a]	% radioactivity
Thymus				
None	1.32	-	nd	-
Guo	1.25	100.0	nd	0
Guo + 8-AGuo	0.86	100.0	nd	0
dGuo	0.69	7.5	0.53	92.5
dGuo + 8-AGuo	0.95	3.2	1.51	96.8
Spleen				
None	0.78	-	nd	-
Guo	1.01	100.0	nd	0
Guo + 8-AGuo	0.69	100.0	nd	0
dGuo	0.94	98.2	nd	1.8
dGuo + *-AGuo	0.47	81.3	nd	18.7

*nd - Not detectable

[a]Units nmole/6x10^6 cells

kinase activity of thymocytes over spleen cells. The addition of 8-amino-guanosine markedly increased deoxyguanosine nucleotide concentration in thymocytes but did not induce their formation in spleen lymphocytes. Similar results with 8-aminoguanosine have been reported for T cell but not B cell lines (14) and without 8-aminoguanosine for T cell lines and thymocytes (6,8-10).

These results document significant differences in deoxyguanosine metabolism between thymocytes and the peripheral lymphocytes of spleen. In these cells two enzymes are competing for a common substrate. Thymocytes exhibit high deoxyguanosine kinase activity and low levels of PNP and the reverse situation occurs in spleen lymphocytes. Therefore, the predominant role of deoxyguanosine metabolism in thymocytes is initiated by deoxyguanosine kinase activity and in spleen lymphocyte metabolism via PNP is the major pathway. In the absence of PNP, formation of dGTP from deoxyguanosine is amplified in thymocytes and remains insignificant in spleen lymphocytes. In thymocytes, the prevention of dGTP accumulation by deoxycytidine and the increase in dGTP accumulation after PNP inhibition support this concept. Thus, in the absence of PNP activity only thymocytes would preferentially accumulate dGTP.

The fact that the spontaneously proliferating thymocytes were uniquely sensitive to the effects of an induced PNP deficient state is important in understanding the stages of T cell differentiation affected by PNP deficiency. Most of the spontaneously proliferating cells in the thymus are subcapsular cortical thymocytes which are the most ancestral cell population within the thymus (22). In contrast, thymocytes that are mitogen responsive are among the most mature cells in the thymus (22). Subcapsular thymocytes are large, immature cycling cells (23) and studies using radioactive DNA precursors have demonstrated that the subcapsular thymocytes are the major generative compartment for cortical and medullary thymocytes (22, 24). Inasmuch as cortical and/or medullary thymocytes serve as the source of thymus emigrants that populate the peripheral lymphoid tissues as mature, functional T cells (22,24), toxicity of subcapsular thymocytes (or their immediate progeny) would likely have a profound deleterious effect on the development of immunocompetence within the T lymphocyte lineage.

REFERENCES

1. Giblett, E.R., A.J. Amman, D.W. Wara and L.K. Diamond. 1975. Lancet 1:1010.
2. Stoop, J.W., B.J.M. Zegers, G.F.M. Hendricks, L.H. Siegenbeek van Heukelom, G.E.J. Staal, P.K. DeBree, S.K. Wadman and R.E. Ballieux. 1977. New Engl. J. Med. 296:651.
3. Polmar, S.H. 1980. Sem. Hematol. 17:30.
4. Osborne, W.R.A. 1981. Trends Biochem. Sci. 63:80.
5. Hirschhorn, R. 1983. Birth Defects 19:73
6. Mitchell, B.S., E. Mejias, P.E. Dadonna and W.N. Kelly. 1978. Proc. Nat. Acad. Sci. 75:5011.
7. Carson, D.A., J. Kaye and J.E. Seegmiller. 1978. J. Immunol. 121:1726.
8. Cohen, A., J.W.W. Lee, H.-M. Dosch and E.W. Gelfand. 1980. J. Immunol. 121:1726.
9. Fox, R.M., E.H. Tripp, S.K. Piddington and M.H.N. Tattersall. 1980. Cancer Res. 40:3383.
10. Ullman, B., L.J. Gudas, S.M. Clift and D.W. Martin. 1979. Proc. Nat. Acad. Sci. 76:1074.
11. Cohen, A., L.F. Gudas, A.J. Amman, G.E.J. Staal and D.W. Martin. 1978. J. Clin. Invest. 61:1405.
12. Gelfand, E.W., J.W. Lee and H.-M. Dosch. 1979. Proc. Nat. Acad. Sci. 76:1998.
13. Chan, T. 1978. Cell 14:523.
14. Osborne, W.R.A. and C.R. Scott. 1983. Biochem. J. 214:711.
15. Stoeckler, J.D., C. Cambor, V. Kuhns, S.-H. Chu and R.E. Parks. 1982. Biochem. Pharmacol. 31:163.
16. Kazmers, I.S., B.S. Mitchell, P.E. Dadonna, L.L. Wotring, L.B. Townsend and W.N. Kelly. 1981. Science 214:1138.
17. Barton, R.W., F. Martinuik, R. Hirschhorn and I. Goldschneider. 1980. Cell. Immunol. 49:208.
18. Beaudoin, B.J. and I. Goldschneider. 1985.
19. Garrett, C. and D.V. Santi. 1979. Anal. Biochem. 99:268.
20. Osborne, W.R.A., W.P. Hammond and D.C. Dale. 1983. J. Clin. Invest. 71:1348.
21. Gudas, L.J., B. Ullman, A. Cohen and D.W. Martin, Jr. Cell 14:531.
22. Goldschneider, I. 1980. Curr. Top. Dev. Biol. 14:33.

23. Metcalf, D. 1966. <u>In</u> The Thymus: Experimental and Clinical Studies. Edited by G.E.W. Wolstenholme and R. Porter. Ciba Foundation Symposium, Churchill, Ltd., London, pp. 242-263.

24. Shortman, K. and R. Scollay. 1985. <u>In</u> Recognition and Regulation in Cell-Mediated Immunity. Edited by J. Watson. Marcel Dekker, New York.

ACKNOWLEDGEMENT

This work was supported by U.S. Public Health Service Grants HD-17061, HD-0554 and AI-19565.

IMBALANCE IN NUCLEOTIDE POOLS OF MYELOID LEUKEMIA CELLS AND HL-60 CELLS:

CORRELATION WITH CELL CYCLE PHASE, CELL PROLIFERATION AND DIFFERENTIATION

Dirk de Korte,[2] Willem A. Haverkort,[1] Martin de Boer,[2]
Dirk Roos[2] and Albert H. van Gennip[1]

[1]Children's Hospital "Het Emma Kinderziekenhuis", Amsterdam
[2]Central Laboratory of the Netherlands Red Cross Blood Transfusion Service and Laboratory for Experimental and Clinical
Immunology of the University of Amsterdam, Amsterdam, The
Netherlands

INTRODUCTION

Recently, several reports have been published about abnormal purine and
pyrimidine metabolism in leukemic cells as compared to normal peripheral
blood cells. Enzyme activities[1,2] as well as metabolite concentrations[3,4]
have been studied. These studies demonstrated marked differences in the
activities of many enzymes controling purine and pyrimidine metabolism in
leukemic cells, such as adenosine deaminase, purine nucleoside phosphorylase and terminal deoxynucleotidyl transferase. De Abreu et al.[3] determined
significantly increased concentrations of adenine and uracil nucleotides
in the lymphoblasts of patients with non-B, non-T acute lymphoblastic leukemia. Liebes et al.[4] found increased concentrations of nicotinamide adenine nucelotides in lymphoid cells of patients with chronic lymphatic leukemia.

During our own investigations on the purine and pyrimidine nucleotide
content of leukemic vs normal cells, we have found an imbalance in the nucleotide pool of myeloblasts from chronic myeloid leukemia (CML) patients
with a myeloid blast crisis. This imbalance might be associated with differences in cell-cycle phase, cell proliferation, maturation stage and/or
with the transformation process. The composition of the nucleotide pool in
CML cells was comparable to that detected in HL-60 (a promyelocytic cell
line[5]) cells. Dimethyl sulfoxide (DMSO) induces HL-60 to maturate into mye-
locytes and granulocytes, which have acquired functional properties characteristic of these cells.[6] Therefore, we have used the HL-60 cell line to
study the imbalance in the myeloid nucleotide pool in more detail.

METHODS

Nucleotide Analysis

Intracellular nucleotides were extracted from the cells with 0.4 M perchloric acid and after neutralization with K_2CO_3, the nucleotides were
quantitated with an anion-exchange HPLC method as previously described.[7]

Cell Culture

Cultures of HL-60 cells were maintained in humidified air/5% CO_2 at 37°C in RPMI-1640 medium (Flow Laboratories), containing penicillin (100 U/ml) and streptomycin (100 µg/ml), supplemented with 10% heat-inactivated fetal calf serum. For differentiation experiments, the cells were diluted 1:1 from a density of $2x10^6$/ml one day before the addition of DMSO and brought to a density of $5x10^5$/ml just before addition of DMSO (1.25%, v/v).

Elutriation

The elutriation medium was phosphate-buffered saline (PBS) with 13 mM trisodium citrate and 5 mg of human albumin/ml; the medium flow was kept constant at 17 ml/min during the entire run. The elutriation medium was kept at 10°C and the centrifuge at 15°C. When the cells had been introduced to the elutriation chamber (Beckman JE-6 rotor), the speed of the centrifuge was decreased from 4000 to 3000 rpm. Fifteen fractions of 100 ml were collected, with a decrease of the rotor speed with 100 rpm after each fraction.

RESULTS and DISCUSSION

Table 1 shows the nucleotide content of myeloblasts from CML patients, compared to exponentially growing HL-60 cells and neutrophils from healthy individuals. Both myelobasts and HL-60 cells showed an imbalance in the nucleotide pool, compared to normal neutrophils. This imbalance includes decreased purine:pyrimidine and adenine:guanine nucleotide ratios, an increased amount of UDP sugars in proportion to the total nucleotides and a changed composition of these UDP sugars. Based on the similarity between myeloblasts and HL-60 cells, we used the HL-60 cell line for our study.

Table 1. Nucleotide Pattern of Myeloblasts from CML Patients, of Normal Neutrophils, of Exponential Growing HL-60 Cells and of DMSO-Induced and Non-induced HL-60 Cells in Stationary Phase (After 144 h Without Medium Refreshment).

	CML myeloblasts n=11	Normal neutrophils n=11	HL-60 cells		
			Exponential phase n=8	Stationary phase + DMSO -DMSO n=5	
CONCENTRATIONS					
Total nucleotides[a]	1374	1933	3789	2548	2595
UDP sugars[b]	120	109	515	262	393
RATIOS					
Purine:pyrimidine	6.0	10.7	3.6	4.9	5.2
Adenine:guanine	3.0	3.4	2.5	3.5	3.4
Uracil:cytosine	3.8	4.2	2.3	2.6	2.2
UDPNAG:UDP-hexoses[c]	1.9	1.2	4.6	0.9	3.2
$\frac{\text{UDP sugars}}{\text{Total nucleotides}}$ x100	8.6	5.6	13.6	10.3	15.1

Concentrations in pmol/10^6 cells; mean of n determinations, variation coefficients 5-30%.
[a] Adenine + guanine + uracil + cytosine nucleotides.
[b] UDP glucose and/or UDP galactose + UDP-N-acetylglucosamine and/or UDP galactosamine.
[c] UDPNAG, UDP-N-acetylhexosamines.

First, the relationship between the imbalance in the nucleotide pool and a specific phase of the cell cycle was investigated. HL-60 cells were separated into 15 fractions by means of elutriation centrifugation (Fig. 1). The recovery of about 70% is normal for elutriation experiments. The first 6 fractions contained only cell debris and a few dead cells. From

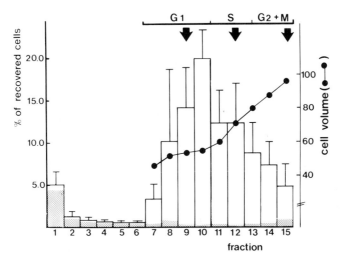

Fig. 1.   Elutriation pattern of HL-60 cells. At a constant medium flow
of 17 ml/min, the rotor speed was step-wise diminished from 3000
rpm to 1600 rpm, yielding 15 fractions. Mean ± S.D. of 3 experiments with a recovery of 70 ± 2%. □, viable cells; ▣, non-viable
cells; cell volume:peak of the size distribution on the Coulter
Channelyzer; ↓, fractions shown in Table 1.

fractions 7 to 15, the mean size of the cells increased, with the volume of
the cells in fraction 15 being more than twice that of the cells in fraction 7. The viability of the HL-60 cells in these fractions was >90%. Mito-
ses were present in fractions 13 to 15. DNA analysis after labeling with
propidium iodide on an Ortho flow cytometer (Spectrum III) showed that
fractions 7 to 10 contained mainly cells in G1 phase, that fractions 11 to
13 were enriched with cells in S phase and that fractions 14 and 15 contained mainly cells in G2+M phase. Thus, elutriation centrifugation with
continuous flow and step-wise decreasing of the rotor speed showed to be
very useful for separation of proliferating cells according to cell-cycle
phase.

Nucleotide analysis of unfractionated HL-60 cells and of the cells in
fractions 9 (G1), 12 (S) and 15 (G2+M) showed an increasing nucleotide content, correlating with cell volume. The composition of the nucleotide pool
was similar in all analyzed cell fractions (Table 2). We conclude that the
imbalance in the nucleotide pool of HL-60 cells, as compared to normal neutrophils, is not associated with a specific cell-cycle phase.

Secondly, the relationship between the imbalance in the nucleotide pool
and the maturation stage and/or the transformed character of the HL-60

Table 2. Nucleotide Pattern of Unfractionated HL-60 Cells
and of HL-60 Cells Enriched in G1, S or G2+M phase

	Unfractionated	G1	S	G2+M
CONCENTRATIONS				
Total nucleotides[a]	4169	3143*	4500*	6105*
UDP sugars[b]	605	462*	676*	967*
RATIOS				
Purine:pyrimidine	3.3	3.1	3.3	3.2
Adenine:guanine	3.0	3.3	3.2	3.3
Uracil:cytosine	2.7	2.5	2.4	2.7

Concentrations in $pmol/10^6$ cells; mean of 3 experiments, with a
variation coefficient of 5-25%.
[a]Adenine + guanine + uracil + cytosine nucleotides.
[b]UDP glucose and/or UDP galactose + UDP-N-acetylglucosamine and/
or UDP galactosamine.
*Significant differences between the cell-cycle phases.

cells was investigated. HL-60 cells were induced to myeloid differentiation
with DMSO (1.25%, v/v). Because differentiation includes loss of proliferation, also non-proliferating HL-60 cells (cells in stationary phase) were
studied to distinguish between changes associated with cell proliferation
or with maturation stage. DMSO induced a decrease in the mean cell volume
within 24 h, whereas in the absence of DMSO the mean cell volume started
to decrease after 72 h (not shown). Both in the absence and presence of
DMSO, exponential growth was observed from 0-72 h. During exponential
growth the increase in cell number was slightly, but significantly faster
in the presence of DMSO. The viability of the HL-60 cells started to decrease, both in the presence and absence of DMSO, after 96 h in culture
without medium refreshment. After 144 h without medium refreshment, the
viability of the cells was under both conditions >90%. The cell number decreased from $1.5x10^6$ cells/ml after 72 h to $1.0x10^6$ cells/ml after 144 h.

Differentiation of the HL-60 cells in the presence of DMSO was proven
by the ability of these cells to show an oxidative burst after stimulation
with phorbol-myristate acetate (PMA) and serum-treated zymosan (STZ), with
a maximum after 144 h. The morphology of the induced HL-60 cells changed
from promyelocytic to myelocytic after 144 h, with already a few band-form
and polymorphic nuclei. Table 1 shows that the nucleotide content of both
induced and non-induced HL-60 cells was decreased after 144 h, correlating
with the decreased cell volume. A significant difference was found between
induced and non-induced HL-60 cells for the concentration and composition
of the UDP sugars. Induced cells contain less UDP-N-acetylhexosamines and
more UDP hexoses, and the amount of UDP sugars in proportion to the total
nucleotides (percentage UDP sugars) was decreased. Thus, the composition
of the UDP sugars in the induced HL-60 cells is more in agreement with
that of normal neutrophils, although the concentration is still higher.
The changes in the ratios of purine:pyrimidine, adenine:guanine and uracil:
cytosine nucleotides are similar for both induced and non-induced cells.
Therefore, the decreased ratios found in exponentially growing HL-60 cells,
compared to normal neutrophils, are associated with proliferation and not
with the immature and/or transformed character.

Lucas et al.[8] described that differentiation of HL-60 cells was associated with an increase in the ratio of adenine:guanine nucleotides. Moreover, they found that differentiation occurred upon treatment of HL-60 cells with inhibitors of IMP dehydrogenase, causing an increase in the adenine:guanine ratio.[9] Our results suggest that there is no causal relationship between the increase in the adenine:guanine ratio and the observed differentiation of HL-60 cells.

The increased concentration of the UDP sugars and the changed composition of these sugar nucleotides seems to be associated with the immature and/or transformed character of the HL-60 cells. Wice et al[10] reported for human colon cancer cells that the only difference between the nucleotide pool of differentiated and undifferentiated non-proliferating cells was found in the concentration of UDP-N-acetylhexosamines (increased in undifferentiated cells). Changes in activities of glycosyltransferases during differentiation have been described for HL-60 by Liu et al[11]; especially the increased activity of N-acetylglucosaminyltransferase might be responsible for the changes in the UDP sugar content. It will be very difficult to distinguish between changes associated with immaturity or transformation, because it is not known whether differentiation is accompanied by loss of the transformed state.

ACKNOWLEDGEMENTS

This study was financially supported by grant number AUKC-EMMA-81-5 from the foundation "Het Koningin Wilhelmina Fonds", Organization for the Fight against Cancer, Amsterdam, The Netherlands.

REFERENCES

1. J. P. R. M. van Laarhoven, G. Th. Spierenburg, and J. A. J. M. Bakkeren, Leukemia Res. 7:407 (1983).
2. H. G. Drexler, G. Gaedicke, and J. Minowada, JNCI 72:1283 (1984).
3. R. A. de Abreu, J. M. van Baal, J. A. J. M. Bakkeren, C. H. M. M. de Bruijn, and E. D. A. M. Schretlen, J. Chromatogr. 227:45 (1982).
4. L. F. Liebes, R. L. Krigel, M. Conklyn, D. R. Nevrla, and R. Silber, Cancer Res. 43:5608 (1983).
5. S. J. Collins, R. C. Gallo, and R. E. Gallagher, Nature 270:347 (1977).
6. J. J. Collins, F. W. Ruscetti, R. E. Gallagher, and R. C. Gallo, Proc. Natl. Acad. Sci. USA 75:2458 (1978).
7. D. de Korte, W. A. Haverkort, D. Roos, and A. H. van Gennip, Clin. Chim. Acta 148:185 (1985).
8. D. L. Lucas, H. K. Webster, and D. G. Wright, J. Clin. Invest. 72: 1889 (1983).
9. D. L. Lucas, R. K. Robbins, R. D. Knight, and D. G. Wright, Bio chem. Biophys. Res. Commun. 115:971 (1983).
10. B. M. Wice, G. Trugnan, M. Pinto, M. Rousset, G. Chevalier, E. Dussaulx, B. Lacroix, and A. Zweibaum, J. Biol. Chem. 260:139 (1985).
11. C- K Liu, R. Schmied, C. Schreiber, A. Rosen, G-X Qian, and S. Waxman, Exp. Hematol. 11:738 (1983).

ENHANCEMENT OF T CELL PROLIFERATION AND DIFFERENTIATION BY

8-MERCAPTOGUANOSINE

Michael G. Goodman and
William O. Weigle

Department of Immunology
Scripps Clinic and Research Foundation
La Jolla, CA

INTRODUCTION

Recent studies of the mechanisms by which cellular interactions govern lymphocyte proliferation and differentiation have resulted in discovery of the lymphokines and elucidation of their modes of action. These molecules are peptides or proteins, whose amino acid sequence and physico-chemical properties have been difficult to characterize due to the difficulty in achieving high degrees of purity in the preparations under study. Recently, however, we have described the potent immunobiologic activity of a novel class of molecule, the C8-subsituted guanine ribonucleosides. The observation that these synthetic guanosine derivatives are mitogenic for murine splenic lymphocytes without participation of cyclic GMP[1] was followed by description of their capacity to induce polyclonal synthesis and secretion of immunoglobulin in both murine and more recently human B lymphocytes. The mitogenic activity of these nucleoside derivatives was shown to be exerted intracellularly[2], predominantly on a subpopulation of mature B lymphocytes and to a lesser degree on a subset of immature B cells[3]. In addition, these agents exhibit powerful adjuvant effects both in vitro and in vivo and provide a T cell-like inductive signal for B cells that enables them to generate antigen-specific responses to T cell-dependent antigens in the absence of T cells[4].

To date, however, little data has appeared regarding the activity of these derivatized nucleosides on T cell function. Therefore, the present experiments were undertaken to investigate the potential effects of 8MGuo on T cell activity.

MATERIAL AND METHODS

Mice. CBA/CaJ, SJL, and BDF_1 mice, 8-12 wks of age, were purchased from the Jackson Laboratory, Bar Harbor, ME.

Culture reagents. Constituents of the serum-free and serum-containing culture media employed in these studies have been described elsewhere[3]. 8MGuo was from the Sigma Chemical Company, St. Louis, MO. Rat IL-2 was from Collaborative Research, Bethesda, MD.

Lymphocyte preparations and culture. Spleen cell, splenic T cell, and thymocyte suspensions were prepared as described previously[3]. Cells were cultured in microculture plates (No. 3596 Costar, Cambridge, MA) at a cell density of 4×10^6 viable cells/ml in a volume of 0.1 ml. Microcultures were incubated at $37^\circ C$ in a humidifed atmosphere of 10% $CO2$ in air. Cultures were fed daily with 8 µl of nutritional cocktail.

Assay for IL-2. The IL-2 dependent murine cell line CTLL was cultured for 24 hrs in a volume of 0.2 ml containing 10^3 cells with the test supernatant or positive control. At that time, 1 µCi per culture of ^3H-TdR was added for the final 24 hrs of culture (5 Ci/mM, Amersham Radiochemicals, Amersham, England). The microcultures were harvested with a Brandel cell harvester (M24V, Biological Research and Development Laboratories, Rockville, MD) onto glass fiber filter strips. Filter disks were transferred to plastic scintillation vials, covered with liquid scintillation cocktail, and counted in the Beckman LS7500 liquid scintillation counter.

Mixed lymphocyte cultures. 1.5×10^5 viable CBA/CaJ thymocytes or SJL spleen cells were cultured with or without an equal number of 2500 r irradiated BDF_1 spleen cells in the presence or absence of IL-2. One µCi ^3H-TdR per culture was added after 90 hrs of incubation for the final 6 hrs. Uptake of tritiated thymidine was assessed as in the assay for IL-2.

Generation of T cell supernatants. 5×10^6 splenic T cells were cultured in 1 ml medium containing 5% FCS with or without 1 mM 8MGuo. One to 4 days later, supernatants were harvested by centrifugation and filtration, and 8MGuo removed by dialysis.

Assessment of T helper activity. Groups of CBA/CaJ mice were immunized i.p. with a low dose of SRBC (5×10^5) previously demonstrated to educate T helper cells preferentially[5]. Three hours later mice received supplemental injections of either carboxymethylcellulose (CMC) alone or variable quantities of 8MGuo suspended in CMC. Five days later, single cell suspensions, prepared from the spleens of these mice, were subjected to 2500 r irradiation and cultured together with antigen in the presence or absence of freshly prepared B cells from antigen-inexperienced animals. Four to five days later, direct PFC to SRBC were determined using a modification of the plaque assay described by Jerne and Nordin[6].

RESULTS

Inability of 8MGuo to stimulate T cell mitogenesis. Table 1 recapitulates our previous experience examining the mitogenic activity of 8MGuo for thymocytes and splenic T cells. Neither cell type responds to 8MGuo, in contrast to their vigorous responses to Con A. However, when thymocytes are pre-stimulated by 24 hr exposure to the mitogenic lymphokine IL-2, they develop responsiveness to co-stimulation by 8MGuo added at this time (Table 2).

Effect of 8MGuo on mixed lymphocyte cultures. In evaluating the effect of 8MGuo on alloantigen-induced proliferation, it is imperative to minimize the simultaneous mitogenic response of B cells to 8MGuo. In the first set of experiments, therefore, thymocytes were chosen as the responder population. These cells are poor producers of IL-2, and their responses can therefore be controlled with IL-2. When thymocytes were stimulated with irradiated allogeneic spleen cells in the presence of exogenous IL-2, the proliferative response of the thymocytes in the presence of 8MGuo increased in a dose-dependent fashion, peaking at 10^{-3} M (Table 3). The

444

Table 1. Effect of 8MGuo on T Cell and Thymocyte Mitogenesis[a]

8MGuo Concentration [M]	[3H]TdR Uptake (cpm/culture)[b]	
	Thymocytes	Spleen T Cells
none	570 $\pm$ 70	3,970 $\pm$ 1,100
10^{-5}	820 $\pm$ 310	4,670 $\pm$ 770
10^{-4}	400 $\pm$ 70	2,870 $\pm$ 350
3 x 10^{-4}	650 $\pm$ 40	5,440 $\pm$ 50
10^{-3}	620 $\pm$ 110	2,440 $\pm$ 170
Con A	19,170 $\pm$ 210	30,310 $\pm$ 1,110

[a]4 x 10^5 viable CBA/CaJ thymocytes or splenic T cells were cultured in 0.1 ml medium containing 5% FCS together with incremental concentrations of 8MGuo.

[b]1 μCi [3H]TdR/well was added after 24 hr of incubation and cultures were harvested 24 hr later. Results are expressed as the arithmetic mean of five replicate cultures $\pm$ S.E.

Table 2. Growth Stimulatory Effect of 8MGuo on Thymocytes Pre-Stimulated with Interleukin-2

8MGuo [M][a]	[3H]TdR Uptake (cpm/culture)[b]	
	-IL-2	+IL-2
none	420 $\pm$ 140	1010 $\pm$ 350
10^{-5}	370 $\pm$ 130	1720 $\pm$ 280
10^{-4}	290 $\pm$ 20	2110 $\pm$ 180
10^{-3}	340 $\pm$ 90	7300 $\pm$ 800

[a]5 x 10^5 CBA/CaJ thymocytes were pre-stimulated with 0.5 U/ml rat IL-2, or cultured in medium as a control, and 24 hr later 8MGuo was added as shown.

[b]One μCi/well [3H]TdR was added at 48 hr, and cultures harvested 24 hr later. Results are expressed as in Table 1.

Table 3. Enhancement of Alloantigen-Induced Thymocyte Proliferation
by 8MGuo

Responder Cells[a]	8MGuo [M]	$[^3H]$TdR Uptake (cpm/culture)[b]	
		- stimulators	+ stimulators
CBA/CaJ	none	680 + 100	6190 + 860
	10^{-5}	460 + 50	5590 + 570
	3×10^{-5}	610 + 30	8250 + 840
	10^{-4}	460 + 30	8340 + 290
	3×10^{-4}	480 + 60	18,640 + 840
	10^{-3}	770 + 60	19,950 + 1110

[a] 1.5×10^5 viable CBA/CaJ thymocytes were cultured with or without an
equal number of 2500 r irradiated BDF_1 spleen cells with incremental
concentrations of 8MGuo together with 5 units rat IL-2.

[b] One μCi $[^3H]$TdR/well was added at 72 hr, and cultures were harvested 24
hr later. Results are expressed as the arithemetic mean of 5 replicate
cultures + S.E.

appropriate controls demonstrated that this was not due to an effect of the
nucleoside on either the thymocytes alone (i.e., Table 1) or on the
irradiated stimulator population. In a second set of experiments, spleen
cells from SJL mice (whose B cells are hyporesponsive to the mitogenic but
not the adjuvant activity of 8MGuo[4]) were stimulated with irradiated
allogeneic cells in the presence of incremental amounts of 8MGuo (Table 4).
This response was similarly enhanced by the nucleoside in a dose-dependent
manner.

Inability of 8MGuo to induce lymphokine production by T cells. We next
investigated the capacity of 8MGuo to induce T cell function, in this case
the elaboration of lymphokines. In experiments designed to determine the
IL-2 content of 8MGuo-stimulated T cell supernatants, we found that in
contrast to concanavalin A, 8MGuo was unable to induce production of this
activity in T cell-enriched populations (Table 5). This was also true for
supernatants produced from unseparated spleen cells, whether activity was
assayed after 24 or 48 hrs of culture (not shown).

Effect of 8MGuo on generation of helper T cells. To evaluate the
effect of 8MGuo on induction of antigen-specific helper T cells, low dose
SRBC priming with or without 8MGuo was carried out in vivo. These educated
spleen cells were irradiated, washed extensively, and cultured with
unprimed B cells in the presence of optimal numbers of SRBC. Anti-SRBC
responses were determined 4-5 days later. The resultant data indicate
clearly that 8MGuo augments either the production or the functional
activity of radioresistent helper T cells in vivo (Table 6).

Table 4. Enhancement of Alloantigen-Induced Proliferation by 8MGuo in SJL Spleen Cells

Responder Cells[a]	8MGuo [M]	[3H]TdR uptake (cpm/culture)[b]	
		− stimulators	+ stimulators
SJL	none	2810 ± 240	$13,360 \pm 450$
	3×10^{-5}	2030 ± 130	$15,610 \pm 610$
	10^{-4}	3090 ± 180	$19,200 \pm 920$
	3×10^{-4}	2740 ± 330	$19,350 \pm 1670$
	10^{-3}	2300 ± 100	$43,440 \pm 1520$

[a]1.5×10^5 viable SJL spleen cells were cultured with or without an equal number of 2500 r irradiated CBA/CaJ spleen cells with incremental concentrations of 8MGuo without supplemental IL-2.

[b]Conditions of labeling and expression of data are as in Table 3.

Table 5. IL-2 Content of 8MGuo-Stimulated T Cell Supernates

Supernate Dilution	[3H]TdR uptake by CTLL cells incubated with:[a]		
	Rat Con A spnt[b]	Control T cell spnt[b]	8MGuo T cell spnt[b]
		(cpm/culture)	
No spnt	718 ± 165[c]	718 ± 165	718 ± 165
1:32	ND	ND	386 ± 12
1:16	ND	ND	336 ± 14
1:8	ND	ND	493 ± 52
1:4	ND	381 ± 36	392 ± 16
1:2	$4,758 \pm 310$	488 ± 75	404 ± 23

[a]10^3 viable CTLL cells were cultures in a volume of 0.2 ml medium containing 5% FCS and 5×10^{-5}M 2-ME.

[b]Supernatants, generated as described in Materials and Methods, were added at onset of culture, yielding the final dilution shown.

[c]One μCi [3H]TdR/well was added after 24 hr of incubation. Cultures were harvested 24 hr later. Results are expressed as the arithmetic mean of five replicate cultures $\pm$ S.E.

Table 6. Modulation of T Helper Activity by 8MGuo

| Helper cells[a] | | | B cells | SRBC | Direct anti-SRBC PFC per culture[b] |
SRBC primed	8MGuo	2500r			
+	none	+	+	+	108 ± 6
+	+	+	-	+	4 ± 2
+	+	+	+	+	1231 ± 141
-	+	+	+	+	18 ± 3

[a]CBA/CaJ mice were immunized with 5 x 10^5 SRBC i.p. followed by an injection of CMC containing 0 or 25 mg 8MGuo in suspension. Four days later spleens were removed and cells from each group were pooled and irradiated with 2500 rads. 5 x 10^6 of these cells were cultured with 4 x 10^6 viable B cells, freshly prepared from naive animals, together with 2 x 10^6 SRBC in 1 ml medium containing 5% FCS.

[b]Direct PFC to SRBC were determined 4 days later. Results are expressed as the arithmetic mean of quadruplicate cultures ± S.E.

DISCUSSION

The data in the current report demonstrate that the C8-substituted guanine ribonucleoside 8MGuo augments proliferation and differentiation of murine T cells only in the presence of other stimulatory signals. In the absence of antigen or a nonspecific signal, comparable immunoenhancement is not observed. Thus, whereas 8MGuo is not mitogenic for thymocytes, it can alter the mitogenic response of thymocytes to IL-2, in cells pre-activated with this lymphokine. Similarly, the proliferative responses of thymocytes or SJL splenocytes to allogeneic stimulator cells can be augmented by this substituted nucleoside. Because mixed thymic (but not splenic) lymphocyte cultures are known to be deficient in IL-2 production[7], the data suggest that susceptibility to the substituted nucleoside is not induced by allogeneic cells acting alone, but requires a cooperative interaction between this signal and IL-2. The data do not speak to the question of whether enhancement of the MLR is due to a direct effect on T cells, or is indirect, as by increasing Ia antigen expression on antigen-presenting cells. The data on augmentation of IL-2 stimulation of thymocytes indicate that 8MGuo may act directly on T cells. However, it is also possible that the MLR is simply a means of generating IL-2 production in culture and that co-stimulation occurs largely by the same mechanism in both cases.

The implication of the response pattern of T cells to 8MGuo is that resting T cells should be uniformly unresponsive to the C8-substituted guanine ribonucleosides, and that such cells must first be activated in order to be responsive to any signal induced by these agents. This was found to be true of T cell function, as well. That is, 8MGuo is unable to cause unstimulated T cells to secrete IL-2, nor does it induce B cell differentiating activity that can restore antigen responsiveness to T-depleted spleen cells (data not shown). Although it does not induce production of T cell-derived lymphokines, 8MGuo does augment differen-tiation of T cells to mature, antigen-specific helper cells. When 8MGuo is administered in vivo shortly after immunization, it enhances the generation of radioresistant functional T helper cell activity. It is currently unknown whether this effect is attributable to clonal expansion of helper cells, recruitment of precursor helper cells, or enhanced efficiency of the helper cells already present.

Mechanistic considerations aside, the present results suggest that 8MGuo can affect both proliferative and differentiative T cell pathways as an adjunct, but not as an inducer. This situation contrasts markedly with that which pertains to normal murine B cells, in which 8MGuo is able to induce both B cell proliferation and differentiation in and of itself. The effect of 8MGuo on T cells is in fact more similar to the situation for B cells in the SJL mouse, in which neither 8MGuo nor 8BrGuo is able to induce early (i.e., inductive) signals. However, in the presence of other, early acting (i.e., antigenic) signals, these nucleosides are capable of exerting adjunctive effects on SJL B cells later in the culture period, resulting in marked enhancement of the primary antibody response in vitro.

ACKNOWLEDGEMENTS

This is publication no. 4032IMM from the Department of Immunology, Scripps Clinic and Research Foundation, La Jolla, CA. This work was supported in part by U.S.P.H.S. grants AI15284 and AI07007, and Biomedical Research Support Grant RRO-5514. Dr. Goodman is a recipient of U.S.P.H.S. Research Career and Development Award AI00374. We thank Lisa Martin, Pat Mullen, Maria Kropp, Pat McLoughlin and Mrs. Terry Else for superb technical assistance, and Alice Bruce for excellent secretarial work.

REFERENCES

1. M. G. Goodman and W. O. Weigle, Activation of lymphocytes by brominated nucleoside and cyclic nucleotide analogues: Implications for the "second messenger" function of cyclic GMP, Proc. Natl. Acad. Sci. USA 78:7604 (1981).
2. M. G. Goodman and W. O. Weigle, Intracellular lymphocyte activation and carrier-mediated transport of C8-substituted guanine ribonucleosides, Proc. Natl. Acad. Sci. USA 81:862 (1984).
3. M. G. Goodman and W. O. Weigle, Activation of lymphocytes by a thiol-derivatized nucleoside: Characterization of cellular parameters and responsive subpopulations, J. Immunol. 130:551 (1983).
4. M. G. Goodman, Inductive and differentiation signals delivered by C8-substituted guanine ribonucleosides, Immunol. Today 5:319 (1984).
5. J. M. Fidler, E. M. McDaniel, and E. S. Golub, Regulation of the immune response. III. Effect of the "accelerated" response on Hapten-carrier responses, Cell. Immunol. 4:29 (1972).
6. N. K. Jerne and A. A. Nordin, Plaque formation in agar by single antibody-producing cells, Science 140:405 (1963).
7. H. Wagner, M. Rollinghoff, K. Pfizenmaier, C. Hardt, and G. Johnscher, T-T cell interactions during in vitro cytotoxic T lymphocyte (CTL) responses, J. Immunol. 124:1058 (1980).

ANTIGEN-SPECIFIC ENHANCEMENT OF THE HUMAN ANTIBODY

RESPONSE BY A SUBSTITUTED NUCLEOSIDE

Michael G. Goodman and
William O. Weigle

Department of Immunology
Scripps Clinic and Research Foundation
La Jolla, CA

INTRODUCTION

Studies from this laboratory have shown that bromination of guanosine endows it with the capacity to induce murine B cells to undergo blast transformation and high level synthesis and secretion of polyclonal immunoglobulin[1,2]. This and certain other C8-substituted guanine ribonucleosides have been demonstrated to transmit T cell-like signals to B cells, allowing the generation of specific antibodies to a T cell-dependent antigen in the absence of T cells. This type of signal is distinct from signals provided by T cells, and in fact displays a strong synergistic interaction with T cell-derived lymphokines[3]. Furthermore, the same nucleosides cause marked augmentation of the magnitude of primary antibody responses both _in vitro_[4] and _in vivo_[5] in the murine model.

Although 8BrGuo is unable to induce B cell proliferation consistently in the human system, it has been found to induce B cells to secrete high levels of immunoglobulin in the absence of a specific eliciting immunogen[6]. The ability of these substituted guanosine compounds to mediate immunologic activity in human lymphoid cells, together with their effects in the murine model, prompted us to investigate their ability to modulate human antibody responses in the presence and absence of lymphokines.

MATERIALS AND METHODS

Materials. 8-mercaptoguanosine was purchased from the Sigma Chemical Co., St. Louis, MO. 7 methyl-8-oxoguanosine was synthesized and generously provided for us by Dr. Roland K. Robins at Brigham Young University. Human IL-2, lot 1464-52, was obtained as a partially purified preparation from Electro-Nucleonics, Inc., Silver Spring, MD. This preparation was found to be free of γ-interferon activity. Human IL-2 (preparation DL-209) was a generous gift of the Cetus Corp. through Dr. David Mark, and was generated through recombinant technology, cloning the IL-2 gene from the Jurkat cell line. The IL-2 had the cysteine at position 125 converted to serine by _in vitro_ mutagenesis. The material received from Cetus was over 96% pure by SDS-page and reverse phase liquid chromatography and contained less than 0.3 ng endoxin/ml.

Cells. Human PBL were prepared from normal heparinized venous blood by ficoll diatrizoate density gradient centrifugation. PBL were depleted of suppressor T cells bearing the histamine type 2 receptor by adhering them to cimetidine-coated petri dishes (Seragen, Boston, MA) and recovering nonadherent cells by panning as described by Cavagnaro and Osband[7]. For convenience, these cells will be referred to as PBL. In some experiments T cells were depleted by single or double rosetting of PBL with neuraminidase-treated SRBC. The B cell populations generated contained about 1-2% OKT3[+] cells by immunofluorescence. Subsequently these populations were depleted of about 90% of their adherent cells by adherence to plastic.

Lymphocyte cultures. The tissue culture medium employed was prepared as described previously[4] except that it contained 10% fresh autologous heat-inactivated plasma. For evaluation of the primary humoral immune response to SRBC, lymphoid cells were cultured at 2×10^6 per ml in either 1.0 ml or 0.1 ml, containing 5×10^6 or 5×10^5 SRBC (respectively) as antigen together with IL-2 and/or derivatized nucleosides.

Assay of plaque-forming cells. Enumeration of PFC that secrete antibodies against SRBC was accomplished after 6 days of culture by a modification of the hemolytic plaque assay of Jerne and Nordin[8]. Cells were brought up in complete medium prior to plaquing, were plaqued in standard low-m_r agarose (Bio-Rad Labs, Richmond, CA), and were incubated in SRBC-absorbed guinea pig complement for 1 hour (rather than 1 1/2 hr) after the usual 1 1/2 hr incubation without complement.

RESULTS

To determine whether C8-substituted guanine ribonucleosides had the capacity to act as adjuvants for the human primary antibody response in vitro, the effects of several of these compounds on the primary antibody response to SRBC were evaluated. At optimal concentrations, 7m8oGuo consistently induced a high degree of enhancement of the underlying response to specific antigen, while 8MGuo induced only modest adjuvant activity (Table 1). The effect of 7m8oGuo was dose-dependent, with its greatest effect around 1 mM.

Table 1. Comparative Adjuvanticity of 8MGuo and 7m8oGuo

		Direct anti-SRBC PFC/10^6	
Antigen	Concentration	+8MGuo	+7m8oGuo
none	none	6 ± 3	6 ± 3
SRBC	none	28 ± 10	28 ± 10
SRBC	10^{-4} M	N.D.	60 ± 18
SRBC	3×10^{-4} M	22 ± 6	182 ± 50
SRBC	10^{-3} M	54 ± 14	582 ± 78

2×10^5 viable H_2R^- human PBL were cultured in a volume of 0.1 ml in the presence or absence of 5×10^5 SRBC with variable concentrations of 8MGuo or 7m8oGuo. All cultures contained 10% Electronucleonics IL-2. Direct PFC to SRBC were determined after 6 days of culture. Results are expressed as the arithmetic mean of 5 replicate cultures $\pm$ S.E.

Table 2. Antigen Specificity of the Primary Response to
Heterologous RBC and Enhancement by 7m8oGuo

		Direct PFC/10^6 against	
Antigen	7m8oGuo	SRBC	HRBC
none	none	1 $\pm$ 1	1 $\pm$ 1
none	1 mM	55 $\pm$ 16	37 $\pm$ 10
SRBC	none	123 $\pm$ 11	8 $\pm$ 3
SRBC	1 mM	583 $\pm$ 85	33 $\pm$ 3
HRBC	none	2 $\pm$ 1	63 $\pm$ 9
HRBC	1 mM	27 $\pm$ 5	270 $\pm$ 24

2 x 10^6 viable H_2R^- human PBL were cultured in a volume
of 1.0 ml medium containing 10% IL-2 in the presence of 5
x 10^6 SRBC, 5 x 10^6 HRBC, or no antigen. 7m8oGuo was
added to culture as shown. Direct PFC to both SRBC and
HRBC were determined for all cultures after 6 days of
incubation. Results are expressed as the arithmetic mean
of 5 replicate cultures $\pm$ S.E.

Antigen specificity of the responses as well as of nucleoside-
mediated immunoenhancement was demonstrated in experiments using HRBC as
a specificity control (Table 2). This antigen is particularly well
suited for use as such a control because it exhibits less than 1%
crossreactivity with SRBC[9]. Thus, SRBC elicited virtually no PFC
specific for HRBC and HRBC elicited virtually no PFC specific for SRBC.
The adjuvanticity of 7m8oGuo was restricted to the SRBC response when
SRBC was used as antigen, and likewise to HRBC when this antigen was
added to culture. These data also demonstrate that the degree of
immunoenhancement is far greater than the sum of the separate polyclonal
and antigen-specific responses.

To evaluate whether the adjuvant activity of 7m8oGuo is exerted at
the time of initial lymphocyte activation or during the course of an
ongoing immune response, the kinetic profile for addition of this
nucleoside was determined. Sensitivity to the adjuvant effect of 7m8oGuo
was observed even when cultures were supplemented with the substituted
nucleoside on day 3 of the 6 day culture period (Table 3). It therefore
appears that the adjuvant activity is not exerted on the initial events,
but rather on those occurring relatively late in the response.

The capacity of purified, recombinant IL-2 (rIL-2) to reconstitute
the in vitro PFC response of human PBL to SRBC was studied next to learn
if contaminant activities might have been responsible for the effects
observed. As increasing doses of rIL-2 are added, the response to SRBC
alone rises to a low peak between 10^2 and 10^3 units/ml. In the presence
of antigen and 1 mM 7m8oGuo, however, rIL-2 elicits a highly significant
increase in antigen-specific PFC. This result substantiates the
existence of synergy between IL-2 and C8-substituted nucleosides (Table
4).

To study the cellular target for the enhancing effects of the C8-
substituted nucleosides, experiments were conducted in which B cell-
enriched populations were cultured with a partially-purified preparation

Table 3. Kinetic Profile for Addition of 7m8oGuo

Antigen	Day of 7m8oGuo Addition	Direct anti-SRBC PFC/Culture
none	none	0 ± 0
SRBC	none	697 ± 191
SRBC	day 0	2017 ± 121
SRBC	day 1	2483 ± 139
SRBC	day 2	2317 ± 145
SRBC	day 3	2500 ± 332
SRBC	day 4	225 ± 57
SRBC	day 5	285 ± 56

2×10^6 viable H_2R^- human PBL were cultured in a volume of
1.0 ml in the presence or absence of 5×10^6 SRBC. All
cultures contained 10% Electronucleonics IL-2. At the
day shown, parallel sets of cultures were supplemented
with 1 mM 7m8oGuo. Direct PFC were determined 6 days
after initiation of culture. Results are expressed as
the arithmetic mean of triplicate cultures ± S.E.

Table 4. Ability of 7m8oGuo and Recombinant IL-2 to
Reconstitute the Human Antibody Response

Antigen	r-IL-2	Direct anti-SRBC PFC/10^6	
		−7m8oGuo	+7m8oGuo
SRBC	none	0 ± 0	20 ± 5
SRBC	10 u/ml	10 ± 8	140 ± 66
SRBC	100 u/ml	24 ± 13	166 ± 54
SRBC	300 u/ml	38 ± 26	400 ± 66
SRBC	1000 u/ml	25 ± 15	26 ± 9

2×10^5 viable H_2R^- human PBL were cultured in a
volume of 0.1 ml in the presence or absence of 5×10^5
SRBC and/or 1 mM 7m8oGuo with incremental concentrations
of recombinant human IL-2. Direct PFC to SRBC were
determined after 6 days of culture. Results are expressed
as the arithmetic mean of 5 replicate cultures ± S.E.

of IL-2 in the presence or absence of 7m8oGuo. Table 5 demonstrates
that this partially-purified lymphokine preparation cannot support a
significant antibody response to antigen by B cell-enriched populations
in its own right, but in the presence of 7m8oGuo a significant response
to SRBC is generated. The same pattern of responses is seen when
purified recombinant material is used as the source of IL-2 (Table 6).

Table 5. Ability of 7m8oGuo to Augment the Response
of Human B Cells (E⁻ Cells) to SRBC

7m8oGuo	Direct anti-SRBC PFC/10^6 culture cells	
	− SRBC[b]	+ SRBC[b]
none	46 ± 11	42 ± 4
10^{-4}M	104 ± 23	179 ± 15
3 x 10^{-4} M	158 ± 18	494 ± 33
10^{-3}M	292 ± 33	1513 ± 145

[a]2 x 10^5 viable E-receptor negative, adherent cell-
depleted human B cells were cultured in a volume of
0.1 ml in the presence of absence of 5 x 10^5 SRBC and
incremental concentrations of 7m8oGuo. All cultures
contained 10% Electronucleonics IL-2.

[b]Direct PFC to SRBC were determined after 6 days of
culture. Results are expressed as the arithmetic
mean of triplicate cultures ± S.E.

Table 6. Ability of 7m8oGuo and Recombinant IL-2 to
Reconstitute the Human B Cell Response to SRBC

		Direct anti-SRBC PFC/10^6	
Antigen	r-IL-2	−7m8oGuo	+7m8oGuo
SRBC	none	10 ± 6	13 ± 3
SRBC	100 u/ml	13 ± 3	440 ± 146
SRBC	300 u/ml	50 ± 20	1095 ± 235
SRBC	1000 u/ml	13 ± 3	227 ± 81

2 x 10^5 viable E-receptor-negative, adherent cell-
depleted human B cells were cultured in a volume of 0.1
ml in the presence or absence of 5 x 10^5 SRBC and/or 1
mM 7m8oGuo with incremental concentrations of recombinant
human IL-2. Direct PFC to SRBC were determined after 6
days of culture. Results are expressed as the arithmetic
mean of triplicate cultures ± S.E.

DISCUSSION

A system is described in which optimal augmentation of the human
antibody response to antigen requires that the B cell receive at least
three distinct signals: 1) the antigenic signal, 2) a T cell-derived
signal, and 3) a nucleoside-dependent signal. This latter signal is
unique in that it is induced by a low m.w. analogue of one of the fundamental building blocks of cellular nucleic acid, a compound that has been
demonstrated to traverse the cell membrane and trigger the cell from an
intracellular site in the murine system[10].

The primary antibody response of antigen-inexperienced PBL to SRBC is dependent upon the presence of the T cell-derived lymphokine, IL-2, in culture. In it absence, PBL fail to generate a significant primary antibody response to antigen. That IL-2 can support this response, however, does not mean that other lymphokines might not also be active in this regard. Indeed, a number of other groups have found it necessary to supplement their cultures with unorthodox additives in order to generate a specific PFC response in vitro[7,11]. The action of these agents is likely mediated by one or more lymphokines.

Enhancement of the primary antibody response in vitro is an antigen-dependent event. In the absence of antigen, nonspecific responses of much lower magnitude are observed. When cultures containing either SRBC or HRBC are supplemented with 7m8oGuo and IL-2, a large increment in the specific PFC response is observed. Antigen-dependency and specificity are reaffirmed by the finding that there is no such change in the PFC response to HRBC when SRBC is used as the eliciting antigen and vice versa; only the underlying polyclonal response, observed also in the absence of antigen, is seen. Consideration of the magnitude of responses involved shows clearly that enhancement does not simply represent the sum of antigen specific and nonspecific (polyclonal) responses.

Augmentation of the antibody response by 8MGuo or 7m8oGuo is highly dose dependent, with optimal effects usually observed at concentrations of approximately 1 mM. Our consistant inability to observe a significant antigen-specific response in populations of unseparated PBL suggests that the histamine released during blood collection and plasma preparation may activate T cells capable of suppressing this response. Removal of these cells appears to unmask the underlying response. The nucleoside-enhanced response, however, can be observed in unseparated cells, albeit of lesser magnitude.

7m8oGuo appears to exert adjuvanticity during the course of an ongoing immune response, insofar as it can be added to culture 3 days after antigen with retention of full activity. These kinetics are in accord with those which pertain in the murine model, suggesting that basically similar mechanisms may be operant in both cases.

Experiments with separated populations of B + accessory cells indicate that an in vitro adjuvant effect can be observed despite depletion of the vast majority of intact T cells. Supplementation of cultures with nucleoside and IL-2-containing preparations enable these cultures to generate antibody responses to antigen. This observation further suggests that the target of the nucleoside may not be the T cell, but rather the B cell or macrophage instead. The observation that B cell-enriched cultures do not contain many more PFC than H_2R^- cultures suggests that 7m8oGuo is not as effective as intact T cells at inducing potentially antigen-reactive B cells to generate antibody. In addition, the observation that IL-2 is effective in T cell-depleted cultures makes it unlikely that production of BCDF by T cells is the primary mechanism of action of IL-2. Although these cultures undoubtedly contain low numbers of residual T cells, these may well be too few to produce sufficient quantities of BCDF to synergize with 7m8oGuo. Further studies are required to resolve this question definitively.

Marked synergy occurs between IL-2 and 7m8oGuo acting on H_2R^- lymphoid cells in the presence of antigen. The identity of the lymphokine involved in synergy with the substituted nucleoside was verified in experiments using IL-2 of recombinant origin, because partially purified IL-2 preparations have been found to contain both B cell growth and B cell differentiation factors. A prerequisite for synergy is that the synergizing elements must, either directly or indirectly, act upon the same cells at some stage of the activation process. The most likely candidate for such a common target cell would be the B cell, with macrophages being somewhat less likely, since B cells bear antigen-specific surface Ig receptors, are the antibody producing cells, and in the murine system appear to be the direct target of action for C8-substituted guanine ribonucleosides. The target for IL-2 in this case is more difficult to resolve. IL-2 receptors have been identified on B cells by a number of investigators, and B cell reactivity to IL-2 has been demonstrated[12]. These observations suggest the possibility that in our system IL-2 might act directly upon B cells after initial activation by antigen. The finding that E⁻ populations remained susceptible to synergy between IL-2 and 7m8oGuo is consistent with this possibility. Several other possibilities, however, are also tenable. Thus, interaction of IL-2 with low numbers of contaminating T cells or with macrophages could conceivably mediate synergy in an indirect manner.

ACKNOWLEDGEMENTS

This is publication no. 4014IMM from the Department of Immunology, Scripps Clinic and Research Foundation, La Jolla, CA. This work was supported in part by U.S.P.H.S. Grants AI15284 and AI07007, and Biomedical Research Support Grant RRO-5514. Dr. Goodman is a recipient of U.S.P.H.S. Research Career Development Award AI00374. We thank Lisa Martin for superb technical assistance, Alice Bruce for excellent secretarial work, and Dr. R.K. Robins for his generous gift of 7-methyl-8-oxoguanosine.

REFERENCES

1. M. G. Goodman and W. O. Weigle, Activation of lymphocytes by brominated nucleoside and cyclic nucleotide analogues: implications for the "second messenger" function of cGMP, Proc. Natl. Acad. Sci. 78:7604 (1981).
2. M. G. Goodman and W. O. Weigle, Induction of immunoglobulin secretion by a simple nucleoside derivative, J. Immunol. 128:2399 (1982).
3. M. G. Goodman and W. O. Weigle, T cell-replacing activity of C8-derivatized guanine ribonucleosides, J. Immunol. 130:2042 (1983).
4. M. G. Goodman and W. O. Weigle, Derivatized guanine nucleosides: a new class of adjuvant for in vitro antibody responses, J. Immunol. 130:2580 (1983).
5. M. G. Goodman and W. O. Weigle, Manifold amplification of in vivo immunity by C8-derivatized guanine ribonucleosides in normal and immunodeficient mice, Proc. Natl. Acad. Sci. 80:3452 (1983).
6. V. Osundwa, C. J. Ledgley, C. M. Martin and H.-M. Dosch, Activation of human B lymphocytes by 8 bromoguanosine (8'BG), Fed. Proc. 42:409 (1983).
7. J. Cavagnaro and M. Osband, Successful in vitro primary immunization of human peripheral blood mononuclear cells and its role in the development of human-derived monoclonal antibodies, Biotechniques Jan./Feb.:30 (1983).
8. N. K. Jerne and A. A. Nordin, Plaque formation in agar by single antibody-producing cells, Science 140:405 (1963).
9. J. M. Fidler, E. M. McDaniel, and E. S. Golub, Regulation of the immune response. III. Effect of the "accelerated" response on hapten-carrier responses, Cell. Immunol. 4:29 (1972).

10. M. G. Goodman, and W.O. Weigle, Intracellular lymphocyte activation and carrier-mediated transport of C8-substituted guanine ribonucleosides, <u>Proc.</u> <u>Nat.</u> <u>Acad.</u> <u>Sci.</u> <u>USA</u> 81:862 (1984).

11. M. K. Hoffmann, Antigen-specific induction and regulation of antibody synthesis in cultures of human peripheral blood mononuclear cells, <u>Proc. Natl. Acad. Sci.</u> USA 77:1139 (1980).

12. M. Tsudo, T. Uchiyama, and H. Uchino, Expression of Tac antigen on activated normal human B cells, <u>J.</u> <u>Exp.</u> <u>Med.</u> 160:612 (1984).

ECTOENZYMES OF NUCLEOTIDE METABOLISM ON HUMAN LYMPHOID CELLS

Wolf Gutensohn and Juliane Rieger

Institut für Anthropologie und Humangenetik der
Universität München
D 8000 München 2, Fed.Rep.Germany

INTRODUCTION

Human lymphoid cells, e.g. normal peripheral blood lymphocytes, leukemic cells and permanent lymphoblastoid cell lines
show extreme variations in the expression of nucleotide degrading ectoenzymes[1,2,3]. Up to now this has mainly been studied
for ecto-5′-nucleotidase (5′-N). On the other hand, a common
expression of 5′-N and the common ALL-antigen (cALLA) frequent-
ly observed in certain forms of acute leukemias [2] is not reflected by a close association of these two surface markers on the
plasmamembrane level[4]. In search for components which might be
more closely connected with 5′-N we studied other ectoenzymes
on the surface of lymphoid cells.

RESULTS

General properties of the enzymes. The following enzymes
were characterized mainly on lymphoid B-cell-lines:
ATPase, ADPase and nucleosidediphosphate-kinase.
ATPase and ADPase are not influenced by either high concentrations of typical phosphatase substrates or by ouabain. Thus
they are clearly distinguished from unspecific phosphatases on
the one side and from Na/K-ATPase on the other. By their metal
ion requirement these enzymes can be defined as Mg^{++}/Ca^{++}-depen-
dent. Their activity is also not decreased by AOPCP, a typical
inhibitor of 5′-N and by polyclonal or monoclonal antibodies
against 5′-N.

When we first observed the nucleosidediphosphate-kinase
activity (NDP-kinase), the question was, whether this was a
separate enzyme or just a sidereaction of the ATPase. We favor
the first possibility:
- Kinetic constants, K_M and K_i resp. of ATP and ADP for the two
  activities differ considerably.
- The two activities are not expressed coordinately on two different B-cell-lines (Raji and BHG-83-1)

All three enzymes are true ectoenzymes by the following
criteria:
- Enzymes can be tested on intact cells with the impermeable

substrates given in the extracellular medium.
- Tests in variable mixtures of broken and viable cells and
  extrapolation to 100% viability gives intercepts on the ordinate, clearly demonstrating the ecto-activity.
- By using detergents only small amounts of cryptic activity
  are released.
- For ATPase we could show partial inhibition by treating intact
  cells with diazotized sulfanilic acid, an impermeable unspecific chemically modifying reagent.

All three enzymes show a rather wide substrate specificity.
Thus the ATPase is rather a nucleoside-triphosphatase. One can
compete for the ATP with dATP > GTP > UTP. Likewise the NDP-kinase
will accept diphosphates other than ADP, for example GDP > dADP >
UDP > CDP and it will accept other triphosphates like dATP, GTP
or UTP as phosphate donors.

All three enzymes can be solubilized from the membranes by
detergents. Nonionic detergents usually have a slightly stimulating effect, whereas ionic detergents show inhibition of enzyme activity. Further purification or separation of the enzymes
starting from a detergent extract have not been attempted.

To study microheterogeneities in membrane composition we
had previously used membrane shedding induced by sublytic concentrations of the detergent ET-12-H, a lysophosphatidyl-choline
analogue. For a null-cell-line (Nalm 1) we could demonstrate
that 5´-N is enriched in the shed membrane material[4]. When we
apply this technique to a B-cell-line (BHG-83-1) and measure the
activities of the other ectoenzymes, we see that shedding of
ATPase, ADPase and 5´-N follow a strictly parallel course
throughout the sublytic concentration range of ET-12-H. From
this we conclude that these enzymes are assembled in membrane
areas specifically prone to shedding under these conditions.

Expression of the enzymes on lymphoid cells. All three ecto
enzymes are found on human peripheral blood lymphocytes (PBL)
and to a lesser degree on granulocytes. Although activities of
ATPase, ADPase and NDP-kinase on PBL are low compared with some
B-cell-lines described below, they are still about one order of
magnitude higher than that of 5´-N on the same cells. NDP-kinase
is usually the most active of the enzymes

On the other hand there is no strict correlation in the expression of ATPase and 5´-N on the same cell. This was shown in
leukemic cells. If we compare cALLA-positive blasts of patients
with common acute lymphocytic leukemia with high or extremely
high 5´-N values on the one hand and cells of T-cell leukemias
with low or absent 5´-N on the other, we find about the same
level of ATPase activity in both groups in a range similar to
activities of normal PBL. Thus ATPase, unlike 5´-N, cannot be
used as a biochemical marker for the differential diagnosis of
subsets of acute leukemias.

When we studied expression of ATPase and 5´-N on various
permanent human lymphoblastoid lines, distinctions were also
not clearcut, although in general activities on T-cell-lines
are much lower than on most B-cell-lines. But there are excep-
tions. One interesting observation was made within the series
of B-cell-lines. In 8 lines established by in vitro infection
of normal PBL with EBV ATPase and 5´-N activities are usually
higher (up to 6 - 8 times) than in normal PBL. In lines derived
directly from Burkitt´s lymphoma (Raji, Daudi, BJAB) ATPase
and 5´-N are very low or even absent.

DISCUSSION

The description and characterization of additional ecto-enzymes of purine metabolism immediately raises the question about the normal physiological function of this enzyme machinery. This question will not be finally answered before our knowledge about possible physiological sources of extracellular purine nucleotides has beeb improved. Nevertheless, ATPase ADPase and 5´-N, do they represent a pathway for the complete breakdown of extracellular ATP to the reusable adenosine ? On the one hand there is indirect evidence from the shedding experiments for a close neighborhood of these enzymes in certain selected membrane areas. But looking at lymphoid cells we certainly do not see a pattern of correlated expression of all 3 enzymes throughout. The best example for this are the blast cells in different forms of acute leukemias. But even in normal PBL the absolute activities of ATPase and ADPase on one side and of 5´-N on the other are so different, that an even flow of substrates through the whole chain of enzymes is difficult to imagine. So most probably the enzymes have independent functions. The possible function of an ecto-NDP-kinase is even less clear.

Recent results from EBV-research have led to the conclusion that cell-lines established by in vitro infection with the virus are different from those derived from Burkitt´s lymphoma, although both are EBV-positive. In a first stage of infection immortalization of the cells is induced. However, additional transformations accompanied or caused by gene rearrangements are required to achieve the final stage of malignancy as represented by the Burkitt´s lymphoma or cell-lines derived from it. In this context consistent phenotypical differences between these B-cell-lines are of interest. If we are able to confirm our results on the ectoenzymes with additional cell-lines this would represent such a phenotypical difference on the cell surface, although the meaning of this is totally obscure.

ACKNOWLEDGEMENTS

This work was supported by the Deutsche Forschungsgemeinschaft (Grant No. Gu 123/13)

REFERENCES

1. L.F.Thompson, A.Saxon, R.D.O´Connor and R.I.Fox, Ecto-5´-nucleotidase activity in human T cell subsets. Decreased numbers of ecto-5´-nucleotidase positive cells from both OKT 4+ and OKT 8+ cells in patients with hypogammaglobulinemia, J.Clin.Invest. 71:892 - 899 (1983)

2. W.Gutensohn, E.Thiel and B.Emmerich, Evaluation of 5´-nucleotidase as biochemical marker in leukemias and lymphomas, Klin.Wochenschr. 61:57 - 62 (1983)

3. W.Gutensohn, L.G.Gürtler, W.Siegert, E.Eichler and S.Ernst, Ectoenzymes on the surface of cells from human lymphoblastoid lines. 5´-Nucleotidase and phosphatase, Blut 41:411 - 420 (1980)

4. W.Gutensohn, B.Wünsch and H.Rodt, Independent expression of the surface markers 5´-nucleotidase and cALLA on leukemic cells, Blut 46:199 - 207 (1983)

SEPARATE MECHANISMS FOR CELLULAR UPTAKE OF PURINE NUCLEOTIDES BY

B- AND T-LYMPHOBLASTS

N. Lawrence Edwards, Annette M. Zaytoun, and
Gail A. Renard

Department of Medicine, University of Florida
and Veterans Administration Medical Center
Gainesville, Florida, USA

INTRODUCTION

Cultured B- and T-lymphocytes have multiple plasma membrane enzymes capable of cleaving extracellular phosphate-containing molecules. These enzymes include a specific 5'-nucleotidase (5'-N, EC 3.1.3.5) whose substrates are primarily purine 5'-nucleoside monophosphates and non-specific phosphatase(s) (NS-phos, EC 3.1.3.1) which has (have) questionable phosphorolytic activity for purine compounds under physiologic conditions. In a tissue culture system, 2 B-cell and 2 T-cell lines were compared for control growth or killing in the presence of 2 members of the cytotoxic 6-mercaptopurine family:6-MPR (nucleoside) and 6-MPRP (nucleotide).

METHODS

The B- and T-lymphoblast cell lines selected for this study had the

TABLE 1.   ECTO-5'-NUCLEOTIDASE AND NON-SPECIFIC PHOSPHATASE IN INTACT CULTURED CELLS

| | B-Lymphoblasts | | T-Lymphoblasts | |
	MGL-8	IM-9	CEM	Molt-4
Ecto-5'-Ampase (5'-N)	33.8 ± 9.5*	18.4 ± 2.1	0.2 ± 0.2	0.2 ± 0.1
% inhibition by 100 μM AMPCP**	100%	100%	100%	100%
Ecto-pNPPase (NS phos)	66.4 ± 36.3*	38.7 ± 9.0	10.6 ± 2.9	17.7 ± 3.4
% inhibition by 100 μM AMPCP	0	0	0	0

*Activity is expressed as nmol/hr · 10^6 cells
**AMPCP = adenosine-α, β-methylene diphosphate

characteristic 5'-nucleotidase (ecto 5'-N) profile—specifically, B-cells
have high ecto-5'-N activity while T-lymphoblasts have low or undetectable
levels. Both B- and T-lymphoblasts have substantial plasma membrane nonspecific phosphatase (ecto-pNPP'ase) activity (Table 1).

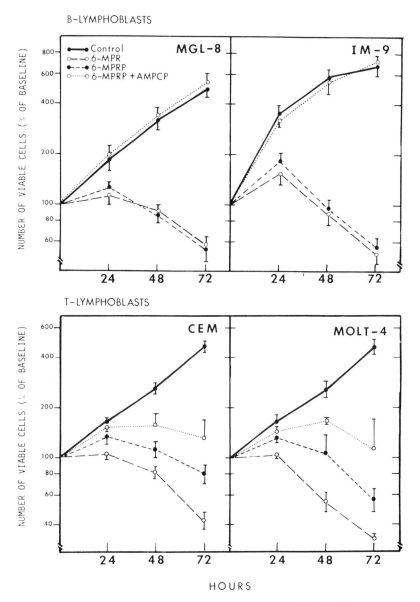

Figure 1. 6-MPR and 6-MPRP toxicity in B- and T-lymphoblasts and the
effect of the specific 5'-nucleotidase inhibitor, AMPCP, on blocking
6-MPRP-induced cytotoxicity.

Cell culture experiments were carried out in RPMI-1640 and 10% fetal calf serum. The fetal calf serum had previously been chromatographed against 5'-AMP sepharose to remove bovine 5'-N activity. Other culture conditions included 40 µM 6-MPR (nucleoside), 40 µM 6-MPRP (nucleotide), 100 µM AMPCP (5'-N inhibitor) and 40 µM 6-MPRP plus 100 µM AMPCP. Incubations were carried out for 72 hours at 37°C.

RESULTS

Control growth was comparable for B- and T-lymphoblasts over 72 hours (Figure 1). 6-MPR and 6-MPRP were equally toxic to B-cells. The addition of the specific 5'-N inhibitor, AMPCP, completely reversed 6-MPRP-induced cytotoxicity but had no effect on 6-MPR toxicity (data not shown). T-lymphoblasts demonstrate delayed sensitivity to 6-MPRP when compared to 6-MPR. AMPCP again had no effect on 6-MPR toxicity and only partially reversed the 6-MPRP toxicity (Figure 1).

DISCUSSION

How purines gain entrance into cells is an area of considerable research interest. The cell surface purine enzyme, 5'-nucleotidase, may control the cellular uptake of adenosine and thus participate in a regulatory signal from the external milieu that drives intracellular responses. The importance of ecto-5'-N in this sequence has been questioned because of the ubiquitous presence of non-specific phosphatase on the cell surface (1, 2).

The cell model described here tests the requirement of ecto-5'-N for the hydrolysis of purine nucleoside monophosphates and the uptake of the resulting nucleosides. Using this model, T cells show delayed cytotoxicity to 6-MPRP (when compared to 6-MPR). This toxicity could theoretically be due to the small amount of AMPCP sensitive AMP'ase activity observed in T-cells. However, when AMPCP is added to the culture system, cytotoxicity is only partially reversed. This differs strikingly with the B-lymphoblast pattern where 6-MPRP toxicity is completely reversed by 5'-N inhibition with AMPCP. Possible explanations for these differences are currently being explored and include: a) the presence in T-cells of an AMPCP resistent nucleotidase or a phosphatase with different substrate specificities than those found in B-cells or (b) a unique T-cell membrane transport mechanism for nucleoside monophosphates not previously described.

REFERENCES

1. W. Gutensohn, L. G. Gurtler, W. Siegert, E. Eichler, and S. Ernst. Ectoenzymes on the surface of cells from human lymphoblastoid lines: 5'-nucleotidase and phosphatase. BLUT 41:411 (1980).
2. M. H. Jensen. Dephosphorylation of purine mononucleotides by alkaline phosphatase. Biochem Biophys Acta 571:55 (1979).

ECTO-5'-NUCLEOTIDASE CAN USE IMP TO PROVIDE THE TOTAL PURINE REQUIREMENTS

OF MITOGEN-STIMULATED HUMAN T CELLS AND HUMAN B LYMPHOBLASTS

Linda F. Thompson

Department of Immunology
Scripps Clinic and Research Foundation
La Jolla, California 92037

Ecto-5'-nucleotidase (ecto-5'-NT)* catalyzes the extracellular
dephosphorylation of purine and pyrimidine ribo- and deoxyribonucleoside
monophosphates to the corresponding ribo- and deoxyribonucleosides (1,2).
This enzyme is located on the external plasma membrane of subpopulations
of human lymphocytes (1). By histochemical stain, 10-25% of adult peripheral T cells and 40-70% of adult peripheral B cells express ecto-5'-NT
activity (3-5). The contribution of this enzyme toward meeting the total
purine requirements of either resting or dividing lymphocytes is unknown.
Similarly, it is unknown whether the relative contributions of purine salvage vs. purine synthesis de novo toward meeting the purine requirements
for replication and cellular metabolism are different in ecto-5'-NT+ and
ecto-5'-NT⁻ lymphocytes. The experiments described here were designed to
determine whether the catalytic activity of ecto-5'-NT is sufficient to
provide the total purine requirements of mitogen-stimulated peripheral T
cells or rapidly-dividing B lymphoblastoid cells from exogenous purine
nucleotides. Inosine 5'-monophosphate (IMP) was chosen as the extracellular purine nucleotide since all other purine nucleotides can be synthesized
from it.

MATERIALS AND METHODS

Isolation of T cells. Peripheral blood mononuclear cells were isolated
from heparinized peripheral blood by Ficoll-Hypaque density gradient centrifugation. T cells were isolated by one cycle of rosetting with neuraminidase-treated sheep red blood cells.

Mitogen stimulation. T cells were cultured at 1 x 10⁶ cells/ml in
25 cm² Corning tissue culture flasks in 6 ml of RPMI 1640 medium with serum-
free supplement (HB 101, Hana Biologics, Berkeley, CA) plus 2 mM glutamine,
1 mM sodium pyruvate, penicillin (100 U/ml), streptomycin (100 µg/ml), and
amphitericin B (0.25 µg/ml) at 37°C in an humidified atmosphere of 5% CO_2
in air. Serum-free medium was used because fetal bovine serum contains 5'-
nucleotidase activity which catalyzes the conversion of IMP to inosine
(HxR). On days 3, 4, 5, 6, and 7 after the initiation of the cultures with

*Abbreviations used in this paper: Ecto-5'-NT, ecto-5'-nucleotidase; IMP,
inosine 5'-monophosphate; HxR, inosine; Hx, hypoxanthine; TdR, thymidine;
Con A, concanavalin A; PRPP, phosphoribosylpyrophosphate.

0.6 μg concanavalin A/ml (Con A, Sigma, St. Louis, MO), and other additions as indicated, quadruplicate 0.2 ml aliquots of each culture were removed, placed in microtiter plates, and pulsed with 1.0 μCi of ^{3}H-thymidine (TdR, 70 Ci/mmol, ICN, Irvine, CA) for 4 hr. The cells were harvested on to glass fiber filters using a multiple automated sample harvester and the filters were counted using scintillation fluid. The results are presented as the mean cpm of ^{3}H-TdR incorporated in 4 hr by 2 x 10^5 cells.

Cell lines. The human B lymphoblastoid cell lines, WI-L2 and #1254, were the gifts of Dr. J. Edwin Seegmiller, University of California, San Diego. WI-L2 was established from the spleen of a patient with hereditary spherocytosis. The cell line #1254 is a spontaneous mutant of WI-L2 which lacks detectable ecto-5'-NT activity; it was isolated fortuitously during routine cloning of WI-L2. Both cell lines were grown in RPMI 1640 with a serum-free supplement as described above, and had a doubling time of approximately 36 hr.

Growth rate experiments. WI-L2 and #1254 were cultured in duplicate at 2 x 10^5 cells/ml with the indicated additions. On each of the next three days, duplicate aliquots of each culture were counted with a Coulter counter. The quadruplicate cell counts agreed within 20%; means were used for calculations and plotting the data. Relative growth rates were calculated by the following formula:

final cell count − 1
initial cell count

Ecto-5'-NT activity. Ecto-5'-NT activity was determined by measuring the ability of whole viable cells to convert ^{14}C-IMP (Amersham, Arlington Heights, IL, diluted to 6 μCi/μmole with carrier IMP) to ^{14}C-HxR as previously described (6).

RESULTS

Ecto-5'-NT activity of T cells, WI-L2 and #1254. Ecto-5'-NT activity in isolated T cells used for these experiments varied between 8.5 and 12.1 nmole/hr/10^6 cells. Ecto-5'-NT activity measured in logarithmically-growing cultures of WI-L2 and #1254 was 18.3 ± 3.5 nmole/hr/10^6 cells for WI-L2 and <0.2 nmole/hr/10^6 cells for #1254.

Ability of IMP, HxR, and Hx to support DNA and protein synthesis in mitogen-stimulated T cells treated with aminopterin or azaserine. Con A stimulated T cells showed a maximum rate of ^{3}H-TdR incorporation into DNA 3 days after the initiation of the culture in purine-free medium (215,213 ± 26,357 cpm/2 x 10^5 cells, Fig. 1). ^{3}H-TdR incorporation was inhibited by 85-95% in mitogen-stimulated T cells treated with 5 μM aminopterin plus 25 μM TdR during the period 3 to 7 days after the addition of Con A. In contrast, the addition of 25 μM TdR plus either 30 μM IMP, HxR, or Hx fully restored the rate of ^{3}H-TdR incorporation in the aminopterin-treated cultures to greater than that of the control cultures (513,823 ± 33,352 cpm, 441,344 ± 22,357 cpm, and 402,187 ± 15,236 cpm/2 x 10^5 cells, respectively). In the rescued cultures, however, maximum rates of ^{3}H-TdR incorporation were delayed to day 4 or 5. This delay in apparent maximal rates of DNA synthesis is most likely due to the inhibition of ^{3}H-TdR incorporation by cold TdR added to the cultures to overcome the simultaneous block in pyrimidine biosynthesis caused by aminopterin.

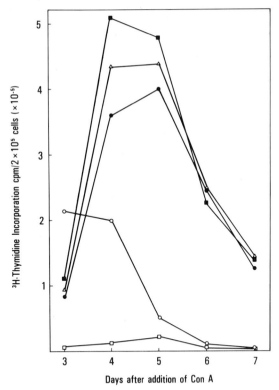

Figure 1. Ability of IMP, HxR, and Hx to support DNA synthesis in mitogen-stimulated T cells treated with aminopterin. Six ml cultures of human peripheral blood T cells at 1 x 10⁶/ml were stimulated to divide with Con A. At the indicated times, quadruplicate 0.2 ml aliquots were removed from each culture, placed in microtiter wells, and pulsed with 1.0 μCi ^{3}H-TdR for 4 hr. ^{3}H-TdR incorporation was measured by liquid scintillation counting after harvesting the cells on to glass fiber filters. The results are expressed as the mean cpm of ^{3}H-TdR incorporated/2 x 10⁵ cells. o, Control, Con A alone; □ , 5 μM aminopterin + 25 μM TdR; ■ , 5 μM aminopterin, 25 μM TdR, 30 μM IMP; Δ, 5 μM aminopterin, 25 μM TdR, 30 μM HxR; ●, 5 μM aminopterin, 25 μM TdR, 30 μM Hx.

Aminopterin inhibited ^{3}H-leucine incorporation into protein by 37% on day 3 in parallel cultures in leucine-free medium (Table 1). This inhibition was reversed completely by the addition of 25 μM TdR plus 30 μM IMP, HxR, or Hx to the cultures. The addition of aminopterin, TdR, and a purine source did not alter the kinetics of Con A-induced ^{3}H-leucine incorporation.

Table 1. Ability of IMP, HxR, and Hx to Support Protein
Synthesis in Mitogen-Stimulated T Cells Treated
With Aminopterin

Culture conditions	^{3}H-leucine incorporation cpm/2 x 10^5 cells
1. Control, −Con A	2,887 ± 154
2. + Con A alone	47,319 ± 2,241
3. Con A + 5 μM aminopterin	30,043 ± 4,043
4. Con A + 5 μM aminopterin + 25 μM TdR	30,202 ± 1,903
5. Con A + 5 μM aminopterin + 25 μM TdR + 30 μM IMP	55,870 ± 2,245
6. Con A + 5 μM aminopterin + 25 μM TdR + 30 μM HxR	53,116 ± 3,225
7. Con A + 5 μM aminopterin + 25 μM TdR + 30 μM Hx	56,334 ± 3,668

Similar experiments were done to evaluate the ability of IMP, HxR, and
Hx to reverse the azaserine-induced inhibition of ^{3}H-TdR incorporation in
Con A-stimulated human peripheral blood T cells. Azaserine alone inhibited
^{3}H-TdR incorporation by 91.5 ± 6.6% in five separate experiments (Table 2).
IMP, HxR, and Hx, at 30 μM, each partially restored the rate of ^{3}H-TdR
incorporation to 51.5 ± 27, 52.1 ± 16, and 45.3 ± 17%, respectively, of
that in the control cultures.

Table 2. Ability of IMP, HxR, and Hx to Support DNA
Synthesis in Mitogen-Stimulated T Cells
Treated With Azaserine

Culture conditions	^{3}H-thymidine incorporation % of response with Con A alone
1. Control, −Con A	0.3 ± 0.3
2. + Con A alone	100
3. Con A + 100 μM azaserine	8.5 ± 6.6
4. Con A + 100 μM azaserine + 30 μM IMP	51.5 ± 27
5. Con A + 100 μM azaserine + 30 μM inosine	52.1 ± 16
6. Con A + 100 μM azaserine + 30 μM hypoxanthine	45.3 ± 17

Inhibition of growth of WI-L2 and #1254 by aminopterin and reversal by
IMP. Parallel cultures of WI-L2 and #1254 were initiated at 2 x 10^5 cells/
ml in the presence and absence of 5 μM aminopterin plus 30 μM TdR. The
control cultures of WI-L2 and #1254 grew at virtually identical rates
(Figure 2) and the growth of both cell lines was inhibited by >95% by 5 μM

aminopterin (plus TdR). The growth of WI-L2 was completely restored to
control levels by the addition of 30 μM IMP (plus TdR) to the culture
medium (Figure 2A), while the addition of the same concentration of IMP
(and TdR) had virtually no effect upon the growth rate of #1254 in the
presence of aminopterin (Figure 2B). Even 300 μM IMP increased the growth
rate of #1254 to only 14% of control values (data not shown). For both
WI-L2 and #1254, the aminopterin-induced growth inhibition was completely
reversed by 30 μM TdR and either 30 μM HxR or 30 μM Hx.

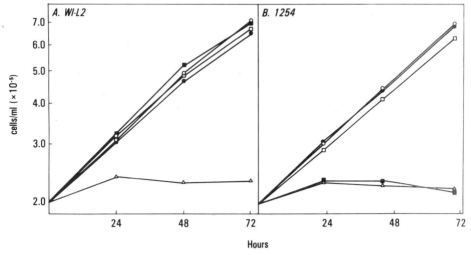

Figure 2.   Growth inhibition of the B lymphoblastoid cell lines, WI-L2 and
#1254, by aminopterin and reversal by thymidine and Hx, HxR, or IMP.
Duplicate 5 ml cultures of WI-L2 or #1254 were initiated at 2 x 10^5/ml.
The cells were counted with a Coulter counter at the indicated times; the
mean cell counts are plotted. A. WI-L2, B. #1254. o, control cultures,
no additions; Δ, 5 μM aminopterin + 30 μM TdR; ●, 5 μM aminopterin, 30 μM
TdR, 30 μM Hx; □ , 5 μM aminopterin, 30 μM TdR, 30 μM HxR; ■ , 5 μM aminopterin, 30 μM TdR, 30 μM IMP.

DISCUSSION

        When purine synthesis de novo is inhibited by aminopterin or azaser-
ine, DNA synthesis and cellular proliferation become dependent upon exogenous purines. Our results demonstrate that the activity of ecto-5'-NT is
sufficient to provide the total purine requirements of cultured human
lymphoid cells when IMP is utilized as the sole exogenous purine nucleotide. In the case of mitogen-stimulated human peripheral blood T cells
treated with aminopterin (and TdR to overcome the simultaneous block in
pyrimidine biosynthesis), inhibition of DNA synthesis as measured by [3]H-
TdR incorporation can be overcome equally well by either IMP, HxR, or Hx.
Rates of ^{3}H-TdR incorporation may not reflect true rates of DNA synthesis
in cultures treated with TdR and aminopterin, however, due to alterations
in intracellular dTTP pools. Therefore, a parallel set of experiments was
done where mitogen stimulation was monitored by ^{3}H-leucine incorporation
into protein. Similar results were obtained; i.e., the aminopterin-induced
inhibition of protein synthesis as monitored by ^{3}H-leucine incorporation
was completely reversed by the addition of 25 μM TdR plus 30 μM IMP, HxR or
Hx. Comparable results were also obtained in a third set of experiments in
which purine synthesis de novo, and thus DNA synthesis, were blocked with
azaserine, a compound which should have no direct effect on pyrimidine
biosynthesis. Therefore, it appears that the ecto-5'-NT activity of human

peripheral T cells is sufficient to supply the total purine requirements for proliferation in response to Con A stimulation.

Similar results were obtained when the ability of ecto-5'-NT to supply the total purines needed for proliferation was studied in WI-L2 and #1254, a pair of rapidly-dividing human B lymphoblastoid cell lines, one positive (WI-L2) and the other negative (#1254) for ecto-5'-NT activity. Proliferation was measured by cell count in these experiments. Aminopterin-induced growth inhibition of WI-L2 was completely reversed by the addition of 30 μM IMP (plus TdR) to the culture medium. However, 30 μM IMP had virtually no effect upon the growth inhibition of #1254 caused by aminopterin. Even 300 μM IMP increased the growth rate of #1254 to only 14% of control values. This modest increase in growth rate was probably due to the non-enzymatic breakdown of some IMP to HxR or to the low 5'-NT activity present in cell line #1254 itself. Since 30 μM HxR (plus TdR) completely reversed the aminopterin-induced inhibition of growth of #1254, the failure of 30 μM IMP (plus TdR) to do so, must be due to the cells' lack of ecto-5'-NT activity and not to subsequent blocks in the purine salvage pathway.

The results of these experiments show that the total purine requirements for mitogen-stimulated T cells or rapidly-growing B lymphoblastoid cells can be met by IMP through the catalytic activity of ecto-5'-NT. These experiments do not prove such a function for this enzyme in human T or B cells in vivo; they do, however, suggest that such a function is feasible, at least under specific conditions. Clearly, ecto-5'-NT activity is not essential for the growth of cells with a capacity for purine synthesis de novo, as shown by the fact that cell line #1254 grows at the same rate as WI-L2 in purine-free media in the absence of aminopterin. However, this enzyme may function in supplying the purines needed for nucleic acid synthesis and cellular metabolism under conditions where purine synthesis de novo is limited, such as in resting lymphocytes (7), or under conditions where extracellular nucleotides are available.

Extracellular nucleotides might accumulate in lymphoid tissues such as the thymus, bone marrow, and spleen where there is rapid cell turnover and massive cell death among specific lymphocyte subpopulations. In mice injected with ^{3}H-TdR, Feinendegen et al. (8) and Shortman et al. (9) have shown that 63-67% of the label incorporated into thymocyte DNA is salvaged and reutilized within the thymus. Since cortical thymocytes have low ecto-5'-NT activity while medullary thymocytes have four- to five-fold higher activity (10), perhaps the function of ecto-5'-NT in medullary thymocytes is to salvage purine (and pyrimidine) nucleotides released by the millions of dying cortical thymocytes. TdR reutilization was also estimated to be 43% in the bone marrow and 38-52% in the spleen (8,9). Thus, ecto-5'-NT could theoretically play a role in purine salvage in these tissues as well. The establishment of a function for ecto-5'-NT will clarify the significance of the marked increases in this enzyme activity which occur during the course of both T and B cell maturation (11,12) and give insight into how particular patterns of purine metabolism might be adapted for lymphocytes at specific stages of differentiation.

ACKNOWLEDGMENTS

L.F.T. thanks Adam Kaufman for cloning cell line #1254. The expert technical assistance of Julie Ruedi and secretarial expertise of Margaret Stone are also gratefully acknowledged. This work was supported by grant AI 18220 and the Biomedical Research Support Program grant RRO 5514 from the National Institutes of Health. This is publication no. 4018 IMM from the Research Institute of Scripps Clinic, La Jolla, CA 92037.

REFERENCES

1. F. Quagliata, D. Faig, M. Conklyn, and R. Silber, Studies on the lymphocyte 5'-nucleotidase in chronic lymphocytic leukemia, infectious mononucleosis, normal subpopulations, and phytohemagglutinin-stimulated cells, Cancer Res. 34:3197 (1974).
2. I.H. Fox, and P.J. Marchant, Purine catabolism in man: characterization of placental microsomal 5'-nucleotidase, Can. J. Biochem. 54:462 (1976).
3. N. Matamoros, D.A. Horwitz, C. Newton, L. Asherson, and A.D.B. Webster, Histochemical studies for 5'-nucleotidase and alpha-naphthyl (nonspecific) esterase in lymphocytes from patients with primary immunoglobulin deficiencies, Clin. Exp. Immunol. 36:102 (1979).
4. D.P. Recker, N.L. Edwards, and I.H. Fox, Histochemical evaluation of lymphocytes in hypogammaglobulinemia, J. Lab. Clin. Med. 95:175 (1980).
5. L.F. Thompson, A. Saxon, R.D. O'Connor, and R.I. Fox, Ecto-5'-nucleotidase activity in human T cell subsets, J. Clin. Invest. 71:892 (1983).
6. L.F. Thompson, G.R. Boss, H.L. Spiegelberg, I.V. Jansen, R.D. O'Connor, T.A. Waldmann, R.N. Hamburger, and J.E. Seegmiller, Ecto-5'-nucleotidase activity in T and B lymphocytes from normal subjects and patients with congenital X-linked agammaglobulinemia, J. Immunol. 123:2475 (1979).
7. R.B. Pilz, R.C. Willis, and G.R. Boss, The influence of ribose 5-phosphate availability on purine synthesis of cultured human lymphoblasts and mitogen-stimulated lymphocytes, J. Biol. Chem. 259:2927 (1984).
8. L.E. Feinendegen, H.J. Heiniger, G. Friedrich, and E.P. Cronkite, Differences in reutilization of thymidine in hemopoietic and lymphopoietic tissues of the normal mouse, Cell Tissue Kinet. 6:573 (1973).
9. D. McPhee, J. Pye, and K. Shortman, The differentiation of T lymphocytes. V. Evidence for intrathymic death of most thymocytes, Thymus 1:151 (1979).
10. D.D.F. Ma, A. Sylwestrowicz, A. Granger, M. Massaia, R. Franks, G. Janossy, and A.V. Hoffbrand, Distribution of terminal deoxynucleotidyl transferase and purine degradative and synthetic enzymes in subpopulations of human thymocytes, J. Immunol. 129:1430 (1982).
11. N.L. Edwards, E.W. Gelfand, L. Burk, H.-M. Dosch, and I.H. Fox, Distribution of 5'-nucleotidase in human lymphoid tissues, Proc. Natl. Acad. Sci. USA 76:3474 (1979).
12. J.F. Bastian, J.M. Ruedi, G.A. MacPherson, H.E. Golembesky, R.D. O'Connor, and L.F. Thompson, Lymphocyte ecto-5'-nucleotidase activity in infancy: increasing activity in peripheral blood B cells precedes their ability to synthesize IgG in vitro, J. Immunol. 132:1767 (1984).

CHANGES OF PURINE METABOLISM DURING DIFFERENTIATION OF RAT HEART MYOBLASTS

Mathias M.Müller,Helmut Rumpold,Gerhard Schopf,and
Peter Zilla

2nd Department of Surgery,Division of Clinical
Biochemistry
University of Vienna,Vienna,Austria

INTRODUCTION

During the last decade studies of cultured cells have become
a tool for biochemists and physiologists. Investigations in cultivated cardiac and skeletal muscle cells have proven to be good
models to study metabolism,drug effects,electrophysiological be-
haviour,and hormone responses of the heart and muscle tissue
(1-4).Most of the cardiac cells obtained from neonatal rat
hearts contain a mixed population of different cells (myocytes,
fibroblasts,and endothelial cells)(4).Therefore it seemed
reasonable to investigate a homogenous clonal cell line.We selected the cell line H9c2 (2-1) derived from embryonic rat heart
tissue.It has been reported that this cell line developed properties of skeletal muscle during differentiation in selective
medium(5).The aim of this study was to characterize this clonal
cell line with respect to purine metabolism and ultrastructural
features.Furthermore it should be investigated if differentiation
from embryonic heart cells (myoblasts) towards differentiated
cells(myocytes)using selective cultivation conditions(5,6)can
result in changes of purine metabolism.

MATERIALS AND METHODS

Cell Cultivation:Embryonic rat heart myoblasts H9c2(2-1)were obtained from the American Type Culture Collection.Cells were grown
in bicarbonate buffered Dulbeccos's minimum essential medium
(DMEM)supplemented with 10% fetal calf serum(Flow Laboratories,
U.K.)and 100mg/l gentamycine(Sigma,FRG)in an atmosphere of 10%
CO_2 and 90% air at 37°C.Differentiation of myoblasts towards
myocytes was performed in DMEM supplemented with only 2% FCS(5)
for one week.Metabolic and SEM studies were performed with monolayers of either myoblasts or myocytes cultivated for one week.
At the beginning of experiments the cells were cloned in order
to obtain a uniform population. The 5 resulting clones were tested
for their CK-and 5'-N activities as well as their ability to
differentiate towards myocytes.All other experiments were performed as triplicates with clone 2 which showed highest enzyme
activities,features of the CK-MB isoenzyme and produced

the highest rate of myocytes after the reduction of FCS to 2%
for one week.

Microscopy:Fixation and preparation of monolayers for scanning electron microscopy (SEM)were performed as described elsewhere (7).After the final rinsing procedure 1 cm^2 samples of
the bottom of the tissue culture flasks were cut out and transfered into a graded acetone series for dehydration.The specimens were dried by the critical point method (CO$_2$),mounted on
copper stubs and coated with 20 nm gold in a sputter coater
using sputter pulse control.Observations of myoblasts and
differentiated myocytes were made with a Jeol JSM T200 SEM.
(In vivo light microscopic investigations were done by means
of phase contrast device using a Zeiss IM 35 microscope.)

Metabolic Studies:Cell flasks containing 1.3 to 1.5x10^6
myoblasts or myocytes were incubated for 1 hour at 37°C with
5.2 μmol/l 8-^{14}C-labeled purine bases (adenine (62 Ci/mol),hypo-
xanthine (53 Ci/mol)) or purine nucleosides (adenosine (55 Ci/
mol),inosine (52 Ci/mol);Amersham International,U.K.))in 15 ml
DMEM supplemented either with 10% or 2% FCS.Following incubation
the monolayers were washed 3-times with 0.9% saline at 4°C,
trypsinized,resuspended in 0.5 ml 0.9% saline and deproteinized
with 40 μl 6 mol/l perchloric acid.After centrifugation the
supernatants were neutralized with 0.5 mol/l potassium hydroxide.
For determination of total purine uptake radioactivities of
supernatants were measured.To investigate the distribution of
radioactive label in purine nucleotides aliquots of neutralized
supernatants were chromatographed on silica gel plates GF 254
(Merck,FRG) with l-propanol/methanol/25% NH$_4$OH (45/15/40)
as solvent.Using a mixture of known standards,spots of nucleotides were identified under ultraviolet light and scraped off.
Their radioactivity was determined in a liquid scintillation
counter.
For the determination of enzyme activities and the concentration of adenine nucleotides monolayers having been cultivated
for one week were immediately subjected to freeze-drying.The
lyophilisate was either resuspended in 0.5 ml 0.9% saline or in
1 ml 0.5 mmol/l perchloric acid.Activities of adenine phosphoribosyltranferase (APRT),adenosine deaminase(ADA),adenosine kinase (AK),adenylate deaminase (AMP-DA),hypoxanthine guanine phosphoribosyltranferase (HGPRT),and purine nucleoside phosphorylase
(PNP) were determined with radiochemical tests (8-10).Activities
of creatine phosphokinase (CK)and 5′-nucleotidase (5′-N) were
measured with UV-tests (11,12)using comercially available kits
(Boehringer Mannheim,FRG,and Sigma,USA).The Paragon electro=
phoresis system (Beckman Instruments,USA) was used for the separation of CK isoenzymes on agarose gel strips.The electrophoretic pattern was visualized under UV-light after incubation with
fluorometric substrate.Protein was determined with Coomassie
Brilliant Blue (13).
Concentrations of adenine nucleotides and creatine phosphate
(CP) were determined in neutralized perchloric acid extracts
enzymatically by means of bioluminometric assays (14).

RESULTS

Clones of H9c2 (2-1) myoblasts:After the first recloning
procedure 5 different cell clones of myoblasts occured which
showed different activities of CK and 5′-N.4 clones proved
rather high enzyme activities,but poor augmentation in enzyme

476

activities when cultivated with 2% FCS for one week.Only one
clone developed 12-fold and 3-fold increases in CK and 5´-N
activities under these conditions.

Table 1.Characteristic features of H9c2(2-1) clones cultivated
in DMEM supplemented with 10% or 2% FCS for one week.
Enzyme activities:nmoles/10^6 cells/hour,mean values of
4 experiments.

Cell clone	Myoblasts 10% FCS		Myocytes 2% FCS		CK-Isoenzymes
	CK	5´-N	CK	5´-N	
1	865	1920	1110	2880	MM
2	384	1560	4740	4800	MM.MB
3	546	792	672	786	MM
4	708	754	851	885	MM
5	816	1324	956	1450	MM

This clone 2 showed differentiaton towards myocytes,whereas the
other 4 clones did not change their morphological features. Therefore clone 2 was chosen for all subsequent experiments.In order
to be sure that this clone was indeed of cardiac origin the CK-
isoenzyme pattern was determined by means of gel electrophoresis.
Homogenates obtained from rat skeletal muscle,heart and brain
tissue served as controls.Two distinct bands due to MM-and MB-
isoenzyme were detected in rat heart tissue and in the cells
derived from clone 2,whereas the other clones only contained
MM-isoenzyme.

SEM appearance of monolayers:Cells cultivated in 10% FCS for
one week(myoblasts):The monolayers of clone 2 myoblasts presented
a uniform picture of flat,polygonal cells which resembled the
cobblestone appearance of endothelial cells.Most of the cell
margins showed a slight degree of overlapping,but the cells
never lost their attachment to the tissue culture flask.The
average cell diameter varied between 75 and 140 micrometers.
The entire surface of each cell displayed a coarse granular
structure,except in the nuclear region.Sparse thin and short
microvilli were visible all over the cell membrane.The nuclear
region was always a sharply defined slightly ellipsoid smooth
area with an average diameter of 23 to 36 micrometers.There
were at least four and at most 7 nucleoli visible per cell.
Cells cultivated in 2% FCS for one week (myocytes):
Following the reduction of the FCS content of the medium,
approximately 60% of the cells became long and spindle-shaped
with a smooth and longitudinally-grooved surface. Most of these
cells divived into two or three main branches with some additional small cell processes.The vast majority were overlapping
and the surface differentiation of microvilli was only rarely
seen.No nuclear region could be identified.The average length
of these differentiated cells was between 190 and 630 micrometers.

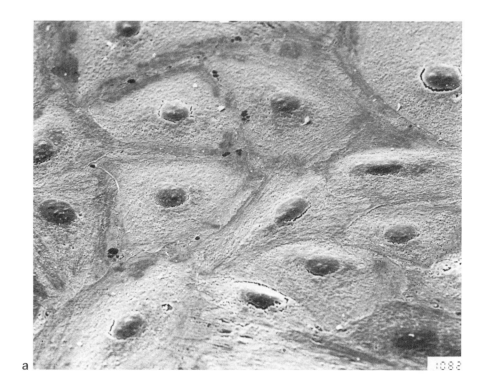

a

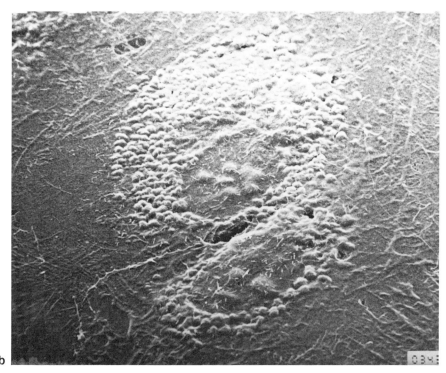

b

Figure 1.a/b: SEM appearance of H9c2(2-1)myoblasts cultivated in
DMEM supplemented with 10% FCS.
Enlargement:a)500x and b)1500x

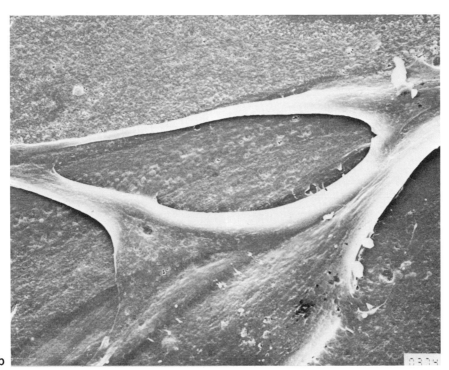

Figure 2.a/b:SEM apearance of H9c2(2-1)myocytes cultivated for
         one week in DMEM supplemented with 2% FCS
         Enlargement:a)500x and b)750x

Metabolic Studies:The contents of adenine nucleotides and CP in both cell populations obtained during cultivation in 10% or 2% FCS are shown in figure 3.There were no significant differences as far as adenine nucleotides were concerned. The concentrations of total adenine nucleotides in myoblasts and myocytes were 60.6 and 62.1 nmoles/10^6 cells with adenylate energy charges of 0.88 and 0.86, respectively. In contrast there was approximately a 2-fold increase of CP during differentiation towards myocytes.

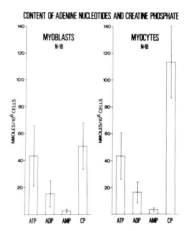

Figure 3.Content of adenine nucleotides and creatine phosphate (CP) in H9c2(2-1) myoblasts and myocytes.

Activities of enzymes involved in purine catabolism and salvage pathway of myoblasts and myocytes are summarized in table 2. Highest activities were obtained in myoblasts for 5′-N which increased during differentiation in 2% FCS 3-fold.AMP-DA also involved in the catabolism of adenine nucleotides showed approximately a 10-fold higher activity in myocytes than in myoblasts.The other enzyme activities measured exhibited only small increases in the cells during cultivation in 2% FCS.

Table 2. Enzyme activities of H9c2 (2-1) myoblasts and myocytes, nmoles/10^6 cells/hour, mean values and standard deviations of 6 experiments.

Enzyme	Myoblasts 10% FCS	Myocytes 2% FCS
CK	384 + 168	4740 + 1920
Purine Catabolism		
AMP-DA	165 + 74	2310 + 820
ADA	82 + 17	114 + 23
5'-N	1560 + 708	4800 + 2700
PNP	265 + 102	460 + 100
Purine Salvage		
AK	142 + 60	202 + 38
APRT	140 + 60	232 + 44
HGPRT	263 + 186	431 + 91

The extent of uptake and incorporation of labeled purine metabolites into intracellular adenine nucleotides of both cell populations is presented in table 3. During 1 hour of incubation the overall uptake of purine nucleosides into myoblasts was higher than that of purine bases. Adenosine was utilized at the highest rate (table 3). Thin layer chromatographic separation of label showed that adenosine, inosine, adenine, and hypoxanthine were incorporated into adenine nucleotides at a rate of 74%, 73%, 71% and 68%, respectively. The differentiation of myoblasts into myocytes was associated with an increase of the uptake which was most distinct for inosine and hypoxanthine. However, the distribution of intracellular label did not change dramatically. With adenosine, inosine, adenine, and hypoxanthine as substrates approximately 76%, 74%, 72%, and 69% of radioactivity was found in adenine nucleotides.

Table 3. Uptake of labeled purine bases and nucleosides into H9c2 (2-1) myoblasts and myocytes and their incorporation into adenine nucleotides (nmoles/10^6 cell/hour, mean values and standard deviations of 6 experiments).

Substrate	Uptake		Incorporation into adenine nucleotides	
	Myoblasts 10% FCS	Myocytes 2% FCS	Myoblasts 10% FCS	Myocytes 2% FCS
Adenosine	11.2 + 4.9	15.0 + 6.2	8.3 + 4.0	11.3 + 4.7
Inosine	3.7 + 2.3	6.7 + 4.2	2.7 + 0.6	5.0 + 3.1
Adenine	3.1 + 1.8	3.8 + 2.4	2.2 + 0.7	2.7 + 1.3
Hypoxanthine	1.6 + 0.6	3.1 + 1.0	1.1 + 0.4	2.1 + 0.7

DISCUSSION

The results of this study show that with selective culture medium the cell line H9c2(2-1),derived from embryonic rat hearts, can be differentiated from myoblasts towards myocytes within one week by reduction of 10% FCS to 2% FCS.It has been reported that the mononucleated myoblasts resemble both skeletal and cardiac muscle myoblasts(5).However,the clone investigated showed the CK-isoenzyme pattern unique for cardiac muscle.Furthermore,as an interesting observation,differentiated myocytes displayed one typical feature of cardiac muscle cells:the cell shape divided into two or three main branches.

It has been reported that the formation of myocytes with myotubes is associated with biochemical alterations indicating selective regulation of specific protein synthesis and enzymatic alterations during the phase of multiplication,fusion and differentiation of myoblasts in vitro(15).The morphological differentiation into myocytes was indeed accompanied by significant increases in the activities of CK,AMP-DA,and $5'$-N.This is in agreement with reports about large changes in activity of CK, adenylate kinase(5) and AMP-DA(16) during differentiation.It is evident that CK holds a predominant position in the maintenance of ATP levels in muscle tissue.The increase of CK-activity was accompanied by a 2-fold increase of CP content indicating a better energy state of myocytes and the availability of a larger energy pool.AMP-DA as well as $5'$-N are also important for the conservation of purine nucleotides in muscles through the operation of purine nucleotide cycles(17).In the contracting muscle AMP formed from ATP may be deaminated by AMP-DA to IMP,which is nondiffusable.This IMP-accumulation will to a fast restoration of depleted ATP(18).On the other hand $5'$-N hydrolyzes AMP and IMP to diffusable adenosine and inosine resulting in a loss of preformed purines.Since both nucleosides are vasodilatators their formation is of importance for the autoregulation of coronary blood flow.Though also a 3-fold increase of $5'$-N activity occure in myocytes the quotient AMP-DA : $5'$-N changed during differentiation from 0.11 to 0.48 indicating that in rat heart myocytes the flow in the interconversion of purine nucleotides is mainly from AMP to IMP as demonstrated with labeled formate in rat skeletal muscle cells(16,19).

The differentiation of myoblasts to myocotes forming myotubes was found to be associated with a marked increase in the overall uptake of purine bases and purine nucleosides.It was striking that there was only little difference to be found in the distribution of label between nucleotides,irrespectively from the subs trate used.In both cell precursors were mainly incorporated into adenine nucleotides at the same percentage.Furthermore the adeny late energy charge of both myoblasts and myocytes was nearly identical.These two facts demonstrate that cells under different environmental and functional conditions organize their metabolis to maintain the identical content of high energy phosphates.The increased uptake of purines as well as the increased rate of purine synthesis de novo(16)in myocytes might reflect a greater demand of precursors for ATP.However,activities involved in the reutilization of bases and nucleosides were only little changed indicating sufficient distribution of these enzymes within both types of muscle cells. An explanation of this phenomenon may be that the specific membrane proteins involved in the transport of purine bases and purine nucleosides(20)are different in myoblasts and myocytes.These proteins might be transformed during differe tiation showing different kinetic characteristics.Further inves tigations will be necessary to study the uptake processes and to determine the rate limiting step in the reutilization of preformed purines.

ACKNOWLEDGEMENTS

The authors gratefully acknowledge the technical assistance of Birgitta Kopf and Gabriele Stückler.This work was supported by a grant from"Jubiläumsfond der österreichischen Nationalbank" (project 2139)

REFERENCES

(1) C.Frelin,A.Pinson,P.Athias,J.M.Surville and P.Padieu,Glucose and palmitate metabolism by beating rat heart cells in culture, Pathol.Biol.27:45(1979)
(2) D.G.Wenzel,J.W.Wheatly,and G.D.Byrd,Effects of nicotine on cultured rat heart cells,Toxicol.Appl.Pharmacol.17:774(1970)
(3) W.C.Claycomb,Cardiac-muscle hypertrophy.Differentiation and growth of the heart cell during development,Biochem.J. 168:599(1977)
(4) N.A.Schroedl and C.R.Hartzell,Myocytes and Fibroblasts exhibit functional synergism in mixed cultures of neonatal rat heart cells,J.Cell.Physiol.117:326(1983)
(5) B.W.Kimes and B.L.Brandt,Properties of a clocal muscle cell line from rat heart,Exp.Cell Res 98:367(1976)
(6) D.Yaffe,Retention of Differentiation potentialities during prolonged cultivation of myogenic cells,Proc.Natl.Acad.Sci. USA 61:477(1968)
(7) D.Schroeter,A procedure for rupture-free preparation of confluently grown monolayers for SEM,J.Scan.El.Mic.Tech. 1:219 (1984)
(8) K.O.Raivio,P.Santavuori,and H.Somer,Metabolism of AMP in muscle extracts from patients with deficient activity of myoadenylate deaminase,Adv.Exp.Med.Biol.165B:431(1984)
(9) E.Kaiser and M.M.Müller,Enzymdefekte des Purinstoffwechsels, Z.med.Labor.-Diagn.20:3(1979)
(10) M.P.Uitendaal,C.H.M.M. de Bruijn, T.L.Oei,and P.Hösli,Ultramicrochemical studies on enzyme kinetics,Adv.Exp.Med.Biol. 76A:597(1977)
(11) G.Szasz,W.Gerhardt,and W.Gruber,Creatine kinase in serum: 4.Differences in substrate affinity among the isoenzymes, Clin.Chem.24:245(1978)
(12) C.L.M.Arkestei jun.,A kinetic method for serum 5′-nucleotidase using stabilised glutamate dehydrogenase,J.Clin.Chem. Clin.Biochem.14:155(1976)
(13) M.M.Bradford,A rapid and sensitive method for the quantitation of microgram quantities of protein utilizing the principle of protein dye binding,Analyt.Biochem.72:248(1976)
(14) R.J.Ellis and C.Gardner,Determination of high-energy phosphosphates using bioluminescence,Analyt.Biochem.105:354 (1980)
(15) F.Bacou and J.Nougués,Alterations of enzymatic activities during red and white muscle differentiation in vitro,Exp.Cell .Res.129:455(1980)
(16) E.Zoref-Shani,A.Shainberg,and O.Sperling,Alterations in purine nucleotide metabolism during muscle diferentiation in vitro,Biochem.Biophys.Res.Commun.116:507(1983)

(17) K.Thornheim and J.M.Lowenstein,The purine nucleotide cycle:
     Interactions with oscillations of glycolytic pathway in musc=
     le extracts,J.Biol.Chem.249:3241(1974).
(18) R.L.Sabina,I.L.Swain,B.M.Patten,T.Ashizawa,W.E.O´Brien,and
     E.W.Holmes,
     J.Clin.Invest.66:1419(1980).
(19) E.Zoref-Shani,A.Shainberg,and O.Sperling,Characterization of
     purine nucleotide metabolism in primary rat muscle cultures,
     Biochem.Biophys.Acta716:324(1982).
(20) M.M.Müller,M.Kraupp,P.Chiba,and H.Rumpold,Regulation of
     purine uptake in normal and neoplastic cells,Adv.Enzyme Reg.
     20:239(1982).

PRODUCTION AND DEGRADATION OF AMP IN CULTURED RAT SKELETAL AND HEART MUSCLE:

A COMPARATIVE STUDY

E. Zoref-Shani, A. Shainberg, G. Kessler-Icekson
and O. Sperling

Department of Chemical Pathology, Tel Aviv University
School of Medicine, Tel Hashomer, Rogoff-Wellcome Medical
Research Institute and Department of Clinical Biochemistry
Beilinson Medical Center, Petah-Tikva and Department of Life
Sciences, Bar-Ilan University, Ramat Gan, Israel

INTRODUCTION

The mechanisms operating for the maintenance of an optimal adenine
nucleotide pool in tissues are interrelated with the metabolism of AMP.
This compound is the parent adenine nucleotide molecule to be produced,
either de novo from nonpurine molecules, or by salvage from preformed
purines. At the same time it is also the substrate for adenine nucleotide
degradation to diffusable nucleosides and bases, which can efflux from the
tissue. The present study was undertaken to clarify the mechanisms operating in the skeletal and heart muscles to produce and degrade AMP. Primary
rat myotube and cardiomyocyte cultures were utilized as models. The activity of the various pathways was gauged in intact cells, under physiological conditions, employing labeled precursors. In addition some of the
key enzymes were assayed in cell extracts.

MATERIALS AND METHODS

Culture media were purchased from Grand Island Biochemicals (USA);
Radiochemicals from Amersham (UK) and fine chemicals from Sigma (USA).

Rat myotube and cardiomyocyte cultures were prepared as described
before (1, 2). Purine synthesis was gauged by the rate of (^{14}C)formate
incorporation into purines (1, 2). Incorporation of labeled purines into
cellular purines was studied in monolayer cultures at 37^{0}C, in fresh Eagle's
MEM growth medium, containing 1.5 μM of the radioactive purine precursor
(1, 2). Enzymes were assayed by radiochemical methods (3, 4).

RESULTS AND DISCUSSION

The cardiomyocyte and myotube cell cultures exhibited capacity for
de novo purine synthesis, indicated by incorporation of (^{14}C)formate into
purines. The rate of (^{14}C)formate incorporation into the cardiomyocyte

Table 1. Purine Nucleotide Synthesis de novo and by Salvage in Muscle Cell Cultures

Cell Culture	de novo nucleotide synthesis: (^{14}C)formate incorporation into total purines (pmol/mg protein/min)	Salvage nucleotide synthesis: Total metabolism of labeled purine precursor by respective salvage enzyme[a] (pmol/mg protein/min)		
		(^{14}C)Adenosine	(^{14}C)Adenine	(^{14}C)Hypoxanthine
Human skin fibroblasts	1.01 ± 0.09 (8)		125.0(4)	64.0(4)
Rat skeletal muscle myotubes	5.59 ± 0.18 (8)	87.4(5)	36.7(4)	6.7(4)
Rat cardiomyocytes	1.46 ± 0.29(13)	21.7(2)	35.0(3)	7.1(4)

[a] For adenosine-labelling in all cellular purines in presence of ADA inhibition, excluding adenosine. For adenine-labelling in all cellular purines excluding adenine. For hypoxanthine-labelling in nucleotides only.

Table 2. Activity of Enzymes of Purine Metabolism in
Extracts of Cultured Cardiomyocytes and Myotubes.

Enzyme	Enzyme Activity[a] (nmol/mg protein/min)		
	Human Fibroblasts	Rat Myotubes	Rat Cardiomyocytes
AMP deaminase	109	560	78
AMP nucleotidase	109	4	11
Adenosine kinase	2	2	7
Adenosine deaminase	19	19	14
IMP nucleotidase	36	1.5	4
Purine nucleoside phosphorylase	15	26	31
HGPRT	4	3	10

[a]Mean of 3 experiments

purines was markedly lower than that in the skeletal muscle myotubes
(Table 1). This finding is compatible with other studies suggesting
that the heart muscle has a relatively limited capacity to produce purine
nucleotides from nonpurine molecules (5, 6). On the other hand, whereas
in the myotubes only a small fraction of the IMP formed was converted to
AMP (ATP), in the cardiomyocytes, adenine nucleotides contained most of
the labelling derived from (^{14}C)formate (1, 2; see below). De novo purine
synthesis in both muscle cell cultures exhibited dependence on the metabolic availability of PRPP and of Rib-5-P. The dependence on PRPP was
indicated by the finding that activation of PRPP synthetase (by increasing the concentration of activator Pi in the incubation medium to 10 mM;
7), increased the rate of purine synthesis in both tissues by 2 to 3
fold. The dependence on the availability of Rib-5-P was indicated by
the finding that addition of 5 mM ribose to the incubation medium accelerated the rate of purine synthesis, in the myotubes by 2 to 3 fold and
in the cardiomyocytes by 1.5 to 2 fold. The dependence of de novo purine
synthesis on PRPP availability is common to many tissues (8,9) but that
on Rib-5-P appears to be characteristic to skeletal and heart muscles
only (7,9,10), probably reflecting the limited capacity of these tissues
to produce Rib-5-P by the oxidative pentose phosphate pathway (1,11).
    The synthesis of AMP by salvage pathways is of great advantage to
the tissue, in being markedly less costly in energy consumption than by
de novo. Indeed, the muscle cultures exhibited a marked capacity to produce purine nucleotides from adenosine, adenine and hypoxanthine. Moreover, this capacity for salvage nucleotide synthesis was found to be
several fold greater than that for the de novo pathway, as gauged by the
rate of incorporation of the purine precursor into nucleotides (Table 1).
The actual-physiological rate of salvage purine nucleotide synthesis in
the intact cell depends on the physiological availability of the purine
precursors. Presently there is no evidence for endogenous adenine formation in the skeletal or heart muscle cell nor on its supply in vivo from
other tissues. On the other hand, as indicated by the results of the
present study (see below), adenosine and hypoxanthine are formed endogenously

in both muscle tissues and therefore are available for nucleotide synthesis.
The results of the present study indicate a better capacity for AMP synthesis from adenosine than from hypoxanthine, but they do not allow quantitative estimation of the actual rate of physiological salvage nucleotide
synthesis in both muscle tissues. The data obtained may however be taken
to suggest that both adenosine kinase (AK) and HGPRT are saturated with
adenosine and hypoxanthine, respectively (see below). The capacity of the
muscle cells to synthesize and degrade AMP, was further evaluated by assay
(in cell extracts) of several enzymes associated with AMP metabolism (Table
2). In comparison to fibroblast extracts, both muscle cells contained
markedly lower nucleotidase activity (for both AMP and IMP), conforming
with nucleotide preservation. ADA activity was similar in all tissues
studies, but AMP deaminase activity was markedly higher in the myotubes.

The extent of activity of the nucleotide interconversion pathways
in the intact skeletal and heart muscle cells was gauged by tracing the
distribution of label from radioactive precursors among the various cellular nucleotides (1, 2). In the contracting skeletal muscle myotube cultures
the main flow was found to be from AMP to IMP, whereas in the cardiomyocytes, the main flow was found to be from IMP to AMP. The above difference
in the fate of AMP reflects the marked difference in the activity of AMP
deaminase between the two tissues. It could also reflect intensive operation of the nucleotide cycle in the myotubes (12). However, in light of
the results of the present study, of a low rate of flow from IMP to AMP
in these cells, the myotubes would have to contain a relatively large
IMP pool for this to be true. This has not yet been verified. The above
differences in the fate of AMP reflect also the rate of phosphorylation
of AMP to ATP (13, 14). In the cardiomyocytes, which possess better capacity for ATP production, a larger proportion of labelled AMP could be
converted to ATP. In contrast, in the myotubes with a less efficient
capacity for ATP production a larger proportion of labelled AMP was deaminated into IMP.

AMP production and degradation in the intact cardiomyocytes and
myotube cells was further studied by tracing the flow of radioactivity
from (^{14}C)adenine into total nucleotides, inosine, hypoxanthine, adenosine and adenine, under conditions at which adenosine deaminase (ADA),
AK, or both were inhibited (Tables 3, 4). ADA inhibition (5µM 2'-
deoxycoformycin (15); a gift from Warner-Lambert Company, Ann Arbor,
Michigan, USA), resulted in both tissues in increased labelling of the
adenosine fraction, associated with decreased labelling of inosine and
hypoxanthine (Tables 3, 4). These effects were markedly greater in the
cardiomyocytes than in the myotubes, indicating that in the cardiomyocytes, under the experimental conditions employed, a greater proportion
of AMP was degraded through adenosine than through IMP, whereas in the
myotubes the ratio between these pathways of AMP degradation was inverse.
In both cell types, the accumulation of adenosine, caused by ADA inhibition, did not increase the labelling in the nucleotide fraction, suggesting that AK is saturated with adenosine at the physiological availability
of this substrate. Inhibition of AK by addition of 50 µM 5'-amino-5'-
deoxyadenosine to the incubation medium (15), caused in the cardiomyocytes
a marked decrease in nucleotide labelling, associated with increased
labelling of the inosine and hypoxanthine fractions (Table 3). Adenosine
did not accumulate unless ADA was inhibited too. In contrast, in the
myotubes, these effects of AK inhibition were minor (Table 4). These
results may be taken to suggest that the futile substrate cycle AMP
adenosine is more active in the cardiomyocytes than in the myotubes.
This difference between the two tissues is probably the reflection of the
difference in AMP nucleotidase activity in the intact cells, being markedly
greater in the cardiomyocytes. Total adenosine formation from AMP could

Table 3. Distribution of Label from (^{14}C) Adenine among Cellular Purines in Cultured Cardiomyocytes.

Fraction labelled	Control (cpm/mg protein)	ADA inhibition[a]	AK inhibition[b]	ADA+AK inhibition
		% of control		
Total nucleotides	35174	109	54	55
Inosine	56697	24_{32}	122_{125}	16_{21}
Hypoxanthine	9459	79	142	51
Adenosine	3250	1470	87	1878
Adenine		undetectable		

Adenine concentration 1.5 μM; incubation 30 min.
[a] 2'-deoxycoformycin 5μM.
[b] 5'-amino-5'-deoxyadenosine 50μM.

Table 4. Distribution of Label from (^{14}C)Adenine among cellular Purines in Cultured Myotubes.

Fraction labelled	Control (cpm/mg protein)	ADA inhibition[a]	AK inhibition[b]	ADA+AK inhibition
		% of control		
Total nucleotides	66750	98	95	89
Inosine	24065	70_{74}	106_{115}	65_{72}
Hypoxanthine	13380	81	131	87
Adenosine	200	4300	100	5000
Adenine		undetectable		

Adenine concentration 1.5 μM; incubation 30 min.
[a] and [b] as in Table 3.

Total radioactivity in incubation medium, excluding adenine, amounted to about 1% of cellular radioactivity.

be assessed under simultaneous inhibition of ADA and AK. Under such conditions, accumulation of labelling from (^{14}C)adenine in adenosine was in
the cardiomyocytes approximately 6 folds that in the myotubes (Tables 3, 4).

In both muscle cell cultures, there was accumulation of labelled
inosine and hypoxanthine under physiological conditions, indicating low
activity of PNP and inefficient salvage of hypoxanthine to IMP. A similar
inefficient activity of HGPRT was reported recently also in the liver (16).
In both cell cultures, under physiological conditions, as well as under
conditions of ADA or AK inhibition, the release of labelled purines into
the incubation medium was found to be very low, ranging between 1-10% of
the cellular labelling. Thus, it seems, that inosine and adenosine do not
efflux freely from the muscle cell. The mechanism of this finding is not
yet clarified, but it could reflect binding of the nucleosides to proteins,
such as binding of adenosine to S-adenosylhomocystein hydrolase (17).
This finding has important implications concerning the conservation of
purines in the skeletal muscle during work and in the heart muscle during
hypoxia.

The study of adenosine metabolism, utilizing adenosine as substrate,
was encountered with difficulties, in view of our finding that the muscle
cultures excrete considerable ADA activity into the medium. A similar
finding was reported before for lymphocytes and lymphoblasts (4). Another
difficulty resulted from our finding that 5'-amino-5'-deoxyadenosine,
used to block intracellular AK activity, interferred with adenosine transport into the cells, preventing interpretations of its effects on adenosine
metabolism. Therefore, incubations in experiments with (^{14}C)adenosine as
substrate were shortened to 10 to 15 minutes and only the effect of ADA
inhibition was studied. The results of these experiments were compatible
with those with (^{14}C)adenine as substrate, as discussed above. In both
muscle cell cultures, inhibition of ADA resulted in accumulation of adenosine, associated with decreased labelling of the inosine and hypoxanthine fractions. The results indicate that in the cardiomyocytes, at
the low concentration of adenosine studied (1.5μ M), about 60% of the
adenosine metabolized in the cell was deaminated by ADA and 40% were
phosphorylated by AK. In contrast, in the myotubes under the same
conditions, only 12% of the adenosine was deaminated and about 88%
phosphorylated.

The myotubes, whether incubated with adenine or adenosine, did
not accumulate adenosine, unless ADA was inhibited. In contrast, the
cardiomyocytes, under the same experimental conditions, were found to
accumulate small amounts of adenosine. These results may be taken to
indicate that under the experimental conditions employed, adenosine
concentration in the cardiomyocyte was greater than that in the myotube,
presumably due to the greater rate of its production from AMP in this
cell. Adenosine, being a strong vasodilator, functions in the heart
as a local hormone, involved in the regulation of O_2 supply. However,
the mechanisms operating in the heart tissue for adenosine production
under hypoxic conditions, whether through increased AMP degradation, or
through inhibition of AK, are not yet fully clarified (12). Preliminary
experiments in our laboratory on the effect of hypoxic conditions on
adenine and adenosine metabolism in the cardiomyocyte suggest that under
such conditions adenine nucleotide degradation is accelerated through
both the IMP and adenosine pathways and that AK activity is inhibited.
These changes were accompanied by a 2-fold increase in the accumulation
of adenosine. The results of the present study indicate clearly that
the most effective mechanism for adenosine accumulation is ADA inhibition,
but if such inhibition takes part in the mechanism leading to adenosine
accumulation in severe hypoxia has not yet been clarified.

ACKNOWLEDGEMENT

Partially supported by the Recanti Foundation, Tel Aviv University (E. Zoref-Shani).

REFERENCES

1. Zoref-Shani, E., Shainberg, A. , Sperling, O. , Biochim. Biophys. Acta 716: 324-330 (1982).
2. Zoref-Shani, E., Kessler-Icekson, G., Wasserman, L., Sperling, O., Biochim. Biophys. Acta 804: 161-168 (1984).
3. Sidi, Y., Livne, P., Brosh, S., Pecht, M., Shilder, D., Pinkhas, J., Trainin, N., Sperling, O., Biomedicine and Pharmacotherapy 38: 455-458 (1984).
4. Snyder, F.F., Mendelsohn, J., Seegmiller, J.E., J. Clin. Invest. 58: 654-666 (1976).
5. Goldthwart, D.A., J. Clin. Invest. 36: 1572-1578 (1957).
6. Zimmer, H.G., Trendelenburg, C., Kammermeier, H., Gerlach, E., Circulation Res. 32: 635-642 (1973).
7. Boer, P., Lipstein, B., De Vries, A., Sperling, O., Biochim. Biophys. Acta 432: 10-17 (1976).
8. Zoref, E., Sivan, O., Sperling, O., Biochim. Biophys. Acta 521: 452-458 (1978).
9. Bashkin, P., Sperling, O., Biochim. Biophys. Acta 538: 505-511 (1978).
10. Brosh, S., Boer, P., Kupfer, B., De Vries, A., Sperling, O., J. Clin. Invest. 58: 289-297 (1976).
11. Krebs, H.A., Eggleston, L.V. in: "Advances in Enzyme Regulation" (Weber, G., Ed.), Vol. 12, pp. 421-434, Pergamon Press, Oxford, (1974).
12. Sabina, R.L., Swain, I.L., Patten, B.M., Ashizawa, T., O'Brien, W.E., Holmes, E.W., J. Clin. Invest. 66: 1419-1423 (1980).
13. Brachfeld, N. in: "New Horizons in Cardiovascular Diseases", vol. 1, part, 1 (Kone, R.J. editor) pp.65-128, Future Publishing Co., N.Y. (1980).
14. Berne, R.M., Rubio, R., Supplement III to Circulation Res., Vols. 34 and 35, : 109-120 (1974).
15. Newby, A.C., Holmquist, C.H., Illingworth, J., Pearson, J.D., Biochem. J. 214: 317-323 (1983).
16. Vincent, M.F., Van Den Berghe, G., Hers, H.G., Biochem. J. 222: 145-155 (1984).
17. Olsson, R.A., Saito, D., Stenhart, C.R., Circ. Res. 50: 617-626, (1982).

DUCHENNE MUSCULAR DYSTROPHY: NORMAL ATP TURNOVER IN CULTURED CELLS

Irving H. Fox, Rachel Shefner, Genaro M.A. Palmieri, and
Tulio Bertorini

Human Purine Research Center, Departments of Internal Medicine
and Biological Chemistry, Clinical Research Center, The University of Michigan, Ann Arbor, Michigan, and University of
Tennessee Center for The Health Sciences, Memphis, Tennessee

SUMMARY

We examined ATP metabolism in cultured muscle cells and fibroblasts
from patients with Duchenne dystrophy. ATP and ADP levels were the same in
cultured cells from normal subjects and patients and there was no difference
in ATP synthesis or degradation. Although there was a significant decrease
in radioactively labelled ATP after incubation with deoxyglucose in Duchenne
muscle cells, there was no difference in ATP concentration or ADP
metabolism.

INTRODUCTION

A reduction of high energy phosphate content has been observed in some
studies of muscle from patients with Duchenne muscular dystrophy,[1-8] but not
in other reports.[9,10] We have found evidence for increased turnover of
adenine nucleotides in Duchenne patients'[11] but the in vivo studies did not
identify the tissue affected by these abnormalities. Therefore, we have now
tested the hypothesis that there is an accelerated rate of ATP turnover in
the cultured muscle cells and fibroblasts from patients with Duchenne
muscular dystrophy.[12]

METHODS

Skin biopsies and needle muscle biopsies were obtained from 9 patients
with Duchenne muscular dystrophy and 10 normal controls who were college-age
males. Skin and muscle biopsies were grown in culture as described.[13-15]

ATP turnover was studied by first labeling the cellular nucleotide pool by incubation with radioactive adenine. Cultured cells are then incubated and harvested as previously described.[12] Radioactive ATP and ADP pools were measured after separation of nucleotides by high-voltage electrophoresis.[12] ATP and ADP concentrations were measured by anion exchange high performance liquid chromatography on a Whatman partisil PXS 10/25-SAX column. Fifty microliters of neutralized extract was injected and ATP and ADP were eluted isocratically in 0.4 M $NH_4H_2PO_4$ pH 4.5 at 1.5 ml/min using standard methods.[16]

Data storage and analysis were performed using CLINFO software (BBN, Cambridge, Massachusetts) and a Vax 11/730 minicomputer. The null hypothesis for the observations of normal cells and cells from Duchenne muscular dystrophy was tested using a two-tailed Student t test.

RESULTS

Adenine Nucleotide Levels

ATP and ADP levels were similar in both muscle cells and fibroblasts from normal subjects and Duchenne patients (Table 1). AMP and IMP levels were at the limits of detection of this system, but did not appear to be different.

ATP Synthesis

ATP synthesis was measured by the incorporation of $[U-^{14}C]$adenine into ATP and ADP (Table 1). Five to 10 times more ATP was formed than ADP. The enrichment of ATP and ADP pools by radioactive isotope was similar as expressed by cpm/nmol. The rates of ATP and ADP formed from radioactive adenine and the enrichment of these pools was the same in cells from controls and patients.

ATP Degradation

ATP degradation was evaluated during a 10-minute incubation in 5.5 mM 2-deoxyglucose or during a 72-hour period of growth. Little or no decrease in ATP levels followed incubation with 2-deoxyglucose, perhaps because the high concentration of glucose (33 mM) in the first incubation period inhibited phosphorylation of deoxyglucose in the second incubation. There was significantly less radioactively-labelled ATP after incubation with deoxyglucose in Duchenne cells than in normal muscle cells (Table 2), but there was no difference in total ATP concentration or ADP concentration or radioactive labelling. There was no difference in ATP or ADP degradation i fibroblasts (Table 2).

494

Table 1. ATP and ADP Levels and Synthesis

Cells were grown in culture as described in methods. After a one hour incubation in tissue culture medium containing [U-^{14}C]adenine the cells were harvested. The ATP, ADP, protein values, and [U-^{14}C]adenine incorporation into ATP and ADP were measured as described in methods. The data are expressed as the mean value plus or minus the standard error of the mean. Statistical analyses showed no difference between normal cells and Duchenne cells.

	ATP			ADP		
	nmol/mg	cpm/mg	cpm/nmol	nmol/mg	cpm/mg	cpm/nmol
Muscle Cells[a]						
Normal	22.2 ± 3.7	7811 ± 1750	359 ± 44	4.9 ± 1.3	1396 ± 439	360 ± 66
Duchenne	30.0 ± 6.1	6889 ± 953	263 ± 38	4.8 ± 1.4	1227 ± 261	304 ± 49
Fibroblasts[b]						
Normal	42.8 ± 5.4	10857 ± 1460	292 ± 73	7.3 ± 0.7	1516 ± 406	203 ± 45
Duchenne	39.9 ± 6.2	12280 ± 2408	305 ± 26	7.1 ± 2.1	1686 ± 294	304 ± 44

a. Cells from 10 normal subjects and 9 patients with Duchenne muscular dystrophy.
b. Cells from 6 normal subjects and 7 patients with Duchenne muscular dystrophy.

Table 2. ATP Degradation Following 2-Deoxyglucose

Cells were grown and incubated as described in Table 1. The cells were washed and then incubated for 10 minutes in medium containing 5.5 mM 2-deoxyglucose. Assays were performed as described in Table 1. The data are expressed as a percentage of the mean baseline values which are listed in Table 1 plus or minus the standard error of the mean. With the exception of ATP radioactivity in muscle cells, statistical analyses show no difference between normal cells and Duchenne cells.

	ATP	ADP	ATP	ADP	ATP	ADP
	cpm/mg (% control value)		nmol/mg (% control value)		cpm/nmol (% control value)	
Muscle Cells[a]						
Normal	94 ± 10[b]	113 ± 17	90 ± 7	94 ± 12	107 ± 11	137 ± 24
Duchenne	62 ± 6[b]	75 ± 10	93 ± 16	107 ± 20	86 ± 14	159 ± 81
Fibroblasts[c]						
Normal	76 ± 11	92 ± 12	82 ± 15	107 ± 7	91 ± 3	86 ± 18
Duchenne	83 ± 6	131 ± 43	89 ± 4	111 ± 18	94 ± 9	86 ± 13

a. Cells from 10 normal subjects and 9 patients with Duchenne muscular dystrophy.
b. p = 0.02
c. Cells from 5 normal subjects and from 4 patients with Duchenne muscular dystrophy.

In longterm study of fibroblasts, radioactively labelled ATP decreased to 50% of control values at 24 hours and about 30% of control values at 72 hours, but there was no difference between normal and Duchenne cells (Table 3).

DISCUSSION

Accelerated turnover of ATP in Duchenne muscular dystrophy has been deduced from the reported decrease in muscle ATP levels[1-8] and increased purine excretion.[11] We found that ATP concentrations were normal in cultured muscle cells and fibroblasts, in accord with other evidence that ATP levels are normal in Duchenne muscle.[9,10] We previously observed increased rates of adenine nucleotide turnover in vivo, but in the current studies there was no evidence of altered ATP synthesis in cultured muscle cells or fibroblasts. ATP degradation in cultured fibroblasts was normal after addition of 2-deoxyglucose or after growth for 72 hours. There was a significantly greater decrease in radiolabeled ATP degradation after

addition of deoxyglucose to cultured muscle cells (Table 3), but this was
not a definite abnormality because other variables were similar to normal
cells and did not suggest increased ATP degradation.

Table 3. ATP Degradation During 72-Hour Incubation
of Fibroblasts[a]

Cells were grown and incubated as described in Table 1. The cells were
washed and then incubated in regular tissue culture medium for up to 72
hours. Assays and expression of data were identical to the procedures
described in Table 2. Statistical analyses show no difference between
normal cells and Duchenne cells.

	24-Hour		72-Hour	
	ATP	ADP	ATP	ADP
cpm/mg (% control value)				
Normal	51 ± 6	80 ± 32	32 ± 3	69 ± 20
Duchenne	50 ± 3	81 ± 23	31 ± 3	48 ± 8
nmol/mg (% control value)				
Normal	89 ± 8	92 ± 7	68 ± 18	91 ± 17
Duchenne	73 ± 13	116 ± 8	67 ± 15	109 ± 18
cpm/nmol (% control value)				
Normal	56 ± 4	89 ± 34	41 ± 6	97 ± 34
Duchenne	89 ± 21	69 ± 18	61 ± 15	57 ± 20

a. Cells from 6 normal subjects and from 7 patients with Duchenne muscular
dystrophy.

Therefore, we found no definite evidence of accelerated ATP turnover in
cultured muscle cells and fibroblasts under the experimental conditions
used. We cannot rule out such a disorder in later stages of cell maturation
or the requirement for specific constituents of the patient environment.
For example, plasma hormones might enhance intracellular calcium
translocation, cause intracellular overload of calcium, and decrease ATP
production.[17-21]

REFERENCES

1. J.A. Callazo, J. Barbudo, L. Torres Der Chemismus des Muskels bei der
   Dystrophia muscularis progressiva (Analyse der Biopsie des
   Deltoideus). Deutsche Med Wchnschr 1936; 62:51-4.
2. P.J. Vignos, J.L. Warner. Glycogen, creatine, and high energy phosphate
   in human muscle disease. J Lab Clin Med 1963; 62:579-90.

3. H. Heyck, G. Laudahn, C.J. Luders. Fermentaktivitatsbestimmugen in der gesunder menschlichen Muskulatur und bei Myopathien. Klin Wschr 1963; 41:500-9.

4. G. Berthillier, D. Gautheron, J.M. Robert. Fractions phosphorylees et nucleotides adenyliques libres de muscle myopathiques chez l'enfant. CR Acad Sci (Paris) 1967; 265:79-82.

5. N.C. Kar, C.M. Pearson. Muscle adenylic acid deaminase activity: Selective decrease in early onset Duchenne muscular dystrophy. Neurology 1973; 23:478-82.

6. L. Stengel-Rutkowski, W. Barthelmai. Muscular energy metabolism in children with progressive muscular dystrophy type Duchenne. Metabolites of the Embden-Meyerhof pathway, the citric acid cycle and high energy phosphates and enzyme activity of alpha-glycerol-oxidase, succinate dehydrogenase and 6-phospho-gluconate-dehydrogenase. Klin Wschr 1973; 51:957-68.

7. W.H.S. Thomson, I. Smith. X-linked recessive (Duchenne) muscular dystrophy (DMD) and purine metabolism: Effects of oral allopurinol and adenylate. Metabolism 1978; 27:151-63.

8. T.E. Bertorini, G.M.A. Palmieri, J. Griffin, C. Chesney, D. Pifer, L. Verling, D. Airozo, I.H. Fox. Chronic allopurinol and adenine therapy in Duchenne muscular dystrophy: Effects on muscle function, nucleotide degradation and muscle ATP and ADP content. Neurology 36:61-65,1985.

9. F.J. Samaha FJ, B. Davis, B. Nagy. Duchenne muscular dystrophy: Adenosine triphosphate and creatine phosphate content in muscle. Neurology 1981; 31:916-19.

10. R.J. Newman, P.J. Bore, L. Chan, D.H. Gadian, P. Styles, D. Taylor, G.K. Radda. Nuclear magnetic resonance studies of forearm muscle in Duchenne dystrophy. Br Med J 1982; 284:1072-74.

11. T.E. Bertorini, G.M.A. Palmieri, D. Airozo, N.L. Edwards, I.H. Fox. Increased adenine nucleotide turnover in Duchenne muscular dystrophy. Pediatr Res 1981; 15:1478-82.

12. I.H. Fox, R. Shefner, G.M.A. Palmieri, T.E. Bertorini. Duchenne muscular dystrophy: Normal ATP turnover in cultured cells. Neurology (In Press).

13. I.H. Fox, I.L. Dwosh, P.J. Marchant, S. Lacroix, M.R. Moore, S. Omura, V. Wyhofsky. Hypoxanthine-guanine phosphoribosyltransferase: Characteristics of a mutation in a patient with gout. J Clin Invest 1975; 56:1239-49.

14. A.F. Miranda, H. Somer, S. DiMauro. Isoenzymes as markers of differentiation. In: Mauro A, ed. Muscle Regeneration. New York: Raven Press, 1979; 453-73.

15. S. Mawatari, A.F. Miranda, L.P. Rowland. Adenyl cyclase abnormality in Duchenne muscular dystrophy: Muscle cells in culture. Neurology 1976; 26:1021-6.

16. L. Rose, R.W. Brockman. Analysis by high pressure liquid chromatography of 9- -D-arabinofuranosyladenine 5'-triphosphate levels in murine leukemia cells. J Chromatogr 1977; 133:335-43.

17. K. Wrogemann, S.K.J. Pena. Mitochondrial calcium overload: A general mechanism for cell-necrosis in muscle diseases. Lancet 1976; 1:672-3.

18. S. Ebashi, H. Sugita. The role of calcium in physiological and pathological processes of skeletal muscle. In: Aguajo AJ and Kapati S, eds. Current Topics in Nerve and Muscle Research. Amsterdam: Excerpta Medica 1979; ICS 475:73-87.

19. G.M.A. Palmieri, D. Nutting, S.K. Phattacharya, T.E. Bertorini, J.C. Williams. Parathyroid ablation in dystrophic hamsters: Effects on Ca content and histology of heart, diaphragm and rectus femoris. J Clin Invest 1981; 68:646-54.

20. T.E. Bertorini, S.K. Bhattacharya, G.M.A. Palmieri, C.M. Chesney, D. Pifer, B. Baker. Muscle calcium and magnesium content in Duchenne muscular dystrophy. Neurology 1982; 32:1088-92.

21. T.E. Bertorini, F. Cornelio, S.K. Bhattacharya, G.M.A. Palmieri, I. Dones, F. Dworzak, B. Brambati. Calcium and magnesium content in fetuses at risk and prenecrotic Duchenne muscular dystrophy. Neurology, (in press).

ACKNOWLEDGMENTS

The authors wish to thank Dr. Armand F. Miranda of Columbia University and Dr. William O. Whetsell, Jr., of Vanderbilt University for their assistance in establishing tissue culture methods, Vera Shively and Barbara Pepper for excellent technical assistance, Stephen Schmaltz for his assistance in statistical analyses using CLINFO, and Holly Gibson and Sharon Demorest for typing the manuscript. This work was supported by grants from the the Muscular Dystrophy Association and the United States Public Health Service (2-R01-AM 19674 and 5-M01-RR00042).

INFERENCES ON THE RAPID METABOLISM OF EXOGENOUSLY APPLIED ATP IN ISOLATED SMOOTH MUSCLE PREPARATIONS USING PHARMACOLOGICAL STUDIES

David Satchell

Department of Zoology
University of Melbourne
Parkville, 3052, Australia

INTRODUCTION

When ATP or adenosine are applied as relaxants to isolated smooth muscle preparations in organ baths despite the presence of large volumes of nutrient solution containing these compounds it has been difficult to ascertain what compounds and what concentrations of compounds are indeed acting on the receptors of the muscle cell surface. For example it is quite possible that applied ATP is rapidly broken down in the extracellular space and that the metabolites themselves may act on muscle receptors in lieu of or in addition to the added substance. The understanding of what happens in this situation may be complicated by the formation of more than one compound, the uptake of metabolites, e.g. adenosine, into the cells and the diffusion of metabolites into the outside medium.

Some insight has been obtained into these problems by two different techniques: Firstly, although ATP and adenosine both relax taenia coli strips the shape of the responses are different and characteristic of each compound. ATP causes a rapid fairly transient relaxation whereas adenosine causes a slower developing relaxation. Adenosine is rapidly taken up into cells and uptake inhibitors such as dipyridamole markedly potentiate responses to adenosine. Thus examination of the shape of the relaxation to ATP in the presence of dipyridamole should reveal whether significant formation of adenosine has occurred during the 60 sec contact time of ATP with the preparation. Secondly, methylene isosteres of ATP are resistant to cleavage at the point of methylene substitution. The relaxant effects of such compounds have been determined on strips of guinea-pig taenia coli and trachea. The ability of α,β and β,γ methylene isosteres to form adenosine during the 60 sec contact time has been assessed using dipyridamole potentiation as an index of adenosine formation. The results may be used to provide information on the site of rapid cleavage of polyphosphate groups of the isosteres and of ATP itself.

METHODS

Taenia coli strips 1.5 cm long and transverse tracheal strips containing 2 to 4 cartilage rings were dissected from guinea-pigs (weights 500-1000 g) of either sex. Preparations were suspended in Krebs solution gassed with 95% O_2 and 5% CO_2 and maintained at 37°C in 50 ml baths.

Tensions of 0.5 and 1.0 g were applied to tracheal and taenia preparations respectively. Muscle activity was registered by means of Gould Statham force-displacement transducers (tracheal strips) and Grass FT03C force displacement transducers (taenia coli strips) coupled to Grass polygraphs. Compounds used were ATP and adenosine (Sigma, St. Louis). Lithium adenosine 5'-α,β-methylene-triphosphate (AOPCPOP) and 5'-adenylylmethylene-diphosphonate (AOPOPCP) (P&L Laboratories Milwaukee), Dipyridamole (Boehringer Ingelheim Australia).

RESULTS

Isolated guinea-pig taenia coli preparations

ATP and adenosine have different actions on taenia coli strips. The relaxation in response to ATP (1.0 µM) was rapid in onset and took 11 sec to reach maximum in contrast to responses to adenosine (2.5 µM which took 30 sec to reach maximum (Fig. 1). The adenosine uptake inhibitor dipyridamole

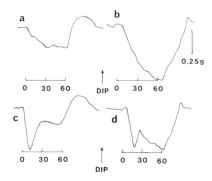

Fig. 1. Relaxations of taenia coli strips to the following: adenosine (2.5 µM), (a) before, (b) 15 min after addition of dipyridamole (1.0 µM). ATP (1.0 µM), (c) before, (d) 15 min after addition of dipyridamole (1.0 µM). The vertical bar refers to tension in gm, the horizontal calibration is in sec.

(1.0 µM) potentiated responses to adenosine. Dipyridamole (1.0 µM) also affected responses to ATP and caused the appearance of a second phase of relaxation: during the 30 to 60 sec contact time (Fig. 1). The second phase could be explained on the basis that some of the ATP entering the tissue rapidly formed adenosine and that a response to this adenosine is only seen under conditions of adenosine uptake inhibition.

AOPCPOP (3 µM) and AOPOPCP (10 µM) both caused rapid relaxations of taenia coli strips consistent with an ATP-like action (Fig. 2). Dipyridamole (1.0 µM) failed to affect responses to AOPCPOP but potentiated responses to AOPOPCP (Fig. 2). This potentiation was due to a second phase of relaxation occurring during the 30 to 60 sec time interval and could be explained in the basis of rapid formation of adenosine.

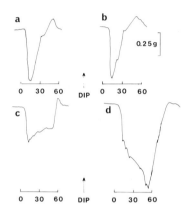

Fig. 2. Relaxations of taenia coli strips to the following: Adenosine 5'-α,β-methylenetriphosphate (AOPCPOP 3 µM), (a) before, (b) 15 min after the addition of dipyridamole (1.0 µM). 5'-adenylylmethylenediphosphonate (AOPOPCP 10 µM); (c) before, (d) 15 min after the addition of dipyridamole (1.0 µM). The vertical bar refers to tension in gm, the horizontal calibration is in sec.

Responses to AOPCPOP may have been unaffected by dipyridamole because this compound could not form adenosine. The reason why AOPOPCP may form adenosine but AOPCPOP is stable may be due to the action of an adenosine triphosphate pyrophoshohydrolase which may also be important in the rapid breakdown of ATP.

Isolated Guinea-pig Tracheal Preparations

The addition of either ATP (60 µM) or adenosine (60 µM) to organ baths containing tracheal strips caused a slow developing relaxation which reached maximum after approximately 60 sec (Fig. 3). Both ATP and adenosine exhibited similar potency and it was not possible to delineate between the actions of the two on the basis of potency or on the shape of the response.

Following treatment with dipyridamole (1.0 μM) the amplitudes of the responses to ATP and adenosine were both potentiated equally (Fig. 3).

These results suggest that in tracheal strips as in taenia coli strips dipyridamole potentiates responses to adenosine due to blockade of uptake of the nucleoside into cells. Moreover the fact that dipyridamole caused a similar potentiation of responses to ATP suggests that a marked breakdown of ATP to adenosine is occurring during the 60 sec contact time.

To further elucidate these findings the effects of AOPOPCP and AOPCPOP were studied on tracheal strips. AOPOPCP (60 μM) caused a slow developing relaxation. Dipyridamole (1.0 μM) caused a marked potentiation of this response (Fig. 4). AOPOPCP was similar to adenosine with respect to the shape of its response and potency before and after dipyridamole. AOPCPOP was inactive up to a concentration of 1000 μM (the limit of its solubility) in the presence or absence of dipyridamole (1.0 μM) (Fig. 4).

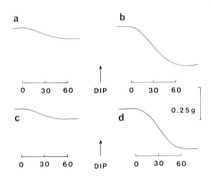

Fig. 3. Relaxations of tracheal strips to the following: adenosine (60 μM), (a) before, (b) 15 min after dipyridamole (1.0 μM). ATP (60 μM), (c) before, (d) 15 min after dipyridamole (1.0 μM). The vertical bar refers to tension in gm. The horizontal calibration is in sec.

The results suggest that AOPOPCP is rapidly broken down to adenosine in trachea as in taenia coli. The inactivity of AOPCPOP in the trachea has several implications: Firstly, it is likely that this compound is stable as in the taenia coli and does not form adenosine during the 60 sec contact time. Secondly, adenosine may need to be formed in order for relaxations to occur. It follows also that not only is AOPCPOP inactive in trachea but that ATP may be also and that relaxations due to ATP result from a very rapid and largely complete breakdown to adenosine during the 60 sec contact time.

DISCUSSION

    Despite the presence of large volumes of applied ATP in the organ bath
rapid metabolism of ATP to adenosine is likely to occur in the extracellular
space of isolated preparations of smooth muscle of the guinea-pig taenia
coli and trachea. The adenosine formed can act on the smooth muscle cell
surface. An important enzyme in the breakdown may be adenosine triphosphate
pyrophoshohydrolase.

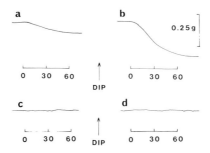

Fig. 4. Relaxations of tracheal strips to the following:
        5'-adenylylmethylenediphosphonate (AOPOPCP 60 µM) (a) before, (b) 15
        min after addition of dipyridamole (1.0 µM) Adenosine 5'-
        α,β-methylenetriphosphate (AOPCPOP 1000 µM). This compound was
        inactive in either the presence or absence of dipyridamole (1.0 µM)
        and this probably signifies an inability to form adenosine. The
        vertical bar refers to tension in gm, the horizontal calibration is
        in sec.

    In taenia coli preparations applied ATP may have a dual action due to
intact ATP and its slower acting adenosine metabolite. In tracheal strips
ATP is likely to be inactive and causes relaxation due to rapidly formed
adenosine. It is likely that separate inhibitory receptors namely P_1 and P_2
are present for adenosine and ATP respectively in the taenia and that P_1
receptors only are present in trachea (see reference 1).

    The status of ADP and AMP in these experiments has not been determined
although in taenia coli ADP has actions similar to ATP and AMP similar to
adenosine[1].

REFERENCES

1. D. G. Satchell, Purine receptors: classification and properties,
        Trends in Pharmacol. Sci. 5:340 (1984).

ENZYME ACTIVITIES OF PURINE CATABOLISM AND SALVAGE IN HUMAN

MUSCLE TISSUE

Gerhard Schopf, Michael Havel, Roland Fasol and
Mathias M. Müller

2nd Department of Surgery, Division of Clinical
Biochemistry
University of Vienna, Vienna, Austria

INTRODUCTION

Metabolic differentiation in muscle is closely related to the
elementary systems of energy supplying metabolism. Most strikingly this is reflected at the level of enzymatic organization(1).
White (fast) muscle is characterized by high capacities of glyco-
genolysis,glycolysis and lactate fermentation,whereas the capacities of glucose phosphorylation,citric acid cycle,oxidative
phosphorylation and fatty acid oxidation are low.Red (slow) muscles,heart and smooth muscles show inverse characteristics.Altera-
tions of enzyme activities during red and white muscle differentiation in vitro suggest the existence of one myogenic cell with
the potential to exhibit those properties which are characteristic for any type of muscle(2). According to this concept the dicho-
tomy between white and red muscles becomes apparent after inner-
vation. Furthermore it was demonstrated that white (fast contracting) muscles can be transformed into red (slow contracting) muscles by longterm electric stimulation with low frequencences(3,4).
The adaptation to the nerval stimulus was reflected by structural
changes and by alterations in the activites of enzymes(4).
The comparative analysis of enzyme activities may give information about the "metabolic programming" of muscle tissue and
should be related to the different functions. Since there is
little information about the distribution of enzymes involved in
purine catabolism and purine salvage,it was reasonable to measure
activities of some key enzymes in human tissue samples.

MATERIALS AND METHODS

10 patients undergoing open heart surgery were informed and
agreed to be selected for this study. Tissue samples were taken
from the heart (Auricula dextra), M. rectus abdominis and M. qua-
driceps femoris, wiped free of adhering blood residues and immediately frozen in liquid nitrogen. Homogenization of the frozen
samples (approximately 45mg wet weight) was performed with a
microdismembrator II (Brown, FRG). Pulverized tissue samples
were resolved in 1 ml 20 mmol/l Tris buffer, pH 7.4, supplemented with 1% Triton-X-100. Activities of adenosine deaminase
(ADA), adenylate deaminase (AMP-DA), purine nucleoside phos-

phorylase(PNP),adenosine kinase(AK),adenine phosphoribosyltrans-
ferase(APRT),and hypoxanthine guanine phosphoribosyltransferase
(HGPRT)were determined using radiochemical tests as described
elsewhere(5-7).An UV-test for measurement of purine 5´-nucleo-
tidase(5´-PNT)activity was used(8). Protein was determined with
Coomassie Brilliant Blue(9).

RESULTS AND DISCUSSION

Enzyme activities of purine catabolism and purine salvage
pathway were determined in homogenates from human muscle tissue.

Table 1.Enzyme activities in human muscle tissue(nmoles/g protein
protein/min,mean values and standard deviations of 10 samples).

Enzyme	Myocardium Auricula dextra	M.rectus abdominis	M.quadriceps femoris
Purine Catabolism			
ADA	5408 + 513	1742 + 414	521 + 76
AMP-DA	2127 + 546	9171 +2137	12182 + 3245
5´-PNT	9555 + 1250	5034 +1771	4861 + 974
PNP	4375 + 646	2381 + 466	2552 + 328
Salvage Pathway			
AK	156 + 38	85 + 21	140 + 42
APRT	242 + 73	383 + 69	266 + 79
HGPRT	134 + 40	175 + 45	280 + 56

In table 1 summarized results show great differences, especially
concerning activities of enzymes involved in purine catabolism.
The highest activities of ADA,5´-PNT and PNP were obtained in
tissues from the right heart auricle,whereas AMP-DA activities
dominated in samples from other sources.The ratio AMP-DA:5´-PNT
in auricles,M.rectus abdominis and M.quadriceps femoris were
0.22,1.82 and 2.51,respectively, indicating that AMP formed during contraction is degraded in different ways.Though only heart
auricles were investigated,AMP might be catabolized in myocard
mainly to hypoxanthine by the subsequent actions of 5´-PNT,ADA
and PNP.This should be of importance for the autoregulation of
coronary blood flow,especially under ischemic conditions,since
adenosine formed and released from the myocard is a potent di-
latator of coronary vessels(10). On the other hand, adenosine might
be phosphorylated to AMP by AK and thus reincorporated into adenine nucleotides(11).In accordance with this concept there are
high activities of AK associated with relatively low activities
of HGPRT.
In contrast to heart auricles in tissue samples from skeletal
muscle(M.quadriceps femoris,M.rectus abdominis)the main route
for AMP degradation should be via AMP-DA yielding IMP.The latter
will be catabolized by 5´-PNT and PNP to hypoxanthine.Furthermore
IMP may be subjected to interconversion through the purine nucleotide cycles having been described in skeletal muscle(12)
Metabolic differentiation of muscle correlates to a large extent with the type of muscle and may be distinguished by differences in distinct enzyme activity ratios(13).Biochemical and histo

logically two types of muscles can be characterized:white(fast) and red(slow)muscles. For this study tissue samples from M.quadriceps femoris,a typical white muscle,from M.rectus abdominis,a red muscle and from the right heart auricle, representative for myocardial tissue, belonging to red persistent muscles, were investigated. The data obtained suggest that differences between myocardium and skeletal muscle,irrespective to their classification,exist with respect to purine catabolism,purine interconversion and purine salvage. However, further experiments will be necessary to correlate the measured enzyme activities with the relative flow of metabolites.

ACKNOWLEDGEMENT

This study was part of project 2139 supported by"Jubiläums fonds der österreichischen Nationalbank".

REFERENCES

(1)   R.A.Lawrie,The activity of the cytochrome system in muscle and its relation to myoglobin,Biochem.J.55:298(1953)
(2)   F.Bacou and J.Nougués,Alterations of enzymatic activities during red and white muscle differentiation in vitro, Exp.Cell Res.129:455(1980)
(3)   D.Pette,W.Müller,E.Leisner,and G.Vrbová,Time dependent effects on contractile properties,fibre population,myosine light chains and enzymes of energy metabolism in intermittently and continously stimulated fast twitch muscles of the rabbit,Pflügers Arch.364:103(1976)
(4)   M.Frey,H.Thoma,H.Gruber,H.Stöhr,L.Huber,M.Havel,and E.Steiner,The chronically stimulated muscle as an energy source for artificial organs,Europ.Surg.Res.16:232(1984)
(5)   K.O.Raivio,P.Santavouri,and H.Somer,Metabolism of AMP in muscle extracts from patients with deficient activity of myoadenylate deaminase,Adv.Exp.Med.Biol.165B:431(1984)
(6)   E.Kaiser,and M.M.Müller,Enzymdefekte des Purinstoffwechsels Z.med.Labor.-Diagn.20:3(1979)
(7)   M.P.Uitendaal,C.H.M.M.de Bruyn,T.L.Oei,and P.Hösli,Ultramicrochemical studies on enzyme kinetics,Adv.Exp.Med.Biol. 76A:597(1977)
(8)   C.L.M.Arkesteijn,A kinetic method for serum 5´-nucleotidase using stabilised glutamate dehydrogenase,J.Clin.Chem. Clin.Biochem.14:155(1976)
(9)   M.M.Bradford,A rapid and sensitive method for the quantitation of microgram quantities of protein utilizing the principle of protein dye binding,Analyt.Biochem.72:248 (1976)
(10) Robert M.Berne,The role of adenosine in the regulation of coronary blood flow,Circ.Res.47:807(1980)
(11) V.T.Wiedermeier,R.Rubio,R.M.Berne,Incorporation and turnover of adenosine-U-14C in perfused guinea pig myocardium,Am.J. Physiol.223:51(1972)
(12) J.M.Lowenstein,Ammonia production in muscle and other tissues: The purine nucleotide cycle,Physiol.Rev.52:382(1972)
(13) A.Bass,D.Brdiczka,P.Eyer,S.Hofer,and D.Pette;Metabolic differentiation of distinct muscle types at the level of enzymatic organization,Europ.J.Biochem.10:198(1969)

PURINE SALVAGE IN RAT HEART MYOBLASTS

Gerhard Schopf, Helmut Rumpold, and Mathias M. Müller

2nd Department of Surgery,Division of Clinical
Biochemistry,University of Vienna
Vienna,Austria

INTRODUCTION

Myocardial metabolism depends on a high level of ATP,which is essential for contraction and relaxation.Under aerobic conditions ADP generated is recycled to ATP mainly by oxidative phosphorylation.However,a small part of ADP is degraded to AMP, thus entering purine interconversion and catabolism.As a consequence purine nucleotides are broken down to adenosine,inosine and hypoxanthine.Since the myocyte membrane is permeable for nucleosides and bases(1),there is a continous loss of preformed purines.Especially during periods of ischemia accompanied by a depletion of myocardial ATP and CP high amounts of these degradation compounds can be detected in coronary venous blood (2,3,4)

Comparing the kinetic data of enzymes involved in purine de novo synthesis and in the salvage pathway,it is most probable that under physiological conditions the reutilization of preformed purines is the main metabolic route for the maintenance of intracellular adenine nucleotide content.It has been reported that for the restoration of the adenine nucleotide pool, after brief ischemia the de novo purine synthesis was limited (5).In primary skeletal rat muscle cultures a limiting availability of phosphoribosyl pyrophosphate(PP-ribose-P)could be demonstrated(6).

In the course of adenine nucleotide degradation a high fraction of adenosine is released into the extracellular space, again taken up by myocardial cells,then rephosphorylated to AMP and reincorporated into myocardial adenine nucleotides(7). So it was interesting to investigate in the present study the reutilization of adenosine,inosine and hypoxanthine as well as of adenine by rat heart myoblasts(8).The information received in this model might be relevant for the metabolism of the heart.

MATERIALS AND METHODS

Cell cultivation:Rat heart myoblasts H9c2(2-1)were obtained from American Type Culture Collection and cultivated in bicarbonate buffered Dulbecco's minimum essential medium(DMEM) supplemented with 10% FCS(Flow Laboratories,U.K.) and 100 mg/l

gentamycine(Sigma,FRG)in an atmosphere of 90% air and 10% CO_2.
Metabolic studies were performed with monolayers of myoblasts
always after 7 days of cultivation.

   Incorporation of labeled purines:one hour before the addition
of labeled purines the medium was changed.Monolyers consisting
of 1.3 to 1.5 x 10^6 cells/flask were incubated at 37^OC for one
hour.Incubations were performed with 5.2 µmol/l ^{14}C-labeled adenosine,inosine,adenine and hypoxanthine at various concentrations
of phosphate(1 to 50 mmol/l).Details of the procedure used are
described in this volume(9).After the incubation period cells
are harvested,purine metabolites extracted with perchloric acid
and separated by means of thin layer chromatography.Radioacti-
vity of labeled purine metabolites was measured in a liquid
scintillation counter.

RESULTS AND DISCUSSION

   During the incubation for one hour in DMEM(1 mmol/l phosphate)
all substrates added were incorporated into purine nucleotides
at various amounts as described in table 1.

Table 1.Uptake and incorporation of adenosine,inosine,adenine,
        and hypoxanthine into H9c2(2-1)myoblasts with 1 mmol/l
        phosphate.Mean values and standard deviations of 6 ex=
        periments.

	Amount of label nmoles/mg 10^6 cells/hour			Distribution of label %
Substrate : ^{14}C-ADENOSINE				
Total uptake	11.2	+	4.9	100.0
ATP+ADP+AMP	8.3	+	4.0	74.4
GTP+GDP+GMP	0.72	+	0.31	6.4
ITP+IDP+IMP	0.12	+	0.04	1.1
Nucleosides	1.48	+	0.50	13.2
Bases	0.37	+	0.15	3.3
Substrate : ^{14}C-INOSINE				
Total uptake	3.7	+	2.3	100.0
ATP+ADP+AMP	2.7	+	0.6	73.0
GTP+GDP+GMP	0.3	+	0.11	8.1
ITP+IDP+IMP	0.1	+	0.03	2.6
Nucleosides	0.37	+	0.09	10.1
Bases	0.19	+	0.05	5.2
Substrate : ^{14}C-ADENINE				
Total uptake	3.1	+	1.8	100.0
ATP+ADP+AMP	2.2	+	0.7	71.0
GTP+GDP+GMP	0.17	+	0.08	5.6
ITP+IDP+IMP	0.04	+	0.02	1.4
Nucleosides	0.16	+	0.08	5.3
Bases	0.5	+	0.20	15.7

Substrate :	^{14}C HYPOXANTHINE	
Total uptake	1.6 ± 0.6	100.0
ATP+ADP+AMP	1.1 ± 0.4	68.0
GTP+GDP+GMP	0.18± 0.05	11.1
ITP+IDP+IMP	0.09± 0.03	5.9
Nucleosides	0.06± 0.02	3.9
Bases	0.16± 0.04	10.1

Adenosine was reutilized at the highest rate,probably because the carrier protein responsible for the uptake of adenosine has the highest affinity compared with the inosine-,adenine-or hypoxanthine-carrier(10).Furthermore intracellular adenosine is metabolized by two enzymes.It is either converted into AMP by adenosine kinase(AK) or catabolized by adenosine deaminase(ADA)to inosine which will be hydrolyzed to hypoxanthine by purine nucleoside phosphorylase(PNP).The latter enzyme is also responsible for the degradation of inosine.Hypoxanthine will be phosphoribosylated to IMP by hypoxanthine guanine phosphoribosyltransferase(HGPRT)and interconverted to AMP.AK exhibits a relatively low reaction velocity in spite of a high affinity.In contrast ADA has low affinity but a 100-fold higher reaction velocity(11,12).Therefore it is likely that adenosine will be metabolized mainly by ADA.The higher amount of labeled nucleosides with adenosine compared to inosine using the same concentrations might reflect the much lower affinity of ADA than PNP for their substrates.Therefore intracellular trapped adenosine will be accumulated even at concentrations of 5.2 μmol/l.Adenine nucleotides were labeled nearly at the same percentage irrespectively of the precursor used(adenosine,inosine,adenine,hypoxanthine). This is an evidence that the myoblast´s metabolism is mainly organized to maintain the equilibrium of adenine nucleotides.

Increase of phosphate concentration in the incubation medium up to 50 mmol/l stimulated the uptake and incorporation of adenosine,inosine,adenine and hypoxanthine into adenine nucleotides (figures 1-3).Under the experimental conditions used nearly a maximal rate of uptake was achieved.Very similar to the curves for the overall uptake,however,on a lower level,were the curves representing incorporation into adenine nucleotides.Taking the kinetics of transport processes and of intracellular enzymes into consideration the latter should be limiting for reutilization of purine bases and purine nucleosides(10).The intracellular radiolabel of other nucleotides went up at a similar percentage in comparison to adenine nucleotides. Since PP-ribose-P synthetase is activated by high inorganic phosphate concentrations (13),these findings indicate that the reutilization of preformed purines by rat heart myoblasts is limited by the availability of PP-ribose-P as reported for fibroblasts(14) and skeletal muscle cells(6).It is evident that under physiological conditions APRT,which converts adenine to AMP,and HGPRT are not saturated with respect to PP-ribose-P.

It was demonstrated that rat heart myoblasts seem to be an excellent model to investigate purine metabolism of the heart.It might be possible to use the biochemical findings obtained with in vitro tests for the development of new strategies in the therapy of myocardial dysfunctions.

PHOSPHATE INDUCED UPTAKE AND INCORPORATION OF ^{14}C-ADENOSINE

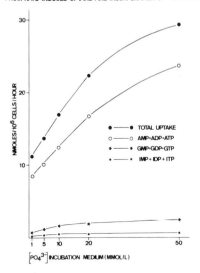

PHOSPHATE INDUCED UPTAKE AND INCORPORATION OF ^{14}C-INOSINE

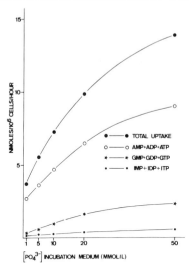

PHOSPHATE INDUCED UPTAKE AND INCORPORATION OF ^{14}C-ADENINE

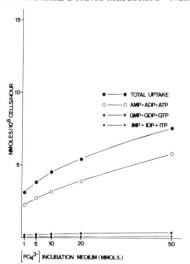

PHOSPHATE INDUCED UPTAKE AND INCORPORATION OF ^{14}C-HYPOXANTHINE

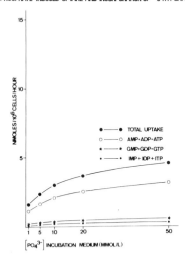

Figure 1.,2.,3.,4.: Phosphate induced uptake and incorporation of ^{14}C-adenosine, ^{14}C-inosine, ^{14}C-adenine and ^{14}C-hypoxanthine into intracellular nucleotides nmoles/10^6 cells/hour; mean value of 3 exp.

ACKNOWLEDGEMENT

This work was supported by a grant of "Jubiläumsfond der österreichischen Nationalbank" (project 2139)

REFERENCES

(1) J.Schrader and E.Gerlach,Compartmentation of cardiac adenine nucleotides and formation of adenosine,Pflügers Archiv 367:129(1976)

(2) W.J.Reeme,J.W.de Jong,and P.D.Verdouw,Effects of pacing-induced myocardial ischemia on hypoxanthine efflux from the human heart,Am.J.Cardiol.40:55(1977)

(3) J.L.Swain,R.L.Sabina,P.A.Mchale,J.C.Greenfield,and E.W.Holmes Prolonged myocardial nucleotide depletion after brief ischemia in the open-chest dog,Am.J.Physiol.242:818(1982)

(4) A.C.Fox,G.E.Reed,H.Heilman,and B.B.Silk,Release of nucleosides from canine and human hearts as an index of prior ischemia,Am.J.Cardiol.43:52(1979)

(5) J.L.Swain,J.J.Hines,R.L.Sabina,and E.W.Holmes,Accelerated repletion of ATP AND GTP pools in postischemic canine myocardium using a precursor of purine de novo synthesis, Circ.Res.51:102(1982)

(6) E.Zoref-Shani,A.Sainberg,and O.Sperling,Characterization of purine nucleotide metabolism in primary rat heart cultures Biochem.Biophys.Acta 716:324(1982)

(7) V.T.Wiedmaier,R.Rubio,and R.M.Berne,Incorporation and turnover of adenosine-U-14C in perfused guinea pig myocardium, Am.J.Physiol.223:51(1972)

(8) B.W.Kimes and B.L.Brandt,Properties of a clonal muscle cell line from rat heart,Exp.Cell.Res.98:367(1976)

(9) M.M.Müller,H.Rumpold,G.Schopf,and P.Zilla,Changes of purine metabolism during differentiation of rat heart myoblasts,in this volume.

(10) M.M.Müller,M.Kraupp,P.Chiba,and H.Rumpold,Regulation of purine uptake in normal and neoplastic cells,Adv.Enzyme Reg. 20:239(1982)

(11) I.H.Fox and W.N.Kelley, The role of adenosine and 2′-deoxyadenosine in mammalian cells,Ann.Rev.Biochem.47:655(1978)

(12) H.P.Schnebli,D.L.Hill,and L.L.Bennett,Jr,Purification and properties of adenosine kinase from human tumour cells, J.biol.Chem.242:1997(1967)

(13) P.Boer,B.Lipstein,A.de Vries,and O.Sperling, The effect of ribose-5-phosphate and 5-phosphoribosyl-5-pyrophosphate availibity on de novo synthesis of purine nucleotides in rat liver slices, Biochim. Biophys.Acta 432:10 (1976)

(14) E.Zoref,O.Sivan and O.Sperling,Synthesis and metabolic fate of purine nucleotides in cultured fibroblasts from normal subjects and from purine overproducing mutants,Biochim.Biophys. Acta,521:452 (1978)

SKELETAL MUSCLE ADENOSINE, INOSINE AND HYPOXANTHINE RELEASE
FOLLOWING ISCHAEMIC FOREARM EXERCISE IN MYOADENYLATE DEAMINASE
DEFICIENCY AND McARDLE'S DISEASE

Sietze Sinkeler, Ed Joosten, Ron Wevers, Rob Bink-
horst and Lian Oei
Departments of Neurology, Physiology and Human
Genetics, University of Nijmegen
Nijmegen, The Netherlands

INTRODUCTION

To assess the value of plasma purine nucleoside and base
measurements in the study of muscle purine nucleotide metabolism, plasma adenosine (Ado), inosine (Ino) and hypoxanthine
(Hx) concentrations were assayed in 7 controls, 5 myoadenylate deaminase deficient (MADD) patients and 6 McArdle patients
before and after ischaemic forearm exercise.

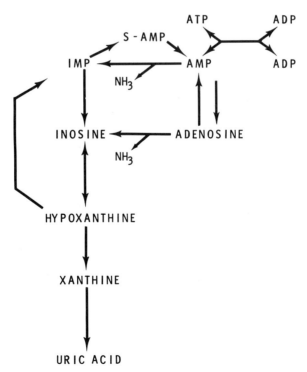

Figure 1. Purine nucleotide catabolism

Methods

The controls, MADD patients and McArdle patients were subjected to a standardized ischaemic isometric forearm test[1].
Plasma Ado, Ino and Hx concentrations were assayed using high performance liquid chromatography (HPLC) essentially according to the method described by Hartwick et al.[2] before and at 3, 5, 7, 9, 11, 15 and 20 minutes in all subjects tested and at 30 minutes in the MADD patients and McArdle patients.

Results and Discussion

The results are presented in Table 1 and 2 and in Figure 2 and 3.
Ado concentrations were low and increased only slightly, if any, in the three test groups and there were no significant differences (Table 1 and 2).
Significantly lower ΔIno and ΔHx was found in the MADD patients as compared to the controls and McArdle patients (Table 1 and 2).
These results do not support the suggestion of Sabina et al.[3] that purine nucleosides and bases diffuse into the vasculature in the absence of IMP accumulation in the exercising MADD muscle (Figure 1), but might be explained by the recent report of Sabina et al.[4], who were not able to find a measurable drop in muscle ATP following exercise in 3 out of 4 MADD patients.
The ΔIno and ΔHx reached the highest values in McArdle no. 6.
The Dixon test showed that there was a significant difference in ΔIno ($P < 0.01$), but not in ΔHx ($P > 0.01$), between this patient and the other McArdle patients.
In either way, however, including or excluding the data from McArdle patient no. 6, the ΔIno and ΔHx were not significantly different in the McArdle patients as compared to the controls (Table 1 and 2).
The ranges of the ΔIno and ΔHx found in the McArdle patients overlap considerably with the control values (Figure 2 and 3).
In the MADD patients, however, the ranges of the ΔHx measured 7 - 15 minutes after ischaemic exercise do not overlap with the control values (Figure 3). The assay of plasma Hx after muscular exertion might, therefore, be of value in the diagnosis of myoadenylate deaminase deficiency, but not in McArdle's disease.

Summary

Plasma adenosine, inosine and hypoxanthine concentrations were assayed in 7 controls, 5 MADD patients and 6 McArdle patients before and after ischaemic forearm exercise. The MADD patients showed a significantly lower increase in plasma inosine and hypoxanthine following exercise as compared to the controls. In the McArdle patients the increase in plasma inosine and hypoxanthine after exercise did not differ significantly from the control values. The plasma adenosine increase was very low in the three test groups and there were no significant differences.

Table 1. Effect of ischaemic forearm exercise on plasma purine nucleoside and base concentrations.

Controls/ Sex		Age	At rest			Maximal value after exercise					
			Ado	Ino	Hx	Ado	Ino	Hx	ΔAdo	ΔIno	ΔHx
no.1	M	52	2.3	0.4	4.0	2.0	6.7	35.2	0.3	6.3	31.2
2	M	29	0.2	0.5	1.7	0.6	17.8	38.5	0.4	17.3	36.8
3	F	22	0.0	0.6	1.6	0.0	11.7	27.2	0.0	11.1	25.6
4	F	23	0.0	0.6	0.6	0.0	15.6	25.6	0.0	15.0	25.0
5	F	33	0.0	0.6	0.6	0.0	9.2	24.3	0.0	8.6	23.7
6	M	34	0.2	1.1	1.3	1.5	4.4	32.9	1.3	3.3	31.6
7	M	32	0.4	0.9	0.9	0.9	8.7	40.5	0.5	7.8	39.6
Median		32	0.2	0.6	1.3	0.6	9.2	32.9	0.0	8.6	31.2
MADD Patients											
no.1	M	47	0.9	0.0	0.7	6.1	0.4	9.4	5.2	0.4	8.7
2	M	24	0.9	0.4	1.1	1.7	0.9	8.7	0.8	0.5	7.6
3	M	67	0.7	0.2	0.9	3.9	0.9	8.3	3.2	0.7	7.4
4	M	14	1.1	0.2	0.4	2.6	1.3	6.3	1.5	1.1	5.9
5	M	29	0.4	0.9	1.1	0.4	0.9	10.4	0.0	0.0	9.3
Median		29	0.9	0.2	0.9	2.6	0.9	8.7	1.5	0.5	7.6
McArdle patients											
no.1	F	19	0.2	0.7	0.4	0.4	4.4	28.5	0.2	3.7	28.1
2	M	37	1.3	0.4	0.2	1.3	5.9	31.5	0.0	5.5	31.3
3	F	27	1.1	0.4	0.7	2.2	17.6	40.5	1.1	17.2	39.8
4	F	8	1.1	0.9	0.7	1.3	3.5	24.8	0.2	2.6	24.1
5	M	24	0.4	0.7	0.4	0.4	11.7	42.0	0.0	11.0	41.6
6	F	25	0.2	0.2	1.0	0.5	39.4	56.5	0.3	39.2	55.5
Median		24,5	0.75	0.55	0.55	0.9	8.8	36.0	0.2	8.25	35.55

Plasma adenosine (Ado), inosine (Ino) and hypoxanthine (Hx): concentrations at rest, maximal concentrations after ischaemic forearm exercise, and the difference between these values (expressed in µM) in controls, myoadenylate deaminase deficient (MADD) patients and McArdle patients.

Table 2.

Maximal value after exercise

	At Rest			Maximal value after exercise					
n = 18	Ado	Ino	Hx	Ado	Ino	Hx	ΔAdo	ΔIno	ΔHx
Kruskal-Wallis	P= 0.12	P= 0.16	P= 0.08	P= 0.09	P=0.006	P=0.005	P= 0.10	P=0.006	P=0.004
Wilcoxon									
Control vs MADD group	–	–	–	–	P=0.003	P=0.003	–	P=0.003	P=0.003
Control vs McArdle group	–	–	–	–	P > 0.1	P > 0.1	–	P > 0.1	P > 0.1
MADD vs McArdle group	–	–	–	–	P=0.004	P=0.004	–	P=0.004	P=0.004
n = 17									
Kruskal-Wallis	P= 0.09	P= 0.14	P= 0.04	P= 0.09	P=0.005	P=0.006	P= 0.11	P=0.006	P=0.006
Wilcoxon									
Control vs MADD group	–	–	P > 0.1	–	P=0.003	P=0.003	–	P=0.003	P=0.003
Control vs McArdle group	–	–	P=0.003	–	P > 0.1	P > 0.1	–	P > 0.1	P > 0.1
MADD vs McArdle group	–	–	P= 0.10	–	P=0.008	P=0.008	–	P=0.008	P=0.008

P values obtained by the one-way analysis of variance according to Kruskal-Wallis and the Wilcoxon rank sum test for 2 groups using the data of Table 1.

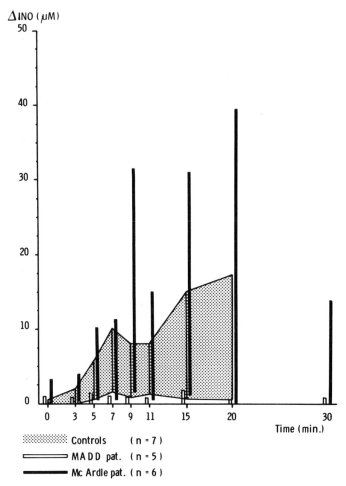

Fig. 2. Increase of plasma inosine (ΔIno)
concentration in controls, myoadenylate deaminase deficient (MADD) patients
and McArdle patients following
ischaemic forearm exercise. The ranges
of the concentrations found at 0, 3, 5,
7, 9, 11, 15, 20 and 30 minutes are
presented by vertical lines and bars.
The wide ranges of the McArdle patients
are due to very high values found in
one of the patients.

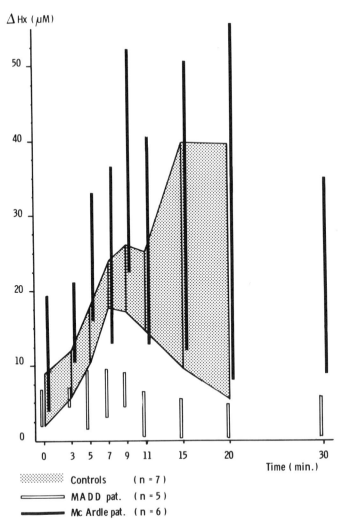

Fig. 3. Increase of plasma hypoxanthine (ΔHx) in
controls, myoadenylate deaminase deficient
(MADD) patients and McArdle patients. The
ranges of the concentrations found at 0,
3, 5, 7, 9, 11, 15, 20 and 30 minutes are
presented by vertical lines and bars. The
wide ranges of the McArdle patients are
due to very high values found in one of
the patients.

References

1. S.P.T. Sinkeler, H. Daanen, R.A. Wevers, T.L. Oei, E.M.G.
   Joosten and R.A. Binkhorst, The relation between blood
   lactate and ammonia in ischaemic handgrip exercise, Muscle
   and Nerve, in press (1985).

2. R.A. Hartwick, S.P. Assenza and P.R. Brown, Identification and quantitation of nucleosides, bases and other UV-absorbing compounds in serum, using reversed-phase high performance liquid chromatography, J. Chromat., 168:647-658 (1979).
3. R.L. Sabina, J.L. Swain, B.M. Patten, T. Ashizawa, W.E. O' Brien and E.W. Holmes, Disruption of the purine nucleotide cycle. A potential explanation for muscle dysfunction in myoadenylate deficiency, J. Clin. Invest., 66: 1419-1423, (1980).
4. R.L. Sabina, J.L. Swain, C.W. Olanow, W.G. Bradly, W.N. Fishbein, S. Dimauro and E.W. Holmes, Myoadenylate deaminase deficiency. Functional and metabolic abnormalities associated with disruption of the purine nucleotide cycle, J. Clin. Invest. 73: 720-730, (1984).

MYOADENYLATE DEAMINASE DEFICIENCY: DIAGNOSIS BY FOREARM ISCHEMIC EXERCISE TESTING

P.A. Valen, D.A. Nakayama, J.A. Veum, and R.L. Wortmann

Department of Internal Medicine, Medical College of Wisconsin and Veterans Administration Medical Center, Milwaukee, Wisconsin 53193, U.S.A.

Myoadenylate deaminase (MADA) (EC 3.5.4.6) catalyzes the deamination of AMP to IMP in skeletal muscle and plays a regulatory role in the purine nucleotide cycle (1). MADA deficiency is associated with a clinical syndrome of easy fatiguability, weakness and post-exercise myalgias (2,3,4). Alterations in purine nucleotide content of skeletal muscle may account for these symptoms in MADA deficiency (5).

Quantitation of lactate and ammonia production after forearm ischemic exercise has been proposed as a screening test for this disorder. A normal response is a several fold increase in both lactate and ammonia concentrations (5,6). In MADA deficient subjects lactate concentrations increase but little or no change is observed in ammonia levels (2,5,6,7). However, false positive results have been seen in normals (7,8), perhaps related to the varied methods of conducting the test.

Forearm ischemic exercise testing was performed in three patients with histochemically defined MADA deficiency and seven healthy volunteers in a standardized fashion to establish the limitations of the test. In addition to lactate and ammonia, plasma purines (adenosine, inosine and hypoxanthine) were measured after exercise.

METHODS

MADA deficient subjects were identified by histochemical analysis (9) of muscle biopsy specimens submitted to the Neurohistochemistry Laboratory at Clement J. Zablocki Veterans Administration Medical Center and Medical College of Wisconsin. The controls were 7 healthy volunteers ages 27 to 40.

Exercise testing was performed as follows. An intravenous line was established in an antecubital vein. Baseline blood samples were obtained and then a blood pressure cuff was placed around the upper arm proximal to the intravenous and inflated to a level 10 to 20 mm of mercury above systolic pressure. Subjects then exercised their forearm by squeezing a grip dynamometer as forcefully as possible at a rate of one grip every two seconds. The dynamometer was attached to a strain gauge and strip recorder so that the amplitude and duration of each grip was recorded. Exercise continued until exhaustion or for a maximum of 90 seconds. Blood samples were collected through the intravenous line at 1,3,5 and 10 minutes after the cuff was deflated.

Ten ml venous samples were collected in a plastic syringe and divided equally into heparinized tubes containing 250 ul of normal saline alone (Tube A) or with 0.1 mM EHNA, an adenosine deaminase inhibitor, and 0.2 mM dipyridamole, an inhibitor of bidirectional nucleoside transport (Tube B). Plasma from Tubes A were immediately analyzed for lactate and ammonia concentrations. Plasma from Tubes B were stored at -70°C for subsequent determination of adenosine, inosine and hypoxanthine concentrations by high performance liquid chromatography (10,11).

A subset of normals repeated the test at least one week later and were asked to exercise at the above rate but with what they considered 50 to 75% of maximum effort. Subject performance was quantitated by summing the areas under the deflection curves generated with each squeeze using a compensating polar planimeter. Planimeter units were expressed in Kg x sec.

RESULTS AND DISCUSSION

Subject performance, or exercise effort, is a critical variable in forearm ischemic exercise testing. Testing seven normal subjects at various levels of performance (Kg x sec) provided correlation coefficients of 0.71, 0.65, and 0.36 for lactate, ammonia and total purines (adenosine plus inosine plus hypoxanthine).

With maximum efforts, measurements of lactate and ammonia concentrations after forearm ischemic exercise clearly distinguished normals from MADA-deficient subjects (Table 1). However with retesting at submaximal exercise levels, five of seven normals failed to generate ammonia. Thus, by exercising with submaximal effort, it is possible for a MADA-normal individual to generate results indistinguishable from those of MADA-deficient subjects (Table 2).

Measurements of the rise in total purines after forearm ischemic exercise provides a better separation between normal and MADA-deficient individuals (Table 1). Despite exercising at submaximum levels, normal subjects increased their total

Table 1: Results of forearm ischemic exercise testing in normal (n=7) and MADA-deficient (n=3) subjects. Values are the range of maximum increase in lactate, ammonia and purines (adenosine plus inosine plus hypoxanthine) concentrations. Normals performed the test on two occasions, once with maximum exercise effort and, one week later, with an effort estimated at 50 to 70% of maximum intensity (Submax Effort).

	Lactate (mEq/L)	Ammonia (umol/L)	Purines (uM)
Normals			
Maximum Effort	4.0-8.4	40-163	23-70
Submax Effort	2.5-6.0	2-92	9.6-20
MADA-deficient	1.2-3.4	5-18	1.5-8.2

Table 2. Results of forearm ischemic exercise testing in normal exercising at different levels of performance and a MADA-deficient patient.

	Normal Subject			MADA Deficient
	100% Effort	91% Effort	58% Effort	100% Effort
LACTATE (mEq/L)				
Baseline	1.2	1.4	1.5	1.1
1 min	9.2	7.2	5.9	2.3
3 min	8.5	6.5	6.0	2.3
5 min	7.3	5.2	2.8	1.8
10 min	4.9	3.0	1.6	0.9
AMMONIA (umol/L)				
Baseline	28	25	26	30
1 min	99	46	25	28
3 min	140	49	33	26
5 min	135	46	26	25
10 min	79	33	25	32

purines 14.9 ± 5.3 uM, levels above those observed for MADA-deficient subjects, 3.5 ± 3.8 uM.

SUMMARY

These results indicate that measuring venous ammonia concentrations after forearm ischemic exercise is an effective means of screening for MADA deficiency but that submaximal exercise performance, whether due to weakness, pain or poor effort, can provide false positive results. Measurements of purine compounds released after exercise may increase the specificity of forearm ischemic exercise testing for MADA deficiency. The low level of purines released after exercise in MADA-deficient subjects supports the hypothesis that disordered purine metabolisms occurs when MADA activity is absent.

REFERENCES

1. Lowenstein JM and Goodman MN: Fed Proc 37:2308-2312, 1978.

2. Fishbein WN, Armbrustmacher VW and Griffen JL: Science 200:545-548, 1978.

3. Shumate JB, Katnik R, Ruiz M, Kaiser K, Frieden C, Brooke MH and Carroll JE: Muscle and Nerve 2:213-216, 1979.

4. Kar NC and Pearson CM: Arch Neurol 38:279-281, 1981.

5. Sabina RL, Swain JL, Patten BM, Ashizawa T, O'Brien WE and Holmes EW: J Clin Invest 66:1419-1423, 1980.

6. Patterson VH, Kaiser KK and Brooke MH: Neurology 33: 784-786, 1983.

7. DiMauro S, Miranda AF, Hays AP, Frank WA, Hoffmann GS, Schoenfeldt RS and Singh N: J Neurol Sc: 47:191-202, 1983.

8. Kelemen J, Rice DR, Bradley WG, Munsat TL, Dimauro S and Hogan EL: Neurology 32:857-863, 1982.

9. Fishbein WN, Griffen JL and Armbrustmacher VW: Arch Pathol Lab Med 104:462-466, 1980.

10. Hartwick RA, Krstulvic AM and Brown PR: J Chromatogr 186:659-676, 1979.

11. Ryan LM, Wortmann RL, Karas B and McCarty DJ: Arthritis Rheum 28:413-418, 1985.

PURINE NUCLEOTIDES IN HUMAN HEARTS DURING OPEN HEART SURGERY

Michael Havel, Werner Mohl, Gerhard Schopf, and
Mathias M.Müller

2nd Department of Surgery, Division of Clinical
Biochemistry
University of Vienna, Vienna, Austria

INTRODUCTION

During open heart surgery aortic clamping is the reason of
a temporary deficiency in oxygen supply and the development of
ischemia.Hypothermia and cardioplegia should protect the myocardium against an irreversible injury.As an effect of ischemia
myocardial content of high-energy phosphates decreases which is
accompanied by the release of degradation products (1,2). In order
to evaluate myocardial purine nucleotide metabolism tissue samples were taken during open heart surgery and the concentrations
of purine necleotides were determined using isotachophoresis.

MATERIALS AND METHODS

Patients
15 patients undergoing coronary artery bypass surgery were
informed and agreed to be selected for this study.The hearts
were arrested with 800 ml St. Thomas's crystalloid cardioplegic
solution which was applied in repetitive doses into the aortic
root.Myocardial biopsies were taken just before aortic clamping
(sample 1),after cardioplegic heart arrest(40 $\pm$ 9 minutes)just
before declamping(sample 2),and after 30 minutes of reperfusion
(sample 3).

Methods
Sample preparation:Tissue samples were wiped free of adhering
blood and immediately frozen in liquid nitrogen.Homogenization
was performed with a microdismembrator II (Brown,FRG).The pulverized tissue samples were extracted with cold 0.5 mmol/l perchloric acid, neutralized with 1 mmol/l potassium hydroxide, and
precipitated potassium perchlorate was removed by centrifugation.
Supernatants were analysed as suplicates for nucleotides by means
of isotachophoresis.Accuracy of these results were checked using
bioluminometric assays.

Capillary isotachophoresis:For determination of creatine phos-
phate(CP),ATP,ADP,AMP,IMP,NAD^+,$NADH/H^+$ and lactate as anions a
LKB 2127 Tachophor(LKB Bromma,Sweden)was used.The neutralized
extracts were directly applied on a teflon capillary tube, length
83 cm, I.D.0.5 mm, thermostated at 20°C. With a constant current of
40μm separation was achieved within 40 minutes using the electro-

lyte system described elsewhere(3).A UV-dectector at 254 nm was used.

Bioluminometric assays: Adenine nucleotides and CP were also determined bioluminometric using a Biocounter M 2000 (Luman, The Netherlands). After incubation at 37° with different enzymes nucleotides were converted to ATP which was determined by means of firefly luciferase (4). For the conversion of ADP to ATP, AMP to to ATP, and CP to ATP pyruvate kinase, pyruvate kinase and myokinase, and creatine kinase were used, respectively. Enzymes and substrates were from Boehringer Mannheim (FRG).

RESULTS

The concentrations of metabolites determined in human myocardial biopsies just before clamping of the aorta are summarized in Table 1. Potassium induced heart arrest for approximately 40 minutes resulted in a pronounced decrease of CP, in contrast to ATP (Fig.1). The drop of CP exhibited 39% of preischemic values. ADP and IMP showed only small increases, whereas the AMP content nearly doubled. A similar pattern showed NAD^+, $NADH/H^+$, and lactate: These metabolites increased during this period approximately 13%, 15% and 53%, respectively.

During reperfusion a decrease of adenine nucleotides especially that of ATP and AMP the latter had accumulated during cardiac arrest occurred (Fig.1). Only 37% ATP, 73% ADP and 75% AMP of preischemic control values were obtained. The only purine nucleotid which increased during reperfusion was IMP: the increase was 33% (preischemic controls = 100%). In contrast no loss/of myocardial CP content during reperfusion was determined. Through NAD^+, $NADH/H$ and lactate had accumulated during the ischemic period, these metabolites showed nearly preischemic values due to washout during 30 minutes of reperfusion.

DISCUSSION

The data obtained in this study show that during potassium induced heart arrest changes in the cellular energy metabolism indicate a tolerable ischemic injury.The dramatic decrease of

Table 1.Myocardial nucleotide content

Nucleotide/Metabolite	Concentration (nmoles/mg protein)
CP	19.9 ± 7.4
ATP	20.8 ± 13.5
ADP	13.1 ± 5.6
AMP	12.0 ± 6.4
IMP	3.4 ± 2.1
NAD^+	9.2 ± 3.0
$NADH/H^+$	5.9 ± 2.8
Lactate	75.3 ± 15.5

myocardial content accompanied by an accumulation of lactate and a more pronounced increase of $NADH/H^+$ are due to decreased myocardial oxygen supply.Since ATP decreased during this period only slightly,the reduced energy demand of the heart might be covered

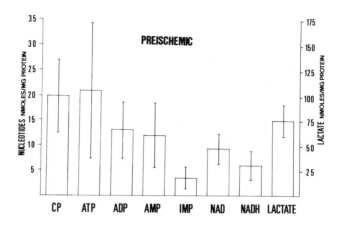

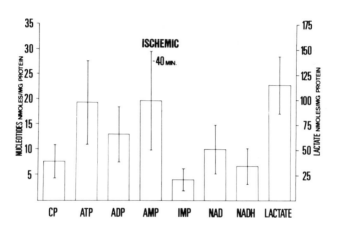

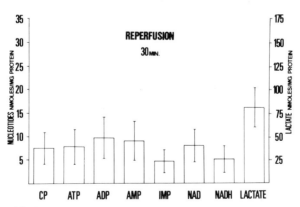

Figure 1.Myocardial nucleotide and lactate content during open
heart surgery.

by CP and anaerobic glycolysis.Thus,CP as an energy resource of the heart is degraded in favour of maintenance of the ATP content.In a number of investigations correlations between duration of ischemia or cardioplegic heart arrest and tissue content of high-energy phosphates have been shown(5-7).It has been proposed that tolerable and reversible ischemias are characterized by small or no fall in tissue ATP.Therefore cardiac arrest for a longer period than in this study should exhaust the CP pool,followed by a decrease of myocardial ATP.

During reperfusion with the blood from the extracorporal circulation at normal temperature a significant decrease of tissue content of nucleotides could be demonstrated.This is in accordance with reports showing a dramatic release of adenosine, inosine and hypoxanthine during initial reperfusion(2),which indicates degradation of purine nucleotides.These degradation products might have been accumulated already during cardioplegic heart arrest inhibiting catabolic enzymes.As a consequence AMP should increase and total adenine nucleotides partially be saved In contrast ATP will be consumed extensively during initial reperfusion when the heart is already beating.Generated ADP will enter purine catabolism by the action of adenylate kinase.In this period the washout of nucleosides and hypoxanthine during reperfusion should decrease substrate inhibition of catabolic enzymes.Therefore purine nucleotides are degraded more rapidly leading to myocardial depletion of high-energy phosphates.Myocardial dysfunction during reperfusion is probably due to this rapid degradation and the loss of preformed purines(2,8).

In open-chest dogs it was demonstrated that after brief occlusion of a segmental coronary artery a prolonged depletion of adenine nucleotides occured(1),which indicates a diminished capability to restore the nucleotide pool.It was suggested that a limited myocardial capacity of de novo purine synthesis and salvage pathway exists due to diminished availability of phosphoribosyl pyrophosphate(PP-ribose-P).If this is true,the utilization of hypoxanthine generated during cardiac arrest and reperfusion by purine salvage pathway should be favoured,since hypoxanthine guanine phosphoribosyltransferase(HGPRT)exhibits a much higher affinity for PP-ribose-P than PP-ribose-P amidotransferase(9).Indeed increased IMP levels were demonstrated. It is most probable that hypoxanthine might be shifted towards IMP by HGPRT thus decreasing de novo synthesis due to PP-ribose-P consumption.

Depletion of adenine nucleotides during open heart surgery should be overcome by modifications of cardioplegic solutions and/or reperfusate.Addition of pentose stimulating PP-ribose-P generation and preformed purines(11,12)will enhance formation of nucleotides mainly via purine salvage.

ACKNOWLEDGEMENT

This study was supported by "Jubiläumsfond der österreichisch Nationalbank",project 2139.

REFERENCES

(1)   J.L.Swain,R.L.Sabina,P.A.Mc Hale,J.C.Greenfield,and E.W.Holmes,Prolonged myocardial nucleotide depletion afte brief ischemia in the open-chest dog,Am.J.Physiol.242:818 (1982)
(2)   A.C.Fox,G.E.Reed,H.Meilman,and B.B.Silk,release of nucleo

sides from canine and human hearts as an index of prior ischemia,Am.J.Cardiol 43:52(1979)

(3)   M.Aomine,M.Arita,S.Imanishi,and T.Kiyosue,Isotachophoretic analyses for metabolites of cardiac and skeletal muscles in four species,Jap.J.Physiol.32:741(1982)

(4)   R.J.Ellis,and C.Gardner,Determination of high-energy phosphates using bioluminescence,Analyt.Biochem.105/354(1980)

(5)   T.C.Vary,E.T.Angelakos,and S.W.Schaffer,Relationship between adenine nucleotide metabolism and irreversible ischemic tissue damage in the isolated perfused rat heart,Circ.Res.45/218(1979)

(6)   R.A.Kloner and E.Braunwald,Observations on experimental myocardial ischemia,Cardiovasc.Res.14/371(1980)

(7)   D.J.Hearse,R.Crome,D.M.Yellon and R.Wyse,Metabolic and flow correlates of myocardial ischemia,Cardiovas.Res.17:452(1983)

(8)   E.R.Rosenkranz and G.D.Buckberg,Myocardial protection during surgical coronary reperfusion,J.Am.Coll.Cardiol.1:1235(1983)

(9)   A.W.Wood,M.A.Becker and J.E.Seegmiller,Purine nucleotide synthesis in lymphoblasts cultured from normal subjects and in a patient with the Lesch-Nyhan syndrome,Biochem.Genetics 9:261(1973)

(10) M.M.Müller,W.Mohl and G.Schopf,Adenosinnukleotidstoffwechsel des Myokards,Z.med.Labor-Diagn.26/71(1985)

(11) M.K.Pasque,T.L.Spray,G.L.Pellom,P.Van Trigt,R.B.Peyton and W.D.Currie,Ribose-enhanced myocardial recovery following ischemia in the isolated working rat heart,J.Thorac.Cardiovasc.Surg.83:390(1982)

(12) D.F.De Witt,E.J.Kenneth and D.M.Behrendt,Nucleotide degradation and functional impairment during cardioplegia:Amelioration by inosine,Circulation 67:171(1983)

CELL CULTURE MODELS FOR THE STUDY OF PURINE

METABOLISM IN HUMAN PLACENTAL TISSUE

Kari O. Raivio and Kim Vettenranta

Children's Hospital and Departments of Obstetrics and Gyne-

cology, University of Helsinki, SF-00290 Helsinki, Finland

INTRODUCTION

The placenta is an organ of crucial importance to the developing fetus. In addition to its function in gas and nutrient transport, it has a variety of biosynthetic capabilities, which have usually been studied in terms of hormone production (stroid hormones, chorionic gonadotropin). Since the placenta develops, grows, matures, and begins to degenerate in the course of 9 months, its biochemical properties will obviously change as a function of gestational age. The study of these properties has been hampered by the presence of several cell types, which are known to have different origins and thus presumably specialized functions.

In vitro methods for the study of placental metabolism all have their drawbacks: tissue slices have defective oxidative metabolism and contain an unknown mixture of cells, malignant cell lines of human placental origin (choriocarcinoma) may be more representative of tumor than of placental cells, and all perfusion methods rely on unphysiological routes of circulation. Primary cultures of placental cells have been successfully set up[1], but interpretation of their metabolic performance depends on the reliable identification of the cells that attach to and grow on the culture plate[2].

The aim of this study was to establish methods for the primary culture of human placental cells with defined morphological and functional, i.e. trophoblastic properties, and to compare them with a choriocarcinoma cell line. With a well-characterized model system, purine metabolism can be studied and correlated with the stage of gestation and with specific placental function. Examples of such applications will also be given.

MATERIALS AND METHODS

A human choriocarcinoma cell line (BeWo) was obtained from the American Type Culture Collection and grown in Ham's F-10 medium containing 10% fetal bovine serum.

Sterile placentae were obtained, by approval of the local ethical committee and after informed consent, either at legal terminations of pregnancy at 8-12 weeks (I trimester), or at elective cesarean sections at 26-31 weeks (early III trimester) or at term (late III trimester). They were rinsed in phosphate-buffered saline at $+0^{\circ}C$, minced with scissors, and incubated for 5 min (I trimester) or 15 min (III trimester) at $+37^{\circ}C$ in MEM containing 0.1% collagenase and 50 IU/ml DNAase. The resulting cell suspension was filtered through sterile cotton gauze into Ham's F-10 medium containing 20% fetal bovine serum. The enzyme treatment was repeated 4 times and the cell suspensions pooled. Erythrocytes were removed by hypotonic lysis or by density gradient centrifugation in 30% (v/v) Percoll. After washing in F-10 medium, the cells were plated out on 35 mm Petri dishes (ca. 0.5×10^6 cells/plate) and grown in an atmosphere of 5% CO_2 in air. Culture medium was collected and replaced daily.

The cells were characterized morphologically by phase contrast microscopy and after staining with May-Grünwald-Giemsa. To verify their origin, immunofluorescent staining for the subunit proteins of cytoskeletal intermediate filaments was used. Cytokeratins, which are expressed in cells of epithelial, i.e. trophoblastic origin, were detected using polyclonal rabbit anti-cytokeratin antibodies[3]. Vimentin, which is characteristic of mesenchymal, i.e. fibroblastic cells, was stained with mouse monoclonal antibodies[4].

To monitor the maintenance of specific trophoblastic function, the culture media were analyzed for human chorionic gonadotropin (HCG) by time-resolved fluoroimmunoassay[5], and for pregnancy-specific beta-1 glycoprotein (SP_1) by radioimmunoassay[6]. These proteins were also localized in cultured cells by immunofluorescence (HCG) or by the immunoperoxidase method (SP_1) using monoclonal antibodies.

Previously described methods were used for studying the incorporation of ^{14}C-adenine[7] and ^{14}C-formate[8] into purine compounds, and for assaying the activity of AMP-deaminase[9], 5'-nucleotidase, and hypoxanthine phosphoribosyltransferase (HPRT)[10].

Because trophoblastic cells tend to form syncytial structures in culture, cell number is not a useful reference basis for biochemical measurements. Therefore, parallel plates were set up for the quadruplicate determination of protein and DNA content.

RESULTS

Characteristics of primary cultures of placental cells

In the first days of culture, over 95% of the cells could be classified into one of four distinct types by morphological criteria:

A. Small round cells with relatively large nuclei and scant cytoplasm.

B. Larger polymorphic cells with abundant cytoplasm and occasional
bi- or trinucleated forms.

C. Multinucleated (up to 20) giant cells.

D. Spindle-shaped fibroblast-like cells.

Cells belonging to types A-C stained positively with anti-cytokeratin
antibodies, indicating their trophoblastic nature, whereas type D cells
expressed vimentin intermediate filaments, typical of mesenchymal cells.
On day 2 of culture, 60-90% of the I trimester and 70-95% of the III
trimester cells were cytokeratin-positive.

The I trimester cultures remained stationary for 5-7 days, after
which type A and type D cells started to multiply, whereas the multinucleated forms disappeared. By day 14 a confluent monolayer was formed,
which contained over 95% vimentin-positive cells. A similar but slower
growth and appearance of vimentin-positivity was seen in early III trimester cultures. In the late III trimester cultures, no growth was seen
during the 14-day observation period, but the number of trophoblastic
cells gradually diminished.

Staining of the cultures with anti-HCG antibodies showed that only
cytokeratin-positive, i.e. trophoblastic cells contained HCG. Expression
of SP_1 was detected not only in trophoblastic but also vimentin-positive
cells, which confirms that the synthesis of this protein is not specific
to the trophoblast.

Secretion of HCG into the medium was significantly higher in the I
than in the III trimester cultures. In all instances, HCG production was
maximal during the first days and then rapidly decreased to undetectable
levels, by day 10 in the I trimester and by day 5 in the III trimester
cultures. SP_1 production remained high in the I trimester but decreased
to undetectable levels by day 5 in the III trimester cultures.

Choriocarcinoma compared to trophoblastic cells

To evaluate, whether malignant cells of human placental origin reflect the metabolism of the normal trophoblast, the incorporation of ^{14}C-
adenine into other purine compounds and production of HCG were compared
in BeWo cells and primary trophoblastic cells from the III trimester on
day 3 of culture (Fig. 1). Choriocarcinoma cells incorporated significantly

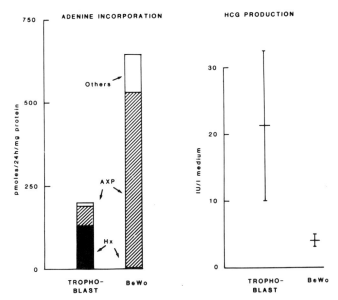

Fig. 1. Incorporation of radioactivity from
^{14}C-adenine (50 uM) into purine compounds and production of human chorionic gonadotropin (HCG) by cultured
trophoblastic cells from a III trimester placenta and by human choriocarcinoma cells (BeWo). AXP=adenine nucleotides, Hx= hypoxanthine.

more adenine than trophoblastic cells, most of it into adenine nucleotides, whereas most of the adenine metabolized by normal placental cells
was recovered as hypoxanthine. In contrast, production of HCG was substantially less in the malignant as compared to normal placental cells.

Purine metabolism in I vs. III trimester cultures

    To assess the capacity for purine catabolism and reutilization, the
activities of some key enzymes were measured in I and III trimester cultures (Table 1). The slight differences in AMP deaminase, 5'-nucleotidase,
and HPRT activity were not statistically significant, suggesting that
these processes do not change noticeably in the course of pregnancy, at
least as judged by enzyme activities alone.

    The capacity of cultured trophoblastic cells for purine synthesis
de novo was assessed in 3 preliminary experiments by measuring the incorporation of ^{14}C-formate into total cellular purines. After 24 h of labelling, the I trimester cultures incorporated an average of 4.1×10^5 cpm/ mg
protein, and late III trimester cultures 1.4×10^5 cpm/mg protein. Although

this suggests a more active purine synthesis early in gestation, the range was wide and partially overlapping, and more experiments are needed before any conclusions can be drawn.

DISCUSSION

Morphological characteristics, expression of intermediate filament proteins, and production of hormones by placental cells indicate that their trophoblastic character is maintained only during the first days of primary culture. Thereafter, specific function gradually decreases, as shown by the diminishing hormone synthesis, and the dominant cell type in the cultures changes from epithelial to mesenchymal. For any studies attempting to characterize the metabolism of the trophoblast, the useful time period in view of "specificity" is thus limited. Whether this period can be extended by the use of appropriate growth factors or by other methods remains to be explored.

The continuous choriocarcinoma cell line (BeWo) used in our experiments seems to be more representative of a malignant cell line in general than of a trophoblastic cell, even though some degree of specific function is retained, as shown by modest HCG production[7].

Our studies suggest a difference in growth potential between I and III trimester placental cells, not only of epithelial but also of mesenchymal origin[2]. This may be a reflection of senescence in an organ nearing the end of its normal life span. A similar decrease in the capacity for HCG and SP_1 synthesis was also demonstrated. Our preliminary observations on purine synthesis de novo do not allow a definite conclusion, whether the high metabolic activity of the trophoblast early in gestation also applies to purine metabolism. If such turns out to be the case, the placenta may play a role in supplying purines to the developing fetus.

Table 1. Enzyme activities in cultured first (I) and third (III) trimester trophoblastic cells.

Enzyme	I	III
	nmoles/min/mg protein (n)	
AMP-deaminase	21.8 $\pm$ 8.8 (9)	24.0 $\pm$ 4.5 (8)
5´-nucleotidase	71.3 $\pm$ 40.5 (15)	43.6 $\pm$ 15.2 (15)
HPRT	0.7 $\pm$ 0.4 (9)	0.6 $\pm$ 0.3 (15)

ACKNOWLEDGEMENTS

We thank Ms. Ritva Löfman for skilled technical assistance. These studies were supported by the Sigrid Jusélius Foundation, The Academy of Finland, and the Foundation for Pediatric Research in Finland.

REFERENCES

1. K. Stromberg, The human placenta in cell and organ culture, in: "Methods in Cell Biology", C.C. Harris, B.F. Trump and G.D. Stoner, eds., Academic Press, New York, Vol 21B (1980).

2. S. F. Contractor, A. Routledge, and S. R. Sooranna, Identification and estimation of cell types in mixed primary cell cultures of early and term human placenta, Placenta 5:41 (1984).

3. H. Holthöfer, A. Miettinen, R. Paasivuo, V-P. Lehto, E. Linder, O. Alfthan, and I. Virtanen, Cellular origin and differentiation of renal carcinomas, Lab. Invest. 49:317 (1983).

4. I. Virtanen, V-P. Lehto, E. Lehtonen, T. Vartio, S. Stenman, P. Kurki, O. Wager, J. V. Small, D. Dahl, and R. A. Badley, Expression of intermediate filaments in cultured cells, J. Cell Sci. 50:45 (1981).

5. K. Pettersson, H. Siitari, I. Hemmilä, E. Soini, T. Lövgren, V. Hänninen, P. Tanner, and U-H. Stenman, Time-resolved fluoroimmunoassay of human choriogonadotropin, Clin. Chem. 29:60 (1983).

6. M. Heikinheimo, H. A. Unnérus, T. Ranta, H. Jalanko, and M. Seppälä, Pregnancy-specific beta-1-glycoprotein levels in cholestasis of pregnancy, Obstet. Gynecol. 52:276 (1978).

7. K. Vettenranta and K.O. Raivio, Purine re-utilization in normal and malignant cells of human placental origin, Placenta 5:315 (1984).

8. M. S. Hershfield and J. E. Seegmiller, Regulation of de novo purine biosynthesis in human lymphoblasts, J. Biol. Chem. 251:7348 (1976).

9. K. O. Raivio, P. Santavuori, and H. Somer, Metabolism of AMP in muscle extracts from patients with deficient activity of myoadenylate deaminase, Advan. Exp. Biol. Med. 165B:431 (1984).

10. J. P. R. M. van Laarhoven, G. Th. Spierenburg, and C.H.M.M. de Bruyn, Enzymes of purine nucleotide metabolism in human lymphocytes, J. Immunol. Meth. 39:47 (1980).

PURINE NUCLEOTIDE SYNTHESIS IN CULTURED RAT EMBRYOS UNDERGOING ORGANOGENESIS

Peter B. Rowe, Sandra E. McEwen and Annette Kalaizis

Children's Medical Research Foundation, The University of Sydney
Camperdown, N.S.W. Australia

INTRODUCTION

The culture of rat embryos undergoing organogenesis (9.5 to 11.5 days gestation) is widely used as a technique for evaluating potentially teratogenic agents and for the study of organ differentiation (1). During culture the embryo develops from a relatively undifferentiated neurula stage to almost the tail bud stage with 24 or 25 somites and the segregation of tissues into the primordia of the neural, cardiac circulatory and hepatic organs. These changes are associated with rapid growth of the surrounding yolk sac. Total cell number increases approximately two orders of magnitude, the embryo containing some $\cong 3$ x 10^6 cells and the yolk sac $\cong 0.7$ x 10^6 cells at 11.5 days gestation (2). Relatively little basic biochemical and physiological data are available however concerning this experimental system. The pre-implantation mouse embryo contain high levels of activity of both purine salvage enzymes, hypoxanthine and adenine phosphoribosyl-transferases and this activity increases in parallel with post-implantation embryonic development (3). In these studies we have shown that (i) de novo synthesis is the sole source of the purine nucleotides required for in vitro rat embryonic growth during organogenesis (ii) the presence of high levels of activity of purine catabolic enzymes in the homologous serum essential for culture prohibits the salvage of purines (iii) the degradation of culture medium serum protein by the yolk sac is the major source of amino acids for embryonic growth and development (iv) while the 3-carbon atom of serine is the major source of one carbon units for purine ring synthesis there is a significant contribution from the 2-ring carbon atom of tryptophan. No one-carbon units are derived from glycine, histidine or choline (v) there is a high level of GTP in both the embryo and its yolk sac which is reflected in the very low ATP/GTP ratios.

METHODS

The rat embryo culture method was based on that of New et al. and has been described in detail elsewhere (4). The embryos are removed at 9.5 days gestation and cultured for 48 hr in 90% heat-treated rat serum containing streptomycin and penicillin. Under these conditions

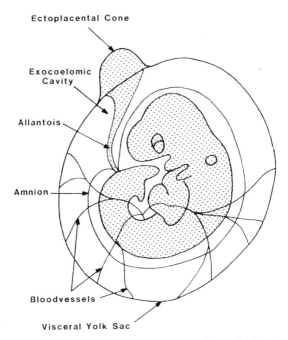

Fig. 1. Rat embryo at 11.5 days gestation after 48 hr in culture.

embryonic development parallels that observed in vivo. At the end of
the culture period (Fig. 1) the exocoelomic fluid and the amniotic
fluid were removed through a micropipette, the yolk sac and embryo
separated by microdissection and processed by the method of Shibko et
al. (5) for the separation of the acid soluble phase, total nucleic
acids and protein. Part of the acid soluble phase nucleotides and the
nucleic acids were hydrolyzed to base level in 12.0 M perchloric acid
for 1 hr at 96° C. The protein fractions from the embryos, yolk sacs
and the culture medium were hydrolyzed under vacuum in constant boiling
HCl for 20 hr at 110°C.

Analyses of purine nucleotide and purine and pyrimidine bases were
carried out by reverse-phase paired-ion HPLC (6). Amino acids were
analysed by gas chromatography after acylation of the amino groups with
hepta-fluorobutyric acid (7). The activity of the embryonic purine
salvage enzymes and the purine catabolic enzymes in the serum-xanthine
oxidase, guanine deaminase, adenosine deaminase and nucleoside
phosphorylase was measured by previously described methods (4).

RESULTS AND DISCUSSION

De Novo Purine Synthesis

The incorporation of [1-14C]glycine into the acid soluble phase,
the nucleic acids and protein of the exocoelomic fluid, the yolk sac
and the embryo throughout a 48 hr culture period is shown on Table 1.
HPLC analysis of the unhydrolyzed acid-soluble phases revealed that
most of the isotope was present as free glycine in the exocoelomic
fluid while it was more evenly distributed between glycine and the
purine nucleotides in the yolk sac and the embryo. This was confirmed
by analysis of the hydrolyzed acid soluble phases (Table 2) and as most
of the adenine and guanine was present as ATP and GTP respectively it
could be seen that the ATP/GTP ratio was close to unity. Although the

Table 1. The incorporation of [1-14C]glycine into embryonic tissue

Tissue	Incorporation nmoles per embryo	Isotope Distribution (%)		
		Acid soluble	Nucleic Acid	Protein
Exocoelomic fluid	3.4	91.8	2.5	5.7
Yolk sac	11.5	30	19	51
Embryo	5.6	30	20	50

The data is from a typical experiment in which 2 flasks each containing three embryos were cultured for 48 hours.

Table 2. Distribution of [14C]-glycine label in hydrolysed acid-soluble phases

Tissue	Distribution (%)		
	Adenine	Guanine	Glycine
Exocoelomic fluid	1.75	1.75	96.5
Yolk sac	28.5	27.8	43.7
Embryo	29.6	38.1	32.4

Part of the acid soluble phase from the experiment shown in Table 1 was hydrolysed in 12N PCA for one hour at 96° and analysed by HPLC.

reason for the high levels of GTP is unknown it may well be related to the relatively undifferentiated state of the embryonic cells. Mycophenolic acid (10 µM) strongly inhibits embryonic growth in vitro. The specific radioactivity both of adenine and guanine in the yolk sac soluble phase was 1/10th that of the medium glycine while in the embryo acid soluble phase it was 1/15th. In the nucleic acid fraction of both the embryo and the yolk sac the isotope was evenly distributed between adenine and guanine and the specific radioactivity was identical with that of their respective soluble phase purines.

Purine Salvage

No salvageable purine bases or nucleosides could be detected by HPLC in the serum used for embryo culture. This was not surprising as high levels of purine catabolic enzyme activity were present. Assayed directly in the culture medium, serum contained (nmol/ml/hr) xanthine oxidase, 696 (with hypoxanthine as substrate) 30 (with adenine as substrate); guanine deaminase, 346; adenosine deaminase, 124; nucleoside phosphorylase, 57. The embryos at the beginning of culture contained significant levels of both purine salvage enzymes (HGPRTase, 1.48 nmol per embryo per hr with hypoxanthine as substrate and APRTase, 0.39 nmol per embryo per hr). After 24 hr culture there was a ten-fold increase in activity.

Table 3.   Adenine incorporation by embryonic tissues (nmol/embryo/3hr)

	At 23 hr		At 26 hr		At 29 hr	
	Soluble phase	Nucleic Acids	Soluble phase	Nucleic Acids	Soluble phase	Nucleic Acids
Yolk sac	1.33	0.31	0.84	0.72	1.03	1.26
Embryo	0.25	0.07	0.20	0.16	0.18	0.21

> In this experiment nine embryos were cultured for 20 hr in serum-based medium. After transfer to RPMI 1640 containing 0.25% bovine serum albumin, they were pulsed for 3 hr with 50 μM [8-14C[adenine. One group of three embryos was then processed while the two other groups of three were cultured for a further 3 and 6 hrs respectively after transfer to the original medium.

Embryos could be cultured briefly in largely serum-free defined medium and a pulse chase study with [8-14C]adenine at 20 hours culture showed that not only was this base readily incorporated but that AMP was cycling through IMP to GMP (Table 3). This study also illustrated that there was no transfer of labelled purines between the yolk sac and the embryo.

Amino acid Metabolism

In the light of the discrepancy between the specific radioactivity of the culture medium glycine and the cellular purine bases and the negligible contribution to the nucleotide pool by purine salvage pathways a more detailed study of amino acid metabolism was under-taken in conjunction with a study of the origin of one-carbon units required for purine nucleotide synthesis.

Amino acid analysis of the culture medium and the exocoelomic fluid showed that there was an overall 4.5 fold-increase in total amino acid concentration in the latter but that there was a significant variation in the extent to which different amino acids were concentrated (Table 4). In studies with [3-14C]serine it could be shown that there was a five-fold decrease in specific radioactivity even within the cells of the yolk sac (Table 5) and this was reflected in the specific radioactivity of the purine bases in the nucleotides of the soluble phase, RNA and DNA (Table 6). These results supported the earlier work of Williams et al. (8) who proposed that there was significant proteolysis of the medium serum proteins by yolk sac lysosomes. The amino acid composition of the exocoelomic fluid did not reflect that of the serum proteins supporting the concept of selectivity of protein degradation (9). If serine were the sole source of one-carbon units for the synthesis of purine rings then their specific radioactivity should be twice that of free intracellular serine. Serine via its 3-carbon atom clearly only supplies about 50% of the one-carbon units required. Independent studies with [2-ring 14C] tryptophan, [2-14C] ring histidine, [2-14C]glycine and [methyl-14C]choline showed that only tryptophan cleavage provided 15-20% of the required one-carbon units.

Table 4. Amino acid concentrations and distribution in the culture
medium and exocoelomic fluid*

Amino Acid	Culture Medium	%	Exocoelomic Fluid	%
Ala	474	18.9	1,763	15.7
Arg	91	3.6	104	0.9
Asp	51	2.0	350	3.1
Glu	275	11.0	933	8.3
Gly	269	10.7	920	8.2
Ile	65	2.6	245	2.2
Leu	194	7.7	833	7.4
Lys	212	8.5	1,302	11.6
Met	15	0.6	40	0.3
Phe	135	5.4	382	3.4
Pro	152	6.1	1,346	12.0
Ser	187	7.5	1,253	11.1
Thr	143	5.7	605	5.4
Tyr	111	4.4	430	3.8
Val	130	5.2	726	6.5
Total	2,504		11,233	

* Concentration is expressed as μM. This experiment
involved 20 embryos which were exposed to [3-14C]serine
throughout the 48 hr culture period.

Table 5. Serine content and specific radioactivity in various phases
of the experimental system.*

Phase	Specific Radioactivity Ci/Mol	Content nmol/20 embryos
1. Culture Medium	1.00	2,390
2. Yolk Sac Acid Soluble	0.24	45
3. Yolk Sac Protein	0.15	786
4. Exocoelomic Fluid	0.12	530
5. Embryo Acid Soluble	0.13	82
6. Embryo Protein	0.11	471

* These values were from the same experiment as in Table 4.

Table 6. Specific radioactivity and content of purine and pyrimidine bases in soluble phase nucleotides and nucleic acids in embryo and yolk sac*

		Adenine		Guanine		Thymine	
		SA	Content	SA	Content	SA	Content
EMBRYO	Nucleotides	0.13	126	0.13	133		
	RNA	0.135	27	0.14	72		
	DNA	0.13	133	0.125	110	0.06	123
YOLK SAC	Nucleotides	0.24	115	0.25	87		
	RNA	0.24	55	0.26	75		
	DNA	0.23	84	0.23	80	0.12	85

*These values were obtained from the same experiment with [3-14C]serine with 20 embryos as for Tables 4 and 5. Specific radioactivity (SA) is expressed as Ci/mole and content as nmol/20 embryos.

REFERENCES

1. D.A.T. New, Whole embryo culture and the study of mammalian embryos during organogenesis, Biol. Rev. 53:81 (1978).
2. N.A. Brown, E.H. Goulding and S. Fabro, Ethanol embryotoxicity: Direct effects on mammalian embryos in vitro, Science 206: 573 (1979).
3. C.J. Epstein, Phosphoribosyltranferase activity during early mammalian development, J. Biol. Chem. 245:3289 (1970).
4. P.B. Rowe and S.E. McEwen, De novo purine synthesis in cultured rat embryos undergoing organogenesis, Proc. Nat. Acad. Sci. (U.S.A.) 80:7333 (1983).
5. S. Shibko, P. Koivistoinen, C.A.Tratnyek, A.R. Newhall and L. Friedman, A method for sequential quantitative separation and determination of protein, RNA, DNA, lipid and glycogen from a single rat liver homogenate or from a subcellular fraction, Anal. Biochem. 19:514 (1967).
6. E. McCairns, D. Fahey, D. Sauer and P.B. Rowe, De novo purine synthesis in human lymphocytes, J. Biol. Chem. 258:1851 (1983).
7. J. Desgres, D. Boisson and P. Padieu, Gas-liquid chromatography of isobutylester N(0)-heptafluorobutyrate derivatives of amino acids on a glass capillary column for quantitative separation in clinical biology, J. Chromatog. 162:133 (1979).
8. K.E. Williams, J.B. Lloyd, M. Davies and F. Beck, Digestion of an exogenous protein by rat yolk sac cultured in vitro, Biochem. J. 125:303 (1971).
9. S.J. Freeman, F. Beck and J.B. Lloyd, The role of the visceral yolk sac in mediating protein utilization by rat embryos cultured in vitro, J. Embryol. exp. Morph. 66:223 (1981).

EVIDENCE FOR A MEMBRANE ADENOSINE RECEPTOR IN LEISHMANIA

MEXICANA MEXICANA (WR 227)

Brian D. Hansen, Peter K. Chiang and
Jose Perez-Arbelo

Walter Reed Army Institute of Research
Washington, D.C. 20307-5100

INTRODUCTION

Leishmania spp., a Protozoan Parasite transmitted to the mammalian
host by the sand fly (genus Phlebotomus), produces a disease manifested
in the cutaneous, mucocutaneous or visceral form. The choice of drugs
has been limited primarily to the pentavalent antimonial compounds.
Successful treatment within the endemic tropical areas of the world
varies due to problems of drug resistance and to strain differences in
Leishmania Spp. These difficulties have prompted the search for new
effective oral antileishmanial agents and ultimately the development of a
vaccine. We are currently studying proteins on the parasite plasma
membrane (enzymes, receptors, transporters) which may be functionally
vital to the organism and amenable to chemotherapeutic attack or elicit a
host immune response. The present study examines evidence for an adenosine receptor on the surface membrane of both the promastigote (insect
stage) and the amastigote form (mammalian stage) and the resultant modulation of adenylate cyclase activity, intracellular cyclic AMP levels,
the rate of cell growth and parasite transformation.

MATERIALS AND METHODS

Promastigotes of Leishmania mexicana mexicana (WR 227) were routinely grown in Schneiders Drosophila Medium supplemented with 30% (V/V)
fetal bovine serum according to the procedure of Hendricks and Wright
(1). Amastigotes were harvested from infected murine macrophages (J774),
isolated and purified after the method of Chang (2). Promastigote
membranes were purified according to the technique of Dwyer et al. (3).

The presence of an adenosine binding protein on the surface of both
the promastigote and amastigote of L. m. mexicana was investigated by
measuring the specific binding of radiolabeled adenosine receptor ligand
agonists (^{3}H-cyclohexyladenosine, CHA; ^{3}H-methyl-2-phenyladenosine,
PIA). Purified promostigote membranes (at 27°C) and isolated amastigotes
(at 4°C) were exposed to the labeled ligands (0-400nM) in the presence
and absence of 1mM adenosine to determine the total, nonspecific and
specific binding. The effect of both 1uM CHA and 1uM PIA on promastigote
plasma membrane adenylate cyclase activity was also tested utilizing the
procedure of Salomon (4). Moreover, following a 5 minute exposure of
promastigotes to 1uM CHA and 1uM PIA, intracellular cyclic AMP levels

were measured. Promastigotes (10^6 cells/ml) grown in vitro were also exposed to 1uM CHA and 1uM PIA for 0-96 hours and the effect of these ligands on the rate of cell replication measured. Finally amastigotes grown in murine macrophages at 34°C were harvested, purified and exposed to 1uM PIA and 1uM CHA at 27°C from 0 to 48 hours. The effect of PIA and CHA on the rate of conversion from the amastigote to the promostigote stage was then measured.

RESULTS AND DISCUSSION

Membrane binding of radiolabeled adenosine receptor ligands

The specific binding of ^{3}H-CHA (0-400nM) and ^{3}H-PIA (0-200nM) to the promastigote plasma membrane was measured utilizing 1mM adenosine to correct for nonspecific binding (Figure 1).

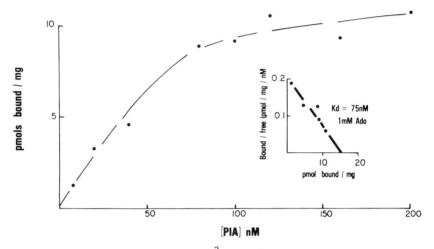

FIGURE 1 The specific binding of ^{3}H PIA (methyl-2-phenyladenosine, 0-200nM) to the promastigote plasma membrane of Leishmania mexicana mexicana. 1mM adenosine was used to correct for nonspecific binding. The inset shows a scatchard plot of the data using linear regression analysis. Each point is the mean of three replicates.

A kd value of 75nM was determined following a "Scatchard Plot" analysis of the data. Similar results were obtained when ^{3}H-CHA binding was determined with a kd value of 156nM. Specific binding of ^{3}H-PIA was also determined in the isolated amastigote form yielding a kd of 0.17nM. These data indicate that these adenosine receptor ligand agonists appear to bind specifically to the promastigote and amastigote surface presumably to an adenosine binding protein. However, the presence of an adenosine receptor can only be demonstated by showing not only specific binding but a resultant cellular physiological response.

The effect of PIA and CHA on promastigote membrane adenylate cyclase activity

Purified plasma membranes of the promastigote stage were pre-incubated with increasing concentrations of PIA and CHA (0-100μM) over a 30 minute period and then monitored for changes in adenylate cyclase activity (Figure 2).

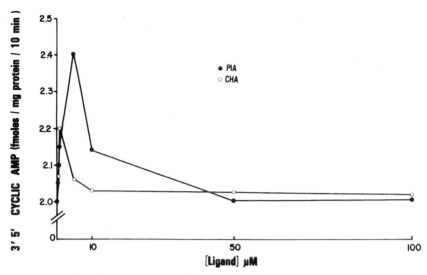

FIGURE 2   The effect of 0 to 100 µM CHA and PIA on plasma membrane
adenylate cyclase activity from the promastigote of
Leishmania mexicana mexicana.

Membrane enzyme activity was maximally stimulated in the presence of 5µM
PIA and 1µM CHA. Numerous workers have demonstrated similar mechanisms
of interaction between adenosine receptor binding and membrane bound
adenylate cyclase in mammalian cells (Wolff,5). A similar system may be
operating on Leishmania Spp.

The effect of CHA and PIA on intracellular cyclic AMP levels, the rate of
promastigote growth and amastigote transformation.

Following a 5 minute exposure of promastigotes to 1µM CHA and 1µM
PIA, a 100% and a 50% increase in intracellular cyclic AMP levels were
noted respectively (Table 1).

Table 1. The effect of 1µM PIA and 1µM CHA on the intracellular
levels of cyclic AMP in promastigotes of Leishmania
mexicana mexicana.

	pmols cyclic AMP/mg protein
CONTROL	11.57 ± 0.59
*CHA	23.05 ± 1.05
*PIA	17.14 ± 0.93

*The cells were exposed to 1µM CHA and 1µM PIA for 5 minutes.

Since these adenosine receptor ligands appear to stimulate membrane adenylate cyclase activity as well as intracellular levels of cyclic AMP and because an increased intracellular concentration of cyclic AMP is known to inhibit the rate of cell proliferation, promastigote growth in the presence of 1μM PIA and 1μM CHA was examined. Figure 3 clearly shows that 1μM PIA will significantly inhibit the rate of cell growth over a period of 96 hours (Figure 3).

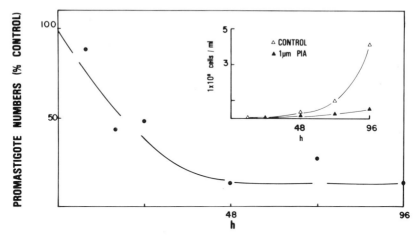

FIGURE 3   The percent of the control promastigote number of Leishmania mexicana mexicana plotted as a function of the exposure time (0-96 hours) to 1μM PIA. The inset presents the same data indicating the rate of cell growth as a function of parasite exposure time to the drug.

Similar results were obtained with 1μM CHA. Interestingly, 1μM PIA and 1μM CHA will also inhibit the rate at which isolated amastigotes will convert back to the promastigote stage (Figure 4).

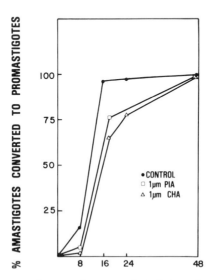

FIGURE 4   The percent of amastigotes of Leishmania mexicana mexicana converting to promastigotes as a function of exposure time to 1μM PIA and 1μM CHA.

CONCLUSIONS

We have demonstrated that two adenosine receptor ligand agonists will bind specifically to the parasite plasma membrane. In addition, these ligands also appear to effect a number of physiological processes of the cell including membrane adenylate cylcase activity, intracellular cyclic AMP levels, the rate of cell replication and the rate of amastigote to promastigote stage transformation. However a number of parameters remain to be examined before this adenosine binding protein can be classified as a receptor. For example, the effect of various adenosine receptor antagonists on the above mentioned physiological responses must be studied. Moreover, we must also determine the effect of these adenosine receptor ligands on cellular cyclic nucleotide phosphodiesterase activity and resultant intracellar cyclic AMP levels. If this adenosine binding membrane protein truly functions as a receptor, however, it may provide a potential target for antileishmanial agents or act as an antigen in mounting a protective host immune response.

REFERENCES

1. L. D. Hendricks, D. E. Wood, and M. E. Hajduk. Haemoflagellates: commercially available liquid media for rapid cultivation, Parasitology 76: 309 (1978).

2. K. P. Chang, Human cutaneous leishmania in a mouse macrophage line: propagation and isolation of intracellular parasites, Science 209: 1240 (1980).

3. D. M. Dwyer, Isolation and partial characterization of surface membranes from Leishmania donovani promastigotes, Journal of Protozoology 27: 176 (1980).

4. Y. Salomon, C. Landos and M. Rodbell, A highly sensitive adenylate cyclase assay, Analytical Biochemistry 58: 541 (1974).

5. J. Wolff, C. Landos, D. M. F. Cooper. Adenosine receptors and the regulation of adenylate cyclase in: "Advances in Cyclic Nucleotide Research", J. E. Dumont, P. G. Greengard and G. A. Robison, ed., Raven Press, New York (1981).

ADENINE PHOSPHORIBOSYLTRANSFERASE-DEFICIENT LEISHMANIA DONOVANI

Kiran Kaur, David M. Iovannisci, and Buddy Ullman*

Department of Biochemistry
University of Kentucky Medical Center
Lexington, Kentucky 40536-0084

SUMMARY

Mutant promastigotes of Leishmania donovani deficient in adenine phosphoribosyltransferase (APRTase) have been isolated in medium containing 4-aminopyrazolopyrimidine. The generation of APRTase-deficient mutants occurred in two discrete steps. In the first step, clones were isolated with 50% of wildtype levels of APRTase activity. These cells were reselected and colonies totally deficient in APRTase were isolated. Partially and totally APRTase-deficient cells exhibited intermediate and complete resistance to cytotoxic adenine analogs, respectively. Nevertheless, wildtype and mutant cells could salvage adenine and utilize adenine as a purine source equally efficiently, suggesting that the adenine deaminase-HGPRTase pathway plays an important role in promastigote adenine metabolism. Kinetic and thermal inactivation studies of purified APRTase and isoelectric focusing of crude extracts from wildtype and partially APRTase-deficient cells suggested that the latter cells possessed wildtype APRTase activity at half the amount found in wildtype parental cells. These data suggest that Leishmania donovani possess two copies of the APRTase structural gene and that these organisms might be diploid for the APRTase locus.

INTRODUCTION

Parasitic protozoa are the etiological agents for a diversity of devastating and often fatal diseases. Rational approaches to chemotherapy of these parasitic diseases have been difficult to devise since most of the major metabolic pathways of these organisms are similar, if not identical to those of their mammalian host. A major metabolic difference between the protozoan parasites and their infected host which has not been exploited chemotherapeutically is the purine pathway. Parasitic protozoa, including Leishmania, are incapable of de novo synthesis of the purine ring and are therefore obligate auxotrophs for purines (1). In order to meet their purine requirements, these organisms have evolved unique purine salvage enzymes not present in mammalian cells. On the basis of metabolic activities identified in cell extracts, two pathways for adenine salvage can be postulated (2,3). Adenine can be phosphoribosylated to AMP via adenine phosphoribosyltransferase (APRTase) or deaminated to hypoxanthine by adenine deaminase. The hypoxanthine formed by this latter reaction can be converted to the nucleotide level via a hypoxanthine-guanine phosphoribosyltransferase activity.

To elucidate the relative roles of these two routes for adenine salvage, we have used biochemical genetic approaches to isolate clonal strains of Leishmania donovani promastigotes genetically deficient in APRTase activity. These studies suggest that the metabolic fate of adenine in these organisms is initiated by deamination. Furthermore, this genetic analysis of adenine metabolism in these mutants suggests that L. donovani promastigotes may be diploid at the APRTase locus.

EXPERIMENTAL PROCEDURES

Cell Culture Conditions   Details of cell culture conditions have been published elsewhere (4). L. donovani can be cloned at virtually 100% efficiency in semi-solid DME-L medium containing 1% Bactoagar (Difco) as described in detail by Iovannisci and Ullman (4). Mutagenized parasites were grown for 7-10 generations under non-selective conditions to ensure phenotypic expression. A clonal cell line, APPB2, was isolated from wildtype cells by virtue of its resistance to 40μM 4-aminopyrazolo[3,4-d]pyrimidine. 4-Aminopyrazolopyrimidine is a cytotoxic substrate of APRTase (2) and is not deaminated in vitro or in situ. APPB2 cells were remutagenized as described above and reselected on semi-solid medium containing 640μM 4-aminopyrazolo-pyrimidine. A clone, APPB2-640A3, was picked from the replated APPB2 cells and characterized further.

Growth Sensitivities to Cytotoxic Agents   The growth inhibitory effects of various purine analogs were determined on wildtype and mutant L. donovani promastigotes in Costar multiwell plates (24 wells) as previously described.

Enzyme Assays   APRTase and HGPRTase activities were assayed in extracts of wildtype and mutant Leishmania in a volume of 0.3 ml containing 50-200 μg protein in 20 mM Tris, pH 7.4, 10 mM NaF, 5 mM MgCl$_2$, 1 mM PRPP, and either 20 μM adenine (9 cpm/pmol) or 20 μM hypoxanthine (9 cpm/pmol), respectively. Assays were performed at 37°. At various time intervals 50 μl aliquots of the reaction mixture were removed, adsorbed to Whatman DE-81 filter paper discs, and placed immediately in 2 l of H$_2$O to terminate the reaction. The discs were pooled, washed twice with 2 l of water, once with 2 l of 1 mM ammonium formate, and twice more with 2 l of water. The discs were then dried and counted in a liquid scintillation counter. All enzymatic assays were performed under conditions that were linear with both time and amount of protein.

Radiolabel Incorporation Experiments and Biochemical Experiments   The rate of uptake of radiolabelled purine nucleobases was determined by the procedures described by Iovannisci et al. (5). The biochemical manipulations performed on the APRTase activities from wildtype and mutant organisms were also carried out as described (5).

RESULTS

Growth Determinations   After isolation in selective semi-solid medium, the abilities of wildtype and mutant cells to grow in suspension medium containing varying concentrations of 4-aminopyrazolopyrimidine were examined. As shown in Figure 1, the APPB2 cell line, mutagenized and selected in a single step against 40μM 4-aminopyrazolopyrimidine, was only slightly less sensitive than the wildtype cell line to the growth

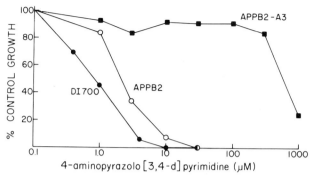

Figure 1. Effect of 4-aminopyrazolopyrimidine on the growth of wildtype and mutant L. donovani promastigotes. The growth of DI700 (●), APPB2 (O), and APPB2-640A3 (■) cells in the presence of various concentrations of 4-aminopyrazolo[3,4-d]pyrimidine was determined. Xanthine at 0.1 mM was the purine in the culture medium.

inhibitory effects of 4-aminopyrazolopyrimidine. In contrast, the APPB2-640A3 cell line which was isolated after two rounds of selection, was several orders of magnitude less sensitive to growth inhibition by 4-aminopyrazolopyrimidine than wildtype cells. The resistant phenotypes were stable after continuous culture of both APPB2 and APPB2-640A3 cells in the absence of selective agent for over 200 generations.

Measurements of APRTase Activities Since 4-aminopyrazolopyrimidine is a substrate for the leishmanial APRTase activity (2), the levels of this enzyme were compared in wildtype and 4-aminopyrazolo-pyrimidine-resistant cells. The results depicted in Figure 2 indicated that extracts of APPB2 cells contained only 50% (280 nmol/h/mg protein at 20μM adenine) of the wildtype APRTase activity (600 nmol/h/mg protein at 20μM adenine) while APPB2-640A3 cells possessed virtually undetectable activity. Mixing experiments with extracts prepared from all three cell lines indicated the absence of a trans-dominant diffusable inhibitor of APRTase activity in mutant cell extracts. As a control, HGPRTase activities were also measured in the same DI700, APPB2, and APPB2-640A3 cell extracts. Extracts from all three cell lines had equivalent amounts of HGPRTase activity.

Incorporation Experiments In order to determine the functional role of APRTase activity in the salvage of adenine by intact cells, the abilities of wildtype cells and cells genetically deficient in APRTase activity to metabolize adenine were compared. As shown in Figure 3, cells with either a partial or total deficiency in APRTase activity salvaged radiolabelled adenine as efficiently as wildtype cells. Thus, it appears that APRTase does not play an important role in the conversion of adenine to the nucleotide level, despite the fact that leishmanial promastigotes possess APRTase activities which are much greater than those found in human tissues (2,5,6).

Biochemical Studies The expression of APRTase activity in APPB2 cell extracts at 50% of wildtype levels could be attributed to either a single

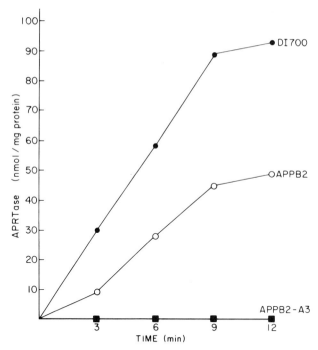

Figure 2. Comparison of APRTase activities in wild-type and mutant cell extracts. The ability of crude cell extracts prepared from DI700 (●), APPB2 (O), and APPB2-640A3 (■) cells to phosphoribosylate [14C]adenine were compared.

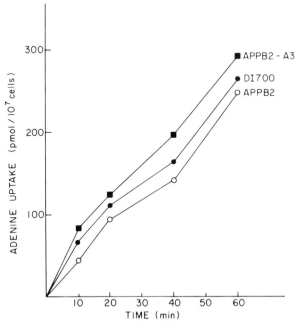

Figure 3. Incorporation of radiolabelled adenine by wild-type and mutant cells. The abilities of DI700 (●), APPB2 (O) and APPB2-640A3 (■) cells to take up [14C]adenine from the cell culture medium are depicted above.

abnormal gene product with altered biochemical properties or to the loss of one half of the genes coding for APRTase activity. Therefore, we attempted to ascertain whether the remaining APRTase activity in APPB2 cells behaved similarily to wildtype enzyme activity with respect to a number of parameters. All parameters including isoelectric point, heat stability and affinity for substrates revealed no differences in APRTase activities prepared from wildtype parental and APPB2 cells.

DISCUSSION

The isolation of APRTase-deficient mutants of Leishmania donovani has allowed a critical examination of the functional role of APRTase in the purine salvage of these organisms. Since wildtype and APRTase-deficient cells could all utilize adenine as a source of cellular purine nucleotides and could convert radiolabelled adenine into nucleotides at similar rates, it appears that the major route of adenine metabolism in L. donovani promastigotes occurs via the adenine deaminase-HGPRTase pathway.

Our data suggest that the genetic loss of APRTase activity in Leishmania donovani promastigotes occurred in two discrete steps, approximately 50% of wildtype APRTase activity being lost in each step. Our biochemical genetic analysis of APRTase-deficient mutants of L. donovani is consistent with a two-gene model for APRTase. Consistent with a two-gene model are the observations that the kinetic and physical parameters of the affinity purified APRTase activity from APPB2 cells are virtually identical to those of the APRTase activity purified from wildtype parental cells, except that the enzyme is present in APPB2 crude extracts at half the amount. Thus APPB2 cells could be considered "presumptive" heterozygotes for APRTase gene expression, possessing one functional and one non-functional APRTase gene copy. While our results are not conclusive evidence for diploidy in Leishmania donovani, they support the hypothesis that there exists two gene copies for APRTase and raises the interesting possibility that these pathogens might be diploid. This has important implications concerning their life cycle, since the possibility exists for genetic exchange through cellular fusion. If such exchange occurs, the generation of cellular variants may be greater than previously anticipated, giving rise to cells resistant to chemotherapeutic agents and thus severely hampering clinical protocols against leishmaniaisis.

ACKNOWLEDGEMENTS

This investigation was supported by grant AI 19866 from the National Institutes of Health National Advisory Allergy and Infectious Diseases Council. B.U. is a recipient of a Research Career Development Award from the National Institutes of Health.

REFERENCES

1. Marr, J.J., Berens, R.L., and Nelson, D.J. (1978) Biochim. Biophys. Acta 544, 360-371.
2. Tuttle, J.V. and Krenitsky, T.A. (1980) J. Biol. Chem. 255, 909-916.
3. Kidder, G.W., and Nolan, L.L. (1979) Proc. Natl. Acad. Sci. U.S.A. 76, 3670-3672.
4. Iovannisci, D.M., and Ullman, B. (1983) J. Parasitol. 69, 633-636.
5 Iovannisci, D.M., Goebel, D., Allen, K., Kaur, K., and Ullman, B. J. Biol. Chem. 259, 14617-14623.
6. Rosenbloom, F.M., Kelley, W.N., Miller, J., Henderson, J.F., and Seegmiller, J.E. (1967) J. Am. Med. Assoc. 202, 103-105.

5'-METHYLTHIOADENOSINE (MTA) PHOSPHORYLASE FROM PROMASTIGOTES OF

LEISHMANIA DONOVANI

George W. Koszalka and Thomas A. Krenitsky

Burroughs Wellcome Co., Research Triangle Park
North Carolina 27709, USA

INTRODUCTION

The activity of various purine analogues against the pathogenic protozoan Leishmania donovani together with the lack of de novo synthesis in Leishmania (sp) has led to more detailed investigations into the purine metabolism of this organism (1,2,3,4). These studies have shown that marked differences in purine metabolism exist between the host and parasite. In a previous report, the naturally occurring purine ribo-nucleosides and 2'-deoxyribonucleosides had been shown to be cleaved in promastigote extracts of L. donovani (5). A detailed investigation revealed the existence of high levels of catabolic enzymes that cleave purine nucleosides in extracts of the promastigote form of this organism. That work also characterized three distinct nucleoside hydrolases. One hydrolyzed purine and pyrimidine ribonucleosides, another cleaved only purine ribonucleosides, while the third hydrolyzed only purine 2'-deoxy-ribonucleosides. None of these enzymes cleaved adenosine and only the purine 2'-deoxyribonucleosidase hydrolyzed 2'-deoxyadenosine. Adenosine was a potent inhibitor of the purine ribonucleosidase (K_{is} = 5 µM), but had little effect on the other two enzymes. In addition, that study identified an adenosine-cleaving activity in crude extracts (5). This report describes further studies on that adenosine-cleaving activity from the promastigote form of Leishmania donovani.

METHODS

All common nucleosides, 5'-methylthioadenosine, ribose-1-phosphate (Rib-1-P), 2'-dRib-1-P and bases were purchased from Sigma Chemicals, St. Louis, MO. 5'-Methylthioribose-1-phosphate was prepared according to established procedures (6).

Growth of Organisms

Cell cultures were supplied by the laboratory of Dr. J. Joseph Marr, University of Colorado, Denver, CO. The promastigotes were grown as previously described (7).

Preparation of Cell Extract

Cell suspensions were subjected to four freeze-thaw cycles in a dry ice-acetone bath. Cell debris was removed by centrifugation at 105,000 x g for 20 min. The resulting supernatant was chromatographed on course G-25 Sephadex (2.5 x 11 cm) and eluted with 50 mM potassium phosphate at pH 7.0. The protein-containing fractions were combined and are referred to as extract.

Enzyme Assays

Nucleoside Synthesis. 5'-Methylthioadenosine phosphorylase was monitored by following the synthesis of adenosine from Rib-1-P and adenine. The spectrophotometric assay was monitored at either 250.5 nm, the isosbestic point between adenine and hypoxanthine ($\Delta\epsilon$ = 0.95 mM^{-1} cm^{-1}) when adenine deaminase was present, or at 256 nm ($\Delta\epsilon$ = 2.1 mM^{-1} cm^{-1}). Assay mixtures (0.3 or 0.5 ml) contained 0.1 mM adenine and 3 mM Rib-1-P in 50 mM potassium Pipes, pH 7. The reaction was initiated by the addition of Rib-1-P.

Nucleoside Phoshorolysis. The rate of phosphorolysis was monitored in extracts by following the adenosine cleavage at 250.5 nm ($\Delta\epsilon$ = 0.95 mM^{-1} cm^{-1}) in 50 mM potassium phosphate, pH 7. Partially purified preparations that contained no adenine deaminase were assayed in the presence of added xanthine oxidase according to established procedures (8).

RESULTS

The rate of adenosine cleavage was dependent on the presence of phosphate. Extracts that had been chromatographed on Sephadex G-25 in potassium Pipes, pH 7, instead of phosphate were void of adenosine-cleaving activity. The addition of 10 mM potassium phosphate, pH 7.0, restored the adenosine-cleaving activity.

The stability of the adenosine-cleaving activity in extracts to heat was dependent on the presence of phosphate. Without phosphate present, 32% of the activity was lost after 5 minutes at 55°C. However, in the presence of 50 mM potassium phosphate, pH 7.0, there was no detectable loss of the adenosine-cleaving activity.

The adenosine-cleaving activity was separated from the three nucleoside hydrolase enzymes and adenase by chromatography on anion exchange resin. A purification scheme was devised that employed the resolution of anion exchange and the enzyme's resistance to thermal inactivation in the presence of phosphate. The procedure resulted in a 92-fold purification of the enzymatic activity and did not contain detectable levels of any nucleoside hydrolase or adenase. The details of this purification procedure will be published elsewhere (G.W. Koszalka and T.A. Krenitsky, manuscript in preparation).

Physical, Chemical and Thermodynamic Properties for the Purified Enzyme.
A molecular weight of 89,000 daltons $\pm$ 10% was estimated by G-150 Sephadex gel filtration chromatography for the purified preparation. Preincubation of the enzyme with EDTA (1 mM) at 25°C for 15 min had no effect on the activity.

The resistance of the adenosine-cleaving activity to thermal inactivation was investigated in more detail by determination of the apparent activation energies in both the phosphorolytic and synthetic directions. The phosphorolysis of adenosine had an apparent activation energy (E_a') of 8600 cal/mole (Figure 1). The Arrhenius plot was linear from 20°C to 60°C. The E_a' for the synthesis of adenosine with this enzyme was higher than that observed for the phoshorolysis of adenosine. In addition, the plot was biphasic over the temperature range studied with a break in the projected slope of the plot occurring at 55°C (Figure 1). This transition temperature is most likely the result of a change in the rate limiting step of the reaction. A drastic decrease in the initial velocity of the promastigote enzyme was apparent at temperatures above 65°C. The addition of unheated enzyme to these reactions, after cooling to 55°C, gave the expected initial velocity for this temperature. Therefore, thermal inactivation of the adenosine- cleaving activity and not substrate stability or depletion was the cause of the lower velocities observed above 65°C.

Catalytic Properties. The substrate and inhibitor specificity of the purine nucleosides for the purified L. donovani enzyme is presented in Table I. Purine ribonucleosides and 2'-deoxyribonucleosides with an amino group in the six position were cleaved by this enzyme in the presence of phosphate. No detectable rate of adenosine phosphorolysis was observed in the absence of phosphate (<0.2 nmol/min/mg of protein). 5'-Methylthioadenosine was the most efficient substrate tested followed by 2'-deoxyadenosine, 5'-deoxyadenosine and adenosine. Adenine arabinoside was neither a substrate nor a potent inhibitor.

The requirement for an amino group in the six position of purine was exemplified by the lack of substrate activity or potent inhibition of the purified enzyme by inosine, 2'-deoxyinosine, guanosine or 2'-deoxyguanosine (Table I).

Nucleoside Synthesis. Purified preparations were able to synthesize adenosine nucleosides when incubated with either Rib-1-P (1 mM), 2'-dRib-1-P (1 mM), or 5'-methylthio-Rib-1-P (0.1 mM) and adenine (0.1 mM). The nucleoside formed with each of the above pentose-1-P analogues had a R_f value identical to the appropriate adenine nucleoside when chromatographed on cellulose and developed in water.

DISCUSSION

This study has indicated the presence of a heat stable nucleoside phosphorylase specific for 6-amino purine nucleosides in extracts of the promastigote form of L. donovani. This activity differs from the nucleoside hydrolases previously identified in promastigote extracts (5). The three other nucleoside-cleaving activities did not synthesize nucleosides from purine bases and Rib-1-P, and none required phosphate for the cleavage of the glycosidic bond. The adenosine-cleaving activity, however, does require phosphate for activity and, in addition, synthesizes adenine ribo-, 2'-deoxyribo- and 5'-methylthioribonucleosides from adenine and the appropriate pentose-1-P.

The substrate specificity of this enzyme suggests it be termed MTA phosphorylase since 5'-methylthioadenosine is the most efficient substrate. However, this protozoal enzyme differs significantly in its substrate specificity from its mammalian counterpart. MTA phoshorylase from S-180 cells cleaves adenosine but with a K_m value 30-fold higher than the L. donovani enzyme (9). In addition, this enzyme from S-180 cells cleaves adenosine with a relative V_{max} 115% that of MTA whereas the L. donovani enzyme has a relative V_{max} for adenosine of 33% (Table I). Another major difference is the observed substrate activity with the promastigote

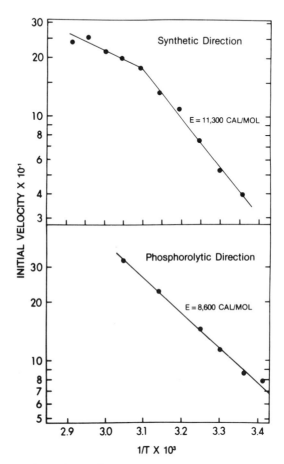

Fig. 1. Arrhenius plots for the synthesis and phosphorolysis of adenosine from L̲. donovani promastigotes.

Table I. KINETIC CONSTANTS FOR PARTIALLY PURIFIED
L. DONOVANI 5'-METHYLTHIOADENOSINE PHOSPHORYLASE

Compound	$K_m{}'{}^a$	Rel. $V_{max}{}^b$	V_{max}/K_m	$K_{is}{}^c$
	(μM)			(μM)
Nucleoside Phosphorolysis				
5'-MTA	0.8±.2	100	26	
2'dAdenosine	20.4	250	12.5	
5'dAdenosine	11.4±1	135	11.8	
Adenosine	47.0±7	33	0.7	
AraA				393±43
Inosine				551±70
2'dInosine				>500
Guanosine				>700
2'dGuanosine				>700

$^a K_m$ values were determined in 50 mM potassium phoshate buffer,
pH 7.
$^b V_{max}$ values are expressed relative to 5'-methylthioadenosine,
640 nmol/min/mg.
^{c}All inhibitors showed competitive inhibition.

enzyme and 2'-deoxyadenosine (Table I). This is in contrast to the
S-180 enzyme which does not react with 2'-deoxyadenosine at 1 mM (9).

The broad substrate specificity for this enzyme and the presence of
nucleoside hydrolases instead of nucleoside phosphorylases, exemplify
fundamental differences in the way purine nucleosides are processed in
L. donovani and man.

REFERENCES

1. J.J. Marr, R.L. Berens, and D.J. Nelson, Biochim. Biophys. Acta
544:360-371 (1978).
2. S.W. LaFon, D.J. Nelson, R.L. Berens, and J.J. Marr, Biochem. Pharm.
31:231-238 (1982).
3. H.F. Hassan, and G.H. Coombs, Exp. Parasitol. 59:139-150 (1985).
4. W.E. Gutteridge and G.H. Coombs, Biochemistry of parsitic protozoa,
MacMillian Press, London (1977).
5. G.W. Koszalka and T.A. Krenitsky, J. Biol. Chem. 254:8185-8193
(1979).
6. D.L. Garbers, Biochim. Biophys. Acta 523:82-93 (1978).
7. R.L. Berens and J.J. Marr, J. Parasitol. 64:160 (1978).
8. R.L. Miller and D. Lindstead, Mol. and Biochem. Parsitol. 7:41-51
(1983).
9. R.E. Parks, Jr., J.D. Stoeckler, C. Cambor, T.M. Savadese,
G.W. Crabree and S. Chu, In Molecular actions and targets for
cancer chemotherapeutic agents, Academic Press, New York, pp.
229-252 (1981).

6-THIOPURINE RIBOSIDE ANALOGS: THEIR TOXICITY AND METABOLISM IN

LEISHMANIA DONOVANI AND MAMMALIAN CELLS

Stephen W. LaFon[1], Naomi K. Cohn[1], Donald J. Nelson[1] and
Randolph L. Berens[2]

[1]Burroughs Wellcome Co., Research Triangle Park, NC   USA
[2]University of Colorado Health Sciences Center
Denver, CO   USA

INTRODUCTION

The purine nucleoside analog, allopurinol riboside, is cytotoxic
to several Leishmania sp.[1]. This nucleoside was metabolized to its
5'-monophosphate by L. donovani which, in turn, was aminated to form
adenine nucleotide analogs. The aminopyrazolopyrimidine 5'-triphosphate
thus formed was incorporated into the RNA of the parasite. Several
purine nucleoside analogs, structurally similar to allopurinol riboside,
have differential cytotoxicity to Leishmania donovani and mouse L cells[2].
The cytotoxic action of these compounds correlates with their conversion
to ATP and/or GTP analogs by these organisms[3].

Thiopurinol riboside, an analog of allopurinol riboside, was
metabolized by L. donovani to its 5'-monophosphate and another unknown
5'-monophosphate metabolite[4]. However, the presence of the sulfhydryl
group in the 4-position of the molecule prevented its amination, as
judged by its failure to yield aminopyrazolopyrimidine nucleotides.
Thiopurinol riboside 5'-monophosphate produced from thiopurinol riboside
was not metabolized to its 5'-di- or 5'-triphosphate. In spite of the
lack of metabolism of thiopurinol riboside to 5'-triphosphate by
L. donovani, this nucleoside inhibited the growth of L. donovani at
concentrations comparable to the concentration of allopurinol riboside
required to inhibit the parasite.

The ability of thiopurinol riboside to inhibit the growth of
L. donovani in the absence of accompanying metabolism to 5'-triphosphate
represents an exception to the so-called "amination hypothesis" of cytotoxicity suggested for certain inosine analogs[3]. This study seeks to
evaluate the cytotoxic activity of thiol-containing compounds structurally similar to thiopurinol riboside and relate these activities to
their metabolism by L. donovani and two mammalian cell lines, mouse
L cells and Detroit 98 cells.

MATERIALS AND METHODS

The cytotoxicity[2] and metabolism[3] studies were performed as previously described. 6-Thio-9-deazapurine riboside was supplied by Dr. Robert Klein of Sloan-Kettering Institute for Cancer Research, Rye, NY and 6-thio-8-azapurine riboside from Dr. John Montgomery of Southern Research Institute, Birmingham, AL. 6-Thiopurine riboside, thiopurinol riboside and 7-thioformycin were synthesized in these laboratories.

RESULTS

Cytotoxicity of 6-Thiopurine Riboside Analogs

A series of 6-thiopurine riboside analogs with modifications in the imidazole portion of the purine ring demonstrated varying inhibitory activity against L. donovani promastigotes (Table 1). Of the compounds studied, 7-thioformycin and thiopurinol riboside had the most potent antileishmanial activities, inhibiting growth by 50% at 0.1 and 1 µM, respectively.

The growth of mouse L cells and Detroit 98 cells was also inhibited by the 6-thiopurine riboside analogs. 6-Thiopurine riboside itself was the most potent growth inhibitor of the series with ED_{50} concentrations of 0.1 µM (L cells) and 0.7 µM (Detroit 98 cells). The relative lack of toxicity of thiopurinol riboside was in contrast to its potent anti-leishmanial activity. 7-Thioformycin was unique among the compounds studied in that it demonstrated a 100-fold difference in potency between the two mammalian cells and was, by far, the most potent antileishmanial agent in this series.

Metabolism of 6-Thiopurine Riboside Analogs and Their Effect on Purine Nucleotide Pools

HPLC analysis of the nucleotide pools of cells incubated with 6-thiopurine riboside analogs revealed that these compounds were metabolized exclusively to 5'-monophosphates by L. donovani and mammalian cells (Figure I). There was no evidence of conversion to 5'-di- or 5'-triphosphates or conversion to other purine analogs. The conversion to 5'-monophosphate was probably the result of direct phosphorylation of the nucleosides since no 6-thiopurine base analogs could be found in the incubation media. The only exception to this was 6-thiopurine riboside which was completely cleaved to 6-thiopurine by both the parasite and the mammalian cells during the incubation. The sulfhydryl moiety of the compounds remained intact throughout the incubations.

With the exception of 6-thio-8-azapurine riboside, all of the nucleoside analogs were metabolized by L. donovani to their respective 5'-monophosphates (Table 2). Of the compounds studied, thiopurinol riboside generated the greatest amount of 5'-monophosphate attaining levels approximately 50% of the intracellular ATP levels. 7-Thioformycin had the greatest antileishmanial activity in this series but was metabolized to only moderate levels of its 5'-monophosphate.

TABLE 1.   EFFECT OF 6-THIOPURINE RIBOSIDE ANALOGS ON THE GROWTH OF L. DONOVANI, MOUSE L CELLS AND DETROIT 98 CELLS.

	ED_{50}[a], μM		
	L. Donovani	L CELLS	DETROIT 98 CELLS
 6-Thiopurine Rib.	40	0.1	0.7
 Thiopurinol Rib.	1	600	>1000
 7-Thioformycin	0.1	100	0.9
 6-Thio-9-Deazapurine Rib.	15	160	10
 6-Thio-8-Azapurine Rib.	>1000	3	4

[a] Concentration at which growth of cells was inhibited by 50%. Incubation time was 72 h.

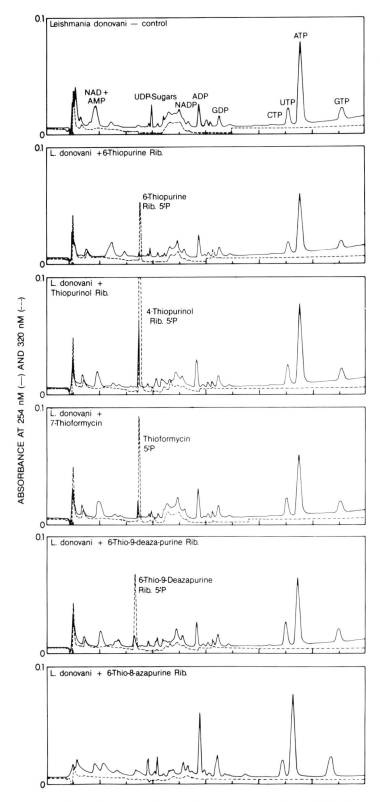

Figure 1. Anion Exchange Chromatogram of perchloric
Acid Extracts of <u>Leishmania</u> <u>Donovani</u> Incubated with 6-Thiopurine Ribosie Analogs.

TABLE 2.   METABOLISM OF 6-THIOPURINE RIBOSIDE ANALOGS BY
L. DONOVANI, L CELLS AND DETROIT 98 CELLS.

	ANALOG 5' - MP	ENDOGENOUS NUCLEOTIDES	
		ATP	GTP
L. DONOVANI			
Control	—	225	49
+ 6-Thiopurine Rib.	6.4	163	31
+ Thiopurinol Rib.	103	220	48
+ 7-Thioformycin	21	162	42
+ 6-Thio-9-Deazapurine Rib.	30	180	52
+ 6-Thio-8-Azapurine Rib.	<1	95	38
L CELLS			
Control	—	3600	450
+ 6-Thiopurine Rib.	1314	4000	131
+ Thiopurinol Rib.	80	8700	1200
+ 7-Thioformycin	0	9400	1700
+ 6-Thio-9-Deazapurine Rib.	0	5700	900
+ 6-Thio-8-Azapurine Rib.	24	5000	120
DETROIT 98 CELLS			
Control	—	4800	860
+ 6-Thiopurine Rib.	27	3800	240
+ Thiopurinol Rib.	0	4400	600
+ 7-Thioformycin	374*	3300	370
+ 6-Thio-9-Deazapurine Rib.	0	3000	490
+ 6-Thio-8-Azapurine Rib.	155	4200	220

Units are in pmol/10^6 cells. Inital nucleoside concentration was 10 μg/ml (35 μM) and incubation times were 4h.

*Two metabolites were produced in this incubation; 7-thioformycin 5'-monophosphate (300 pmol/10^6 cells) and another eluting shortly after 7-thioformycin 5'-monophosphate (74 pmol/10^6 cells).

L cells metabolized 6-thiopurine riboside, thiopurinol riboside and 6-thio-8-azapurine riboside to their respective 5'-monophosphates (Table 2).   The other compounds in this series were not metabolized by these cells.   6-Thiopurine riboside was the most toxic of the series to L cells and, also, generated the greatest amount of 5'-monophosphate. Detroit 98 cells metabolized 6-thiopurine riboside, 7-thioformycin and 6-thio-8-azapurine riboside to their 5'-monophosphate.   Other compounds in this series were not metabolized by the Detroit 98 cells.   7-Thioformycin was unique in that it was converted to another metabolite by Detroit 98 cells which eluted after 7-thioformycin 5'-monophosphate on anion exchange HPLC.   6-Thiopurine riboside was a potent growth inhibitor of the Detroit 98 cells but, unlike in mouse L cells, this nucleoside was metabolized to only small amounts of its 5'-monophosphate.

While having no significant effects on L. donovani, some of these compounds altered the nucleotide pools in both the mouse L cells and the Detroit 98 cells (>50% charge in pool sizes).   For example, in L cells, thiopurinol riboside, 7-thioformycin and 6-thio-9-deazapurine riboside produced substantial increase in both ATP and GTP pools, while 6-thio-purine riboside and 6-thio-8-azapurine riboside caused decreases only in the GTP pools. Although the 6-thiopurine riboside analogs had no major effect on theATP pools of the Detroit 98 cells, it was observed that 6-thiopurine riboside, 7-thioformycin, 6-thio-9-deazapurine riboside and 6-thio-8-azapurine riboside significantly lowered the GTP pool sizes in these cells.

Correlation of Cytotoxicity and Metabolism of 6-Thiopurine Riboside Analogs

In general, the antileishmanial activity of the 6-thiopurine riboside analogs correlated with intracellular conversion to their respective 5'-monophosphate. Thiopurinol riboside was metabolized to its 5'-monophosphate much more extensively than the other nucleosides (Table 2) and was also one of the most potent antileishmanial agents studied (Table 1). On the other hand, 6-thiopurine riboside and 6-thio-8-azapurine riboside showed little or no metabolism by the para-site and were also only slightly effective against the organism. The one exception to this correlation was 7-thioformycin which was extremely toxic to the parasite while being metabolized to only moderate levels of its 5'-monophosphate. The toxicity of these compounds to L cells and Detroit 98 cells also appears to be related to their conversion to 5'-monophosphate. In mammalian cells, this correlation can be taken one step further, in that, those compounds (in their monophosphate form) which decreased the GTP levels were most toxic to the cells.

DISCUSSION

6-Thiopurine riboside analogs were selectively cytotoxic to L. donovani. Thiopurinol riboside and 7-thioformycin were the most potent antileishmanial agents with ED$_{50}$ concentrations 15- to 10,000-fold lower than the other compounds studied. As well, these compounds were approximately 10-fold more potent than their 6-oxo congeners, allopurinol riboside and formycin B[2]. On the other hand, 6-thio-9-deazapurine riboside and 6-thio-8-azapurine riboside were sub-stantially less potent than their 6-oxo analogs (ED$_{50}$ concentrations of 15 µM versus 1 µM and >1000 µM versus 35 µM, respectively). A similar study with L. tropica-infected macrophages found 6-oxypurine ribosides to have superior antileishmanial activity to their 6-thio congeners[5].

Previous studies with a series of 6-oxypurine riboside analogs revealed a direct correlation between their toxicity to L. donovani and L cells and their conversion to ATP and/or GTP analogs[3]. However, the thiopurine riboside analogs represent a unique class of compounds that demonstrate cytotoxicity to both L. donovani and mammalian cells without accompanying metabolism to 5'-triphosphate analogs. As previously observed for thiopurinol riboside[4], this series of compounds is metab-olized only to their 5'-monophosphate. The lack of metabolism to 5'-triphosphates would appear to preclude their incorporation into nucleic acids and interference at that level. These compounds have no major effect on the purine nucleotide pools of L. donovani although they do alter the purine triphosphate pools of both the L cells and the Detroit 98 cells. It is of interest to note that those compounds which lowered GTP levels in the mammalian cells were also the compounds most cytotoxic to these cells. The inhibitory effect of at least one of the nucleotide metabolites identified in this study, 6-thiopurine riboside 5'-monophosphate, on IMP dehydrogenase has been demonstrated[8]. Inhibition of this enzyme in vivo could explain the lower GTP pools in the mammalian cells exposed to these compounds. The lack of effect on the purine nucleotide pools in <u>L</u>. <u>donovani</u> is interesting since some of these compounds (in their 5'-monophosphate form) inhibit several of the enzymes responsible for purine biosynthesis in this organism[4,9,10]. This finding suggests a unique ability of this organism to maintain purine nucleotide levels under conditions where their synthesis is inhibited <u>in</u> <u>vitro</u>. Although the mechanism by which these 6-thiopurine riboside analogs inhibit parasite growth is not understood, it appears that it is not dependent upon the formation of analog 5'-triphosphate or alterations of the endogenous nucleotide pools.

ACKNOWLEDGEMENTS

The authors wish to thank Dr. Robert Klein for supplying the 6-thio-9-deazapurine riboside, Dr. John Montgomery for the 6-thio-8-azapurine riboside and Drs. G. B. Elion, T. A. Krenitsky, J. L. Rideout, S. Daluge and G. W. Koszalka for their efforts in the synthesis of the other nucleosides used in this study. We also thank Ms. Rita Hendricksen, Ms. Bevery Raab and Mr. Ernest Dark for their technical assistance.

REFERENCES

1. D. J. Nelson, S. W. LaFon, J. V. Tuttle, W. H. Miller, R. L. Miller, T. A. Krenitsky, G. B. Elion, R. L. Berens, and J. J. Marr, Allopurinol ribonucleoside as an antileishmanial agent, J. Biol. Chem., 254:11544 (1979).
2. J. J. Marr, R. L. Berens, N. K. Cohn, D. J. Nelson, and R. S. Klein, Biological action of inosine analogs in Leishmania and Trypanosoma spp., Antimicr. Agents and Chemother., 25:292 (1984).
3. S. W. LaFon, D. J. Nelson, R. L. Berens, and J. J. Marr, Inosine analogs: Their metabolism in mouse L cells and Leishmania donovani, J..Biol. Chem. in press (1985).
4. J. J. Marr, R. L. Berens, D. J. Nelson, T. A. Krenitsky, T. Spector, S. W. LaFon and G. B. Elion, Antileishmanial action of thiopurinol and its ribonucleoside, Biochem. Pharmacol., 31:143 (1982).
5. J. D. Berman, L. S. Lee, R. K. Robins, and G. R. Revankar, Activity of purine analogues in vitro, Antimicr. Agents and Chemother., 24:233 (1983).
6. "6-Mercaptopurine" in Annals of the N. Y. Acad. of Sci., R. W. Miner, ed. 60:183 (1954).
7. J. D. Brockman, Mechanism of resistance to anticancer agents, Advances Cancer Res., 7:129 (1963).
8. R. L. Miller and D. L. Adamczyk, Inosine 5'-monophosphate dehydrogenase from sarcoma 180 cells-substrate and inhibitor specificity, Biochem. Pharmacol., 25:883 (1976).
9. T. Spector, T. E. Jones, and G. B. Elion, Specificity of adenylosuccinate synthetase and adenylosuccinate lyase from Leishmania donovani, J. Biol. Chem., 254:8422 (1979).
10. T. Spector and T. E. Jones, GMP reductase from Leishmania donovani: A possible chemotherapeutic target, Biochem. Pharmacol., 31:3891 (1982).

PURINE SALVAGE ENZYMES IN TRICHOMONAS VAGINALIS

Richard L. Miller and Wayne H. Miller

Wellcome Research Laboratories
Research Triangle Park, N.C. 27709

INTRODUCTION

Trichomonas vaginalis is one of four trichomonad species of protozoal parasites that infect humans. Of these, it is the only species which appears to possess pathogenic strains.[1]

As noted for other parasitic protozoa, T. vaginalis is incapable of de novo purine synthesis and is thus dependent upon purine salvage[2]. Unlike mammalian systems and most other protozoa, T. vaginalis requires an exogenous source of adenine and guanine for growth; hypoxanthine can not serve as a replacement for either compound[3]. In addition, the organism has no purine phosphoribosyltransferase activity and thus appears to be dependent upon a phosphorylase/kinase pathway for the salvage of purines and their nucleosides[4]. The purine nucleoside phosphorylase and the purine nucleoside kinase have been purified and characterized from the Bushby strain of T. vaginalis. Both enzymes have substrate specificities which differ from their mammalian counterparts.

RESULTS AND DISCUSSION

The purine nucleoside phosphorylase (mol. wt. 95,000) has been purified approximately 450-fold by a combination of DEAE-cellulose, agarose-Blue A and agarose-adenosine gel chromatography. A summary of the purification is presented in Table 1. Throughout the purification, the ratio of adenosine/guanosine/inosine activity remained constant indicating that all three activities are catalyzed by the same enzyme. The purified enzyme was free of interfering enzyme activities. Initial velocity studies indicated that the enzyme catalyzed the synthesis and cleavage of the nucleosides by a sequential mechanism. Kinetic constants derived from these studies are presented in Table 2. Unlike the mammalian purine nucleoside phosphorylase, the T. vaginalis enzyme catalyzed the efficient synthesis and cleavage of adenosine as well as guanosine and inosine.

Table 1. Purification of Purine Nucleoside Phosphorylase

Fraction	Total Units	Sp. Act. (μmol/min/mg)	% Recovery
Crude Extract	70	0.86	100
DE-52	102	3.58	145
Blue A	92	11.6	131
Agarose-adenosine	90	400	128

Table 2. Kinetic Constants for Purified
Purine Nucleoside Phosphorylase

Substrate	K_m (μM)	V_{max} (μmol/min/mg)	V_{max}/K_m
Phosphorolysis			
Inosine	49	390	8.0
Guanosine	17	81	4.8
Adenosine	54	16	0.3
Phosphate	390	-	-
Synthesis			
Hypoxanthine	25	236	9.4
Guanine	24	590	24.6
Adenine	21	363	17.3
Ribose-1-phosphate	15	-	-

The purine nucleoside kinase proved to be extremely labile thus allowing only the limited purification presented in Table 3. The purified enzyme (mol. wt. 16,000) was free of interfering enzyme activities. Kinetic studies indicated that the enzyme was capable of catalyzing the phosphorylation of purine and pyrimidine nucleosides (Table 4).

2'-Deoxyadenosine, 2'-deoxycytidine and 2'-deoxythymidine did not serve as substrates for this enzyme. Neither AMP, GMP nor p-nitrophenylphosphate could substitute for ATP as the phosphate donor. The uridine phosphorylating activity associated with the purine nucleoside kinase has been shown to be distinct from the specific uridine kinase found in extracts.[5]

Table 3. Purification of Purine Nucleoside Kinase

Fraction	Total Units	Sp. Act. (nmol/min/mg)	% Recovery
Crude Extract	20	1.3	100
Sephadex G-25	5.3	0.40	27
Agarose-ATP	5.1	25	26

Table 4. Kinetic Constants for Purified Purine Nucleoside Kinase

Substrate	K_m (μM)	Rel. V_{max}	V_{max}/K_m
Guanosine	1	100	100
2'-Deoxyguanosine	20	2	0.1
Adenosine	200	111	0.6
Inosine	20	67	3.4
Uridine	30	67	2.2
Cytidine	200	4	0.02

The finding that both the purine nucleoside phosphorylase and the purine nucleoside kinase catalyze reactions involving the common purine nucleosides, guanosine and adenosine, suggests that they may act as a coordinated set of enzyme activities to salvage purines and purine ribonucelosides in T. vaginalis.

REFERENCES

1. B.M. Honigberg, Trichomonads of Importance in Human Medicine, in: "Parasitic Protozoa", Vol. 2, J.P. Kreier, ed., pp. 275-454, Academic Press, New York (1978).
2. P.G. Heyworth and W.E. Gutteridge, Further Studies on the Purine and Pyrimidine Metabolism in Trichomonas vaginalis, J. Protozool. 25:9B (1978).
3. P.G. Heyworth, W.E. Gutteridge, and C.D. Ginger, Purine and Pyrimidine Metabolism in Trichomonas vaginalis, Parasitol. 77:xi (1978).
4. R.L. Miller and D. Linstead, Purine and Pyrimidine Metabolizing Activities in Trichomonas vaginalis Extracts, Mol. Biochem. Parasitol. 7:41-51 (1983).
5. W.H. Miller and R.L. Miller, unpublished results.

BIOCHEMICAL BASIS FOR DEOXYADENOSINE AND 2-CHLORODEOXYADENOSINE TOXICITY TO RESTING HUMAN LYMPHOCYTES

Shiro Seto, Carlos J. Carrera, D. Bruce Wasson and Dennis A. Carson

Scripps Clinic and Research Foundation, Department of Basic and Clinical Research, La Jolla, California

INTRODUCTION

An Inherited deficiency of adenosine deaminase (ADA; adenosine aminohydrolase EC 3,5,4,4) results in a combined immunodeficiency disease (1). The selective lymphopenia seen in ADA deficient children has been attributed to the toxic effects of deoxyadenosine (dAdo) metabolites (2). Micromolar concentrations of dAdo are toxic in vitro to ADA-inhibited human resting peripheral blood lymphocytes (PBL). The ADA-resistent dAdo congener, 2-chlorodeoxyadenosine (CdA) is similarly cytotoxic to resting human T cells (3). Since the toxic mechanisms of dAdo have not been clearly elucidated, we have examined the metabolic changes that follow exposure of PBL to dAdo plus deoxycoformycin (dCF), as well as to CdA (4).

As reported previously (4,5), alkali sensitive DNA single strand breaks accumulated in PBL early after incubation with dAdo plus dCF, or with CdA. The two deoxynucleosides also inhibited the repair of radiation induced DNA lesions. The metabolic changes, caused by the DNA damage, were closely linked to cell death (5). DNA strand breaks trigger poly(ADP-ribose) synthesis which consumes NAD (6,7). Thus, NAD and ATP levels in lymphocytes decreased sequentially after incubation with dAdo or CdA. Importantly, the addition of nicotinamide, a precursor for NAD and an inhibitor of poly(ADP-ribose) synthesis, protected partially PBL from dAdo toxicity, at least in short time cultures. In this context, the relationships between dAdo toxicity and NAD metabolism were examined in more detail in the present experiments.

METHODS

Lymphocytes were isolated from heparinized blood and maintained in RPMI 1640 medium supplemented with 20 % autologous plasma, as described previously (4). Cultures were performed at 37° C in an atmosphere of 5% CO_2. Viable cells were counted microscopically after the addition of Erythrosin B. Alkali-sensitive DNA strand breaks in the various lymphocyte suspensions were measured by the method of Birnboim et al. (8) with some modifications (4). In this assay, the amount of double-stranded DNA remaining after alkali treatment is quantitated fluorometrically, using the DNA binding dye, ethidium bromide. In control experiments, the DNA unwinding rate was correlated with the number of DNA strand breaks. In the Results, data from DNA unwinding assays is shown as the percent of control cultures without deoxynucleosides or irradiation. To measure DNA repair, PBL were irradiated on ice with gamma rays from a 137Cs source at a dose rate of 500 rads/min. The DNA unwinding assay was performed immediately, and at intervals after irradiation. Intracellular NAD and ATP levels were determined as described previously (4).

Table 1. METABOLIC CHANGES IN LYMPHOCYTES TREATED WITH DEOXYADENOSINE OR 2-CHLORODEOXYADENOSINE

Assay	Compound	Incubation time					
		4h		8h		24h	
		%control		%control		%control	
ds-DNA	dAdo+dCF	90 ± 5	*	74 ± 6	***	41 ± 4	***
	CdA	90 ± 2	*	78 ± 1	***	37 ± 3	***
NAD	dAdo+dCF	95 ± 10	NS	81 ± 6	**	31 ± 4	***
	CdA	95 ± 5	NS	81 ± 7	*	43 ± 8	***
ATP	dAdo+dCF	99 ± 5	NS	93 ± 3	NS	61 ± 1	***
	CdA	92 ± 8	NS	92 ± 7	NS	64 ± 6	***

DeoxyAdo, dCF and CdA were added into cultures at concentrations of 10 μM, 1 μM and 1 μM, respectively.
ds-DNA; the amount of double-stranded DNA remianing after alkali treatment.
Data represent the mean ± SEM of four determinations.
The nucleoside-treated cultures were compared with control cultures by the two-tailed paired t test; NS not significant, * p <0.05, ** p <0.01, *** p <0.005.

RESULTS

As shown in Table 1, an increase in DNA strand breaks was apparent in cultures as early as four hours after exposure to the deoxynucleosides. Thereafter, the disintegration of DNA proceeded progressively. At 24 hours of incubation, 10 μM dAdo plus 1 μM dCF or 10 μM CdA produced DNA strand breakage (or alkali labile sites), almost equivalent to the amount caused by 500 rads gamma irradiation. The DNA damage was followed by a significant fall in intracellular NAD levels at 8 hours and a drop in ATP pools at 24 hours. The depletion of these essential metabolites preceded cell death.

The inhibitory effects of deoxynucleosides on the repair of radiation induced DNA lesions are summarized in Table 2. Lymphocytes were capable of rejoining DNA strand breaks quickly. Most DNA lesions, including alkali labile sites, were repaired within two hours after 500 rads irradiation. Preincubation of lymphocyte cultures for one hour with 10 μM dAdo inhibited DNA repair moderately. However, longer exposure hampered severely the repair capacity of PBL. Under the same conditions, 10 μM CdA inhibited DNA repair as strongly as 100 μM ara-C, a known potent inhibitor of DNA polymerization.

Table 2. Effects of deoxynucleosides on the repair of radiation induced DNA lesions

Compound	Incubation time before irradiation	% of control double stranded DNA	
		0 h	2h
Exp.1			
–	–	45	91
10 μM dAdo + 1 μM dCF	1 h	46	71
	4 h	43	59
	8 h	34	42
Exp.2			
–	–	36	91
1 μM CdA	1 h	32	58
10 μM CdA	1 h	30	39
100 μM ara-C	1 h	32	34

DNA unwinding assays were performed at 0 h and 2 h after 500 rads irradiation. The amounts of double-stranded DNA after alkali treatment were compared with that of an unirradiated control sample.

Regular RPMI medium contains 8 μM nicotinamide. The further addition of nicotinamide to the lymphocyte cultures raised NAD levels remarkably, and prevented fall in intracellular NAD levels caused by dAdo (Table 3). Additionally, nicotinamide protected partially cells from dAdo toxicity in a dose dependent manner.

Table 3. Effects of nicotinamide supplementation
on lymphocyte cultures exposed to 10 µM
deoxyadenosine

Supplemental nicotinamide	% survival at 72 hours	% control NAD levels at 24 hours
0 µM	12 %	40 %
40 µM	20 %	55 %
200 µM	37 %	80 %
1 mM	48 %	185 %
5 mM	63 %	308 %

Control NAD levels were 97.6 ± 21.7 pmol in 10^6 cells
at 24 hours.

DISCUSSION

Human resting lymphocytes examined in vitro have a sizable number of DNA strand breaks (9). Conceivably, the nicking and rejoining of DNA strands takes place even during the prolonged G_0 phase of the lymphocyte cell cycle. Both dAdo and CdA inhibited DNA repair in gamma irradiated lymphocytes. CdA inhibited DNA repair more effectively than ara-C. Ara-CTP is a chain terminator that interrupts the elongation of newly synthesized DNA chains (10). Possibly, CdA also functions as a prompt and potent DNA chain terminator, after phosphorylation and conversion into 2-chloro-dATP. In contrast, the inhibition of DNA repair caused by dAdo was much more time-dependent, and apparently was mediated by a different process. The exact mechanism remains to be clearly elucidated. Perhaps the continued accumulation of dATP, and the associated deoxynucleotide pool imbalance, gradually retards DNA polymerization. The net result is an increasing difference between the rates of breaking and rejoining of DNA, with a progressive increase in strand breaks.

Pyridine nucleotide metabolism is intimately connected with the toxic effects of dAdo toward PBL. Lymphocytes form minimal NAD from tryptophan. Rather, NAD is synthesized and degraded through a pyridine nucleotide cycle, as outlined in Figure 1. This metabolic cycle is active in human resting lymphocytes. NAD is utilized as a substrate for poly(ADP-ribose) synthesis. The poly(ADP-ribose) synthesis induced by massive DNA damage is known to cause the rapid degradation of NAD (6,7). Our preliminary data indicate that poly(ADP-ribose) synthesis was enhanced in dAdo- treated lymphocyte cultures. The increased consumption of NAD, if not balanced by a rise in NAD synthesis, could account for the decline in intracellular NAD levels. Nicotinamide could lessen the fall in NAD levels in two ways: (a) by accelerating NAD synthesis, and (b) by inhibiting NAD degradation mediated by poly(ADP-ribose) synthesis. It is still not clear whether nicotinamide protects the dAdo-treated cells only through the preservation of NAD pools. Many nuclear proteins are poly(ADP-ribosyl)ated after DNA damage (11). The excessive poly(ADP-ribosyl)ation induced by dAdo exposure might inactivate essential nuclear proteins as part of a programmed suicide reaction, which occurs in cells with DNA damage (12). Inhibition of poly(ADP-ribosyl)ation by nicotinamide might therefore prevent the "programmed cell death" induced by dAdo and CdA.

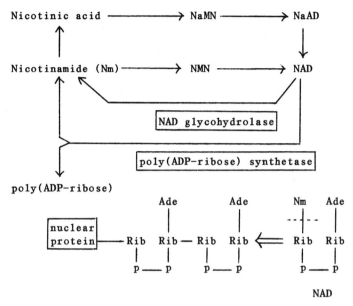

Fig. 1. The synthetic pathway of NAD and the pyridine nucleotide cycle

ACKNOWLEDGMENTS

This research was supported by grants GM 23200, AM 07144, RR 00833 and CA 35048 from the National Institutes of Health.

REFERENCES

1. Giblett, E.R., J.E. Anderson, F. Cohen, B. Pollara and H.J. Meuwissen. Adenosine-deaminase deficiency in two patients with severely impaired cellular immunity. Lancet II:1067-1069. (1972)

2. Carson, D.A., J. Kaye and J.E. Seegmiller. Lymphospecific toxicity in adenosine deaminase deficiency and purine nucleoside phosphorylase deficiency: possible role of nucleoside kinase(s). Proc.Natl.Acad.Sci.USA. 74:5677-5681. (1977)

3. Carson, D.A., D.B. Wasson, R. Taetle and A. Yu. Specific toxicity of 2-chlorodeoxyadenosine toward resting and proliferating human lynphocytes. Blood 62:737-743. (1983)

4. Seto, S., C.J. Carrera, M. Kubota, D.B. Wasson and D.A. Carson. Mechanism of deoxyadenosine and 2-chlorodeoxyadenosine toxicity to nondividing human lymphocytes. J.Clin.Invest. 75:377-383. (1985)

5. Brox, L., A. Ng, E. Pollock and A. Belch. DNA strand breaks induced in human T-lymphocytes by the combination of deoxyadenosine and deoxycoformycin. Cancer Res. 44:934-937. (1984)

6. Skidmore, C.J., M.I. Davis, P.M. Goodwin, H. Halldorsson, P. Lewis, S. Shall and A-A. Zia'ee. The involvement of poly(ADP-ribose) polymerase in the degradation of NAD caused by gamma radiation and N-methyl-N-nitosourea. Eur. J. Biochem. 101:135-142. (1979)

7. Rankin, P.W., M.K. Jacobson, V.R. Mitchell and D.L. Busbee. Reduction of nicotinamide adenine dinucleotide levels by ultimate carcinogens in human lymphocytes. Cancer Res. 40:1803-1807. (1980)

8. Birnboim, H.C. and J.J. Jevecak. Fluorometric method for rapid detection of DNA strand breaks in human white blood cells produced by low doses of irradiation. Cancer Res. 41:1889-1892. (1981)

9. Greer, W.L. and J.G. Kaplan. Regulation of repair of naturally occurring DNA strand breaks in lymphocytes. Biochem. Biophys. Res. Commun. 122:366-372. (1984)

10. Major, P.P., E.M. Egan, D.J. Herrick and D.W. Kufe. Effect of ara-C incorporation on deoxyribonucleic acid synthesis in cells. Biochem. Pharmacol. 31:2937-2940. (1982)

11. Hayaishi, O. and K. Ueda. Poly- and Mono(ADP-ribosyl)ation reactions: their significance in molecular biology, in "ADP-ribosylation reactions," O.Hayaishi and K.Ueda, ed., Academic Press, London, (1982)

12. Berger, N.A. Symposium: Cellular response to DNA damage: the role of poly(ADP-ribose). Radiat. Res. 101:4-15. (1985)

2-CHLOROADENOSINE IS PHOSPHORYLATED AND

INCREASES THE PRODUCTION OF HYPOXANTHINE IN HUMAN CELLS

Hisashi Yamanaka*,   Naoyuki Kamatani,   Tsutomu Nobori,
Kusuki Nishioka, Yutaro Nishida**, and Kiyonobu Mikanagi

Institute of Rheumatology, Tokyo Women's Medical College
2-4-1 NS bldg. Nishi-Shinjuku, Shinjuku-ku, Tokyo, Japan
* Present Address: Basic and Clinical Research, Scripps
Clinic and Research Foundation, 10666 North Torrey Pines Rd.
La Jolla, CA, 92037, USA
** Department of Medicine and Physical Therapy, School of
Medicine, University of Tokyo, 7-3-1 Hongo, Bunkyo-ku
Tokyo, Japan

INTRODUCTION

2-Chloroadenosine (ClAdo) has been reported not to be a substrate
for either adenosine deaminase(1) or adenosine kinase(2) and for this
reason, has been widely used as a ligand for the adenosine receptor
assay(3,4). However, since it is quite toxic to human growing cells, we
suspected it might be, as most other toxic nucleosides, phosphorylated
by some mammalian enzymes.   We cultured a human B cell line with this
adenosine analog and found that human cells are able to phosphorylate
ClAdo, and, in the course of its phosphorylation, this compound induces
the overproduction of hypoxanthine in the culture medium.

MATERIALS AND METHODS

Cell Culture

A variant human splenic B lymphoblastoid cell line WI-L2 deficient
in HGPRT, and secondary variant selected from the former cell line for
the deficiency of adenosine kinase were cultured in RPMI1640 medium
supplemented with 10% fetal calf serum, 2mM glutamine, 100units
penicillin per ml,100ug streptomycin per ml at 37 C in the atmosphere
containing 5% CO_2. Contamination of mycoplasma was routinely checked by
the method of McGarrity, et al(5).   Human T-lymphocytes growing under
the presence of interluekin 2 were cultured by the method described
before(6).

Enzyme Assay

Hypoxanthine guanine phosphoribosyltransferase(HGPRT) or adenosine
kinase was determined by the method previously described(7).

Identification of nucleotide analogs of ClAdo within the cells

Each of the adenosine kinase positive or negative cell line was suspended at 10^6 cells per ml in 30ml medium containing 40 or 100μM ClAdo and cultured for 5 hours. After the incubation, cells were washed with chilled phosphate buffered saline, then lysed by adding 0.4N cold perchloric acid. After extracting perchloric acid, the samples were applied to the HPLC system. The measurements of the intracellular nucleotide analogs of ClAdo were performed by the application of the method described previously(8). After complete dephosphorylation by alkaline phosphatase, the nucleoside was quantitated by HPLC system with C_{18} μBondapak reversed-phase column. Same experiment was done using human T-lymphocytes cultured in the presence of interleukin 2.

Determination of hypoxanthine excreted into culture medium and of intracellular nucleotide concentrations

30ml of exponentially growing HGPRT-deficient WI-L2 cells(10^6 per ml) was incubated with the indicated concentrations of ClAdo and was harvested at 1 and 6 hours by centrifugation. The supernatants were prepared for the quantitation of hypoxanthine. They were treated with charcoal as described before(8),lyophilized, resuspended and applied to the HPLC system. Cells were prepared for the determinations of the intracellular nucleotides as described above, and quantified by the method described before(9).

RESULTS

Enzyme activities

Adenosine kinase activities in the cell extracts from the adenosine kinase positive and negative cell lines were 0.02 and less than 0.001 nmol/min. per mg protein, respectively. HGPRT activities from these cell extracts were less than 0.001 nmol/min. per mg protein.

Cytotoxic effects of ClAdo on each cell line

WI-L2 cells were incubated for three days with ClAdo or 6-methylmercaptopurine riboside(MMPR), and percent control viable cell numbers were compared between adenosine kinase positive and negative cells. Adenosine kinase negative cells were resistant to as high as 80 μM MMPR, while the enzyme positive cells were more than 100-fold less resistant to the same nucleoside, suggesting that the phosphorylation of this nucleoside is mediated by adenosine kinase and the cytotoxicity is dependent on the phosphorylation. Unlike this, ClAdo inhibited the growth of even adenosine kinase negative cells at concentrations higher than 80μM. IC50 values for ClAdo of adenosine kinase negative or positive cells were approximately 65 or 14μM, respectively. Therefore, approximately 4.6-fold higher concentration of ClAdo was required to achieve the same level of growth inhibitory effects on adenosine kinase negative cells than positive cells. These data suggest that adenosine kinase participates, at least some extent, in both of phosphorylation and cytotoxicity of ClAdo.

Identification of nucleotide analogs of ClAdo within the cells

In order to examine the participation of adenosine kinase in the phosphorylation more precisely, we determined the phosphorylated analogs

584

of ClAdo within the cells by dephosphorylation of the nucleotide analogs with alkaline phosphatase after culturing with ClAdo. Without alkaline phosphatase treatment, cell extract from adenosine kinase positive cells cultured with 40µM ClAdo contained less than 0.01nmol ClAdo per 10^7 cells, however, after dephosphorylation, 2.61nmol of ClAdo was detected. ClAdo nucleotides were also detected in the cell extracts of adenosine kinase negative cells cultured with 40 and 100µM ClAdo in amounts of 0.52 and 1.04 nmol per 10^7 cells, respectively. These data suggest that there should be another enzyme(s) converting ClAdo into nucleotide analogs in human cells. These phosphorylations of ClAdo were observed not only in human B-cell line cells but in cultured human peripheral T-lymphocytes as well.

Determination of hypoxanthine excreted into culture medium and of intracellular nucleotide concentrations

After the incubation with 5µM of ClAdo for one and six hours, concentrations of hypoxanthine excreted into the culture medium were increased to 5.1 and 5.2 µM/ml, respectively, compared to that in the drug-free control, 3.0µM/ml. These increased productions of hypoxanthine were not observed in the culture using the adenosine kinase deficient WI-L2 cells (data not shown).

The changes of intracellular concentrations of nucleotides are as follows:

Incubation : 1 hour		IMP	ADP	GDP	ATP	GTP
2-Chloroadenosine	0uM	0.216*	0.64	0.24	7.58	2.08
	5	0.232	0.63	0.29	6.55	1.77
	40	0.308	0.73	0.31	8.65	2.50

Incubation : 6 hours		IMP	ADP	GDP	ATP	GTP
2-Chloroadenosine	0uM	0.123	1.81	0.48	15.32	2.59
	5	0.217	1.48	0.36	15.28	2.69
	40	0.262	0.76	0.31	9.27	2.79

* nmole per 10 cells

In one hour incubation with ClAdo, intracellular concentration of IMP was increased, on the other hand, concentrations of other nucleotides were not changed so much. At longer incubation time or with higher concentration of ClAdo, the enlargement of IMP pool was enhanced, however, concentrations of other nucleotides, especially those of ATP, were diminished.

DISCUSSION

Our data exhibit that both cytotoxicity and phosphorylation of ClAdo depend on adenosine kinase, at least to some extent. ClAdo is known as a potent inhibitor of S-adenosylhomocysteine hydrolase(10), the cytotoxicity of ClAdo in adenosine kinase negative cells might be involved in this effect. However, because we detected the phosphorylated analogs of this compound in the cell extracts, our data really show that there should be another enzyme converting ClAdo into nucleotide analogs within human cells. It is unlikely that this phosphorylation is attributable to the residual activity of adenosine kinase in "adenosine kinase negative" cells. The enzyme activity of adenosine kinase of this cell line was essentially zero, and besides that, these cells were quite resistant to MMPR, which is a much better substrate of adenosine kinase than ClAdo.

ClAdo is also a good substrate of adenosine phosphorylase. Although human cells do not have this enzyme, mycoplasma contaminated cultures are known to express this enzyme activity. However, we have checked the contamination of mycoplasma routinely, and never detected this enzyme activity. Therefore, the possibility that ClAdo was phosphorylized by adenosine phosphorylase in these experiments is also unlikely.

In addition, ClAdo induced overproduction of hypoxanthine accompanied by the enlargement of intracellular IMP concentration. These changes were observed even in such culture conditions that intracellular ATP concentration did not fall. Because of the deficiency of HGPRT, hypoxanthine is a final product of purine metabolites in these cells. The overproduction of hypoxanthine should have a relationship with consumption of ATP in the course of phosphorylation of ClAdo, because these accerelations of hypoxanthine production were dependent on the presence of adenosine kinase. And, consumption of ATP should require cells to increase intracellular IMP for the purpopse to maintain intracellular ATP level. Large amount of IMP should be made via de novo purine biosynthesis, and a lot of hypoxanthine should be excreted into culture medium as a result of elevated IMP pool. There are numbers of reactions within the body that consume ATP, and accerelation of those reactions may be associated with the mechanism of human hyperuricemia.

REFERENCES

1) Huang M, Daly JW : Adenosine elicited accumulation of cyclic AMP in brain slices : Potentiation by agents which inhibit uptake of adenosine. Life Sci. 14:489 (1974)
2) Schnebli HP, Hill DL, Bennett LLJr : Purification and properties of adenosine kinase from human tumor cells of type H.Ep.No.2 J. Biol.Chem. 242:1997 (1967)
3) Williams M, Risley EA : Biochemical characterization of putative central purinergic receptors by using 2-chloro[3H]adenosine, a stable analog of adenosine. Proc.Natl.Acad.Sci.USA 77:6892 (1980)
4) Daly JW : Adenosine receptors : Targets for future drugs. J.Med.Chem. 25:197 (1982)
5) McGarrity GJ, Carson DA : Adenosine phosphorylase-mediated nucleoside toxicity. Exp.Cell Res 139:199 (1982)
6) Kamatani N, Yamanaka H, Nishioka K, Nakamura T, Nakano K, Tanimoto K, Mizuno T, Nishida Y : A new method for the detection of Lesch-Nyhan heterozygotes by peripheral blood T cell culture using T cell growth factor. Blood 63:912 (1984)

7) Carson DA, Kaye J, Seegmiller JE : Lymphospecific toxicity in adenosine deaminase deficiency and purine nucleoside phosphorylase deficiency : Possible role of nucleoside kinase(s). <u>Proc. Natl. Acad.Sci.USA</u> 74:5677 (1977)

8) Ullman B, Wormsted MA, Cohen MB, Martin DW Jr : Purine oversecretion in cultured murine lymphoma cells deficient in adenylsuccinate synthetase: Genetic model for inherited hyperuricemia and gout. <u>Proc.Natl.Acad.Sci.USA</u> 79:5127 (1982)

9) Brown PR : Current high-performance liquid chromatographic methodology in analysis of nucleotides, nucleosides, and their bases.1. <u>Cancer Invest.</u> 1:439 (1983)

10) Chiang PK, Guranowski A, Segall JE : Irreversible inhibition of S-adenosylhomocysteine hydrolase by nucleoside analogs. <u>Arch. Biochem.Biophys.</u> 207:175 (1981)

MECHANISM OF CYTOTOXICITY OF 2-CHLORO AND 2-BROMODEOXYADENOSINE FOR

A HUMAN LYMPHOBLASTIC CELL LINE, CCRF-CEM

Raymond L. Blakley, Min-Chi Huang, Richard A. Ashmun and Rainer Koob

Divisions of Biochemical and Clinical Pharmacology and
Hematology/Oncology
St. Jude Children's Research Hospital
Memphis, TN 38101

INTRODUCTION

The considerable cytotoxicity of 2-chloro-2'-deoxyadenosine (2-CldAdo) and 2-bromo-2'-deoxyadenosine (2-BrdAdo) is accounted for by their relatively high resistance to adenosine deaminase (1) and consequently more prolonged effect on cell metabolism compared with that of 2'-deoxyadenosine (dAdo) unless the latter is protected by an inhibitor of the deaminase like EHNA [erythro-9-(2-hydroxy-3-nonyl)adenine]. Significant therapeutic activity for mice with L1210 leukemia has been reported for both 2-CldAdo (2) and 2-BrdAdo (3), and in a phase I clinical trial in patients with intractable malignancies 2-CldAdo had significant antileukemic activity (4).

The cytotoxicity of dAdo itself is attributed to intracellular accumulation of dATP (5-8), which inhibits the reduction of all substrates by ribonucleotide reductase (9-12). dAdo is therefore assumed to deplete pools of dCTP, dGTP and dTTP and thus limit DNA synthesis. We have investigated the effect of dAdo, 2-CldAdo and 2-BrdAdo on cell kinetics, the intracellular conversion of the nucleosides to triphosphates, the inhibition of ribonucleotide reductase by these triphosphates, and the effects of the nucleosides on intracellular pools of dNTPs.

METHODS

Clonogenic assay was performed on RPMI 1640 medium containing 10% fetal bovine serum and 1.0% methylcellulose. Cells were exposed to nucleoside for either 4 hr or 18 hr before washing three times and suspending in the plating medium. Colonies were examined after 11 days. Cell kinetic studies were performed with an EPICS-V flow cytometer (Coulter Electronics) with excitation by a Spectra-Physics 164-05 argon laser. The frequency histogram data were transferred to a Data General MV/8000 computer for calculation of percentages of cells in G_1, S and G_2-M phases of the cell cycle (13,14). Ribonucleotide reductase was prepared from L1210 cells (15) and assayed by published methods (10,16, 17). Deoxyribonucleotide pools were determined by HPLC by the method of Garret and Santi (18).

RESULTS

   As seen in Table 1 the concentrations of 2-CldAdo to cause 50%
inhibition of cell proliferation, and to decrease clonogenicity by 50%
during an 18 hr exposure are similar and quite close to the concentration required in an 18 hr exposure to inhibit thymidine incorporation
into DNA by 50%. This is also true for 2-BrdAdo and for dAdo in the
presence of 5 µM EHNA. However, in all three measurements 2-CldAdo is
slightly more inhibitory than 2-BrdAdo which in turn is considerably
more inhibitory than dAdo + EHNA. These results suggest that the loss
of viability and inhibition of proliferation are closely linked to inhibition of DNA synthesis. The inhibitory effects of 2-BrdAdo on DNA
synthesis are much greater than its effects on RNA synthesis or protein
synthesis (results not shown). It is noteworthy that when the decline
in viability with increase in nucleoside concentration is examined
(Figure 1), the slope of the curve is much greater for dAdo than for the
halo analogues, suggesting differences in metabolism or metabolic
action.

Table 1.   Effect of dAdo, 2-BrdAdo and 2-CldAdo on Proliferation,
            Clonogenicity and DNA Synthesis in CCRF-CEM cells

	Concentration to produce 50% decrease (µM)		
	dAdo[a]	2-BrdAdo	2-CldAdo
IC_{50}[b] (48 hr exposure)	0.9	0.068	0.045
EC_{50}[c] (4 hr exposure)	4.8	3.0	0.40
EC_{50}[c] (18 hr exposure)	0.68	0.13	0.052
DNA synthesis[d] (4 hr exposure)	1.8	0.05	0.014
DNA synthesis[d] (18 hr exposure)	0.85	0.052	0.031

[a]5 µM EHNA also present.
[b]Inhibition of increase in cell numbers during 48 hr.
[c]Inhibition of colony formation in clonogenic assay.
[d]Inhibition of [³H]thymidine incorporation during a 1 hr incubation
after the period of exposure to the nucleoside indicated.

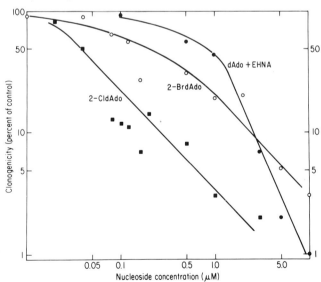

Figure 1. Effect of exposure of CCRF-CEM to nucleosides for 18 hr on
cell viability.

The effects of various concentrations of the nucleoside on the cell cycle kinetics of CCRF-CEM cells are summarized in Table 2. It is clear that concentrations of the order of the EC_{50} for 18 hr exposure cause accumulation of cells in S phase. At concentrations approximating EC_{95} the great majority of cells are accumulated at the G_1-S border. Evidence based on cell counts indicates that under the latter conditions cells complete S, G_2 and M before becoming arrested at the G_1-S border.

Table 2. Effect of Nucleosides on Cell Cycle Kinetics of CCRF-CEM

| Nucleoside | Concentration (μM) | Fraction in Cycle Phase | | | Decrease in viability[a] % |
		G_1	S	G_2-M	
None		0.378	0.482	0.140	0
dAdo	2.0	0.287	0.677	0.036	84
	4.0	0.019	0.974[b]	0.007	95
None		0.426	0.453	0.121	0
2-BrdAdo	0.08	0.304	0.622	0.073	34
	0.32	0.080	0.895	0.025	60
	5.0	0.015	0.960[b]	0.025	95
None		0.323	0.536	0.142	0
2-CldAdo	0.04	0.055	0.745	0.200	53
	0.08	0.035	0.928	0.038	74
	0.48	0.020	0.954[b]	0.026	94

[a]From Figure 1.
[b]At G_1-S border.

When CCRF-CEM cells were incubated with 0.15 μM [8-^{3}H]2-CldAdo (Moravek Biochemicals, 0.5 Ci/mmole) formation of 2-CldATP could be readily detected by HPLC. The intracellular concentration was 1-2 μM after 24 hr and 2-3 μM after 48 hr. A similar experiment with 0.15 μM 2-BrdAdo indicated accumulation of 2-BrdATP at about half the levels found for 2-CldATP. These concentrations of the triphosphates were significantly inhibitory for CDP reduction, but less so for ADP reduction, when a crude preparation of L1210 reductase was used (Table 3). 2-BrdATP is a significantly better inhibitor of the reductase than 2-CldATP, and both are considerably better than dATP and araATP and significantly better than 2-F-araATP.

Table 3. Inhibition of L1210 Ribonucleoside Diphosphate Reductase by 2-BrdATP and 2-CldATP

| Nucleotide | Concentration (μM) necessary to obtain 50% inhibition of enzyme activity | |
	CDP reduction	ADP reduction
dATP	270	32
2-CldATP	3	15
2-BrdATP	0.85	5.5
araATP	600	>1000
2-F-araATP	95[a]	15[a]

[a]Data from White et al. (15).

When the effect of 2-BrdAdo on nucleotide pools was examined it was found that 0.1 μM nucleoside (which reduces viability to 59% in an 18 hr exposure) causes a reduction of all four deoxyribonucleoside triphosphates to 45-60% of controls by 8 hr, but by 24 hr the values were close to controls. Although 0.5 μM 2-BrdAdo (which decreases viability to 33%) in 12 hr lowers dGTP and dTTP to 50% of controls and dCTP and dATP to 20-30%, all pools were above controls by 24 hr. It was necessary to raise the 2-BrdAdo concentration to 2.5 μM in order to maintain any of the pools at a low value for 24 hr. At this concentration dCTP remained below 10% of controls and dATP had recovered only to about 40% at 24 hr, but at this time, dGTP and dTTP were 3 and 4 times control values, respectively. Hence, although 2.5 μM 2-BrdAdo decreases viability to 10% in 18 hr it does not deplete dATP completely and causes accumulation of dGTP and dTTP.

Similar results were obtained with 2-CldAdo. Thus, 0.1 μM 2-CldAdo (which reduces viability to 22%) produces a marked decrease in intracellular dCTP (<5% of controls) and dATP (≈10% of controls) at 8 hr, but these pools recover to a considerable extent by 24 hr. There is little effect on dTTP and dGTP undergoes only a transitory decrease to 50% of controls at 8 hr. When cells are exposed to 0.5 μM 2-CldAdo the dATP, dCTP and dGTP pools are all low at 8-12 hr. However, dATP and dCTP remain low at 24 hr whereas dGTP fully recovers. dTTP is little affected.

The results with dAdo were surprisingly different from those with the halo analogues. At a concentration of 3.5 μM (which reduces viability to 6% in 18 hr) only dGTP was significantly decreased. As expected dATP was increased (8-fold), but dCTP and dTTP were also increased, markedly so by 24 hr.

DISCUSSION

The results provide quite clear evidence that although 2-CldAdo and 2-BrdAdo behave like dAdo in many aspects of their effects on cells there are also significant differences. Not only are the analogues more toxic, but they have quite different effects on deoxyribonucleotide pools and the curve relating residual viability to drug concentration during exposure is less steep than in the case of dAdo.

Although the 2-haloadenine deoxyribonucleoside triphosphates are much better inhibitors of ribonucleotide reductase than dATP it is not clear whether the consequent depletion of dNTP pools is severe enough to account for cytotoxicity. Although cytotoxicity seems to be related to interference with DNA replication, this may arise from other mechanisms in addition to depletion of dNTP pools. If depletion of dNTP pools plays a part it must be due more to an imbalance in the deoxyribonucleotide pools rather than to overall depletion. Completion of the cell cycle and accumulation at the G_1-S border by cells exposed to high nucleoside concentrations suggests that some other important mechanism is at work which prevents initiation of S phase.

Acknowledgements. This work was supported by research grant GM 30423 from the National Institute of General Medical Sciences, USPHS CORE Center grant P30-CA 21765 from the National Cancer Institute and by American Lebanese Syrian Associated Charities.

REFERENCES

1. J. A. Montgomery, The chemistry and biology of nucleosides of purines and ring analogs, in: "Nucleosides, Nucleotides and Their Biological Applications," J. Ridout, D. W. Henry and L. M. Beacher III, ed., Academic Press, New York (1984).

2. D. A. Carson, D. B. Wasson, J. Kaye, B. Ullman, D. W. Martin Jr., R. K. Robins and J. A. Montgomery, Deoxycytidine kinase-mediated toxicity of deoxyadenosine analogs toward malignant human lymphoblasts in vitro and toward murine L1210 leukemia in vivo. Proc. Natl. Acad. Sci. USA 77:6865 (1980).

3. M.-C. Huang, T. L. Avery, R. L. Blakley, J. A. Secrist III and J. A. Montgomery, Improved synthesis and antitumor activity of 2-bromo-2'-deoxyadenosine. J. Med. Chem. 27:800 (1984).

4. D. A. Carson, D. B. Wasson and E. Beutler, Antileukemic and immunosuppressive activity of 2-chloro-2'-deoxyadenosine. Proc. Natl. Acad. Sci. USA 81:2232 (1984).

5. M. S. Coleman, J. Donofrio, J. J. Hutton, L. Hahn, A. Davod, B. Lampkin and J. Dyminski, Identification and quantiation of adenosine deoxynucleotides in erythrocytes of a patient with adenosine deaminase deficiency and severe combined immunodeficiency. J. Biol. Chem. 253:1619 (1978).

6. A. Cohen, R. Hirschhorn, S. D. Horowitz, A. Rubinstein, S. H. Polmar, R. Hong and D. W. Martin Jr., Deoxyadenosine triphosphate as a potentially toxic metabolite in adenosine deaminase deficiency. Proc. Natl. Acad. Sci. USA 75:472 (1978).

7. B. S. Mitchell, E. Mejas, P. E. Dadonna and W. N. Kelley, Purinogenic immunodeficiency diseases; selective toxicity of deoxyribonucleosides for T-cells. Proc. Natl. Acad. Sci. USA 75:5011 (1978).

8. D. A. Carson, B. D. Wasson, E. Lakow and N. Kamatani, Possible metabolic basis for the different immunodeficient states associated with genetic deficiencies of adenosine deaminase and purine nucleoside phosphorylase. Proc. Natl. Acad. Sci. USA 79:3848 (1982).

9. E. C. Moore and R. B. Hurlbert, Regulation of mammalian deoxyribonucleotide biosynthesis by nucleotides as activators and inhibitors. J. Biol. Chem. 241:4802 (1966).

10. E. C. Moore, Mammalian ribonucleoside reductase. Methods Enzymol. 12:155 (1967).

11. S. Eriksson, L. Thelander and M. Ackerman, Allosteric regulation of ribonucleoside diphosphate reductase. Biochemistry 18:2948 (1979).

12. L. Thelander and P. Reichard, Reduction of ribonucleotides. Ann. Rev. Biochem. 48:133 (1979).

13. P. N. Dean, A simplified method of DNA distribution analysis. Cell Tissue Kinet. 13:299 (1980).

14. P. N. Dean, Simplified methods of analyzing DNA distributions from perturbed cell populations. Cell Tissue Kinet. 13:672 (1980).

15. E. L. White, S. C. Shaddix, R. W. Brockmann and L. L. Bennett, Comparison of the actions of 9-β-D-arabinofuranosyl-2-fluoroadenine and 9-β-D-arabinofuranosyladenine on target enzymes from mouse tumor cells. Cancer Res. 42:2260 (1982).

16. J. R. Steeper and C. D. Steuart, A rapid assay for CDP reductase activity in mammalian cell extracts. Anal. Biochem. 34:123 (1970).

17. J. G. Cory, F. A. Russell, and M. M. Mansell, A convenient assay of ADP reductase activity using Dowex-1-Borate columns. Anal. Biochem. 55:449 (1978).

18. C. Garret and D. V. Santi, A rapid and sensitive HPLC assay for deoxyribonucleoside triphosphates in cell extracts. Anal. Biochem. 99:268 (1973).

GENETIC AND BIOCHEMICAL CHARACTERISTICS OF THREE DIFFERENT TYPES OF

MUTANTS OF MAMMALIAN CELLS AFFECTED IN ADENOSINE KINASE

Radhey S. Gupta and Kamal D. Mehta[*]

Department of Biochemistry
McMaster University
Hamilton, Canada L8N 3Z5

INTRODUCTION

Purine nucleoside analogs constitute an important group of anti-metabolites, many of whom possess very useful biochemical and medicinal properties (1,2). For the past several years we have been investigating the metabolism and mechanism of cellular resistance to a number of different purine nucleoside analogs viz. toyocamycin, tubercidin, 6-methyl mercapto-purineriboside (6-MeMPR), pyrazofurin, formycin A and formycin B in cultured Chinese hamster cells using a combined genetic and biochemical approach (3-7). In this approach cellular mutants resistant to various analogs are initially selected and then their cross resistance patterns towards other nucleoside analogs and the affected cellular functions are examined. Our studies with mutants resistant to the above purine nucleoside analogs in Chinese hamster ovary (CHO) cells show that all of the mutants obtained are affected in the enzyme adenosine kinase (AK), thus providing strong evidence regarding the central role of AK in the cellular metabolism and toxicity of these analogs. Studies with these mutants which are reviewed here show that at least three different types of mutants, which show novel differences in the biochemical properties of the enzyme and in their cross resistance patterns towards N- and C-purine nucleoside analogs have been isolated. (C-Nucleosides are those in which the purine base is linked to ribose via a C-C bond rather than the usual N-C linkage found in adenosine and most other nucleosides (1,2)).

SELECTION OF MUTANTS RESISTANT TO PURINE NUCLEOSIDE ANALOGS

The parental CHO cell line which is auxotrophic for proline (8) is referred to as WT in our work (3-7). We and others have previously reported that stable mutants which are highly resistant to the 7-deazaadenosine analogs toyocamycin (Toy[r] mutants) and tubercidin (Tub[r] mutants) are spontaneously obtained at a very high frequency (between 1×10^{-3} to 1×10^{-4}) in CHO cells (3,4,9,10). Similar results have been obtained when other nucleoside analogs such as 6-MeMPR (Mpr[r] mutants) or pyrazofurin (Pyr[r] mutants) were employed in the mutant selection (4). The various properties (e.g. cross resistance pattern, AK activity, etc.) of the Toy[r], Tub[r], Mpr[r] and Pyr[r] mutants of CHO cells suggest that they all involved

[*] Present address: Department of Biochemistry, University of British Columbia, Vancouver, Canada V6T 1W5

very similar genetic lesions (see later sections and ref. 4). Toy[r]4 and Toy[r]5 are two representative mutants of this kind, which have been studied in detail by us. The frequency of the above kind of mutants in CHO cells increases a concentration-dependent manner upon treatment with a wide variety of mutagens, and the optimal conditions for selection of these mutants have been previously described (4,11). In contrast to CHO cells, the frequency of Toy[r] or Tub[r] mutants in other Chinese hamster, mouse or human cell lines have been found to be at least 2 to 3 logs lower (3,12-18). Although the precise basis of the observed differences in the mutant frequency remains unclear, this may be related to the number of functional AK genes that are present in different cell lines (3). Mutants resistant to formycin A (Fom[R] mutants), which is a C-adenosine analog, were selected in single step in medium containing 10 μg/ml of the drug (5). The frequency of the Fom[R] mutants in mutagen (ethyl methanesulfonate; EMS) treated cultures was found to be about 1×10^{-5}, and the two mutants, Fom[R]2 and Fom[R]4, which we have studied in detail both show stable drug-resistance and other characteristics (5). Mutants resistant to formycin B (Fom[r] mutants), which is a C-nucleoside analog of inosine, were obtained (spontaneous mutant frequency between 1- to 5- x 10^{-6}) by selection in presence of 10 μg/ml of the drug. The two mutants, Fom[r]10 and Fom[r]12, which have been further investigated, both show stable drug-resistant phenotype.

CROSS RESISTANCE OF MUTANTS TOWARDS OTHER NUCLEOSIDE ANALOGS

Table 1 shows the results of cross resistance studies with the Toy[r], Fom[R] and Fom[r] mutants towards a number of different nucleoside analogs. In these experiments, cellular toxicity of different nucleoside analogs towards various cell lines was determined from their plating efficiencies in presence of different concentrations of the drugs, as described previously (4-6). From the D_{10} values (drug concentrations which reduce plating efficiency to 10% of control) of the nucleoside analogs for different cell lines, the degree of resistance of the mutant cell lines in comparison to the parental sensitive cell line (i.e. WT) was calculated. As seen from Table 1, the two Toy[r] mutants exhibited high degree of cross resistance to various adenosine analogs examined which included both N-nucleosides (viz. toyocamycin, tubercidin, 6-methyl aminopurineriboside (6-MeAPR), 6-MeMPR, 8-azaadenosine) as well as a number of different C-nucleosides (viz. formycin A, N[7]-benzyl-formycin A (Bbb-73), N[7]-(Δ[2]-isopentenyl)formycin A (Bbb-85), 9-deazaadenosine (18-19). The Toy[r] mutants also exhibited high degree of cross resistance to the structurally distinct nucleoside pyrazofurin, which is also known to be phosphorylated via AK (20). These mutants also showed slight cross resistance to formycin B, which is an inosine analog. A number of Tub[r], Mpr[r] and Pyr[r] mutants which have been examined, showed similar cross resistance patterns as seen with the Toy[r] mutants. In contrast to the above mutants, the two Fom[R] mutants, Fom[R]2 and Fom[R]4 showed very slight to no cross resistance to different N-nucleosides, but were highly resistant to various C-nucleosides examined. The Fom[r] mutants which have been selected in presence of formycin B also exhibited cross resistance to both N- and C-nucleosides, but their degree of resistance was lower in comparison to the Toy[r] mutants. The above mutants, however, showed no cross resistance towards tiazofurin, which is a C-nucleoside analog that is structurally related to pyrazofurin but is not known to be phosphorylated via AK (21), or to the purine bases 8-azaadenine and 6-thioguanine, which are phosphorylated via other enzymes (results not shown). Similar results for these nucleoside analogs for these cell lines have been obtained in at least two independent experiments.

CELLULAR UPTAKE AND PHOSPHORYLATION OF ADENOSINE ANALOGS IN DIFFERENT MUTANT

Earlier studies on purine and pyrimidine nucleoside analogs have shown that the most common mechanism by which cells become resistant to nucleoside analogs involves an alteration or deficiency in the phosphorylation

TABLE 1
CROSS RESISTANCE PATTERNS OF THE TOY^r, FOM^R AND FOM^r
MUTANTS TOWARDS VARIOUS NUCLEOSIDE ANALOGS

Nucleoside Analogs	D_{10} value for[a] the WT cells (ng/ml)	Relative resistance of the mutant cell lines[b]					
		Toy^{r}4	Toy^{r}5	Fom^{R}2	Fom^{R}4	Fom^{r}10	Fom^{r}12
N-Nucleosides							
Toyocamycin	0.4	500	400	1	2	3	5
Tubercidin	2.0	1000	1000	2	3	2	3
8-Azaadenosine	3.0	>1000	>1000	1	1	N.D.[d]	N.D.[d]
6-MeAPR	200	> 500	> 500	2	2	50	70
6-MeMPR	250	> 400	> 400	3	3	15	30
C-Nucleosides							
Formycin A	1.0	> 500	> 500	70	150	15	20
Formycin B	2500	2	2	3	8	5	7
Bbb-73[c]	2000	50	50	> 50	> 50	10	55
Bbb-85[c]	20	1250	>1000	100	350	25	70
Pyrazofurin	15	1000	1000	20	45	5	10
9-Deazaadenosine[c]	0.5	> 200	> 200	30	80	5	15
Tiazofurin	200	1	1	1	1	1	1

[a] The D_{10} values represent the concentration of the analog which reduces plating efficiency of the cells to 10% of that observed in the absence of any drug. Plating efficiencies of various cell lines in presence of different concentrations of the nucleoside analogs were determined as described previously (4-6).

[b] Assuming the D_{10} value of an analog for WT cells as 1, the relative degrees of resistance of the mutant cell lines were determined from the ratios of the D_{10} values of the mutant cell lines compared to the WT cell.

[c] The formycin A analogs Bbb-73 (N^7-benzylformycin A) and Bbb-85 (N^7-(Δ^2-isopentenyl)formycin A) which have been synthesized by Robins and Trip (19), were obtained through Dr. R.H. Hall of McMaster University. 9-Deazaadenosine was kindly supplied by Dr. R.S. Klein of Sloan Kettering Institute for Cancer Research, Rye, NY.

[d] N.D. - not determined.

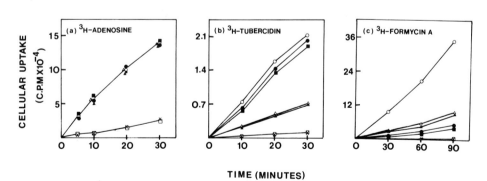

Figure 1. Cellular uptake of [^{3}H]adenosine, [^{3}H]tubercidin and [^{3}H]formycin A by various cell lines. The experiment was carried out as described previously (5). (a) [^{3}H]adenosine; (b) [^{3}H]tubercidin; (c) [^{3}H]formycin A. Symbols: o —— o, WT; ● —— ●, Fom^{R}2; ■——■, Fom^{R}4; x —— x, Toy^{r}4; □——□, Toy^{r}5; Δ —— Δ, Fom^{r}10; ▲ —— ▲, Fom^{r}12.

step (see 3,9,12-17,22). Therefore, to investigate whether these mutants were affected in this regard, cellular uptake and phosphorylation of [^{3}H]-adenosine, [^{3}H]tubercidin and [^{3}H]formycin A in various cell lines was determined. From the results of these studies shown in Fig. 1, it is evident that in case of the Toy^r mutants, cellular uptake of all three adenosine derivatives, i.e. [^{3}H]adenosine, [^{3}H]tubercidin and [^{3}H]formycin A was greatly reduced. In contrast to the Toy^r mutants, both Fom^{R}2 and Fom^{R}4 lines showed nearly normal uptake of the two N-nucleosides, i.e. [^{3}H]adenosine and [^{3}H]tubercidin, but the cellular uptake of [^{3}H]formycin A in both these mutants was found to be greatly reduced ($\approx$10% of the parental WT cell lines). These results indicate that the Fom^R mutants are specifically affected in the phosphorylation of the C-nucleosides. The behavior of the Fom^r mutants differed from both the Toy^r and the Fom^R mutants. These mutants showed normal uptake of [^{3}H]adenosine, but uptake of both [^{3}H]tubercidin and [^{3}H]-formycin A was found to be much reduced. However, the uptake of the latter nucleosides in the Fom^r mutants was not reduced to the same extent as that seen with the Toy^r mutants. In other experiments, whose results are not shown, we have observed that of the total radioactivity which was taken up by various cell lines, more than 90% was in the forms of phosphorylated derivatives. These results thus show that all of the above mutants are affected in the phosphorylation of adenosine analogs and that the observed phosphorylation deficits are in accordance with the results of cross resistance studies.

ADENOSINE KINASE ACTIVITY IN THE MUTANT CELL EXTRACTS

Since adenosine and its derivatives are phosphorylated via the enzyme adenosine kinase, in view of the reduced phosphorylation of adenosine analogs in the mutants, the activity of AK in the parental and mutant cell extracts was examined. Results of these studies based upon three different experiments are shown in Fig. 2. As can be seen, extracts from both the Toy^r and the Fom^R mutants contained no measurable activity of AK, as measured by conversion of [^{3}H]adenosine into [^{3}H]AMP (3,13). In view of the normal incorporation of [^{3}H]adenosine and [^{3}H]tubercidin in the Fom^R mutants, the complete absence of AK activity in the Fom^R mutant extracts

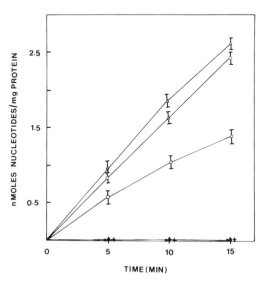

Figure 2. AK activity in the extracts of WT and mutant cells. The activity of AK was measured by conversion of [^{3}H]adenosine into [^{3}H]AMP as described in earlier studies (4,5). o ---- o, WT; Δ —— Δ, Fom^{r}10; $\square$ —— $\square$, Fom^{r}12; $\blacktriangle$ —— $\blacktriangle$, Toy^{r}4; $\blacksquare$ —— $\blacksquare$, Toy^{r}5; $\bullet$ —— $\bullet$, Fom^{R}12; $\diamondsuit$ —— $\diamondsuit$, Fom^{R}4.

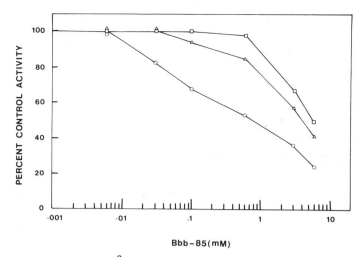

Figure 3. Competition of [^{3}H]adenosine phosphorylation by the adenosine analog, Bbb-85 in the WT and mutant cell extracts. Equal amounts of AK activity from wild-type or resistant cells were incubated with a constant amount of [^{3}H]adenosine and increasing amounts of adenosine analog Bbb-85 and the amount of [^{3}H]adenosine which was phosphorylated during 15 mins of incubation under the standard AK assay conditions was determined. The AK activity (i.e. amount of [^{3}H]adenosine phosphorylated) in presence of Bbb-85 is expressed as a percentage of the activity in the absence of any competing drug. o ——— o, WT; Δ ——— Δ, Fom^{r}10; □ ———□ , Fom^{r}12.

has been very surprising. However, all attempts to obtain AK activity in these cell extracts under various conditions have so far proven unsuccessful (5). The possibility that the mutant cell extracts may contain an inhibitor of AK has also been excluded by mixing experiments using cell extracts from the parental and the mutant cells (5). Similar to the Toy^r mutants, extracts from various Tub^r, Mpr^r and Pyr^r mutants also contained no measurable activity of AK (4). In contrast to the above mutants, cell extracts from both Fom^{r}10 and Fom^{r}12 contained appreciable amounts of AK activity. However, in case of Fom^{r}10, the specific activity of AK was found to be slightly higher (≈10-15% more than WT cells), whereas Fom^{r}12 consistently showed about 35-40% less AK activity in comparison to the parental cells. The AK activity in the Fom^{r}12 extracts also differed from the WT and Fom^{r}10 enzymes with regards to the kinetics of AMP formation, which did not show a linear time course for this mutant.

Since the Fom^r mutants showed normal phosphorylation of [^{3}H]adenosine, but reduced phosphorylation of the adenosine analogs (see Figure 1), the possibility that the enzyme from the mutant cells may show reduced affinities for adenosine analogs was investigated. To examine this aspect, competition experiments between [^{3}H]adenosine and an adenosine analog, Bbb-85 to which the Fom^R mutants showed high degree of cross resistance (Table 1), were carried out in cell extracts of the WT and the mutant cells. In these experiments equivalent amounts of AK activity from the WT and the mutant cell extracts were incubated with a constant amount of [^{3}H]adenosine and varying concentrations of Bbb-85, and the inhibition of [^{3}H]adenosine phosphorylation by Bbb-85 was determined. From the results of these studies, shown in Fig. 3, it is evident that Bbb-85 was a more effective competitor of [^{3}H]adenosine for the WT enzyme, as compared to the enzyme from the two Fom^r mutants. For example, in case of the WT enzyme, 50% inhibition of [^{3}H]-adenosine phosphorylation was observed at 0.8 mM concentration of Bbb-85. However, similar competition of [^{3}H]adenosine phosphorylation in the Fom^{r}10

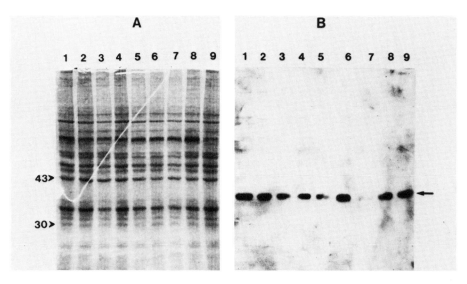

Figure 4. Cross reactivity of AK antibody with proteins from cell extracts of different AK⁻ mutants. Equivalent amounts of total cellular proteins from the indicated cell lines were applied onto two SDS-polyacrylamide slab gels, which were run in parallel. One gel (A) was stained with Coomassie brilliant blue, while the other was blotted on nitrocellulose and then successively treated with 1:100 dilution of the AK antiserum and then with [^{125}I]labeled goat-antirabbit IgG, following the procedure of Towbin et al. (26). (A) Coomassie brilliant blue staining of the applied proteins. (B) Fluorograph of the proteins which cross react with the AK antiserum. The cell lines whose proteins were applied in different lanes are: 1. WT; 2. Toy^{r}17; 3. Toy^{r}18; 4. Tub^{r}17; 5. Tub^{r}18; 6. Mpr^{r}12; 7. Mpr^{r}11; 8. Pyr^{r}14; 9. Pyr^{r}18.

and Fom^{r}12 cell extracts was observed at 4 mM and 6 mM concentrations of Bbb-85. Similar differences between the WT and the mutant enzymes have been observed in two independent experiments. These results indicate that AK from the Fom^r mutants has lower affinity for Bbb-85 in comparison to the enzyme from the parental WT cells.

PRESENCE OF AK CROSS REACTING MATERIAL IN THE MUTANT CELL EXTRACTS

The unusually high spontaneous mutant frequency of the Toy^r, Tub^r, Mpr^r and Pyr^r mutants in CHO cells (henceforth referred to as Class A mutants) together with the complete loss of AK activity in all such mutants raises the question regarding the nature of the genetic lesion responsible for this mutant phenotype (3,9). Such a phenotype could conceivably result from either loss, deletion or silencing of the AK gene (23,24), or it could involve a high frequency mutation within the structural gene for AK that leads to complete loss of the enzymic activity. To gain insight into the nature of the genetic lesion responsible for these mutants, AK from CHO cells has been purified to homogeneity using affinity chromatography and specific antibodies to it have been raised (25). This antibody pulls out AK activity and in immunoblots of WT cell extracts, specific cross reaction with a protein that has the same electrophoretic mobility and M_r value (≈38,000) as purified AK is observed (25).

The presence of AK cross reacting material (CRM) in different mutants was examined by the immunoblotting technique (26). These studies were carried out with a large number of independently selected Toy^r, Tub^r, Mpr^r and Pyr^r mutants, as well as the two Fom^R and Fom^r mutants. Results of these studies for the WT cells and two representative Toy^r, Tub^r, Pyr^r and

TABLE 2

Genetic and Biochemical Characteristics of Different Types of Mutants Affected in AK

Mutant Class	Representative Mutant	Selective Agent	Mutation Frequency	AK Activity in Cell Extracts	Cross-Resistance to Adenosine Analogs	Phosphorylation of Adenosine Analogs In Vivo	Behaviour in Hybrids (Drug^R x Drug^S)
A	Toy^{r}4	Toyocamycin, tubercidin, 6-MeMPR, pyrazofurin	10^{-3} – 10^{-4}	<0.5%	High degree of cross-resistance to both N- and C-nucleosides	Greatly reduced phosphorylation of both N- and C-nucleosides	Recessive
B	Fom^{R}4	Formycin A/ formycin B	1×10^{-5}*	<0.5%	Cross-resistance to only C-nucleosides	Reduced phosphorylation of only C-nucleosides	Codominant
C	Fom^{r}12	Formycin B	2×10^{-5}*	Contains 60 to 110% of biochemically altered AK activity	Lower degree of cross-resistance to both N- and C-nucleosides in comparison to the class A mutants	Reduced phosphorylation of both N- and C-nucleotides	Recessive

* Frequency in EMS-treated cells.

Mpr^r mutants are shown in Fig. 4. As seen from Fig. 4A, the cell extracts from the WT and the mutant cell lines showed similar staining of protein bands and no differences were observed between them. In the parallel gel-blot probed with the AK antibody, the various cell lines (except Mpr^{r}11) showed the presence of a major cross reacting protein that has a M_r of ≈38,000. The cross reacting band has identical electrophoretic mobility in different cell lines, however, its amount differed in various mutants, and in one of the mutants, Mpr^{r}11, very little CRM was observed. A large number of other Toy^r, Tub^r, Pyr^r and Mpr^r mutants which have been examined also showed the presence of substantial amounts of CRM of identical electrophoretic mobility (results not shown). Similarly, cell extracts from Fom^{R}2, Fom^{R}4, Fom^{r}10 and Fom^{r}12 were also found to contain nearly equivalent amounts of AK cross reacting material of similar electrophoretic mobility as seen in the WT cells (results not shown).

DISCUSSION: PROPERTIES OF VARIOUS MUTANTS AFFECTED IN ADENOSINE KINASE

Table 2 summarizes the properties of the three types of mutants affected in purine nucleoside phosphorylation which have been identified by our studies. The Class A mutants which can be selected using a number of different adenosine analogs and are represented here by the Toy^r mutants, are obtained spontaneously at unusually high frequency in CHO cells (3,4,9). These mutants exhibit high degree of cross resistance to various N- as well as C-adenosine analogs and show greatly reduced phosphorylation of adenosine and its various analogs. These mutants show recessive behavior in cell hybrids and cell extracts of these mutants contain no measurable activity of the enzyme AK (3,4,9). Based on their reported characteristics, mutants resistant to various adenosine analogs which have been previously isolated, all appear to be of this kind (3,4,9,10,12-17,22). The presence of AK-cross reacting material in cell extracts from all Class A mutants provides strong evidence that these mutants do not involve loss or failure of expression of the AK gene. The similar electrophoretic mobilities in SDS-gels of the major cross reacting band, which has the same relative molecular mass as purified AK, also argue strongly against the possibility of a large deletion within the AK gene, or for a nonsense type of premature chain termination mutation. These results instead strongly suggest that the genetic lesions in the Class A mutants are missense mutations within the structural gene for AK (occurring at a presumed mutational hot-spot).

The Class B (i.e. Fom^R) mutants which have been selected in presence of formycin A exhibit a novel genetic and biochemical phenotype (5). These mutants exhibited increased resistance to various C-nucleosides, but no appreciable cross resistance was observed to the various N-nucleosides examined. In accordance with their cross resistance pattern, the Fom^R mutants showed reduced phosphorylation of C-adenosine analogs (viz. [^{3}H]-formycin A) but not of N-adenosine derivatives. The lack of cross resistance of the Fom^R mutants to tiazofurin, a C-nucleoside which is not phosphorylated via AK (21), further suggests that the lesion in these mutants is specific for only those C-nucleosides which are substrates for AK. Very interestingly, the cell extracts from the Fom^R mutants also contained no activity of AK under various conditions and they exhibited codominant behavior in cell hybrids (5). As discussed elsewhere (5), it is highly unlikely that the Fom^R mutants involve more than one genetic lesion. Instead, various properties of the Fom^R mutants suggest that the genetic lesion in these directly affects AK in a manner that specifically affects the phosphorylation of C-nucleosides and alters the behavior of the enzyme in cell extracts. It is of interest in this regard that in contrast to various N-adenosine derivatives which normally exist in anti-conformation, the C-nucleosides such as formycin A and pyrazofurin are predominantly present in syn-conformation (1,27). Because of this difference in their preferred conformations, it is possible that N-nucleosides interact with certain specific region or domain of AK that is not important in interaction with N-nucleosides and which may be altered in the Fom^R mutants.

The Fom^r mutants which are described in the present studies constitute the third type (Class C) of mutants which are affected in purine nucleoside phosphorylation. These represent the only mutants resistant to adenosine analogs which contain large amounts of AK activity in cell extracts. (The only exception in this regard were the mutants resistant to adenosine isolated by McBurney and Whitmore (28) which were originally reported to contain AK activity, however, a later study (11), found no detectable AK activity in these mutant cell extracts). The genetic lesion in the Class C mutants seems to directly affect AK in a manner that leads to production of an altered form of the enzyme that has lower affinity for adenosine analogs. The lower spontaneous mutant frequencies of the Class B and C mutants indicate that the genetic lesions which result in such phenotypes are highly specific and relatively rare, in comparison to those which produce Class A phenotype.

REFERENCES

1. Suhadolnik, R.J. (1979) Prog. Nucl. Acid Res. Mol. Biol. 22: 193–291.
2. Bloch, A. (1975) Ann. N. Y. Acad. Sci. 255: 576–596.
3. Gupta, R.S. and Siminovitch, L. (1978) Som. Cell Genet. 4: 715–735.
4. Gupta, R.S. and Singh, B. (1983) Mut. Res. 113: 441–454.
5. Mehta, K.D. and Gupta, R.S. (1983) Mol. Cell. Biol. 3: 1468–1477.
6. Gupta, R.S. and Mehta, K.D. (1984) Biochem. Biophys. Res. Commun. 120: 88–95.
7. Gupta, R.S. and Moffat, M.R.K. (1982) J. Cell. Physiol. 111: 291–294.
8. Kao, F.T. and Puck, T.T. (1967) Genetics 55: 513–524.
9. Rabin, M.S. and Gottesman, M.M. (1979) Som. Cell Genet. 5: 571–583.
10. Plagemann, P.G.W. and Wohlhueter, R.M. (1983) J. Cell. Physiol. 116: 236–246.
11. Singh, B. and Gupta, R.S. (1983) Environ. Mut. 5: No. 6, 871–880.
12. Bennett, L.L. Jr., Schnebli, H.P., Vail, M.H., Allan, P.W. and Montgomery, J.A. (1966) Mol. Pharmacol. 2: 432–443.
13. Chan, T.S., Ishii, K., Long, C. and Green, H.G. (1973) J. Cell. Physiol. 81: 315–322.
14. Debatisse, M. and Buttin, G. (1977) Som. Cell Genet. 3: 497–511.
15. Gupta, R.S. and Hodgson, M. (1981) Exp. Cell Res. 132: 496–500.
16. Murray, W. and Gupta, R.S. (1984) Can. J. Genet. Cytol. 26: 576 (Abst.).
17. Thacker, J. (1980) Mut. Res. 74: 37–59.
18. Robins, M.J, and Trip, E.M. (1973) Biochemistry 12: 2179–2187.
19. Zimmerman, T.P., Deeprose, R.D., Wolberg, G., Stopford, C.R., Duncan, G.S., Miller, W.H., Miller, R.L., Lim, M.I., Ren, W.Y. and Klein, R.S. (1983) Biochem. Pharmacol. 32: 1211–1217.
20. Dix, D.E., Lehman, C.P., Jakubowski, A., Moyer, J.D. and Handschumacher, R.E. (1979) Cancer Res. 39: 4485–4490.
21. Saunders, P.P., Kuttan, R., Lai, M.M. and Robins, R.K. (1983) Mol. Pharmacol. 23: 534–539.
22. Astrin, K.H. and Caskey, C.T. (1976) Arch. Biochem. Biophys. 176: 397–410.
23. Simon, A.E. and Taylor, M.W. (1983) Proc. Natl. Acad. Sci. USA 80: 810–814.
24. Bradley, W.E.C. (1979) J.Cell. Physiol. 101: 325–340.
25. Mehta, K.D. and Gupta, R.S. (1985) Mol. & Cell. Biol. (Submitted for publication).
26. Towbin, H., Staehelin, J. and Gordon, J. (1979) Proc. Natl. Acad. Sci. USA 76: 4350–4354.
27. Daves, G.D. and Cheng, C.C. (1976) Prog. Med. Chem. 13: 303–349.
28. McBurney, M.W. and Whitmore, G.F. (1975) J.Cell Physiol. 85: 87–99.

CHARACTERIZATION OF ARABINOSYLGUANINE RESISTANCE

IN A LYMPHOBLASTOID CELL LINE

Donna S. Shewach and Beverly S. Mitchell

University of Michigan Medical Center

Ann Arbor, MI 48109

INTRODUCTION

In recent years, cancer chemotherapists have been searching for drugs that would selectively destroy malignant cells while sparing healthy host tissue. The discovery that dAdo and dGuo are more toxic to T compared to B lymphoblasts through the actions of their 5'-triphosphate derivatives (1) has spurred interest in developing analogs of these compounds that would exhibit a similar specificity of action, without being degraded by the purine catabolic enzymes adenosine deaminase (ADA) and purine nucleoside phosphorylase (PNP). Using this rationale, 2-fluoroadenine arabinoside was developed as an ADA-insensitive analog of adenine arabinoside, and it is presently undergoing clinical trials as an antileukemic agent (2). A recent report (3) indicated that araG was resistant to cleavage by PNP, and exhibited selective toxicity for T relative to B lymphoblasts. We have previously compared the effects of araG in human lymphoblasts to those of araC, another arabinosyl antimetabolite that is in clinical use an an antileukemic agent (4). While araG was less potent than araC, it was far more selective in its cytotoxic effects. Based on these findings, araG shows promise as an antileukemic agent and merits greater study. In this paper, we report on the further characterization of the metabolism of araG in human cell lines exhibiting varying levels of sensitivity to araG.

RESULTS AND DISCUSSION

The metabolism of araG in cultured cells has been studied by several investigators with conflicting results. An early report indicated that araG was phosphorolyzed by PNP (5) whereas a more recent report stated that araG was not a substrate for this enzyme (3). We have studied the catabolism of araG in a variety of ways, as illustrated

Table 1. Metabolism of araG

Sample	Drug Concentration	% Initial Drug Remaining		Incubation Period
		AraG	dGuo	
Calf Spleen PNP	1, 10, 100 or 1000 μM	>95	<10	12 hr
MOLT-4 T Lymphoblasts	20 μM	92	0	72 hr
Human Whole Blood Normal	100 μM	26	0	4 hr
HGPRT-deficient	100 μM	57	0	4 hr

All samples were incubated at 37°C. The amount of drug remaining was analyzed by high pressure liquid chromatography, employing a reverse-phase column and eluting with 5% MeOH in 0.05M KH_2PO_4, pH 5.4.

in Table 1. Under the conditions studied, no significant degradation of araG was observed during incubation with either calf spleen PNP or MOLT-4 T lymphoblasts in culture.

These conditions resulted in the complete phosphorolysis of dGuo. In contrast, a 4 hr incubation with 100 μM araG in diluted human whole blood resulted in the disappearance of 74% of the initial amount of araG. In this system, dGuo was completely phosphorolyzed within 1 hr. The decrease in the amount of araG or dGuo present in the serum was associated with a more than 5-fold increase in the intraerythrocytic GTP pools. In contrast, a similar incubation with araG and dGuo in blood from a patient who was deficient in the enzyme hypoxanthine-guanine phosphoribosyltransferase (HGPRT) resulted in a lesser degree of phosphorolysis of the drugs. Guanine accumulated in the serum of the drug-treated HGPRT-deficient blood and less than a 2-fold increase in the intraerythrocytic GTP pools was observed after 4 hr. These data indicate the araG is indeed a substrate for human PNP although it is degraded at a slower rate than dGuo. The resulting araG phosphorolytic cleavage product, guanine, can be salvaged by HGPRT to effect an increase in the intraerythrocytic guanine nucleotide pools.

To study the mechanism of action of araG in more detail, several clonal lymphoblast cell lines were developed that were resistant to the growth inhibitory effects of araG. The cell line (designated line 24B3) showing the greatest level of resistance to araG (IC_{50} value for araG increased 600-fold compared to wild type cells) was characterized further in order to determine the mechanism responsible for the observed drug resistance. As indicated in Table 2, the 24B3 cell line was highly resistant to growth inhibition by araG and dGuo, but showed little resistance to araC. Of the biochemical parameters tested, the greatest

Table 2. Inhibition of growth of human lymphoblastoid and myeloid cells.

Cell line	AraG	dGuo[a]	AraC
		IC_{50} Values (μM)	
MOLT-4 (wild type)	0.15	2.0	0.015
24B3	90.0	72.0	0.05
HL-60	21.0	28.0	N.D.[b]

Cells were incubated in the presence of varying concentrations of each drug for 72 hr as previously described.

[a]In the presence of 500 μM 8-aminoguanosine, an inhibitor of PNP.

[b]Not determined.

Table 3. Accumulation of araGTP in MOLT-4, 24B3 and HL-60 cell lines.

Cell line	AraG Concentration	pmol araGTP/5×10^6 cells Incubation period	
		6 hr	24 hr
MOLT-4 (wild type)	20 μM	227[a]	1,132
24B3	100 μM	43	7
HL-60	100 μM	4	9

AraGTP was detected and quantitated from acid-soluble cell extracts by gradient high pressure liquid chromatography using a strong anion exhange column as previously described (4).

[a]Represents the amount of araGTP at 4 hr.

difference between the 24B3 and parental cell lines was in their abilities to accumulate araGTP and dGTP. At equimolar concentrations of araG and dGuo, the mutant cell line accumulated araGTP and dGTP to less than 11% of the level observed in the wild type cell line after 4 hr. The accumulation of araCTP was affected to a lesser degree, as the 24B3 cells were able to accumulate this nucleotide to greater than 30% of the wild type level. Similar results were obtained using cell lysates, indicating that the decreased ability to accumulate nucleotides in the 24B3 cell line cannot be explained by decreased membrane transport of the corresponding nucleosides.

In order to better characterize the metabolism and mechanism of action of araG, we have examined its effects on other cell lines as well. Compared to the effect of araG on the growth of T lymphoblasts, the HL-60 promyelocytic cell line shows relative resistance to this drug. The IC_{50} value for araG in HL-60 cells is more similar (28 μM) to that determined in the araG-resistant 24B3 lymphoblast line than that observed in the wild type lymphoblasts. Furthermore, dGuo is toxic to HL-60 cells only in the presence of an inhibitor of PNP (IC_{50} = 21 μM). The HL-60 cells are also similar to the 24B3 line in that araGTP accumulates to less than 1% of the level observed in the MOLT-4 (wild-type) lymphoblast line after 24 hr, as indicated in Table 3. The relative resistance of the HL-60 promyelocytes and the 24B3 T lymphoblasts to the cytotoxic effects of araG are not due to increased degradation of the nucleoside since no phosphorolysis of araG was observed during a 72 hr incubation period in either of these two cell lines.

The 24B3 cell line may have acquired resistance to araG through increased catabolism of araG nucleotides; however, preliminary data suggests that this is an unlikely explanation. Based on the lack of cross resistance to araC and the lesser accumulation of araGTP and dGTP compared to araCTP in the araG-resistant cell line, it is possible that araG resistance in the 24B3 cells has occurred through the deficiency of a specific kinase. This mechanism would suggest that the kinase involved in the phosphorylation of araG and dGuo differs from that responsible for araC phosphorylation. Alternatively, regulation of the phosphorylation of araG may be altered in the 24B3 cell line. In mammalian cells, araC is phosphorylated by dCyd kinase, and it has been suggested that this kinase is also responsible for the phosphorylation of dGuo (6). The development of a mutant cell line resistant to araG may aid in delineating the substrate specificities of the deoxyribo-nucleoside kinases.

These data suggest that the biochemical pathways involved in the intracellular accumulation of araGTP and dGTP differ, in part, from those involved in the accumulation of araCTP. In addition, these biochemical differences appear to be cell-specific. We are presently investigating the nature of the biochemical mechanism that is responsible for the differential accumulation of araGTP in human cells.

REFERENCES

1.   Martin D.W. Jr. and Gelfand E.W.   Biochemistry of diseases of immunodevelopment.   Ann. Rev. Biochem. 40:845-877, 1981.

2.   Hutton J.J., Von Hoff D.D., Kuhn J., Phillips J., Hersh M. and

Clark G. Phase I clinical investigation of 9-β-D-arabinofuranosyl-2-fluroadenine 5′-monophosphate (NCS 312887), a new purine antimetabolite. Cancer Res. 44:4183-4186, 1984.

3. Cohen A., Lee J.W.W. and Gelfand E.W. Selective toxicity of deoxyguanosine and arabinosyl guanine for T-leukemic cells. Blood 61:660-666, 1983.

4. Shewach D.S., Daddona P.E., Ashcraft E. and Mitchell B.S. Metabolism and selective cytotoxicity of 9-β-D-arabinofuranosylguanine in human lymphoblasts. Cancer Res. 45:1008-1014, 1985.

5. Elion G.B., Rideout J.L., de Miranda P., Collins P. and Bauer D.J. Biological activities of some purine arabinosides. Ann. N.Y. Acad. Sci. 55:468-480, 1975.

6. Krenitsky T.A., Tuttle J.V., Koszalka G.W., Chen I.S., Beacham L.M. III, Rideout J.L. and Elion G.B. Deoxycytidine kinase from calf thymus. Substrate and inhibitor specificity. J. Biol. Chem. 251;4055-4061, 1976.

THE SYNTHESIS OF PHOSPHORIBOSYLPYROPHOSPHATE FROM GLUCOSE DECREASES

DURING AMINO ACID STARVATION OF HUMAN LYMPHOBLASTS

Renate B. Pilz and Gerry R. Boss

Department of Medicine
University of California, San Diego Medical Center
San Diego, CA

INTRODUCTION

Phosphoribosylpyrophosphate (PP-Rib-P) is a key regulator of de novo purine synthesis and because of the kinetic properties of glutamine amido-phosphoribosyltransferase, small changes in the intracellular concentration of PP-Rib-P can cause major changes in rates of de novo purine synthesis (1). We recently showed that ribose 5-phosphate (Rib-5-P), the immediate precursor of PP-Rib-P may also be rate-limiting for de novo purine synthesis in human lymphoblasts, since its intracellular concentration is well below the K_m of PP-Rib-P synthetase for this substrate at physiological phosphate concentrations (2,3).

Rib-5-P and thus PP-Rib-P may be produced from glucose via the oxidative or nonoxidative pentose phosphate pathway. The question which of these two pathways provides most of the PP-Rib-P needed for nucleotide synthesis is controversial and little is known about the regulation of Rib-5-P synthesis from glucose (4,5,6).

We used amino starvation of cultured human lymphoblasts as a model system to study these questions. When lymphoblasts are starved 3 hours for a single essential amino acid, purine nucleotide synthesis via the de novo pathway decreases by about 80% and via the salvage pathway by about 60%. These effects are rapid and fully reversible (7). Purine nucleotide synthesis is significantly less affected in mutant lymphoblasts with either an increased capacity to generate PP-Rib-P via superactive PP-Rib-P synthetase or with an expanded PP-Rib-P pool due to HPRT deficiency (3). There is no change in the specific activities of PP-Rib-P synthetase or amidophosphoribosyltransferase during amino acid starvation. In addition, there is no change in intracellular concentrations of phosphate or purine nucleotides, both known regulators of PP-Rib-P synthetase. But, there is a 40% decrease in intracellular concentrations of Rib-5-P and PP-Rib-P. Moreover, inhibitors of IMP-dehydrogenase and adenylosuccinate synthetase reverse the effect of amino acid starvation on de novo purine synthesis by increasing the intracellular Rib-5-P and PP-Rib-P concentrations through increased endogenous ribose reutilization from IMP. Thus, amino acid starvation appears to decrease purine nucleotide synthesis by decreasing PP-Rib-P synthesis from glucose (3).

METHODS

The experimental procedures are described in detail elsewhere (8).

RESULTS AND DISCUSSION

Glucose transport was unaffected after amino acid starvation. However, total glucose consumption decreased by approximately 25% from 135 ± 11 nmol/h/10^6 cells to 101 ± 9 nmol/h/10^6 cells. There was a small statistically insignificant decrease in lactate production and total anion excretion into the medium.

The effect of amino acid starvation on the oxidative pentose phosphate pathway was examined by measuring $^{14}CO_2$ production from [1-^{14}C]glucose. $^{14}CO_2$ release from [1-^{14}C]glucose decreased by 18% from 3.7 ± 0.4 nmol/h/10^6 cells to 3.0 ± 0.4 nmol/h/10^6 cells after 3 hours of starvation. Similar results were obtained when the carbon flow through the oxidative pentose phosphate pathway was calculated according to Katz and Wood (9). None of the enzymes involved in the oxidative pentose phosphate pathway, glucose 6-phosphate dehydrogenase, 6-phosphogluconate dehydrogenase, or ribulose 5-phosphate 3-epimerase and phosphoriboisomerase changed in specific activity after amino acid starvation. When cells were incubated with the electron acceptor methylene blue which reduces NADPH-inhibition of glucose 6-phosphate dehydrogenase, the $^{14}CO_2$ release from [1-^{14}C]glucose was stimulated approximately 3-fold to values that were nearly identical in the control and the starved state. These results suggest that the synthetic capacity of the oxidative pentose phosphate pathway remains intact and that amino acid starvation affects the carbon flow through the pathway by altering the oxidative state of NADP.

Since [1-^{14}C]glucose loses its radioactive carbon on passage through the oxidative pentose phosphate pathway, the incorporation of radioactivity from [1-^{14}C]glucose into PP-Rib-P as well as into the ribose moiety of nucleotides can be used as a measure of nonoxidative pentose phosphate pathway activity. In control experiments using glucose labelled in the 6 and 2 position, we ruled out 1) any significant exchange of the glucose carbon atoms in position 1 and 6 due to isomerization at the triose phosphate level, and 2) any significant reflux from pentoses to the hexose phosphate pools which would have interfered with our measurement (8). We could, therefore, measure the effect of amino acid starvation on the nonoxidative pentose phosphate pathway by following [1-^{14}C]glucose incorporation into PP-Rib-P and found that it decreased by approximately 55% from 3.1 ± 0.3 pmol/h/10^6 cells to 1.4 ± 0.2 pmol/h/10^6 cells after 3 h of amino acid starvation. Similarly, the incorporation of [1-^{14}C]glucose into ATP and GTP decreased by approximately 51% from 86.4 ± 6.1 pmol/h/10^6 cells to 44.1 ± 3.9 pmol/h/10^6 cells after starvation. Interestingly, none of the enzymes involved in the nonoxidative pentose phosphate pathway (including transketolase, transaldolase, aldolase, phosphofructokinase, and fructose diphosphatase) changed in specific activity during amino acid starvation. Therefore, the activity of the nonoxidative pentose phosphate pathway seems to be regulated by a metabolite which changes in concentration during amino acid starvation; the nature of this regulator is currently under investigation.

The incorporation of [6-^{14}C]glucose into PP-Rib-P in cells in complete media was only about 30% greater than that of [1-^{14}C]glucose and it decreased after amino acid starvation by a similar amount as found for [1-^{14}C]glucose. Since the incorporation of the 6 label could arise from either oxidative or nonoxidative pentose phosphate pathway activity, these data suggested to us that the oxidative pentose phosphate pathway provides little PP-Rib-P for purine nucleotide synthesis. In order to further examine this question, we studied the two enzymes responsible for the conversion of ribulose 5-phosphate to xylulose 5-phosphate and to Rib-5-P, respectively, to determine whether one of the reactions might be favored over the other. The specific activities of ribulose 5-phosphate epimerase and phosphoriboisomerase were very similar at about 1.1 nmol/min/mg of protein. However, the K_m value of the epimerase was about one-quarter that of the isomerase for the shared substrate ribulose 5-phosphate (0.36 $\pm$ 0.04 mM and 1.4 $\pm$ 0.1 mM, respectively). The intracellular ribulose 5-phosphate concentration was well below both K_m values (8). Therefore, ribulose 5-phosphate is more likely to be converted to xylulose 5-phosphate rather than to Rib-5-P. Although this model suggests that carbon flow through the oxidative pentose phosphate pathway mostly results in xylulose 5-phosphate which may be returned to glycolysis, it does not exclude the possibility that increasing the carbon flow through this pathway, e.g. by electron acceptors, can increase the Rib-5-P and PP-Rib-P concentrations and thus increase rates of purine nucleotide synthesis as has been reported by other authors (10,11).

CONCLUSIONS

We conclude that 1) decreased purine nucleotide synthesis in lymphoblasts during amino acid starvation is largely secondary to decreased PP-Rib-P synthesis from glucose; 2) carbon flow through the oxidative pentose phosphate pathway decreases by approximately 18%, whereas the carbon flow through the nonoxidative pentose phosphate pathway decreases by about 55% during amino acid starvation; 3) $^{14}CO_2$ release from $[1-^{14}C]$glucose is a measure of ribulose 5-phosphate production by the oxidative pentose phosphate pathway; this product seems to be converted preferentially to xylulose 5-phosphate rather than to ribose 5-phosphate. Our data, therefore, suggest that the nonoxidative pentose phosphate pathway is the major source of PP-Rib-P for purine nucleotide synthesis in cultured human lymphoblasts; this pathway appears to be regulated by a metabolite which changes in concentration during amino acid starvation.

REFERENCES

1. M. A. Becker, K.O. Raivio and J. E. Seegmiller, Phosphoribosylpyro-phosphate in mammalian cells, Adv. Enzymol. 49:281-306 (1979).

2. R. B. Pilz, R. C. Willis and G. R. Boss, The influence of ribose 5-phosphate availability on purine synthesis of cultured human lymphoblasts and mitogen-stimulated lymphocytes, J. Biol. Chem. 259:2927-2935 (1984).

3. G. R. Boss, Decreased phosphoribosylpyrophosphate as the basis for decreased purine synthesis during amino acid starvation of human lymphoblasts, J. Biol. Chem. 259:2936-2941 (1984).

4. K. A. Steer, M. Sochor, A. Gonzalez and P. McLean, Regulation of pathways of glucose metabolism in kidney. Specific linking of pentose phosphate pathway activity with kidney growth in experimental diabetes and unilateral nephrectomy, FEBS Letters 150: 494-498 (1982).

5. M. Sochor, N. Z. Baquer and P. McLean, Regulation of the pathways of glucose metabolism in kidney. The effect of experimental diabetes on the activity of the pentose phosphate pathway and the glucuronate-xylulose pathway, Arch. Biochem. Biophys. 198:632-646 (1979).

6. K. Brand and K. Deckner, Quantitative relationship between the pentose phosphate pathway and the nucleotide synthesis in ascites tumor cells, Hoppe-Seyler's Z. Physiol. Chem. 351:711-717 (1970).

7. G. R. Boss and R. W. Erbe, Decreased purine synthesis during amino acid starvation of human lymphoblasts, J. Biol. Chem. 257: 4242-4247 (1982).

8. G. R. Boss and R. B. Pilz, Phosphoribosylpyrophosphate synthesis from glucose decreases during amino acid starvation of human lymphoblasts, J. Biol. Chem. 260:6054-6059 (1985).

9. J. Katz and H. G. Wood, The use of $^{14}CO_2$ yields from glucose-1- and -6-^{14}C for the evaluation of the pathways of glucose metabolism, J. Biol. Chem. 238:517-523 (1963).

10. K. O. Raivio, C. S. Lazar, H. R. Krumholz and M. A. Becker, The phosphogluconate pathway and synthesis of 5-phosphoribosyl-1-pyrophosphate in human fibroblasts, Biochim. Biophys. Acta 678: 51-57 (1981).

11. G. C. Yeh and J. M. Phang, Pyrroline-5-carboxylate stimulates the conversion of purine antimetabolites to their nucleotide forms by a redox-dependent mechanism, J. Biol. Chem. 258:9774-9779 (1983).

EFFECT OF FRUCTOSE ON THE CONCENTRATION OF PHOSPHORIBOSYLPYROPHOSPHATE IN
ISOLATED HEPATOCYTES

M.F. Vincent, G. Van den Berghe and H.G. Hers

Laboratory of Physiological Chemistry, International Institute of
Cellular and Molecular Pathology and University of Louvain
Avenue Hippocrate 75, B-1200 Brussels, Belgium

INTRODUCTION

Fructose loads provoke a depletion of the hepatic purine nucleotides
(1, reviewed in 2) which is followed by a compensatory enhancement of their
"de novo" synthesis, as evidenced by an increased rate of incorporation of
$[^{14}C]$glycine into urinary uric acid after oral (3) and intravenous (4)
administration of the ketohexose in man. A conversion of glutamine-
phosphoribosylpyrophosphate amidotransferase, the limiting enzyme of the
"de novo" pathway, from its large inactive to its small active form, has
been shown to play a role in this enhancement (5). This conversion most
likely results from the fructose-induced decrease of the purine nucleotides, which are inhibitory of the enzyme, combined with an increase in
PRPP, which is stimulatory. Indeed, an approx. twofold increase in hepatic
PRPP has been demonstrated 30 min after a fructose infusion in mice (5). In
the present work, the mechanism of the elevation of PRPP induced by fructose has been investigated in isolated hepatocytes.

METHODS

The preparation and incubation of isolated hepatocytes from fed rats,
and the methods used for the extraction and determination of nucleotides
have been given (6). PRPP was measured in heat-treated extracts (7) as
previously described (8). Ribose 5-phosphate was determined in the same
extracts by the method of Becker (9). Intracellular Pi was measured in
hepatocytes that had been centrifuged through a 0.5 ml layer of silicone
into 0.5 ml of ice-cold 10 % $HClO_4$ as given in (10), by the method of
Itaya and Ui (11). Fructose 1-phosphate was measured as given in (12).

RESULTS

(1) Influence of fructose on the concentration of PRPP and other
metabolites in isolated hepatocytes

Addition of 5 mM fructose to the cell suspension provoked, after a 5
to 15 min latency and within a similar time interval, a 5- to 10-fold
increase in the concentration of PRPP above its basal value of around 10
nmol per g of cells. The concentration of PRPP reached its maximum between

15 and 30 min after the addition of fructose and thereafter decreased progressively to its basal level, which was reached at around 60 min. As shown in Fig. 1, depicting a representative experiment, fructose decreased the concentration of ATP to 30 % of its initial value within 2 minutes and that

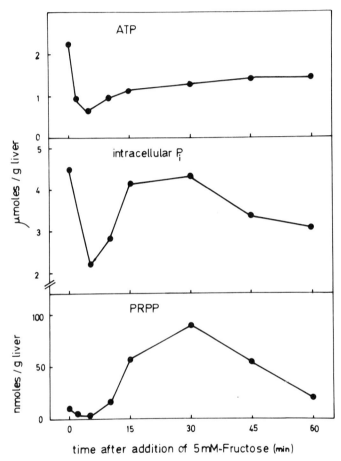

Fig. 1 Influence of fructose on the concentrations of ATP, Pi and PRPP in isolated rat hepatocytes.

of Pi to 50 %. Both decreases preceded the rise in PRPP. Strikingly, the elevation of PRPP occured in close parallelism with the rebound of the concentration of intracellular Pi. The latter returned to the normal range approx. 15 min after the addition of fructose. Together with the increase in PRPP and Pi, there was also a slower, progressive increase in the concentration of ATP. This elevation proceeded reproducibly in two stages : a more rapid initial increase, lasting for 10 to 15 min, followed by a slower phase.

As shown in Fig. 2, the effect of fructose on the concentration of PRPP at 15 min displayed a dose-dependency which mirrored that on the level of ATP 5 min after the addition of the ketohexose. As expected, fructose provoked a dose-dependent accumulation of fructose 1-phosphate; it also induced an up to approx. 2-fold elevation of ribose 5-phosphate.

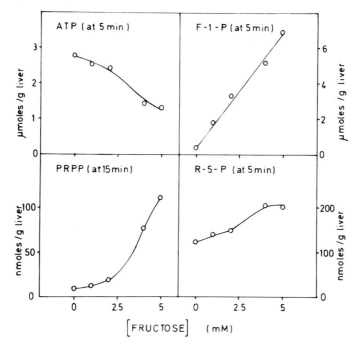

Fig. 2 Influence of increasing concentrations of fructose on the levels of ATP, fructose 1-phosphate, PRPP and ribose 5-phosphate in isolated rat hepatocytes.

(2) Effect of xylitol and tagatose on the concentration of PRPP

Other compounds are known to provoke a depletion of ATP when added at millimolar concentrations to liver preparations, among them xylitol (13) and tagatose (14). With 5 mM xylitol, ATP decreased by 43 % as compared to 70 % with the same concentration of fructose. The time course and the extent of the depletion of Pi was similar with both compounds, but xylitol elevated PRPP only to approx. 50 % of the level reached with fructose (not illustrated). As compared to fructose, tagatose, also at the concentration of 5 mM, provoked a slightly less rapid but equally pronounced degradation of ATP, 65 % of which was lost after 15 min (Fig. 3). Both ketohexoses decreased the intracellular concentration of Pi to the same extent, in parallel with that of ATP. However, following the addition of tagatose, intracellular Pi reincreased at a markedly slower rate than after fructose, its intracellular concentration reaching only approx. 50 % of the normal value after 45 min. With both ketohexoses the rise in PRPP evolved in striking parallelism with that of intracellular Pi.

(3) Influence of the elevation of the concentration of PRPP on the incorporation of hypoxanthine in the cellular nucleotides

We have shown that the incorporation of hypoxanthine in the purine mononucleotides of isolated hepatocytes is severely limited, due to a

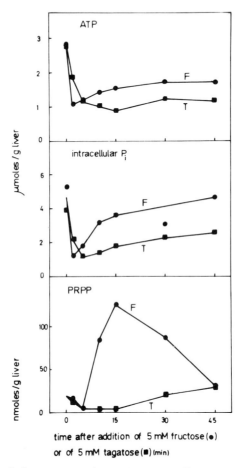

Fig. 3 Influence of fructose and tagatose on the concentrations of ATP, Pi and PRPP in isolated rat hepatocytes.

potent inhibition of HGPRT under physiological conditions, which renders the enzyme dependent upon the concentration of PRPP prevailing intracellularly (8). The influence of the addition of fructose on the concentration of PRPP prompted therefore a study of its effect on the incorporation of hypoxanthine in the purine mononucleotide pool of the cells.

618

As shown in Fig. 4, in the absence of fructose, approx. 2.5 % of 10 µM hypoxanthine added to the hepatocyte suspension was incorporated into the purine nucleotides within 30 sec. This result was not significantly modified when hypoxanthine was added 2 min after fructose. However, when hypo-

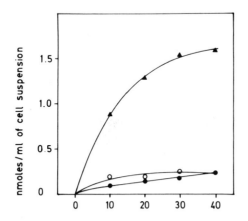

time after addition of 10 µM - hypoxanthine
(seconds)

Fig. 4 Influence of fructose on the incorporation of hypoxanthine into liver nucleotides. Hypoxanthine was added before (O), 2 min (●) or 30 min (▲) after the addition of 5 mM fructose.

xanthine was added 30 min after fructose, when the concentration of PRPP was elevated, approx. 15 % of the purine base was incorporated into the purine mononucleotide pool.

DISCUSSION

PRPP synthetase, the enzyme that generates PRPP from ATP and ribose 5-P, is subjected to a complex regulation (reviewed in 15). Similarly to glutamine-PRPP amidotransferase, it is inhibited by various purine nucleotides and, moreover, displays an absolute dependence upon its stimulator Pi. Our results suggest that the addition of fructose to isolated hepatocytes, by its degradative effect on the purine nucleotide pool, renders PRPP synthetase more sensitive to Pi. Because of the accompanying initial decrease in the concentration of Pi, resulting from the accumulation of fructose 1-phosphate, this increased sensitivity is not immediately apparent. It becomes manifest when, due to the further metabolism of fructose 1-phosphate, the concentration of Pi returns to its normal intracellular level. This rebound of the intracellular concentration of Pi, which we recorded in isolated hepatocytes, has also been demonstrated in rat liver in vivo (1). The mechanism proposed for the fructose-induced elevation in PRPP is corroborated by the dose-effect of the ketohexose and by the observation that other compounds that produce a degradation of hepatic ATP and a temporary depletion of Pi have a similar effect on the concentration of PRPP. In peculiar, the magnitude of the effect appears correlated with the extent of the degradation of ATP, whereas its time-course is determined by the elevation of intracellular Pi. Xylitol, which at the same concentration as fructose depletes ATP to a smaller extent than the ketohexose, brings about a smaller elevation of PRPP. Tagatose, which induces a longer lasting depletion of hepatic Pi than fructose, most likely because its phosphorylation product tagatose 1-phosphate is split at a lower rate than fructose 1-phosphate by liver aldolase, has a markedly delayed effect on the concentration of PRPP.

The reason why PRPP synthetase becomes more sensitive to Pi following ATP-depletion is not immediately apparent since this nucleotide, a substrate of the enzyme, has no inhibitory effect on its activity. Moreover, the concentration of ADP, the most potent inhibitor of the enzyme, is barely modified following fructose loads (1). Several other factors may, nevertheless, be involved. An increase in free Mg^{++}, likely to occur in conjunction with ATP-depletion, has been shown to stimulate the activity of purified PRPP synthetase from rat liver (16). An additional factor may be the elevation of ribose 5-phosphate, recorded in the presence of fructose (Fig. 2), since the Km of the enzyme for this substrate is approx. 0.3 mM (16). The decrease in UTP following fructose loads (17) may also play a role, since this nucleotide is inhibitory of liver PRPP synthetase (18).

That the transient increase in the concentration of PRPP recorded after the administration of fructose, besides stimulating the "de novo" synthesis of purines, also results in an enhanced activity of the salvage pathway, is shown by the 6-fold increase in the rate of incorporation of hypoxanthine into the purine mononucleotides (Fig. 4). This increased purine salvage may explain the biphasic time-course of the elevation of ATP following its depletion by the addition of fructose. Indeed, the initial more rapid elevation of ATP, occurs more or less simultaneously with the increase in PRPP. Our results thus demonstrate that several mechanisms attempt to compensate the catabolism of the purine nucleotide pool provoked by fructose.

ACKNOWLEDGEMENTS

The authors wish to express their appreciation to Mrs Thérèse Timmerman for expert technical assistance.

Supported by grant nr 3.4563.82 of the "Fonds de la Recherche Scientifique Médicale". M.F. Vincent is "Chargé de Recherches" and G. Van den Berghe "Maître de Recherches" of the Belgian "Fonds National de la Recherche Scientifique".

REFERENCES

1. P.H. Mäenpää, K.O. Raivio & M.P. Kekomäki, Liver adenine nucleotides : fructose-induced depletion and its effect on protein synthesis, Science 161:1253 (1968).

2. G. Van den Berghe, Metabolic effects of fructose in the liver, Curr. Top. Cell. Regul. 13:97 (1978).
3. B.T. Emmerson, Effect of oral fructose on urate production, Ann. Rheum. Dis. 33:276 (1974).
4. K.O. Raivio, M.A. Becker, L.J. Meyer, M.L. Greene, G. Nuki & J.E. Seegmiller, Stimulation of human purine synthesis de novo by fructose infusion, Metabolism 24:861 (1975).
5. M. Itakura, R.L. Sabina, P.W. Heald & E.W. Holmes, Basis for the control of purine biosynthesis by purine ribonucleotides, J. Clin. Invest. 67:994 (1981).
6. G. Van den Berghe, F. Bontemps & H.G. Hers, Purine catabolism in isolated rat hepatocytes. Influence of coformycin, Biochem. J. 188:913 (1980).
7. M.A. Becker, L.J. Meyer, & J.E. Seegmiller, Gout with purine overproduction due to increased phosphoribosylpyrophosphate synthetase activity, Am. J. Med. 55:232 (1973).
8. M.F. Vincent, G. Van den Berghe & H.G. Hers, Metabolism of hypoxanthine in isolated rat hepatocytes, Biochem. J. 222:145 (1984).
9. M.A. Becker, Patterns of phosphoribosylpyrophosphate and ribose-5-phosphate concentration and generation in fibroblasts from patients with gout and purine overproduction, J. Clin. Invest. 57:308 (1976).
10. J.M. Whelan & A.S. Bagnara, Factors affecting the rate of purine ribonucleotide dephosphorylation in human erythrocytes, Biochim. Biophys. Acta 563:466 (1979).
11. K. Itaya & M. Ui, A new micromethod for the colorimetric determination of inorganic phosphate, Clin. Chim. Acta 14:361 (1966).
12. L.V. Eggleston, D-Fructose-1-phosphate, in "Methods of enzymatic analysis" (H.U. Bergmeyer, ed.) pp. 1308-1313, Academic Press, New York and London (1974).
13. H.F. Woods and H.A. Krebs, Xylitol metabolism in the isolated perfused rat liver, Biochem. J. 134:437 (1973).
14. G. Van den Berghe, Biochemical aspects of hereditary fructose intolerance, in "Normal and pathological development of energy metabolism" (F.A. Hommes and C.J. Van den Berg, eds) pp. 211-228, Academic Press, New York and London (1975).
15. M.A. Becker, K.O. Raivio & J.E. Seegmiller, Synthesis of phosphoribosylpyrophosphate in mammalian cells, Adv. Enzymol. 49:281 (1979).
16. D.G. Roth, E. Shelton & T.F. Deuel, Purification and properties of phosphoribosylpyrophosphate synthetase from rat liver, J. Biol. Chem. 249:291 (1974).
17. H.B. Burch, O.H. Lowry, L. Meinhardt, P. Max Jr & K-J. Chyu, Effect of fructose, dihydroxyacetone, glycerol and glucose on metabolites and related compounds in liver and kidney, J. Biol. Chem. 245:2092 (1970).
18. D.G. Roth & T.F. Deuel, Stability and regulation of phosphoribosylpyrophosphate synthetase from rat liver, J. Biol. Chem. 249:297 (1974).

THE MECHANISM OF INSULIN-INDUCED INCREASE OF THE RATE OF DE NOVO PURINE

BIOSYNTHESIS IN PRIMARY CULTURED RAT HEPATOCYTES

Masami Tsuchiya, Mitsuo Itakura and Kamejiro Yamashita

Division of Endocrinology and Metabolism
Institute of Clinical Medicine
University of Tsukuba, Ibaraki 305, Japan

SUMMARY

The effect of insulin on the rate of de novo purine biosynthesis was studied in freshly isolated hepatocytes from adult male rats. Insulin started to increase the rate of de novo purine biosynthesis at 1.5×10^{-10} M and plateaued at 1.5×10^{-8} M with the magnitude of increase of about 280%. Insulin at 1.5×10^{-9} M or the higher concentration increased the 5-phosphoribosyl 1-pyrophosphate (PRPP) availability for purine ribonucleotide synthesis to about 230%. Insulin increased ATP concentration to 127% and decreased AMP, ADP, GMP and GDP concentration to 73, 69, 73 and 69% in association with the increased adenylate energy charge of 0.90 in comparison to the control level of 0.83. But the total adenine and guanine nucleotide concentration in the cell was not significantly changed. In addition insulin increased the specific activity of amidophosphoribosyltransferase (ATase) with the increased Vmax and the unchanged Km for PRPP.

Based on these results it is concluded that insulin increases the rate of de novo purine biosynthesis by increasing ATase activity and PRPP availability for purine ribonucleotide synthesis and that the increased adenylate energy charge induced by insulin is regarded as the basis for the acceleration of multiple metabolic pathways by insulin.

INTRODUCTION

The growth of liver in vivo after the partial hepatectomy[1] and that of hepatocytes in vitro[2] are under the control of various hormones but the mechanism by which the quiescent hepatocytes are brought to the proliferative stage is not well understood. In order to elucidate the mechanism of cell growth the effect of insulin, one of the typical growth factors in vitro on the rate of de novo purine biosynthesis, one of the typical anabolic pathways was studied in primary cultured rat hepatocytes. The mechanism of the increased rate of de novo purine biosynthesis was also studied from the viewpoints of the substrate availability, the concentration of feedback inhibitors and the specific activity of ATase, the supposed rate limiting enzyme of the purine biosynthetic pathway.

MATERIALS AND METHODS

Materials

7 week old male Wistar rats were purchased from Shizuoka
Agricultural Cooperative Association for Laboratory Animals, Shizuoka
Japan. Their body weights ranged from 180 to 230 g.

Eagle's minimum essential medium was purchased from Nissui Pharmaceutical Co. Ltd., Tokyo Japan, Williams' medium E and newborn calf serum
were from Life Technologies, INF. New York, U.S.A., actorapid insulin was
from Novo Co. Denmark, penicillin G potassium salt was purchased from
Banyu Pharmaceutical Co. Ltd., Tokyo, Japan and streptomycin was from
Meiji Pharmaceutical Co. Ltd., Tokyo, Japan. Culture plastic dishes were
purchased from Corning, New York, U.S.A. Dexamethasone was purchased from
Towa Pharmaceutical Co. Ltd., Osaka Japan. Collagenase, tris-n-octylamine,
1,1,2-trichlorotrifluoroethane and scintisol EX-H were purchased from Wako
Pure Chemical Industries Ltd., Tokyo, Japan, Dowex 50 W-H$^+$ was purchased
from Bio Rad Laboratory. 5-phosphoribosyl-1-pyrophosphate (PRPP)
magnesium salt, AMP, ADP, ATP, GMP, GDP and GTP were purchased from Sigma.
Other chemicals were of the highest quality available.

[^{14}C]formate (55.0 µCi/µmol), [^{14}C]adenine (54.8 µCi/µmol) and
[^{14}C]glutamine (258-292 µCi/µmol) were purchased from New England Nuclear.

Methods

Isolation and culture of hepatocytes. Hepatocytes were isolated
from male Wistar rats, weighing 180 to 230 g by the two-step collagenase
perfusion technique of Seglen[3] with a slight modification. After the
perfusion, the hepatocytes were suspended in Eagle's minimum essential
medium (MEM) and nonparenchymal cells were removed by washing 4 times
with MEM and by centrifugation (50 g, 1 min). The viability of the
hepatocytes determined by trypan blue exclusion test at the time of
isolation was more than 90%.

Hepatocytes were cultured in Williams' medium E with 10% newborn calf
serum, 10^{-7} M insulin, 10^{-6} M dexamethasone to facilitate the attachment
of the cells to the surface of culture dishes, and with 100 u/ml of
penicillin and 100 µg/ml of streptomycin. The initial cell density was
6×10^5 cells/35 mm Corning plastic dish. To determine the unmodified
effect of insulin, insulin was administered to the hepatocytes which were
cultured for 20 hours beforehand without serum and hormones following the
initial plating culture for 4 hours.

Assay of the rate of de novo purine biosynthesis. The rate of de
novo purine biosynthesis was determined according to the published
methods[4] by measuring the incorporation of [^{14}C]formate to adenine
nucleotides in hepatocytes. [^{14}C]formate at the final concentration of
9.5 µM was added and the hepatocytes were incubated for 30 min. The
reaction was stopped by ice-cold phosphate buffered saline and purine
ribonucleotides were extracted in 1 ml of 2 N perchloric acid for 60 min
on ice. The extraction was centrifuged at 3,000 g for 2 min. The
supernatant was heated at 100°C for 60 min, cooled on ice for 10 min and
applied to Dowex 50 W-H$^+$ columns that were preequilibrated with 0.01 N
HCl. The columns were washed with 5.0 ml of 1 N HCl and the adenine
fraction was eluted with 7.5 ml of 6 N HCl. A 1 ml aliquot of this
eluate was introduced to 10 ml of Scintisol EX-H and 1 ml of H$_2$O and the
radioactivity was measured by liquid scintillation counter. The rate of
de novo purine biosynthesis was expressed as the specific activity of
adenine fraction in the unit of dpm/µmol adenine.

Assay of PRPP availability. PRPP availability for nucleotide synthesis was determined according to the published methods[5] by measuring the incorporation of [^{14}C]adenine into purine ribonucleotides. [^{14}C]adenine at the final concentration of 4.7 μM was added and the hepatocytes were incubated for 30 min. The reaction was stopped by rinsing with ice-cold phosphate buffered saline and the purine ribonucleotides were extracted by 12% trichloroacetic acid on ice for 30 min. The extraction was centrifuged and neutralized with buffer containing 1 N Tris pH 7.4 and 6 N KOH at the ratio of 6.5 to 1. A 20 μl aliquot was applied to the high voltage paper electrophoresis to separate purine ribonucleotides from adenine. The spot of purine ribonucleotides was determined by the UV lamp and the radioactivity was measured by the liquid scintillation counter. The PRPP availability was expressed as the radioactivity of purine ribonucleotide in 10^5 cells in the unit of dpm/10^5 cells.

Assay of ATase activity. After rinsing hepatocytes with ice-cold phosphate buffered saline they were disrupted by freeze-thawing and harvested with a rubber policeman and ATase was extracted in 25 mM KPi buffer pH 7.4 containing 60 mM beta-mercaptoethanol. The homogenate was centrifuged at 3,000 g for 30 min and the supernatant was introduced to the enzyme assay. The methods for the assay were those previously reported[6]. The reaction mixture contained in final concentration 5 mM PRPP, 4 mM [^{14}C]glutamine (0.15 μCi/μmol), 5 mM MgCl$_2$, 1 mM dithiothreitol, 50 mM potassium phosphate buffer pH 7.4 and the enzyme preparation. The reaction was carried out at 37°C for 60 min and stopped with ice-cold 200 mM EDTA with the final concentration of 20 mM. A 20 μl aliquot was applied to the high voltage paper electrophoresis to separate the formed [^{14}C]glutamate from [^{14}C]glutamine. The radioactivity was measured by the liquid scintillation counter and the PRPP dependent glutaminase activity was regarded as ATase activity. The amount of DNA was measured by the method of Schneider[7] and cellular protein was measured by the method of Lowry[8]. The specific activity of ATase was expressed as nmol glutamine deaminated/hr/μg DNA or nmol glutamine deaminated/hr/mg protein.

Assay of purine ribonucleotide concentrations in the cell. Hepatocytes were incubated for 8 hours with or without 10^{-7} M insulin. After rinsing the cells with ice-cold phosphate buffered saline, purine ribonucleotides in the cell were extracted by ice-cold 12% trichloroacetic acid for 30 min. Acid insoluble material was removed by centrifugation at 3,000 g for 2 min. The methods for the assay of purine ribonucleotides by HPLC system are those reported previously[9,10]. The supernatant containing the free purine ribonucleotide was neutralized by the equal volume of 0.5 N tris-n-octylamine/1,1,2-trichlorotrifluoroethane and centrifuged at 3,000 g for 2 min. A portion of the upper layer was applied to the high performance liquid chromatography containing SAX 10 anion exchange column. The sample was eluted over a linear gradient starting at 5 mM NH$_4$PO$_4$ pH 2.8 and ending at 750 mM NH$_4$PO$_4$ pH 3.9 at a rate of 2 ml/min. The nucleotides were identified and quantitated by the appropriate standards.

Statistical significance. Statistical significance was tested by Student's unpaired t test and the marks of * or ** were used to show the P less than 0.05 or 0.01.

RESULTS

Effect of insulin on the rate of de novo purine biosynthesis

The incorporation of [^{14}C]formate into adenine nucleotides was linear for at least 30 min with or without hormones and serum (data not shown). Accordingly 30 min incorporation period in the [^{14}C]formate incorporation assay was used in all of the following experiments as the index of the rate of de novo purine biosynthesis. The time course of the effect of insulin on the rate of de novo purine biosynthesis is shown in the upper panel of Fig. 1. Insulin at 10^{-7} M increased the rate of de novo purine biosyntheis time-dependently with the first effect observed 30 min after the administration. The rate of de novo purine biosynthesis plateaued after 4 hours and reached the maximal value of $24.5 \pm 1.8 \times 10^3$ dpm/μmol adenine 12 hours after the administration which is 270–280% of the starting level. The dose-dependency of the effect of insulin on the rate of de novo purine biosynthesis is shown in the upper panel of Fig. 2. The effect of insulin was dose-dependent and reached the near maximal level at 1.5×10^{-8} M.

(a) RATE OF PURINE BIOSYNTHESIS DE NOVO

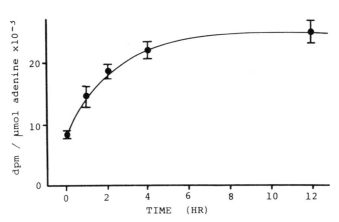

(b) PRPP AVAILABILITY

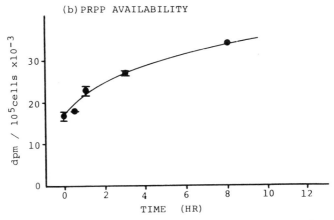

Fig. 1. The time course of the rate of de novo purine biosynthesis and PRPP availability. The time courses of the rate of de novo purine biosynthesis and PRPP availability for purine ribonucleotides at 1.5×10^{-7} M of insulin are shown respectively in the upper and lower panel. The time after the administration in hour is shown on the abscissa. Each point is the mean $\pm$ S.E.M. of 4 culture dishes.

Effect of insulin on PRPP availability

The incorporation of $[^{14}C]$adenine into purine ribonucleotides was linear for at least the first 1 hour with or without hormones (data not shown). Accordingly the incubation period of 30 min was used in all of the subsequent experiments as the index of PRPP availability. The time course of the increase of PRPP availability is shown in the lower panel of Fig. 1. With insulin at 10^{-7} M the effect of insulin on PRPP availability was firstly observed at 1 hour and the time course of the increase of PRPP availability was in parallel with that of the rate of de novo purine biosynthesis. After the incubation with insulin at the concentration of 10^{-7} for 8 hours, PRPP availability was $34.4 \pm 0.3 \times 10^3$ dpm/10^5 cells which is 208% of the control. The dose-dependency of the increase of PRPP availability induced by isulin is shown in the lower panel of Fig. 2. The effect was already observed at the concentration of 1.5×10^{-9} M and the concentration above this did not further increase PRPP availability.

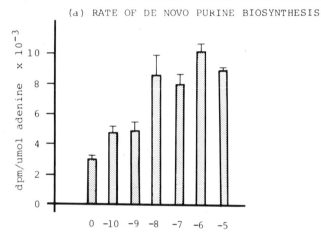

(a) RATE OF DE NOVO PURINE BIOSYNTHESIS

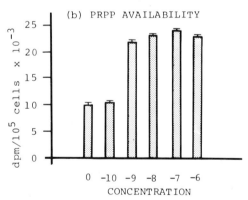

(b) PRPP AVAILABILITY

CONCENTRATION

Fig. 2. Dose-dependency of the rate of de novo purine biosynthesis and PRPP availability on insulin concentration: The dose dependencies of the rate of de novo purine biosynthesis and PRPP availability for purine ribonucleotede synthesis after 8 hours' incubation with insulin are shown respectively in the upper and the lower panel. The concentration of insulin is shown on the abscissa by the logarithmus of (molar concentration/1.5). In other words -10 in the figure denotes 1.5×10^{-10} M of insulin. Each bar represents the mean ± S.E.M. of 4 culture dishes.

Effect of insulin on the specific activity of ATase

The specific activity of ATase per unit amount of DNA and protein after the incubation for 8 hours in the presence or the absence of insulin at 10^{-7} M are respectively 13.85 ± 0.41 or 7.88 ± 0.61 nmol glutamine deaminated/hr/μg DNA x 10 and 176.62 ± 5.99 or 96.17 ± 9.99 nmol glutamine deaminated/hr/mg protein. The increases both to about 180% are statistically significant with P less than 0.01. The kinetic analysis of ATase activity by Hanes-Woolf plot[11] are shown in Fig. 3. The plots were linear and the Km for (PRPP) obtained from the intercept on the abscissa was 0.65 mM both in the absence or the presence of insulin. The Vmax values obtained from the inclination were 2.89 and 1.87 nmol glutamine deaminated/hr/μg DNA respectively in the presence or the absence of insulin.

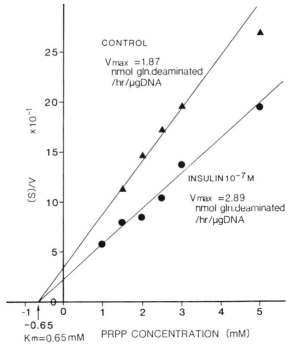

Fig. 3. The kinetic analysis by Hanes-Woolf plot of ATase
in the presence and absence of insulin. Hepatocytes
were incubated for 8 hours in the presence (●) or
absence (▲) of insulin at 10^{-7} M. PRPP concentration
in mM is plotted on the abscissa and [PRPP]/V x 10^{-1}
is plotted on the ordinate.

Effect of insulin on purine ribonucleotide   concentrations in the cell

The intracellular concentration of purine ribonucleotides 8 hours

in the presence or absence of insulin at 10^{-7} M are summarized in Table 1.
ATP concentration significantly increased to 127%. By contrary AMP, ADP,
GMP and GDP concentration significantly decreased by respectively to 73,
69, 73 and 69% in comparison to the control. Although the concentration
of several purine ribonucleotides were changed, the total adenine and
guanine nucleotides concentration did not show significant changes by the
administration of insulin. The adenylate energy charge was 0.90 and 0.83
respectively in the presence or the absence of insulin.

Table 1. Concentration of purine ribonucleotides, total
adenine and guanine nucleotides (TAGN) and
adenylate energy charge (AEC) in the hepatocytes
after the incubation for 8 hours in the presence
and the absence of insulin at 10^{-7} M.

	CONTROL		INSULIN
AMP	1.00 ± 0.04	*	0.73 ± 0.08
ADP	2.65 ± 0.08	**	1.82 ± 0.13
ATP	10.40 ± 0.39	*	13.25 ± 0.67
GMP	0.15 ± 0.01	*	0.11 ± 0.01
GDP	0.35 ± 0.03	*	0.24 ± 0.03
GTP	1.84 ± 0.07		1.82 ± 0.11
TAGN	16.33 ± 0.56		17.19 ± 1.00
AEC	0.83 ± 0.00	**	0.90 ± 0.00
(N)	(4)		(5)

(Each number is a mean ± S.E.M. in the unit of x 10^{-10}
mol/10^5 cells. (N) stands for the number of culture dishes.)

DISCUSSION

Our study has convincingly demonstrated that insulin as one of the
typical growth factors in the culture system increases the rate of de
novo purine biosynthesis, one of the typical anabolic pathways in the
primary cultured hepatocytes. As the mechanisms of the increased rate of
de novo purine biosynthesis, our study has firstly demonstrated that the
specific activity of ATase both per unit amount of DNA and protein are
increased by insulin. The kinetic analysis of the increased activity of
this enzyme showed the increase of Vmax without changing the Km for [PRPP].
These results suggest that the mechanism of the increased specific
activity of this enzyme is due to the induction of this enzyme. The idea
of induction of this enzyme is in accord with other examples of enzymes
induced by insulin, for example ornithine decarboxylase in the primary
cultured hepatocytes[12] or tyrosine aminotransferase in cultured hepatoma
cells in association with the increase of the enzyme specific messenger
RNA activity[13].

Our data have secondly demonstrated that insulin increased the PRPP
availability for purine ribonucleotide synthesis. Since PRPP availability is in parallel with PRPP concentration[14], the observed increased
PRPP availability for purine ribonucleotide synthesis is regarded as
another factor to explain the increased rate of de novo purine
biosynthesis.

Thirdly the role of the decreased feedback inhibition by purine
ribonucleotides on ATase was unlikely. Although insulin significantly
changed the concentration of several purine ribonucleotides including the

increase of ATP and the decrease of AMP, ADP and GDP concentration, the total inhibitory potential on ATase by purine ribonucleotides was not significantly changed. Thus it is concluded that the changes of the concentrations of purine ribonucleotides induced by insulin resulted in the unchanged feedback inhibition on ATase as a whole.

The increased adenylate energy charge in the hepatocytes cultured with insulin is the first demonstration in our knowledge. This is regarded as the basis for the potency of insulin in the culture system to activate multiple metabolic pathways leading to the cell growth, since intracellular pool size of ATP has been reported to accelerate RNA and DNA synthesis[15,16].

Based on these results it is concluded that insulin increases the rate of de novo purine biosynthesis by increasing the specific activity of ATase and by increasing the PRPP availability for purine ribonucleotide synthesis. In addition the increased adenylate energy charge produced by insulin is regarded as the basis of the potency of insulin for the multiple metabolic activations.

REFERENCES

1. N. L. R. Bucher, and M. N. Swanfield, Regulation of hepatic regeneration in rats by synergistic action of insulin and glucagon. Proc. Nat. Acad. Sci. USA, 72:1157 (1975).
2. R. A. Richman, T. H. Claus, S. J. Pilkis, and D. L. Friedman, Hormone stimulation of DNA synthesis in primary cultures of adult rat hepatocytes. Proc. Nat. Acad. Sci. USA, 73:3589 (1976).
3. P. O. Seglen, Chapter 4. Preparation of isolated rat liver cells. Methods Cell Biol. 13:29 (1976).
4. G. R. Boss, Decreased phosphoribosylpyrophosphate as the basis for decreased purine synthesis during amino acid starvation of human lymphoblasts. J. Biol. Chem. 259:2936 (1984).
5. C. B. Thomas, J. C. Meade, and E. W. Holmes, Aminoimidazole carboxamide ribonucleoside toxity: A model for study of pyrimidine starvation. J. Cell. Physiol. 107:335 (1981).
6. M. Itakura, and E. W. Holmes, Human amidophosphoribosyltransferase. An oxygen-sensitive iron-sulfur protein. J. Biol. Chem. 333:999 (1979).
7. W. C. Schneider, Phosphorus compounds in animal tissues. III. A comparison of methods for the estimation of nucleic acids. J. Biol. Chem. 164:747 (1946).
8. O. H. Lowry, N. J. Rosenbrough, A. L. Farr, and R. J. Randall, Protein measurement with the Folin reagent. J. Biol. Chem. 193:265 (1951).
9. L. M. Rose, and R. W. Brockman, Analysis by HPLC of 9-beta-D arabinofuranosyladenine-5'-triphosphate level in murine leukemia cells. J. Chromatogr. 133:335 (1977).
10. M. Itakura, R. L. Sabina, P. W. Heald, and E. W. Holmes, Basis for the control of purine biosynthesis by purine ribonucleotides. J. Clin. Invest. 67:994 (1981).
11. C. S. Hanes, CLXVII. Studies on plant amylases. I. The effect of starch concentration upon the velocity of hydrolysis by the amylase of germinated barley. Biochem. J. 26:1406 (1932).
12. Y. Tomita, T. Nakamura, and A. Ichihara, Control of DNA synthesis and orinithine decarboxylase activity by hormones and amino acids in primary cultures of adult rat hepatocytes. Expt. Cell Res. 135:363 (1981).
13. R. E. Hill, K. Lee, and F. T. Kenney, Effects of insulin on messenger RNA activities in rat liver. J. Biol. Chem. 256:1510 (1981).

14. P. Bashkin, and O. Sperling, Some regulatory properties of purine biosynthesis de novo in long-term cultures of epithelial-like liver cells. Biochem. Biophy. Acta 538:505 (1978).
15. I. Grummt, and F. Grumt, Control of nucleolar RNA synthesis by the intracellular pool sizes of ATP and GTP. Cell 7:447 (1976).
16. E. Rapaport, M. A. Garcia-Blanco, and P. C. Zamecnik, Regulation of DNA replication in S phase nuclei by ATP and ADP pools. Proc. Nat. Acad. Sci. 76:1643 (1979).

HORMONE-INDUCED STIMULATION OF PHOSPHORIBOSYLPYROPHOSPHATE

AND OF PURINE SYNTHESIS IN MOUSE LIVER IN VIVO

P. Boer, S. Brosh and O. Sperling

Department of Clinical Biochemistry, Beilinson Medical Center
Tel Aviv University Medical School, Petah-Tikva, Israel

INTRODUCTION

A possible role for glycogenolytic hormones in the regulation of purine metabolism was indicated by the finding in normal mice that glucagon and epinephrine increased liver content of PRPP (1) and that epinephrine stimulated purine synthesis (2). Furthermore, increased glucagon level was suggested recently to underly the purine overproduction common in glycogen storage disease type I (3).

In the present communication we report our studies concerning the stimulatory effect of glycogenolytic hormones on the rate of PRPP and purine synthesis in mouse liver in vivo (4, 5).

MATERIALS AND METHODS

For the study of the effect of the hormones on liver PRPP content, the mice (males; 18-22 gm) were killed (by cervical dislocation) 15 min following the administration (I.P.) of the hormones and the livers rapidly removed and assayed for PRPP, as described previously (6). The rate of de novo purine nucleotide synthesis in the liver tissue in vivo was gauged by the rate of incorporation of precursor (^{14}C)formate into total tissue purines, during 4 min. 5μCi (^{14}C)formate were injected intraveneously 11 min following the administration of the effectors. 4 min subsequently, the mice were killed and livers removed as described above and the total tissue purines precipitated and counted as described before (6). For the purification of PRPP synthetase, livers were homogenized (7) and the extracted enzyme was partially purified by DEAE (8). PRPP synthetase was assayed radiochemically by the one step procedure (9).

RESULTS AND DISCUSSION

Glucagon, epinephrine, vasopressin, oxytocin, angiotensin and dibutyryl cAMP caused a rapid 2 to 3 fold increase in liver PRPP content and in the rate of purine synthesis de novo (Table 1).

The effect of epinephrine and glucagon on purine synthesis was evident within minutes following the administration of hormones. Maximum stimulation was observed around 1 hour later, followed by rapid decrease. However, significant acceleration of the rate of purine synthesis could be observed

Table 1.   Effect of Hormones and of cAMP on Liver Content of PRPP and on the Rate of de novo Purine Nucleotide Synthesis.

Effector (dose)	PRPP content (nmoles/g wt)	Rate of de novo Purine Synthesis: (^{14}C)Formate Incorporation into Total Tissue Purines (dpm/g wt)
Control	5.4 + 2.3 (10)[a]	8791 + 4491 (15)
Glucagon (2mg/kg)	10.8 + 2.1 (3)	28260 + 19988 (3)
L-Epinephrine (1 mg/kg)	14.0 + 1.3 (4)	19965 + 9427 (9)
Dibutyryl cAMP(40 mg/kg)	16.3 + 5.5 (4)	17636 + 3507 (3)
Lys vasopressin (25U/kg)	13.2; 24.2 (2)	18911 + 2145 (4)
Oxytocin (25U/kg)	17.2; 31.6 (2)	19064 ; 11435 (2)
Angiotensin (65 µg/kg)		19624 ; 18184 (2)

[a]Mean + S.D. Numbers in parentheses represent number of groups of 4 mice each.

also 12 hours following hormone administration (Fig. 1).

The hormones studied, although representing different physiological tasks, share in the liver a common glycogenolytic effect, through either cAMP-dependent or calcium-dependent mechanisms. The observed effects of

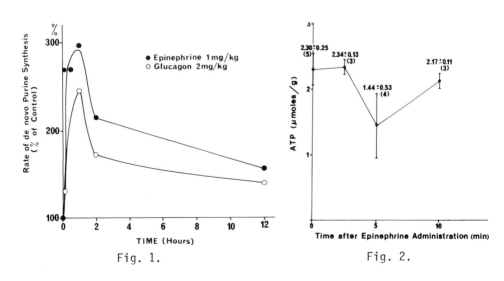

Fig. 1.

Fig. 2.

Fig. 1.   Effect of time between administration of hormones and the assay of the rate of purine synthesis on the effect of the hormones on the rate of purine synthesis.
Fig. 2.   Effect of epinephrine administration on liver ATP content.

Table 2.   Specific Activity of PRPP Synthetase in
            Liver Extracts from Control and Hormone-treated Mice.

Mice Group	Spinco Extract (100000 g)	DEAE Preparation
Control	1.42 nmol/mg prot/h	11.92 nmol/mg prot/h
L-Epinephrine	92% (3)[a]	92% (4)
Glucagon	113% (4)	89% (3)
dbcAMP	115% (3)	97% (2)

[a]% of control; numbers in parentheses represent number of experiments.

the hormones on PRPP and purine metabolism could be secondary to the gly-
cogenolysis, or reflect a direct effect of the hormones on PRPP synthetase,
e.g. phosphorylation through the activity of protein kinase. The amounts
of hormones administered into the mice in the present study were several
fold greater than the physiological levels, probably resulting in formation of glucose-6-P in excess of the activity of glucose-6-phosphatase.
The accumulation of the latter substrate could result in stimulation of
both the pentose shunt and glycolysis, each of which being able theoretically to increase purine synthesis. Stimulation of the pentose shunt
could lead to increased production of ribose-5-P, a substrate for PRPP,
the limiting substrate for purine synthesis. However, this mechanism
seems unlikely in view of the common knowledge that the pentose shunt is
regulated by the NADP/NADPH ratio rather than by glucose-6-P availability
(10). Furthermore, several studies done in our laboratory, have suggested

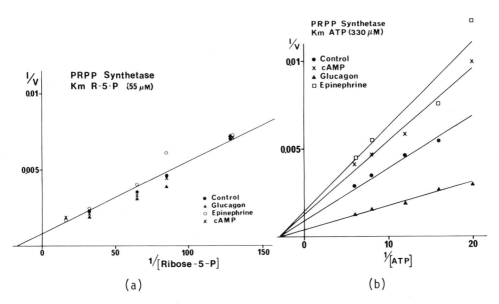

Fig. 3.   Effect of hormones on Km of PRPP synthetase for ribose-5-P (a)
          and for ATP (b).

that in several tissues, including the liver, under physiological concentration of inhibitors and activators, the physiological availability of ribose-5-P is saturating for PRPP synthetase (11, 12). On the other hand, the possibility that glycogenolysis-induced acceleration of glyco-lysis, is involved in the mechanism by which the glycogenolytic hormones affect PRPP and purine synthesis, seems more likely. Rapid phosphorylation

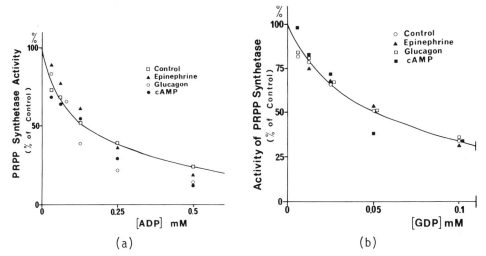

Fig. 4.   Effect of hormones on sensitivity of PRPP synthetase to inhibition by ADP (a) and by GDP (b).

of glucose-6-P into fructose 1,6-diphosphate is associated with depletion of ATP and of Pi (13, 14), the latter activating adenylate deaminase which increases AMP degradation (15). This decrease in ATP and AMP concentration probably leads to additional alterations in the relative concentration of the various nucleotides which could diminish the feedback inhibition exerted on PRPP synthetase and on PRPP amidotransferase, accelerating the rate of this pathway (16, 17). In support of this mechanism, we found that epine-phrine administration to mice resulted in prompt decrease in liver ATP content (Fig. 2). Nevertheless, the decrease was found to last only several minutes, whereas the stimulation of purine synthesis was found to last several hours (Fig. 1). Evidently, more data is needed on the effect of the other hormones on liver ATP content and especially on the effect of these hormones on the liver content of the other purine nucleotides, in order to clarify this mechanism.

Extensive studies were carried in our laboratory concerning the possibility of a direct effect of the hormones on PRPP synthetase. How-ever, until today no such evidence could be obtained. The specific acti-vity of the enzyme (Table 2), its affinity to substrates ribose-5-P and ATP (Fig. 3), as well as its sensitivity to inhibition by ADP and GDP (Fig. 4), were not altered in preparations from livers of hormone-treated mice in comparison to control mice.

REFERENCES

1. Lalanne M and Henderson F: Canad. J. Biochem 53: 394, 1975.
2. Bauman N. and Wyngaarden JB: Fed. Proc. 23: 324, 1964.
3. Cohen JL, Vinik A, Faller J and Fox IH: J. Clin. Invest. 75:251,1985.
4. Brosh S, Boer P and Sperling O: Biomedicine, 35: 50, 1981.
5. Boer P, Brosh S and Sperling O: Israel J. of Med. Sci. 20: 564, 1984.
6. Boer P, Lipstein B, de Vries A and Sperling O: Biochim. Biophys. Acta,
    432: 10, 1976.
7. Oliver JM: Biochem. J. 128 :771, 1972.
8. Fox IH and Kelley WN: J. Biol. Chem. 246: 5739, 1971.
9. Sperling O, Boer P, Persky-Brosh S, Canarek E and de Vries A:
    Europ. J. Clin. Biol. Res. 17: 703, 1972.
10. Krebs HA and Eggleston LV: Adv. Enzyme Regulation, 12: 421, 1974.
11. Bashkin P and Sperling O: Biochim. Biophys. Acta, 538:505, 1978.
12. Sperling O: in Purine and Pyrimidine Metabolism (Elliot K and
    Fitzsimons DW, eds) Ciba Foundation Symposium 48 (New Series)
    Elsevier, Amsterdam, p347, 1977.
13. Green HL, Wilson FA, Hefferan P, Terry AB, Roberto Moran J, Slonim AE,
    Claus TH and Burr IM: J. Clin. Invest. 62: 321, 1978.
14. Roe TF and Kogut MD: Pediat. Res. 11: 664, 1977.
15. Van den Berghe G, Bronfman M, Vanneste R and Hers HG: Biochem. J. 162:
    601, 1977.
16. Hershko A, Razin A and Mager J : Biochim. Biophys. Acta 184: 64, 1969.
17. Holmes EW, McDonald JA, McCord JM, Wyngaarden JB and Kelley WN: J. Biol.
    Chem, 248: 144, 1973.

COBALAMIN INACTIVATION DECREASES PURINE AND METHIONINE SYNTHESIS IN CULTURED HUMAN LYMPHOBLASTS

Gerry R. Boss

Department of Medicine
University of California, San Diego Medical Center
San Diego, CA

INTRODUCTION

Cobalamin (vitamin B_{12}) is required by two enzymes in mammalian cells, specifically methionine synthetase (N^5-methyltetrahydrofolate-homocysteine methyltransferase, EC 2.1.1.13) and methylmalonyl CoA mutase (EC 5.4.99.2). Cobalamin deficiency in humans leads to megaloblastic anemia which is probably secondary to decreased methionine synthetase activity and to subacute combined degeneration of the spinal cord which appears to be secondary to decreased methylmalonyl CoA mutase activity. A decrease in methionine synthetase activity should cause 5-methyltetrahydrofolate to accumulate because this folate is a major storage form of intracellular folate and its production from 5,10-methylene tetrahydrofolate is essentially irreversible in vivo. These were the data that originally formulated the methylfolate trap hypothesis to describe cobalamin deficiency. Trapping intracellular folates as 5-methyltetrahydrofolate should decrease intracellular purine synthesis and thus lead to decreased DNA synthesis and ultimately to megaloblastosis. However, not all of the data concerning B_{12} deficiency are totally explainable by the folate trap and animal studies only partially support this mechanism.

Nitrous oxide has been known for some time to induce megaloblastic changes in bone marrow cells when used as an anesthetic. Experimentally it has been used recently to induce a cobalamin deficiency by oxidizing the active cobalamin I to inactive cobalamin III. I, therefore, decided to use nitrous oxide to induce a functional cobalamin deficiency in cultured human lymphoblasts to study the sequelae of such inactivation (1).

METHODS

Intracellular rates of methionine and serine synthesis were measured as previously described employing a [^{14}C]formate label (2). Similarly, rates of de novo purine synthesis were measured as previously described, also using a [^{14}C]formate label (3).

RESULTS

Methionine and Serine Synthesis

Nitrous oxide rapidly reduced rates of methionine synthesis in cultured human lymphoblasts so that by 30 minutes there was an approximate 50% reduction and by four hours almost a 90% reduction. Rates of serine synthesis were reduced at a somewhat slower rate, such that by 30 minutes there was an approximate 18% reduction and by four hours an approximate 50% reduction.

Purine Synthesis at Variable Folate Concentrations

In untreated control cells, the rates of purine synthesis were independent of the folate concentration in the medium. However, when the cells were treated with nitrous oxide, rates of purine synthesis decreased with time and the rate of decrease was greater in the absence of folate, such that by four hours there was an approximate 20% decrease.

Purine Synthesis at Variable Methionine Concentration

As previously reported, rates of purine synthesis were dependent on the methionine concentration in the medium. This is secondary to decreased phosphoribosylpyrophosphate production by the pentose phosphate pathway. Nitrous oxide reduced rates of purine synthesis more in the absence of folate than in the presence of folate as the methionine concentration was reduced. Of note, nitrous oxide decreased purine synthesis most significantly at the physiologic methionine concentrations of 10 and 30 μM. Specifically, at these concentrations purine synthesis was decreased by approximately 37% and 57%, respectively, in the absence of folate.

Purine Synthesis in the Presence of Homocysteine

Nitrous oxide reduced rates of purine synthesis even more dramatically when homocysteine replaced methionine in the medium, such that by 30 minutes of nitrous oxide exposure rates of purine synthesis were decreased by 56% and after four hours exposure by 75%.

DISCUSSION

In these studies we show that inactivation of cobalamin by nitrous oxide decreases intracellular methionine synthesis from homocysteine. Of greater note, this inactivation significantly decreases intracellular purine synthesis and this decrease in purine synthesis is dependent on both the folate concentration in the medium and on the methionine concentration. The dependence on the folate concentration further suggests that there is 5-methyltetrahydrofolate trapping as previously suggested. The dependence on the methionine concentration suggests that ample methionine is required for full rates of purine synthesis.

Decreased intracellular methionine could decrease purine synthesis by at least three different mechanisms. First, a decreased methionine concentration could decrease S-adenosylmethionine synthesis and since this compound is a potent inhibitor of 5,10-methylene tetrahydrofolate reductase a decrease in its intracellular concentration could increase 5-methyltetrahydrofolate trapping. Moreover, S-adenosylmethionine is the main intracellular methyl donor and a decrease in its concentration could affect many enzymatic reactions. Second, methionine may be a major provider of cellular formate and thereby methionine may serve as a precursor of the purine ring. And third, decreased methionine availability could lead to decreased phosphoribosylpyrophosphate synthesis since a decrease in any intracellular essential amino acid will decrease production of phosphoribosylpyrophosphate.

These data are consistent with the previous clinical observations

that cobalamin deficiency may be associated with homocystinuria and that
methionine administration to cobalamin deficient patients does decrease
the urinary excretion of formiminoglutamic acid which is a marker of
cobalamin deficiency.

REFERENCES

1. G. R. Boss, Cobalamin inactivation decreases purine and methionine
   synthesis in cultured lymphoblasts, J. Clin. Invest In Press.

2. G. R. Boss and R. W. Erbe, Decreased rates of methionine synthesis by
   methylene tetrahydrofolate reductase-deficient fibroblasts and
   lymphoblasts, J. Clin. Invest. 67:1659-1664 (1981).

3. G. R. Boss, Decreased phosphoribosylpyrophosphate as the basis for
   decreased purine synthesis during amino acid starvation of human
   lymphoblasts, J. Clin. Invest 259:2936-2941 (1984).

5'-DEOXY-5'-METHYLTHIOADENOSINE PHOSPHORYLASE DEFICIENCY IN LEUKEMIA:

GENETICS AND BIOCHEMICAL ASPECTS

Carlos J. Carrera, Erik H. Willis, Robert R. Chilcote*,
Masaru Kubota, and Dennis A. Carson

Department of Basic and Clinical Research
Scripps Clinic and Research Foundation
La Jolla, California 92037

*Department of Pediatrics
Los Angeles County – USC Medical Center
Los Angeles, California 90033

INTRODUCTION

In mammalian cells, 5'-deoxy-5'-methylthioadenosine (MTA) derives from decarboxylated S-adenosylmethionine during spermidine and spermine synthesis.[1] To a lesser extent, MTA is also produced following the aminocarboxypropyl group transfer from S-adenosylmethionine to certain tRNA uridine residues[2] (Fig. 1). Although MTA can inhibit polyamine aminopropyl transferase reactions[3], the thioether nucleoside does not accumulate in normal cells but is rapidly cleaved to adenine and 5-methylthioribose 1-phosphate by the enzyme MTA phosphorylase. As shown in Figure 1, MTA phosphorylase is important not only for the balanced synthesis of polyamines, but also for the economic intracellular salvage of adenine nucleotides and methionine.[4,5] The enzyme is present in all normal tissues studied thus far. Recently, we have assigned the gene for MTA phosphorylase to the 9pter-->9q12 region of human chromosome 9 by analysis of mouse-human somatic cell hybrids.[6]

Several murine and human malignant cell lines have been found to lack MTA phosphorylase.[7,8] Nearly all MTA phosphorylase-deficient cell lines are of leukemic origin, and in 1982 Kamatani et al. found that freshly isolated malignant cells from two of 20 leukemia patients had markedly decreased MTA phosphorylase activity.[9] Both MTA phosphorylase-deficient samples were from patients with acute lymphoblastic leukemia (ALL). Recent cytogenetic studies in this disease have identified a distinct subgroup of patients with abnormalities of chromosome 9 in their leukemic cells.[10,11] These patients often present with lymphomatous clinical features and have a poor prognosis.[11] We have now identified 2 additional patients from this subgroup of lymphomatous ALL whose leukemic cells contain very low MTA phosphorylase activity.

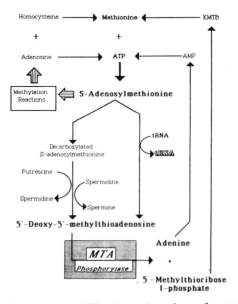

Figure 1. Metabolic role of MTA phosphohorylase in mammalian cells.
Abbreviations: URNA = 3-(3-amino-3-carboxypropyl) uridine modified tRNA; KMTB = 2-keto-4-methylthiobutyrate.

Cell lines deficient in MTA phosphorylase cannot metabolize MTA but excrete the nucleoside into the culture medium.[12] Because elevated plasma or urine MTA might be a useful marker for MTA phosphorylase-deficient leukemia in patients, we have developed a mouse model for measuring plasma MTA in the presence of MTA phosphorylase-negative tumor growth. We present here evidence that plasma MTA does indeed rise dramatically in animals bearing MTA phosphorylase-deficient mutant tumor cells but not in those with enzyme replete wild-type cells.

METHODS

MTA Phosphorylase Assay of ALL Samples

Patients were classified as having lymphomatous ALL if they presented with two of the following three clinical features: (a) lymphadenopathy greater than 3 cm in diameter (or matted lymph nodes greater than 5 cm), (b) splenomegaly, and (c) radiologic evidence of superior mediastinal enlargement, adjusted for patient age.[11] Mononuclear leukocytes were isolated within 24 hr from blood or marrow samples by Ficoll-Hypaque density gradient centrifugation. Extracts were prepared from patients' cells and from cell lines by 5 cycles of freeze-thaw disruption followed by centrifugation at 10,000 xg. MTA phosphorylase activity in cell extracts was assayed by a radiochemical method.[4] Enzyme assays were also performed on several leukemic cell samples which had been cryopreserved up to two years in liquid nitrogen.

Immunoadsorption of MTA Phosphorylase

BALB/c mice were immunized with human MTA phosphorylase which had been partially purified by sequential chromatographic procedures. Cell extracts at varying dilutions were mixed with antiserum for 3 hr at 5° C prior to the addition of formalin-treated _S. aureus_ Cowan strain I cells (Calbiochem, La Jolla, CA). Following removal of the _S. aureus_ and bound antibody by centrifugation, the supernatants were assayed for remaining MTA phosphorylase activity.

Measurement of Plasma MTA

The wild-type murine T lymphoma cell line R1.1, which contains MTA phosphorylase, was provided by Dr. Robert Hyman (Salk Institute, La Jolla, CA). Mutant clone H cells lacking MTA phosphorylase were derived from wild-type R1.1 by a tritium suicide method[12] employing 5'-deoxy-5'-chloro-[2,8-^{3}H] adenosine, an alternate enzyme substrate. C58/J mice were injected intraperitoneally with 0.5-1 x 10^7 R1.1 wild-type or clone H cells and were bled at intervals for 22 days.

Plasma samples were spiked with 5'-deoxy-5'-chloroadenosine as an
internal standard, and the nucleosides were concentrated by passage over
Sep-Pak C$_{18}$ cartridges (Waters Associates, Milford, MA). Eluted
compounds were lyophilyzed, reconstituted in water, and the adenine-
containing nucleosides were derivatized by reaction with chloroacetaldehyde. The resulting fluorescent etheno-MTA derivative was then
quantitated by reverse-phase HPLC.

RESULTS

MTA phosphorylase activity was measured in 9 fresh and in 5 cryopreserved leukemic cell samples from patients with ALL. In all but 2
samples, the enzyme activity was within a normal range from 0.72-4.69
nmol/min per mg protein (Fig. 2). Both samples with low MTA phosphorylase activity (0.041 and 0.12 nmol/min per mg protein) were from patients
with lymphomatous ALL. However, one other patient with lymphomatous
clinical features had leukemic cells with normal enzyme activity.

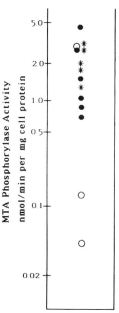

Figure 2. MTA phosphorylase activity
in leukemia cells from
patients with ALL. (●) = ALL, (o)
= lymphomatous ALL, (*) = cryopreserved ALL samples. MTA phosphorylase activity was determined by
radiochemical assay.[4]

Cytogenetic studies on cells from one of the MTA phosphorylase-deficient leukemia patients have been reported in detail.[11] The cells have a complex karyotype with a deletion involving the short arm of chromosome 9 [del(9)(p21p22)]. Peripheral blood mononuclear cells from this patient during remission, as well as cells obtained from several family members, contain normal MTA phosphorylase activity (1.25-2.11 nmol/min per mg protein).

MTA phosphorylase activity can be efficiently removed from cultured human lymphoblast extracts by immunoadsorption with murine antisera plus S. aureus cells (Fig. 3). Serum from non-immunized mice was without effect. The addition of extracts from seven MTA phosphorylase-deficient human cell lines did not affect the antiserum binding titration, suggesting that these enzyme-negative cell lines do not synthesize significant amounts of cross-reactive material.

The median survival was approximately 22 days for mice inoculated either with R1.1 wild-type or with MTA phosphorylase-deficient R1.1 clone H cells. Large tumor masses developed in both groups. Figure 4 shows that elevated plasma MTA was detected by 13 days in mice bearing R1.1 clone H tumors, rising to nearly 1 μM pre-terminally. In contrast, no increase in plasma MTA above that of healthy animals (less than 80 nM) was detectable in mice injected with R1.1 wild-type cells despite fatal tumor progression. In other experiments, plasma MTA also increased in mice injected with L_{1210} leukemia which, like R1.1 clone H, lacks MTA phosphorylase.

DISCUSSION

MTA phosphorylase activity was markedly decreased in leukemic cells from two of 20 patients in a previous study (16 with ALL),[9] and the enzyme activity is low in two of 14 ALL samples reported here. Interestingly, the incidence of chromosome 9 deletion abnormalities is especially common among the approximately 10% of patients with ALL who present with lymphomatous clinical features.[11] Localization of the MTA phosphorylase gene to chromosome region 9pter-->9q12 strongly suggest that efficiency of the enzyme in leukemic cells may be a useful biochemical marker for this karyotypically distinct clinical subgroup of ALL. However not all patients with lymphomatous ALL have cells with chromosome 9 deletions [11] or which lack MTA phosphorylase (Fig. 2). Biochemical and cytogenetic studies must be performed in additional cases of ALL to confirm any correlations between lymphomatous clinical features, chromosome 9 deletions, and MTA phosphorylase deficiency.

The mechanism producing deficiency of MTA phosphorylase in patient samples and in certain malignant cell lines is unknown. Low but detectable enzyme activity in the two patient samples reported here probably reflect a small number of normal cells in the leukocyte preparations from peripheral blood. However, low levels of MTA phosphorylase (partial deficiency) or a heterogeneous enzyme deficiency in the leukemic cell population cannot be excluded. MTA phosphorylase is recessive in both inter- and intraspecies somatic cell hybrids.[6,13]

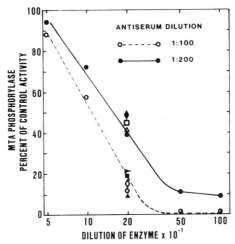

Figure 3. Immunoadsorption of MTA phosphorylase from WI-L2 lymphoblasts in the absence (●,○) or presence of equal amounts of cell protein from CEM (◆), K562 (□), DHL-9 (▲), Reh (➤), EW 36 (■), Jurkat (◇), or NALL-1 (▲) cell lines. Normal mouse serum was used in control assays.

By the immunoadsorption assay, no cross-reactive, inert enzyme protein could be detected among seven MTA phosphorylase-deficient human cell lines. Together, these findings suggest that either gene deletion or synthesis of defective messenger RNA occurs in the enzyme deficient cells. Competitive immunoadsorption studies have not yet been performed on MTA phosphorylase deficient ALL samples.

Our results with the murine model show that MTA phosphorylase-deficient malignant cells excrete MTA into the extracellular space in vivo as well as in vitro.[12] Chheda et al. identified MTA sulfoxide as one of two novel nucleosides in the urine of a patient with chronic myelogenous leukemia;[14] unfortunately, leukemic cells from this patient are not available for enzyme assay. Plasma MTA measurements appear to reflect accurately the growth of MTA phosphorylase-deficient tumors in mice. Thus, the finding of elevated plasma MTA in certain patients with ALL may have prognostic importance as a signal of relapse or disease progression.

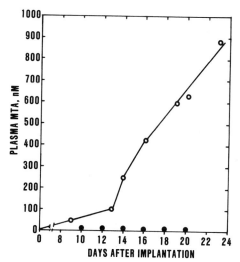

Figure 4. Plasma MTA concentrations in mice bearing either R1.1 wild-type (●) lymphoma, which contains MTA phosphorylase, or R1.1 clone H lymphoma (o), which has mutational deficiency of the enzyme. $0.5 - 1 \times 10^7$ cells were injected intraperitoneally at day 0. Plasma MTA was measured by HPLC following chloroacetaldehyde derivatization.

ADCKNOWLEDGEMENTS

This work was supported in part by grants CA 35048, GM 23200 and AM 07022 from the National Institutes of Health.

REFERENCES

1. A. E. Pegg and H. G. Williams-Ashman, On the role of S-adenosyl-L-methionine in the biosynthesis of spermidine by rat prostate, *J. Biol. Chem.* 244:682 (1969).

2. A. G. Saponara and M. D. Enger, The isolation from ribonucleic acid of substituted uridines containing alpha-aminobutyrate moieties derived from methionine, *Biochim. Biophys. Acta* 349:61 (1974).

3. R.-L. Pajula, A. Raina, and T. Eloranta, Polyamine synthesis in mammalian tissues. Isolation and characterization of spermine synthesis from bovine brain, *Eur. J. Biochem.* 101:619 (1979).

4. N. Kamatani and D. A. Carson, Dependence of adenine production upon polyamine synthesis in cultured human lymphoblasts, *Biochim. Biophys. Acta* 675:344 (1981).

5. P. S. Backlund, Jr. and R. A. Smith, Methionine synthesis from 5'-methylthioadenosine in rat liver, *J. Biol. Chem.* 256:1533 (1981).

6. C. J. Carrera, R. L. Eddy, T. B. Shows, and D. A. Carson, Assignment of the gene for methylthioadenosine phosphorylase to human chromosome 9 by mouse-human somatic cell hybridization, *Proc. Natl. Acad. Sci. USA* 81:2665 (1984).

7. J. I. Toohey, Methylthio group cleavage from methylthioadenosine. Description of an enzyme and its relationship to the methylthio requirement of certain cells in culture, *Biochem. Biophys. Res. Commun.* 78:1273 (1977).

8. N. Kamatani, W. A. Nelson-Rees, and D. A. Carson, Selective killing of human malignant cell lines deficient in methylthioadenosine phosphorylase, a purine metabolic enzyme, *Proc. Natl. Acad. Sci. USA* 78:1219 (1981).

9. N. Kamatani, A. L. Yu, and D. A. Carson, Deficiency of methylthioadenosine phosphorylase in human leukemic cells in vivo, *Blood* 60:1387 (1982).

10. J. Kowalczyk and A. A. Sandberg, A possible subgroup of ALL with 9p-, *Cancer Genet. Cytogenet.* 9:383 (1983).

11. R. R. Chilcote, E. Brown, and J. D. Rowley, Lymphoblastic leukemic with lymphomatous features associated with abnormalities of the short arm of chromosome 9, *N. Engl. J. Med.* 313:in press (1985).

12. M. Kubota, N. Kamatani, and D. A. Carson, Biochemical genetic analysis of the role of methylthioadenosine phosphorylase in a murine lymphoid cell line, *J. Biol. Chem.* 258:7288 (1983).

13. N. Kamatani, M. Kubota, E. H. Willis, and D. A. Carson, 5'-Methylthioadenosine phosphorylase deficiency in malignant cells: recessive expression of the defective phenotype in intraspecies (mouse x human) hybrids, *Adv. Exp. Med. Biol.* 165B:279 (1984).

14. G. B. Chheda, H. B. Patrzyc, A. K. Bhargava, S. K. Sethi, P. F. Crain, J. A. McCloskey, and S. P. Dutta, Characterization of two novel nucleosides isolated from chronic myelogenous leukemia (CML) urine, *Proc. Am. Assoc. Cancer Res.* 25:22 (1984).

METABOLISM OF 5'-METHYLTHIOADENOSINE IN METHIONINE-DEPENDENT AND

METHIONINE-INDEPENDENT CELLS

Laurence Christa, Joëlle Kersual, Jean-Louis Pérignon
and Pierre Hubert Cartier

Laboratoire de Biochimie, INSERM U 75, Faculté de Médecine
Necker-Enfants Malades, Paris, France

INTRODUCTION

Two metabolic pathways are known to result in the formation of methionine (Met) in mammalian cells: one is the methylation of L-homocysteine (Hcy) by methyltetrahydrofolic acid in the presence of vitamin B12, or by betaine; the other one is the salvage of the ribose moiety of the purine nucleoside 5'-deoxy-5'-methylthioadenosine (MTA). In an attempt to elucidate the intermediate steps of this last pathway, we tried to grow cells in media were Met was replaced by MTA. Unfortunately, our attempts were unsuccessfull. However, Hoffman and Erbe (1) have reported that some cell lines ("methionine-dependent") were unable to grow when Met was replaced by Hcy in spite of the fact that they were perfectly able to synthesize Met from Hcy; we thus hypothesized that the inability to grow on MTA may not merely reflect the inability to metabolize MTA into Met. In the present work, we studied a) the ability of CCL 39 (Met-dependent) and Raji (Met-independent) cells to grow on MTA or Hcy; b) the metabolism of exogenous, preformed, Met compared to that of endogenous Met (newly synthesized from MTA or Hcy).

MATERIALS AND METHODS

S-adenosyl-L- [methyl-^{14}C]methionine, L-[methyl-^{14}C]methionine and 5-[^{14}C]methyl-tetrahydrofolic acid were purchased from the Radiochemical Center (Amersham, U.K.); [6-^{3}H]thymidine was from CEA (Saclay, France). [^{14}C-methyl] MTA was prepared from S-adenosyl-L-[methyl-^{14}C]methionine by acid hydrolysis (2). MTA, Met, 2-ketomethylthiobutyrate, S-adenosyl-L-methionine (AdoMet), cyanocobalamin, folic acid, L-homocysteine thiolactone, were from Sigma (St Louis, MO). 5'-deoxy-5'-methylthioinosine was synthesized from MTA by nitrous acid deamination (3). CCL 39, a line of female chinese hamster lung fibroblast was a gift of Dr. M. Buttin (Institut Pasteur, Paris, France). Met-free media were reconstructed RPMI 1640 and Dulbecco's modified Eagle's media.

Raji cells were grown on RPMI 1640 medium supplemented with 10% fetal calf serum (FCS), and the proliferation was evaluated by ^{3}H thymidine incorporation (4). CCL 39 cell line was maintained in Dulbecco's modified Eagle's medium supplemented with 10% FCS; cells were counted in a Coulter counter. For salvage experiments, Met-free media were used and FCS was dialysed for 6 hours at 4°C against 3 changes of 154mM NaCl. For CCL 39 cells, MTA salvage experiments were made in the presence of 50 μM uridine to palliate the toxic effect of adenine. For Hcy requirement experiments, Met-free media were supplemented with 100 μM folic acid and 8 μM cyanocobalamin. The effectors were added at culture initiation.

Table I. Effects of different Met precursors on the proliferation of CCL 39 and Raji cells in a Met-free medium

	CCL 39 percent of control	Raji percent of control
Met 100 μM	100	100
KMTB* 100 μM	95	96
MTA 15 μM	5	95
MTA 100 μM	5	40
{ Hcy 100 μM { Vitamin B12 8 μM { Folic acid 100μM	4	100
{ Hcy 100 μM { Betaine 100 μM	5	nd

* KMTB: 2-ketomethylthiobutyrate

For the metabolic studies, CCL 39 cells were plated at 10^6 cells/well in 6-well microtiter plates and grown for 48h to obtain about 5.10^6 cells/well. The medium was then removed and the cells washed with Met-free medium. Raji cells were washed in Met-free medium and plated at 2.10^6 cells/ml just before experiments. Incorporation of 25 μM radiolabeled Met or MTA was conducted during 6h at 37°C in 1 ml of Met-free medium. The incorporation of [^{14}C methyl] tetrahydrofolic acid (34 μM) was conducted for 24h at 37° C in 1 ml of a Met-free medium supplemented with 100 μM Hcy thiolactone and 8 μM cyanocobalamin.. After labeling, aliquots of the medium were counted and extracted with 0.8 N cold PCA. The cells were cooled on ice, washed with ice-cold Dulbecco's PBS and extracted in 0.4 N cold PCA. Nucleic acids were solubilized by heating the acid precipitate at 90°C for 15 min in 1 ml of 5% TCA. The pellets were hydrolysed in 6 N HCl for 24 h at 110°C. Radioactivity of the acid-soluble fraction and of the hydrolysate was measured in a Packard liquid scintillation spectrometer. Aliquots of the acid extracts were neutralized with tri-N-octylamine/freon (5), then evaporated and analysed by either of 2 chromatographic systems: a) a paper chromatography on Whatman 3MM in n-butanol/acetic acid/water (5/1/4) separated AdoMet, Met, Met sulfoxide and MTA; after vizualisation (UV light and Ninhydrin spray), the spots were cut out and counted; b) a HPLC chromatography on a μ Bondapak C18 column, with isocratic elution by 20 mM KH_2PO_4, pH 4.2, during 15 min, followed by linear gradient elution (0% to 40% methanol in 20 mM KH_2PO_4, pH 4.2, in 80 min); 1 ml fractions were collected and counted. The counts of Met and Met sulfoxide were added and expressed as Met.

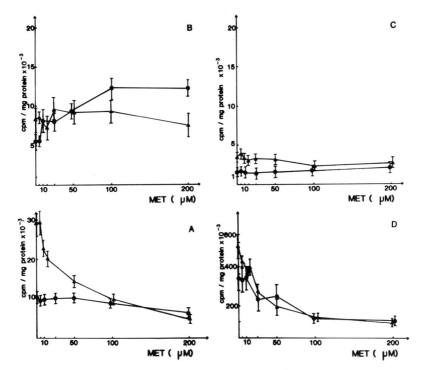

Fig. 1. Effect of Met on the metabolism of $[^{14}C\text{-methyl}]$ MTA (25 uM) in CCL 39 (●) and Raji (▲) cells. The cells were incubated with labeled MTA (0.62 μCi/ml) as described in "Materials and Methods". Values are cpm/mg protein and represent means ± SD of five (CCL 39) and four (Raji) separate culture experiments.
A: AdoMet in the acid soluble fraction
B: Met in the acid soluble fraction
C: MTA in the acid soluble fraction
D: Protein Met.

RESULTS

Growth of cell lines in Met-free media

To compare the quantitative importance of the various Met biosynthetic pathways, we measured the cellular proliferation in a Met-free medium supplemented with different Met precursors (Table I). The proliferation of CCL 39 cells reached a plateau at 50 μM Met. 2-ketomethylthiobutyrate could support the growth of the cells, whereas MTA, methylthioinosine and Hcy (even after preincubation in a complete medium supplemented with 8 μM cyanocobalamin for 4 days), could not. In contrast, Raji cells could grow on MTA and on Hcy.

Metabolism of $[^{14}C\text{-methyl}]$ MTA in CCL 39 and Raji cells

In order to determine why CCL 39 cells cannot proliferate when MTA replaces Met, we compared $[^{14}C\text{-methyl}]$ MTA metabolism into Met, proteins and AdoMet in CCL 39 and Raji cells (Fig.1) Both types of cells

Table II. Metabolism of $\begin{bmatrix} ^{14}C\text{-methyl} \end{bmatrix}$ MTA in CCL 39 and Raji cells

		CCL 39	Raji
Intracellular metabolites (cpm.mg protein^{-1})	AdoMet	10,313 + 2,970	29,231 + 9,058
	Met	5,616 + 2,761	6,613 + 1,433
	MTA	1,410 + 813	3,445 + 1,295
	Protein Met	377,407 + 198,619	589,040 + 137,405
Extracellular metabolites (cpm.mg protein^{-1})	AdoMet	248,657 + 48,746	81,873 + 16,214
	Met	209,500 + 57,858	440,531 + 76,237
	MTA	146,832 + 35,298	1,189,896 + 277,213

$\begin{bmatrix} ^{14}C\text{-methyl} \end{bmatrix}$ MTA (25µM, 0.62 uCi/ml) was added in Met-free medium. Each point is the mean + SD of five (CCL 39) and four (Raji) cell cultures.

metabolized MTA into Met and AdoMet in the acid-soluble fraction and incorporated Met synthesized from MTA into proteins. Exogenous Met, up to 200 µM, did not reduce the $\begin{bmatrix} ^{14}C\text{-methyl} \end{bmatrix}$ Met pools, but inhibited the incorporation into proteins of Met newly synthesized from MTA. Exogenous Met exhibited an important ability to inhibit the synthesis of AdoMet from MTA in Raji cells, whereas it only moderately affected it in CCL 39 cells (Fig. 2, A). To estimate the rate of MTA metabolism by the two cell lines, we also analysed the culture media (Table II): on a mg protein basis, 85% MTA were metabolized by CCL 39 cells, versus 49% by Raji cells, thereby indicating paradoxically a higher metabolic rate of MTA in the former. Furthermore, CCL 39 cells excreted a noticeable amount of AdoMet, the radioactivity of extracellular $\begin{bmatrix} ^{14}C\text{-methyl} \end{bmatrix}$ AdoMet corresponding to 25% of the total radioactivity, vs 3.5% for Raji cells.

Table III. Comparative study of the metabolism of preformed and endogenously synthesized Met in CCL 39 and Raji cells.

		CCL 39	Raji
Labeled precursor			
Extracellular AdoMet	$\begin{bmatrix} ^{14}C\text{-methyl} \end{bmatrix}$ MTA	0.834 + 0.331	0.316 + 0.018
	$\begin{bmatrix} ^{14}C\text{-methyl} \end{bmatrix}$ Met	0.031 + 0.006	0.085 + 0.013
Intracellular AdoMet	$\begin{bmatrix} ^{14}C\text{-methyl} \end{bmatrix}$ MTA	0.041 + 0.003	0.062 + 0.021
	$\begin{bmatrix} ^{14}C\text{-methyl} \end{bmatrix}$ Met	0.038 + 0.017	0.086 + 0.027
Protein Met	$\begin{bmatrix} ^{14}C\text{-methyl} \end{bmatrix}$ MTA	1.142 + 0.779	1.161 + 0.231
	$\begin{bmatrix} ^{14}C\text{-methyl} \end{bmatrix}$ Met	1.446 + 0.337	1.582 + 0.681

The cells were cultured in 25 µM MTA plus 25 µM Met as described in "Materials and Methods" with 0.62 µCi/ml labeled MTA or 0.62 µCi/ml labeled Met. Each point is the mean + SD of five (CCL 39) or four (Raji) cell cultures incubated with $\begin{bmatrix} ^{14}C\text{-methyl} \end{bmatrix}$ MTA, and two CCL 39 and Raji cell cultures incubated with $\begin{bmatrix} ^{14}C\text{-methyl} \end{bmatrix}$ Met. Results are expressed as nmol.mg^{-1} protein.h^{-1}.

Table IV. Metabolism of 5-$[^{14}C]$ methyltetrahydrofolic acid in CCL 39 and Raji cells.

		CCL 39	Raji
Intracellular metabolites (cpm.mg protein^{-1})	AdoMet	1,812 + 153	715 + 275
	Met	3,766 + 459	2,940 + 307
	Nucleic acids	14,849 + 234	18,372 + 375
	Protein Met	101,737 + 12,332	32,544 + 3,520
Extracellular metabolites (cpm.mg protein^{-1})	AdoMet	undetectable	undetectable
	Met	750,000 + 250,000	1,790,000 + 150,000

Cells were incubated for 24h at 37 C with 5-$[^{14}C]$ methyltetrahydrofolic acid (34µM, 2 µCi/ml) in a Met-free medium supplemented with 8 µM vitamin B12 and 100 µM L-homocysteine thiolactone. Each point is the mean + SD of triplicate determinations.

Comparative study of the synthesis and disposition of AdoMet metabolized from exogenous or endogenous Met

To further investigate the paradox of Met-dependent growth despite active MTA metabolism in CCL 39 cells, we tried to determine if there were differences between the metabolism of exogenous and endogenous Met. Accordingly, the cells were incubated with 25 µM $[^{14}C\text{-methyl}]$ Met plus 25 µM MTA or with 25 µM $[^{14}C\text{-methyl}]$ MTA plus 25µM Met. The 25 µM Met concentration chosen allows 80 and 100% of the growth in optimal Met concentrations for CCL 39 and Raji cells, respectively. Table III shows that the incorporation of radioactive Met into proteins is the same in the two types of cells, and with the two labeled precursors. On the contrary, the synthesis and disposition of AdoMet depends both of the type of the cell and of the precursor used. With $[^{14}C\text{-methyl}]$ MTA as the precursor, AdoMet is excreted by CCL 39 cells to a large extent, with ratios of extracellular to intra cellular AdoMet of 20.3 and 5.1 for CCL 39 and Raji cells, respectively. With $[^{14}C\text{-methyl}]$ Met as the precursor, AdoMet is much less excreted by both types of cells, with ratios of extracellular to intracellular AdoMet of 0.8 and 1.0 for CCL 39 and Raji cells, respectively.

The other biosynthetic pathway of Met, i.e. the cobalamin-dependent methylation of Hcy by methyltetrahydrofolic acid, was similarly studied in both cell lines (Table IV). The present results indicate that CCL 39 and Raji cells incorporate Hcy into Met, AdoMet and macromolecules. In these conditions, no radioactive extracellular AdoMet was found with either cell type.

DISCUSSION

Growth experiments revealed major differences between CCL 39 and Raji cells: Raji cells, which are malignant human B lymphoid cells, can grow on MTA, as previously described (4), and also on Hcy, contrary to the findings of Kano et al. (6), who probably used unsufficient folic acid and vitamin B12 concentrations. CCL 39, which are non-malignant hamster fibroblasts, do not grow either on MTA or on Hcy. Our findings that CCL 39 cells can, however, synthesize Met and AdoMet from Hcy, and incorporate the endogenously synthesized Met into proteins, are in accordance with the results obtained by Stern et al. in "Met-dependent" cells (7). The fact that Met dependence is not, as initially suggested by Hoffman and Erbe (1), limited to malignant cells, has been also reported by Carson et al.(8). Our study of MTA metabolism in CCL 39 cells is the first demonstration that "Met-dependence" can be observed with both the metabolic precursors of Met. This study also clearly indicates that AdoMet synthesis is not limiting in Met-dependent CCL 39 cells, which is in accordance with the report of Oden et al. (9) that the activity of AdoMet synthetase is normal in Met-dependent cells, ; they rather suggest an abnormal disposition of AdoMet synthesized from endogenous Met. Indeed, the comparative study of AdoMet metabolism in cells incubated with Met and MTA shows that in CCL 39 cells, the excretion of AdoMet synthesized from MTA is 25.4 fold higher than that synthesized from preformed Met, and only 5.1 higher in Raji cells. Furthermore, the ability of exogenous unlabeled Met to inhibit the incorporation of C-methyl MTA into intracellular AdoMet is 4-fold higher in Raji cells than in CCL 39 cells. Taken together, these results would suggest the existence, in CCL 39 cells, of a metabolic compartmentation between exogenous Met and Met endogenously synthesized from MTA. Wether this compartmentation may be correlated with Met dependence, as recently suggested by Tisdale (10), is questionable, since it was not found with the other Met precursor (Hcy). Alternative hypotheses should be also examined, such as the toxicity of MTA on cell proliferation via the inhibition of transmethylations or of polyamine metabolism (11).

ACKNOWLEDGEMENT

We are grateful to Mrs. Joëlle Augé for expert technical assistance.

REFERENCES

1 Hoffman RM, Erbe RW. High in vivo rates of methionine biosynthesis in transformed human and malignant rat cells auxotrophic for methionine. Proc Natl Acad Sci USA 1976; 73: 1523-1527.
2 Schlenk F, Ehninger DJ. Observations on the metabolism of 5'-methyl-thioadenosine. Arch Biochem Biophys 1964; 106: 95-100.
3 Savarese TM, Ghoda LY, Dexter DL, Parks RE. Conversion of 5'-deoxy-5'-methylthioadenosine and 5'-deoxy-5'-methylthioinosine to methionine in cultured human leukemic cells. Cancer Res 1983; 43: 4699-4702.
4 Christa L, Thuillier L, Munier A, Pérignon JL. Salvage of 5'-deoxy-methylthioadenosine into purines and methionine by lymphoid cells and inhibition of cell proliferation. Biochim Biophys Acta 1984; 803: 7-10.
5 Khym JX. An analytical system for rapid separation of tissue nucleotides at low pressures on conventional anion exchangers. Clin Chem 1975; 21: 1245-1252.
6 Kano Y, Sakamoto S, Kasahara T, Kusumoto K, Hida K, Suda K, Ozawa K, Miura Y, Takaku F. Methionine dependency of cell growth in normal and malignant hematopoietic cells. Cancer Res 1982; 42: 3090-3092.
7 Stern PH, Wallace CD, Hoffman RM. Altered methionine metabolism occurs in all members of a set of diverse human tumor cell lines. J Cell Physiol 1984; 119: 29-34.
8 Carson DA, Wilis EH, Kamatani N. Metabolism to methionine and growth stimulation by 5'-methylthioadenosine and 5'-methylthioinosine in mammalian cells. Biochem Biophys Res Commun 1983; 112: 391-397.
9 Oden KL, Carson K, Mecham JO, Hoffman RM, Clarke S. S-adenosylmethionine synthetase in cultured normal and oncogenically-transformed human and rat cells. Biochim Biophys Acta 1983; 760: 270-277.

10 Tisdale MJ. Utilization of preformed and endogenously synthetized methionine by cells in tissue culture. Br J Cancer 1984; 49: 315-320.
11 Williams-Ashman HG, Seidenfeld J, Galletti P. Trends in the biochemical pharmacology of 5'-deoxy-5'-methylthioadenosine. Biochem Pharmacol
1982; 31:277-288.

INHIBITION OF DE NOVO PURINE SYNTHESIS BY METHYLTHIOADENOSINE

Ross B. Gordon and Bryan T. Emmerson

University of Queensland Department of Medicine

Princess Alexandra Hospital, Brisbane 4102, Australia

INTRODUCTION

The synthesis of the polyamines, spermidine and spermine is accompanied by the stoichiometric formation of the nucleoside, 5'-deoxy-5'-methylthioadenosine (MTA) (1). The concentration of MTA in normal mammalian cells is extremely small. The presence of a specific phosphorylase (methylthioadenosine phosphorylase, MTAP) cleaves MTA to adenine and methylthioribose-1-phosphate (MTR-I-P). Adenine can be salvaged by the purine salvage enzyme, APRT, and pathways exist for the conversion of MTR-I-P to methionine (2, 3). Via such pathways the cell has the potential to recover the purine and methionine moieties of S-adenosylmethionine consumed during the synthesis of polyamines.

A number of human malignant cells in culture have been found to be deficient in MTAP activity(4) and these cells excrete large amounts of MTA into the culture medium (5). We have observed that two human cell lines which are MTAP-deficient (MTAP-) have increased rates of de novo purine synthesis when assayed by the incorporation of a radio-labelled precursor into a purine pathway intermediate. This prompted us to investigate the possible relationship between polyamine and purine metabolism in dividing mammalian cells. We show that MTA in the culture medium can inhibit de novo purine synthesis in MTAP+ cells but not MTAP- cells. This inhibition appears to be dependent on the phosphorolytic cleavage of MTA by the MTAP enzyme.

METHODS

The human MTAP- cells were the T-cell leukemic line, CCRF-CEM and an erythroleukemic line, K562. The MTAP+ cells were the human T-cell leukemic line, MOLT-4 and cultures derived from EBV-transformed B-lymphocytes obtained from normal individuals and from HGPRT-deficient patients. All cultures were grown in RPMI-1640 medium containing 2 mM glutamine, 20 mM Hepes (pH 7.2), penicillin (100 U/ml) and streptomycin (100 μg/ml) and supplemented with 10% fetal calf serum. Cultures were maintained in logarithmic growth in a 37°C humidified incubator in a 5% CO_2 atmosphere. Cultures were regularly checked for mycoplasma contamination.

Table 1. Rates of De Novo Purine Synthesis in MTAP⁻ and MTAP⁺ Cells

Lymphoblast Cell Line	De Novo Purine Synthesis pmole/h/10^6 cells
K562 (MTAP⁻)	1881 ± 415 (6)
T-cell CCRF-CEM (MTAP⁻)	1100, 1046
T-cell MOLT-4 (MTAP⁺)	350
B-cell (HGPRT⁺, MTAP⁺)	347 ± 154 (5)
B-cell (HGPRT⁻, MTAP⁺)	1253 ± 260 (6)

De novo purine synthesis was assayed by the rate of incorporation of [^{14}C]-formate into the pathway intermediate, FGAR, essentially as described by Gordon et al (6) except that the incubation medium was RPMI-1640 containing 2 mM glutamine, 20 mM Hepes (pH 7.4), 1 mM NaHCO$_3$ and 0.4% bovine serum albumin. MTAP activity was assayed by the procedure of White et al (7) using 5'-[^{14}C-methyl] methylthioadenosine as substrate.

RESULTS AND DISCUSSION

The cell lines K562 and CCRF-CEM were completely deficient in MTAP activity. The T-cell line MOLT-4 had activity comparable to B-cell lymphoblasts. All the MTAP⁺ cell lines exhibited activity similar to reported values (4). Suitable mixing experiments verified that extracts from the MTAP⁻ cell lines did not contain any inhibitor of the MTAP enzyme. Activities of other purine metabolic enzymes were as expected for all these cell lines. The concentration of PRPP in the MTAP⁻ cells was not elevated above that in MTAP⁺ cells.

The rate of de novo purine synthesis was increased considerably in the MTAP⁻ cell lines and was similar to values obtained for HGPRT-deficient cell lines (Table 1). In MTAP⁺ cells de novo purine synthesis was inhibited by the presence of MTA in the assay medium; 10 μM concentration produced > 90% inhibition. Purine synthesis in the MTAP⁻ cells was resistant to inhibition by MTA. Adenine in the assay medium inhibited de novo purine synthesis in the MTAP⁺ and MTAP⁻ cells to a similar degree.

The presence of MTA or adenine in the medium affected the growth of MTAP⁺ and MTAP⁻ cells. Following 72 hours of culture in the presence of either inhibitor, a concentration of 100 μM MTA or 400 μM adenine was required to achieve 50% inhibition of growth. Since the complete inhibition of de novo purine synthesis in MTAP⁺ cells could be elicited by concentrations of adenine or MTA of the order of 10 μM, it seems unlikely that inhibition of purine synthesis is secondary to the inhibition of growth.

The above results suggest that the inhibition of de novo purine synthesis by MTA in MTAP⁺ cells is mediated by the prior phosphorolytic cleavage of MTA to adenine. The adenine is readily salvaged to AMP by APRT in a reaction which utilises PRPP. This salvage could result in less PRPP being available for de novo purine synthesis. In a number of different cell types, it is well documented that the PRPP concentration can be a factor regulating purine synthesis (8, 9, 10).

In addition, the adenine nucleotides formed from the salvage might result in feedback inhibition of PRPP amidotransferase (11) (the rate-limiting step of de novo purine synthesis) and in feedback inhibition of PRPP synthesis (12). In the case of MTAP⁻ cells these potential regulatory mechanisms would not be available.

Kamatani and Carson (13) have shown recently that the major source of intracellular adenine in mammalian lymphoblasts is from MTA derived from polyamine synthesis. Whether adenine so derived normally contributes to the in vivo regulation of de novo purine synthesis via mechanisms outlined above has not been determined. We are currently examining other cell lines which are MTAP⁻ to ascertain (i) if increased rates of de novo purine synthesis are solely a character-istic of MTAP-deficiency and (ii) whether differences in the concentrations of various purine nucleotides exist between MTAP⁺ and MTAP⁻ cells.

Acknowledgement

This study was supported by a Special Project Grant from the University of Queensland.

References

1.   H. G. Williams-Ashman, J. Seidenfeld, and P. Galletti, Trends in the biochemical pharmacology of 5'-deoxy-5'methylthioadenosine, Biochem. Pharmacol. 31:277 (1982).

2.   P. S. Backlund Jr., and R. A. Smith, Methionine synthesis from 5'methylthioadenosine in rat liver, J. Biol. Chem. 256:1533 (1981).

3.   T. M. Savarese, L.C. Ghoda, D.L. Dexter, and R.E. Parks Jr., Conversion of 5'deoxy-5'-methylthioadenosine and 5'-deoxy-5'-methylthioinosine to methionine in cultured human leukemic cells. Cancer Res. 43:4699 (1983).

4.   N. Kamatani, W. A. Nelson-Rees, and D. A. Carson, Selective killing of human malignant cell lines deficient in methylthio-adenosine phosphorylase, a purine metabolic enzyme, Proc. Natl. Acad. Sci. USA 78:1219 (1981).

5.   N. Kamatani, and D.A. Carson, Abnormal regulation of methylthio-adenosine and polyamine metabolism in methylthioadenosine phos-phorylase-deficient human leukemic cell lines, Cancer Res. 40:4178 (1980).

6.   R. B. Gordon, A. C. Counsilman, S. M. Cross, and B. T. Emmerson, Purine synthesis de novo in lymphocytes from patients with gout, Clin. Sci. 63:429 (1982).

7.   M. W. White, A. A. Vandenbark, C. L. Barney, and A. J. Ferro, Structural analogs of 5'-methylthioadenosine as substrates and inhibitors of 5'-methylthioadenosine phosphorylase and as inhibitors of human lymphocyte transformation, Biochem. Pharmacol. 31:503 (1982).

8.   M. A. Becker, Regulation of purine nucleotide synthesis : effects of inosine on normal and hypoxanthine guanine phospho-ribosyltransferase-deficient fibroblasts, Biochim. Biophys. Acta 435:132 (1976).

9.   P. Boer, B. Lipstein, A. de Bries, and O. Sperling, The effect of ribose-5-phosphate and 5-phosphoribosyl-1-phrophosphate availability on de novo synthesis of purine nucleotides in rat liver slices, Biochim. Biophys. Acta 432:10 (1976).

10. R. B. Gordon, L. Thompson, L. A. Johnson, and B. T. Emmerson, Regulation of purine de novo synthesis in cultured human fibro-blasts : The role of P-ribose-PP, Biochim. Biophys. Acta 562:162 (1979).

11. E. W. Holmes, J. B. Wyngaarden, and W. N. Kelley, Human glutamine phosphoribosylpyrophosphate amidotransferase : two molecular forms interconvertible by purine ribonucleotide and phosphoribosylpyrophosphate, J. Biol. Chem. 248:6035 (1973).

12. A. S. Bagnara, A. A. Letter, and J. F. Henderson, Multiple mechanisms of regulation of purine biosynthesis de novo in intact tumor cells, Biochim. Biophys. Acta 374:259 (1974).

13. N. Kamatani, and D. A. Carson, Dependence of adenine production upon polyamine synthesis in cultured human lymphoblasts, Biochim. Biophys. Acta 675:344 (1981).

DETERMINATION OF S-ADENOSYLHOMOCYSTEINE IN TISSUES FOLLOWING PHARMACOLOGICAL INHIBITION OF S-ADENOSYLHOMOCYSTEINE CATABOLISM

S. Helland* and P.M. Ueland

*Department of Dermatology and Department of Pharmacology
University of Bergen, 5000 Bergen, Norway

INTRODUCTION

S-Adenosylhomocysteine (AdoHcy) hydrolase (EC 3.3.1.1.), the enzyme responsible for the metabolic degradation of the endogenous transmethylase inhibitor, AdoHcy, is a target enzyme for various adenosine analogues. Some of these compounds have oncostatic and antiviral properties. Adenosine analogues, including 9-ß-D-arabino - furanosyladenine (ara-A), function as inactivators or inhibitors of AdoHcy hydrolase, and these analogues induce a massive accumulation of intracellular AdoHcy. Other adenosine analogues serve as substrates for AdoHcy hydrolase, and are metabolized to their corresponding nucleosidylhomocysteine (1).

The potential role of AdoHcy hydrolase as a chemotherapeutic target enzyme has demanded the development of methods for determination of AdoHcy in tissues. Hoffman et al. reported that the amount of AdoHcy in rat liver increased rapidly following death of the animal (2). We have recently confirmed and extended this observation by showing that the amount of AdoHcy in various tissues of mice under physiological conditions was less than 30% when the organs were frozen in vivo using liquid nitrogen, compared with values obtained with organs rapidly removed and frozen after the animal was put to death (3).

The increase in AdoHcy after death may be related to production of AdoHcy from methyl-transfer reactions still operating, and failure to metabolize the resultant AdoHcy because of accumulation of adenosine and/or homocysteine (2). Alternatively, the postmortal increase in AdoHcy may be related to reversal of the AdoHcy hydrolase reaction induced by accumulation of adenosine during tissue anoxia (4-6). However, we observed only minimal change in the amount of free homocysteine in tissues of mice following death of the animal (7). This observation led us to investigate whether postmortal increase in AdoHcy may contribute to the massive accumulation of AdoHcy in tissues of animals treated with inhibitors of AdoHcy catabolism (1). The present report focuses on the AdoHcy content in various tissues of mice treated with the drug combination ara-A plus 2'-deoxycoformycin (dCF), which nearly totally inactivates AdoHcy hydrolase (8,9).

METHODS

Mice were given repetitive i.p. injections with the drug combination ara-A plus dCF. The first injection was 50 mg ara-A and 0.16 mg dCF per kg, followed by ara-A (25 mg/kg) plus dCF (0.16 mg/kg) each hour for 8 hours. The drugs were dissolved in 0.9% sodium chloride. The animals were put to death 1 hour after the (last) injection.

Two procedures were used for isolation of organs:

1) The animals were put to death by decapitation, and the liver, kidney, brain, heart, lung and spleen were immediately removed and placed in liquid nitrogen. The liver was frozen within 15 seconds, the brain within 25 seconds, and the other organs within 30 seconds.

2) The animals were anesthesized with ether, and the abdominal cavity was opened to expose the liver, kidney and spleen. The skin and tissue were removed from the skull bone. The animals were then allowed to breath in air containing no ether, and were then submerged in liquid nitrogen just before they recovered. The organs were isolated while still frozen, and kept at -80°C until analysis.

Frozen tissue was homogenized (1:4, w/v) in perchloric acid, which was neutralized as described previously (8,9). AdoHcy was determined by high performance liquid chromatography on a Partisil 10 SCX column or a 3um ODS Hypersil column, as described (9).

RESULTS AND DISCUSSION

The amount of AdoHcy in several tissues has recently been determined in our laboratory (3). In accordance with data published by others (2, 10, 11), we observed a rapid increase in AdoHcy (from 13 to 26 nmol/g in liver) in several tissues following death of the animal. This phenomenon could be avoided by freezing the organs in vivo using liquid nitrogen (3). If the same amount of AdoHcy accumulates after death in tissues of mice treated with ara-A plus dCF as that observed with non-treated mice, this postmortal phenomenon will be totally obscured under conditions of the high AdoHcy content induced by ara-A plus dCF (8,9).

In the present paper data are presented suggesting that a large fraction of AdoHcy in several tissues of mice treated for 8 hours with the drug combination stems from postmortal increase in AdoHcy. The amount of AdoHcy in the liver, and to a lesser degree in kidney, heart and lung was significantly lower when the organs were frozen in vivo compared with the amounts determined when the organs were isolated and frozen after the animal was put to death. In the liver of treated mice, the AdoHcy content increased from 150 nmol/g to 370 nmol/g within 15 seconds after death. In brain and spleen, the amount of AdoHcy was essentially the same whether the organs were frozen in vivo or isolated and frozen after decapitation (Table 1).

Four mice treated with ara-A plus dCF for 8 hours were anesthesized with ether and operated, as described in a preceding paragraph. The animals were killed by decapitation, and the organs removed and frozen immediately after death. The amount of AdoHcy in liver, kidney, heart, brain, lung and spleen of these animals was the same as the amount in these organs of treated mice not anesthesized and operated (data not shown). These data suggest that the low AdoHcy content in organs frozen in vivo (Table 1) is not induced by the anesthesia and/or the operative trauma.

Table 1

AdoHcy Content in Various Tissues of Mice Injected with ara-A plus dCF for 8 Hours

Mice were given i.p. injections of the drug combination ara-A plus dCF each hour for 8 hours, as described in the text.

Tissue	Concentration of AdoHcy (nmol/g wet wt)		p-values[c]
	Organs frozen after death[a]	Organs frozen in vivo[b]	
Liver	370±36	146±12	<0.0001
Kidney	644±53	374±42	<0.001
Heart	24.5±5.2	11.1±0.9	<0.01
Brain	29.0±1.8	24.5±1.7	<0.05
Lung	45.7±2.4	25.2±2.4	<0.001
Spleen	69.6±5.5	51.3±5.5	<0.05

[a] Mean + S.E.M. of 9 animals. The animals were killed by decapitation and the organs were immediately removed and frozen.

[b] Mean +S.E.M. of 12 animals. The exposed organs were frozen in vivo using liquid nitrogen, as described in the text.

[c] The p-values for the difference between AdoHcy content in organs frozen in vivo and frozen after death, were calculated using Student's t-test for two means.

Treatment of mice with ara-A plus dCF almost completely inactivates AdoHcy hydrolase in several tissues of mice (9). Therefore it seems unlikely that the postmortal increase in AdoHcy content in tissues of treated mice is related to further inhibition of AdoHcy catabolism by adenosine or to condensation of endogenous homocysteine with adenosine formed in response to postmortal tissue anoxia (4-6).

There was essentially no postmortal increase in AdoHcy in brain of mice injected with ara-A plus dCF (Table 1). Alternatively, the data obtained (Table 1) may be interpreted as failure to prevent postmortal increase of AdoHcy by freezing the brain in vivo. There are consistent reports on a rapid, massive accumulation of adenosine in brain during anoxia (4-6, 12), and the postmortal accumulation of this nucleoside can be avoided by freezing the brain in vivo through the intact skull bone (6, 12).

CONCLUSION

A large fraction (from 50 to 75%) of the AdoHcy content in liver, kidney, heart and lung of mice treated with ara-A plus dCF may rapidly accumulate following death of the animal. The postmortal increase in AdoHcy in brain and spleen was less pronounced. This source to erratic results can be avoided by freezing the organs in vivo, using liquid nitrogen. The mechanism(s) behind the postmortal elevation of AdoHcy content in tissues is unknown. Nevertheless, attention should be paid to this phenomenon when determining the metabolic response to compounds interfering with AdoHcy catabolism.

REFERENCES

1. Ueland, P.M. , Pharmacological and biochemical aspects of S-adenosylhomocysteine and S-adenosylhomocysteine hydrolase, Pharmacol. Rev. 34: 223 (1982).

2. Hoffman, D.R., Cornatzer, W.E., and Duerre, J. A., Relationship between tissue levels of S-adenosylmethionine, S-adenosylhomocysteine, and transmethylation reactions, Can. J. Biochem. 57: 56 (1979) .

3. Helland, S. and Ueland,P.M., Effect of 2'-deoxycoformycin infusion on S-adenosylhomocysteine hydrolase and the amount of S-adenosylhomocysteine and related compounds in tissues of mice, Cancer Res. 43:4142 (1983).

4. Arch, J.R.S., and Newsholme, E. A., The control of the metabolism and the hormonal role of adenosine, Essays Biochem. 14: 82 (1978) .

5. Wojcik, W. J., and Neff, N. H. , Adenosine measurement by a rapid HPLC-fluorometric method: induced changes of adenosine content in regions of rat brain, J. Neurochem. 39:280 (1982).

6. Helland, S., Broch, O. J., and Ueland, P. M., Neurotoxicity of deoxycoformycin: effect of constant infusion on adenosine deaminase, adenosine, 2'-deoxyadenosine and monoamines in the mouse brain, Neuropharmacol. 22: 915 (1983).

7. Ueland, P.M., and Helland,S., Homocysteine in tissues of the mouse and rat, J. Biol. Chem. 259:2360 (1984).

8. Helland, S., and Ueland, P. M., Inactivation of S-adenosylhomocysteine hydrolase by 9-ß-D-arabinofuranosyladenine in intact cells, Cancer Res. 42: 1130 (1982).

9. Helland, S., and Ueland, P. M., S-Adenosylhomocysteine and S-adenosylhomocysteine hydrolase in various tissues of mice given injections of 9-ß-D-arabinofuranosyladenine, Cancer Res. 43:1847 (1983).

10. Hoffman, D. R., Haning, J. A., and Cornatzer, W. E., Microsomal phosphatidylethanolamine methyltransferase: inhibition by S-adenosylhomocysteine, Lipids 16: 561(1981) .

11. Finkelstein, J.D.,Kyle,W.E., Harris, B.J., and Martin, J.J.,Methionine metabolism in mammals: concentration of metabolites in rat tissues, J. Nutr. 112:1011 (1982).

12. Nordstrøm, C. H.,Rehncrona, S., Siesjø, B.K., and Westerberg, E., Adenosine in rat cerebral cortex: its determination, normal values, and correlation to AMP and cyclic AMP during shortlasting ischemia, Acta Physiol. Scand.101:63 (1977) .

4'-THIOADENOSINE AS A NOVEL INHIBITOR OF S-ADENOSYLHOMOCYSTEINE HYDROLASE AND AN INDUCER FOR THE DIFFERENTIATION OF HL-60 HUMAN LEUKEMIA CELLS

George A. Miura, Richard K. Gordon, John A. Montgomery*, and Peter K. Chiang
Division of Biochemistry, Walter Reed Army Institute of Research, Washington, DC 20307-5100; and *Southern Research Institute, Birmingham, AL

Analogs of adenosine with modifications in the ribofuranosyl moiety such as (±)aristeromycin (1), 3-deaza-(±)aristeromycin (2-4) and neplanocin A (5) are excellent inhibitors of S-adenosylhomocysteine (AdoHcy) hydrolase with competitive K_i and irreversible K_I values in the nanomolar ranges. In addition to being inhibitors, some of these modified compounds can also substitute for adenosine as alternative substrates, yielding S-nucleosidinylhomocysteine (NucHcy) analogs (4). Inhibitors of AdoHcy hydrolase will cause elevations in the cellular levels of AdoHcy and/or the production of NucHcy analogs. The main biochemical effect is the inhibition of transmethylation reactions (6-9), normally accompanied by a rise in the S-adenosylmethionine (AdoMet)

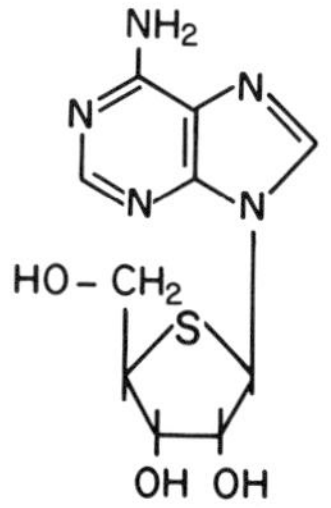

Fig. 1. Structure of 4'-thioadenosine.

level. The correlative biological effects are many (4), and chief among them is the induction of differentiation of cell lines. For example, 3-deaza-adenosine enhances the transformation of 3T3-L1 fibroblasts to fat cells upon confluency (10), and 3-deaza-(±)aristeromycin induces the differentiation of HL-60 human promyelocytic leukemia cells to neutrophils (11).

Analysis of extracts from HL-60 cells incubated with 4'-SAdo by high pressure liquid chromatography (14) revealed the formation of 4'-thio-adenosylmethionine (Fig. 3), and this supports the contention that one of the new nucleotide peaks (Fig.2), presumably peak (#4), was the tri-phosphate of 4'-SAdo. Concomitant with the formation of 4'-thio-adenosylmethionine, there was a dose-dependent increase in AdoMet (Fig. 4). Although purified AdoHcy hydrolase from hamster liver (3) could synthesize 4'-thioadenosylhomocysteine (4'-SAdoHcy) (Fig. 5), the formation of the latter in HL-60 cells was not detected.

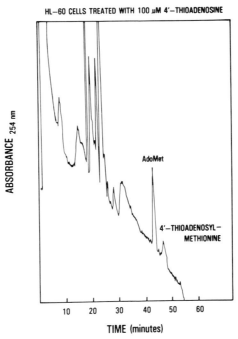

Fig. 3. Formation of 4'-thioadenosylmethionine in HL-60 cells incubated with 100 μM of 4'-SAdo for 2 h.

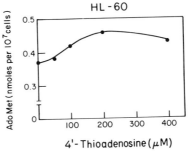

Fig. 4. Increase in cellular AdoMet levels in HL-60 cells incubated with 4'-SAdo for 2 h.

Table 1. Effect of 4'-thioadenosine (4'-SAdo) on levels of nucleotides in HL-60 cells after 2 hours of treatment

4'-SAdo (μM)	ATP	ADP (nmol/10^7 cells)	AMP	GTP	GDP	GMP
0	46 ± 3.0	10.4 ± 0.7	3.4 ± 0.5	9.3	3.2	0.3
50	48 ± 0.8	11.5 ± 1.4	3.3 ± 0.2	8.9	3.0	0.6
100	24 ± 1.3	6.7 ± 0.9	3.6 ± 0.1	3.2	1.1	0.6
200	20 ± 1.0	6.3 ± 1.0	2.4 ± 0.6	3.3	0.6	0.9
400	11 ± 1.3	3.6 ± 0.3	1.9 ± 0.2	0.3	0.2	0.6

4'-Thioadenosine (4'-SAdo) is an adenosine analog with the 4' bridge oxygen in the ribose replaced by a sulfur atom (Fig. 1). When the HL-60 cells were incubated with 4'-SAdo, there was a dose-dependent decrease in the cellular concentrations of GDP, GTP, ADP, ATP and to a smaller extent AMP (Table 1).   At the same time, three new and prominent nucleotide peaks appeared (Fig. 2): one behind ADP (#2), one following GDP (#3), and the largest one trailing GTP (#4). A small noticeable peak was generated right after the AMP peak (#1).   The obvious conclusion was that 4'-SAdo was phosphorylated to the 4' nucleotides, presumably the mono-, di-, and tri-phosphates.   Chemical synthesis of 4'-SAdo 5'phosphate has been reported (12). There was, however, no change in UDP and UTP.

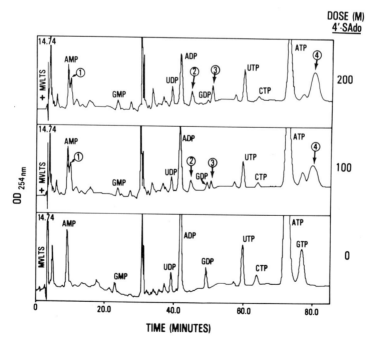

Fig.2. High pressure liquid chromatography of nucleotides (13) in HL-60 cells treated with 4'-SAdo.

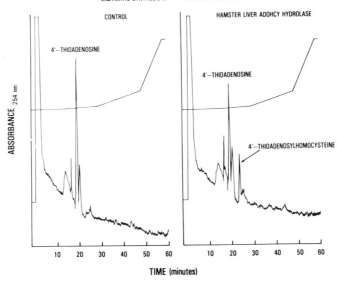

Fig. 5.   Formation of 4'-thioadenosylhomocysteine by purified hamster liver AdoHcy hydrolase.

The activity of AdoHcy hydrolase of HL-60 cells was inactivated irreversibly by 4'-SAdo (Fig. 6). The inactivating constant K_I was 0.2 µM, and this is a potent K_I in comparison with other nucleosides (4, 15). A K_I of 0.06 µM was obtained for the inactivation of purified AdoHcy hydrolase from hamster liver (not shown).

Next, the effect of 4'-SAdo on HL-60 cells was examined. Up to 15 µM, 4'-SAdo was cytostatic (Fig. 7); cytotoxicity was observed at 20 µM. After four days of culture with 15 and 20 µM of 4'-SAdo, 22 ± 3 and 22 ± 2 percent of the cells were able to reduce p-nitroblue tetrazolium, respectively. In comparison, no more than 2 percent of the control cells reduced p-nitroblue tetrazolium. No α-naphthylacetate esterase activity could be detected in the HL-60 treated with 4'-SAdo. Thus, biochemically and morphologically, the cells treated with 4'-SAdo underwent differentiation to become more like neutrophils and not monocytes.

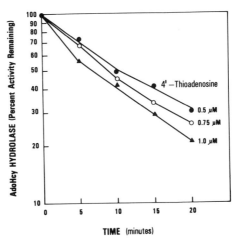

Fig. 6.   Inactivation of AdoHcy hydrolase of HL-60 cells by 4'-SAdo.

670

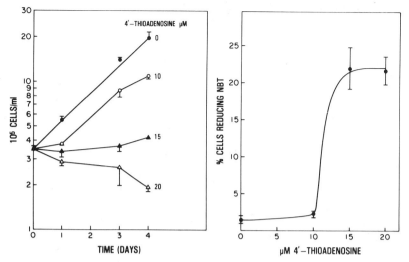

Fig. 7.   Growth and induction of HL-60 cells treated with 4'-SAdo for 4 days.

Our present investigation shows that 4'-SAdo is a potent irreversible inactivator of AdoHcy hydrolase. The formation of 4'-thio-adenosylmethionine was most likely due to its ability to be phosphorylated, and subsequently becomes 4'-SAdo triphosphate, which then reacts with methionine to form 4'-thioadenosylmethionine, a reaction catalyzed by AdoMet synthetase (16). It is not known at present whether the induction of the differentiation of the HL-60 cells was due to an aberration in the transmethylation of crucial macromolecules, such as DNA, RNA, phospholipids, or proteins. The appearance of 4'-thio-adenosylmethionine and the increase in cellular AdoMet suggest that transmethylation reactions might be inhibited. Although 4'-SAdoHcy was not detected in the HL-60 cells, a transient formation of 4'-SAdoHcy cannot be precluded at present. 4'-SAdoHcy may inhibit transmethylation reactions like AdoHcy or S-3-deaza-adenosylhomocysteine. On the other hand, the formation of novel 4'-SAdo nucleotides could also contribute to the induction of differentiation of the HL-60 cells treated with 4'-SAdo. 4'-SAdo has been reported to inhibit the growth of P388 murine leukemia cells (17). In conclusion, inhibitors of AdoHcy hydrolase and/or nucleosides that can form AdoMet analogs are excellent candidates for leukemia inhibitory agents.

REFERENCES

1. A. Guranowski, J. A. Montgomery, G. L. Cantoni and P. K. Chiang, Adenosine analogues as substrates and inhibitors of S-adenosyl-homocysteine hydrolase, Biochemistry 20: 110 (1981).
2. J. A. Montgomery, S. J. Clayton, H. J. Thomas, W. M. Shannon, G. Arnett, A. J. Bodner, I.-K. Kim, G. L. Cantoni and P. K. Chiang, Carbocyclic analogue of 3-deazaadenosine: a novel antiviral agent using S-adenosylhomocysteine hydrolase as a pharmacological target, J. Med. Chem. 25, 626 (1982).
3. I.-K. Kim, C.-Y. Zhang, P. K. Chiang and G. L. Cantoni, S-Adenosyl-homocysteine hydrolase from hamster liver: purification and kinetic properties, Archiv. Biochem. Biophys. 226: 65 (1983).
4. P. K. Chiang, S-Adenosylhomocysteine hydrolase: measurement of activity and use of inhibitors, Methods Pharmacol. 6: 127 (1985)

5. J. Aarbakke, R. K. Gordon, A. S. Cross, G. A. Miura and P. K. Chiang, Correlation between DNA hypomethylation and differentiation of HL-60 promyelocyticcells induced by neplanocin A and 3-deaza nucleosides. Fed. Proc. 44, 313 (1985).

6. P. K. Chiang and G. L. Cantoni, Perturbation of biochemical transmethylation by 3-deazaadenosine in vivo, Biochem. Pharmacol. 28: 1897 (1979).

7. S. Sariban-Sohraby, M. Burg, W. P. Wiesmann, P. K. Chiang and J. P. Johnson, Methylation increases sodium transport into A6 apical membrane vesicles: possible mode of aldosterone action, Science 225, 745 (1984).

8. W. P. Wiesmann, J. P. Johnson, G. A. Miura and P. K. Chiang, Aldosterone-stimulated transmethylations are linked to sodium transport, Am. J. Physiol. 248: F43 (1985).

9. J. P. Bader, N. R. Brown, P. K. Chiang and G. L. Cantoni, 3-Deazaadenosine, an inhibitor of adenosylhomocysteine hydrolase, inhibits reproduction of Rous sarcoma virus and transformation of chick embryo cells, Virology 89: 494 (1978).

10. P. K. Chiang, Conversion of 3T3-L1 fibroblasts to fat cells by an inhibitor of methylation: effect of 3-deazaadenosine, Science 211: 1164 (1981).

11. P. K. Chiang, D. L. Lucas and D. G. Wright, Induction of differentiation of HL-60 human promyelocytic leukemia cells by 3-deaza purines, N. Y. Acad. Sci. 435: 126 (1984).

12. D. J. Hoffman and R. L. Whistler, Synthesis and properties of nucleotides containing 4-thio-D-ribofuranose, Biochemistry 9, 2367 (1970).

13. R. A. Hartwick and P. R. Brown, The performance of microparticle chemically bonded anion-exchange resins in the analysis of nucleotides, J. Chromatogr. 112: 651 (1975).

14. G. A. Miura, J. R. Santangelo, R. K. Gordon and P. K. Chiang, Analysis of S-adenosylmethionine and related sulfur metabolites in animal tissues, Anal. Biochem. 141: 161 (1984).

15. I.-K. Kim, C.-Y. Zhang, G. L. Cantoni, J. A. Montgomery and P. K. Chiang, Inactivation of S-adenosylhomocysteine hydrolase by nucleosides, Biochim. Biophys. Acta, 829, 150 (1985).

16. P. K. Chiang and G. L. Cantoni, Activation of methionine for transmethylation: purification of the S-adenosylmethionine synthetase of bakers' yeast and its separation into two forms, J. Biol. Chem. 252, 4506 (1977).

17. R. H. Adamson, D. W. Zaharevitz and D. G. Johns, Enhancement of the biological activity of adenosine analogs by the adenosine deaminase inhibitor 2'-deoxycoformycin, Pharmacology 15: 84 (1977).

IN VIVO INHIBITION OF MOUSE LIVER METHYLTRANSFERASE ENZYMES

FOLLOWING TREATMENT WITH 2'-DEOXYCOFORMYCIN AND

2'-DEOXYADENOSINE

Jane Renshaw and Kenneth R. Harrap

Dept. Biochemical Pharmacology
Institute of Cancer Research
Sutton, Surrey, England

INTRODUCTION

Treatment of BDF_1 mice with 2'-deoxyadenosine (AdR) in combination with the adenosine deaminase inhibitor 2'-deoxycoformycin (dCf) leads to acute hepatic dysfunction with cumulative dose-related hepatocellular necrosis (1). Toxicity may derive from AdR-induced inhibition of S-adenosylhomocysteine (SAH) hydrolase since dramatic increases in liver SAH levels were observed following AdR/dCf treatment (2). S-adenosyl methionine (SAM) levels also increased and the SAM:SAH ratios fell from 15 to less than 1. Further investigations were undertaken to determine whether the pattern and/or degree of perturbation of SAM:SAH ratios correlated with the manifestation of hepatotoxicity. In addition, creatinine levels in both urine and plasma were monitored in order to assess, indirectly, the activity of hepatic methyltransferase enzymes; the last step in the biosynthesis of creatinine is the methylation of guanidinoacetic acid (GAA), a reaction which takes place almost exclusively in the liver. Inhibition of GAA methyltransferase will be reflected by reduced plasma and urinary creatinine levels. Blood urea levels were also measured in order to rule out the possibility of kidney toxicity.

METHODS

Animal Studies

Male BDF_1 mice received dCf (0.27mg/kg, once daily x5) and AdR (267mg/kg, twice daily x5) as described previously (2). Animals were anaesthetised with pentobarbital (100mg/kg i.p.) and sections of liver freeze-clamped for SAM and SAH estimations. Urine was collected by direct aspiration from the bladder and blood was removed via the axillary vessels.

Metabolic Estimations

Both sample preparation, and high pressure liquid chromatographic determination of SAM and SAH levels were carried out as described previously (2). Plasma and urinary creatinine and blood urea estimations were achieved using a Beckman Clinical Autoanalyser.

RESULTS AND DISCUSSION

Mouse liver SAM:SAH ratios during AdR/dCf treatment are shown in Table 1.

TABLE 1: Mouse liver SAM:SAH ratios during treatment with AdR/dCf

	SAM:SAH Ratios	
	$-30min^a$	$+6hrs^a$
Day 1	15.4 ± 3.3^b	0.5 ± 0.2
Day 2	9.7 ± 0.7	0.6 ± 0.2
Day 3	15.7 ± 1.8	0.5
Day 4	3.1 ± 2.4	0.5 ± 0.2
Day 5	1.3 ± 0.2	0.7 ± 0.4

[a]Time related to the first of each days' AdR injections
[b]Control animals: Mean+S.D. (n=10)
All other values: mean+range (n=2)

Ratios were reduced to less than 1 at 6 hours following the start of each days treatment. Recovery to control levels was observed following the first two days treatment but not thereafter. Mice treated for up to 3 days are able to recover from drug induced hepatotoxicity while those treated for the full 5 days suffer irreversible hepatocellular necrosis. These data indicate a correlation between the duration of depression of SAM:SAH ratios and the manifestation of hepatocellular necrosis.

Evidence of inhibition of hepatic methyltransferase enzymes as a result of reduced SAM:SAH ratios was obtained (Table 2).

TABLE 2: Creatinine levels in mouse urine and plasma during AdR/dCf Treatment

| | Creatinine Levels % Control | | |
	Urine	Plasma	Urine:Plasma
Day 1	63	92	66
Day 3	63	71	97
Day 5	23	N.D.	N.D.

N.D. Not Determined

Urinary creatinine levels were reduced to 60% control levels for the first three days treatment. A further reduction to 23% control levels was seen by day 5. The degree of inhibition of creatine formation appears to correlate with the duration of depression of SAM:SAH ratios (see Table 1). Plasma creatinine levels were also reduced and no consistant alteration in urine to plasma creatinine ratios was observed indicating a lack of kidney toxicity. This was confirmed by blood urea levels which were reduced to 70% control levels from day 1, (data not shown) which probably reflects the contribution of adenosine deamination to the production of urea.

These data support the hypothesis that the hepatotoxicity seen in both animals and man following dCf treatment either alone, or in combination with AdR, is due to AdR-induced inhibition of SAH hydrolase activity with subsequent inhibition of hepatic methylation reactions.

REFERENCES

1   R.M. Paine, B.J. Weston, H.McD. Clink, J. Kohn, A.M. Neville, K.G. McGhee, and K.R. Harrap, Toxicity and Immunosuppressive activity of binary combinations of 2'-deoxycoformycin and 2'-deooxyadenosine. Cancer Treat. Rep., 65:259 (1981)

2   J. Renshaw and K.R. Harrap. Purine Deoxyribonucleoside Induced Hepatotoxicity in The Mouse. Adv. Exp. Med. Biol. 165B:363 (1984)

ALTERATION OF RIBONUCLEOTIDE AND DEOXYRIBONUCLEOTIDE METABOLISM

BY INTERFERON IN HUMAN B-LYMPHOBLASTOID CELLS

Jerzy Barankiewicz, Chaim Kaplinsky and Amos Cohen

Division of Immunology
Research Institute
The Hospital for Sick Children
Toronto, Ontario, Canada   M5G 1X8

Inhibition of cell growth and proliferation by interferon has been studied intensively[1-3]. Although it was believed that the suppression of cell proliferation by interferon is associated with inhibition of thymidine kinase activity, recent studies using thymidine kinase deficient cells showed that activity of this enzyme is not crucial for interferon action[4]. Therefore, it was of great interest to study effects of interferon on other reactions of both ribonucleotide and deoxyribonucleotide metabolism, which possibly could explain inhibitory action of interferon on nucleic acid synthesis.

Ribonucleotide metabolism

Both pathways of ribonucleotide biosynthesis, e.g. biosynthesis de novo and salvage pathways were markedly inhibited by interferon. Incorporation of formate into ribonucleotides and nucleic acids as well as salvage of purine and pyrimidine bases and nucleosides were markedly inhibited in interferon treated cells (Table 1).

The inhibition by interferon of ribonucleotide synthesis is probably due to significant reduction of the intracellular level of P-ribose-PP (Table 2) a common substrate for both ribonucleotide biosynthesis de novo and salvage pathways. In contrast nucleotide catabolism is significantly increased by interferon (Table 3).

B-lymphoblastoid cells incubated with interferon excrete significantly more inosine and hypoxanthine as the end products of ATP catabolism. Other reactions of ribonucleotide metabolism including ribonucleotide interconversions catalyzed by, e.g., AMP synthetase and lyase, IMP dehydrogenase seem to be not affected by interferon. Only reductive deamination of GMP seems to be somewhat inhibited (results not shown). Inhibition of ribonucleotide synthesis is associated with inhibition of both RNA (10% of inhibition of uridine incorporation) and DNA synthesis (70% of inhibition of thymidine incorporation) due to limitation of intracellular levels of ribonucleotides. Indeed intracellular levels of ribonucleotides are reduced 40-60% (Table 2).

Deoxyribonucleotide metabolism

Deoxyribonucleotide biosynthesis de novo via ribonucleotide reduction is 50-70% decreased by interferon (Table 4). Also alternative deoxyribonucleotide biosynthesis via salvage of deoxyadenosine, deoxyguanosine, thymidine and deoxycytidine is inhibited by 40-50% (Table 4). Inhibition of deoxyribonucleotide biosynthesis reduces intracellular levels of deoxyribonucleotides by 20-70% (Table 2) and therefore suppresses DNA synthesis.

The study of interferon effect on nucleic acids precursors showed that interferon specifically blocks pathways leading to synthesis of both ribonucleotides and deoxyribonucleotides. This inhibition together with elevation of nucleotide degradation can

Table 1. Effect of interferon of ribonucleotide metabolism

Substrate	Ribonucleotides	Nucleic Acids
	(% of Control)	
Formate	66.6	42.5
Adenine	44.9	29.1
Hypoxanthine	55.2	33.6
Guanine	40.3	34.4.
Uracil	59.9	N.S.
Adenosine	65.1	40.3
Uridine	70.0	N.S.
Cytidine	56.2	N.S.

B-lymphoblastoid cells (0.25×10^6 per ml) were incubated for 24 hours with recombinant α-interferon (50 units per ml) in RPMI medium containing 10% of heat inactivated fetal calf serum. Cells were then washed and incubated (2×10^6 cells in 0.1 ml) with radioactive purine and pyrimidine bases or nucleosides (Sp. act. between 20-50 Ci/mmol) at concentrations 0.5-2 µM. Incorporation of radioactivity into ribonucleotides and into nucleic acids was measured after 1 or 3 hrs respectively[5]. Control cells were incubated in the absence of interferon and incorporation of radioactive substrates to ribonucleotides and nucleic acids in pmoles/10^6 cells were respectively: formate 2898 and 604; adenine 4273 and 1200; hypoxanthine 1531 and 93; guanine 810 and 867; uracil 617 and N.S.; adenosine 1261 and 676; uridine 180 and N.S.; cytidine 128 and N.S. N.S. - non studied.

Table 2. Effect of interferon on intracellular nucleotide levels and
P ribose PP

Nucleotide	% of Control
ATP	58.8
GTP	46.9
CTP	60.2
UTP	61.4
dATP	82.2
TTP	37.7
dCTP	26.6
dGTP	33.3
P-ribose-PP	55.9

Cells (10 x 10^6) incubated with interferon were
extracted with 0.1 ml of 0.4 M perchloric acid for 10 min at
0°C. Neutralized cell extract with Alamine 336/Freon[6] were
analyzed by HPLC on Partisil-10 SAX anion exchange column with
0.5 M ammonium phosphate (pH 3.4)[7]. P-ribose-PP was
determined in reaction with radioactive orotic acid, orotate
diphospholipase and orotidine-5'-phosphate decarboxylase[8].
Control cells were incubated without interferon and levels of
nucleotides in pmoles/10^6 cells were respectively: ATP
3642, GTP 801, CTP 136, UTP 218 dATP 107, TTP 45, dCTP 1.5
dGTP 0.3 PRPP 18.4. For details see Table 1.

Table 3. Effect of interferon on ATP catabolism

| Time (hrs) | + interferon | | − interferon | |
	inosine	hypoxanthine pmoles/10^6 cells	inosine	hypoxanthine
1	920	320	190	90
2	1250	750	320	100
3	1610	1010	480	190
4	1700	1260	590	220
5	1750	1590	750	290

Cells (10 x 10^6) treated with interferon were incubated
for 1 hr with 2 μCi of radioactive adenine to synthesize
radioactive ATP. Unincorporated adenine was then washed out
and cells were resuspended in RPMI medium. Excretion of
radioactive inosine and hypoxanthine was measured[7] in medium
during 5 hours incubation. For details see Table 1.

Table 4. Effect of interferon on deoxyribonucleotide metabolism

Substrate	Deoxyribonucleotides (% of control)
Adenine	35.7
Guanine	50.9
Deoxyadenosine	45.6
Deoxyguanosine	44.6
Thymidine	42.6
Deoxycytidine	66.6

Cells (2×10^6) treated with interferon were incubated in 0.1 ml with radioactive substrates at concentrations 0.5-2.0 μM. (sp. act. between 15-50 Ci/mmol) for 1 or 2 hours. Prior to incubation with deoxyadenosine cells were preincubated for 30 min. with 40 μM deoxycoformycin. Incorporation of radioactivity into deoxyribonucleotides was measured[9]. Control cells were incubated in the absence of interferon and incorporation of radioactive substrates into deoxyribonucleotides in pmoles/10^6 cell were respectively: adenine 1.1; guanine 0.5; deoxyadenine 3.1; deoxyguanosine 1.1; thymidine 0.6; deoxycytidine 0.5 pmoles/10^6 cells. For details see Table 1.

markedly limit the availability of nucleotides for both RNA and DNA synthesis and in consequence arrest cellular multiplication. It was found that interferon blocks human B-lymphoblastoid cells in Go/Gl phase (results not shown) and this result agrees with data of others[3].

REFERENCES

1. E. Knight Jr., Antiviral and cell growth inhibitory activities reside in the same glycoprotein of human fibroblast interferon, Nature (London) 262:302-303 (1976).
2. F. Balkwill, Cancer Immunol. Immunother. 7:7-14, (1977).
3. I. Gresser, and M.G. Tovey, Biochem. Biophys. Acta 516:231-247 (1978).
4. D.R. Gewert, A. Cohen, and B.R.G. Williams, The effect of interferon on cells deficient in nucleoside transport or lacking thymidine kinase activity, Biochem. Biophys. Res. Comm. 118:124-130, (1984).
5. J.F. Henderson, J.H. Fraser, and E.E. McCoy, Methods for the stud of purine metabolism in human cells in vitro, Clin. Biochem. 7:339-358 (1974).
6. J.X. Khym, An analytical system for rapid separation of tissue nucleotides at low pressure on conventional anion exchange. Clin. Chem. 21:1245-1251 (1975).

7. J. Barankiewicz and A. Cohen, Evidence for distinct catabolic pathways of adenine ribonucleotides and deoxyribonucleotides in human T lymphoblastoid cells. <u>J. Biol. Chem</u>. 259: 15178–15181 (1984).

8. A.S. Bagnara, L.W. Brox and J.F. Henderson, Kinetics of amidophosphoribosyltransferase in intact tumor cells. <u>Biochem. Biophys. Acta</u> 350:171–182, 1974.

9. D. Hunting, J. Hordern, and J.F. Henderson, Quantitative analysis of purine and pyrimidine metabolism in Chinese hamster ovary cells, <u>Can. J. Biochem.</u> 59:838–847 (1981).

INHIBITOR OF 2', 5'-OLIGOADENYLATE SYNTHETASE INDUCED IN HUMAN T
LYMPHOBLASTOID CELL LINE TREATED WITH DEOXYADENOSINE, DEOXYCOFORMYCIN
AND INTERFERON

T. Heike, K. Katamura, M. Kubota, K. Shinomiya and H. Mikawa

Department of Pediatrics
Kyoto University
Kyoto, Japan

INTRODUCTION

The biochemical mechanism for the antiviral effect on interferon (IFN)
is being elucidated. Among the biochemical changes which IFN causes, the
induction of 2',5'-oligoadenylate (2-5A) synthetase has been considered to
play a central role in the antiviral activity (1,2). This enzyme is activated by double-stranded RNA and catalyzes the polymerization of ATP into
a series of (2',5')-linked oligo(adenylate) molecules with 5'-terminal triphosphate. When cells are infected by viruses, double-stranded RNA is
detected in viral replicative complexes. This viral double-stranded RNA
can activate 2-5A synthetase (3). Thus, 2-5A synthetized through this
process activates an latent endonuclease and viral mRNA is specifically
cleaved by this endonuclease.

On the other hand, severe or recurrent viral infections are common
in patients with severe combined immunodeficiency disease (SCID). Recently, Williams et al. reported that in vitro IFN treatment of peripheral
blood mononuclear cells from patients with SCID failed to induce 2-5A
synthetase and suggested the responsibility for their increased susceptibility to viral infections (4). In SCID, heterogeneous groups of disorders
are comprised, and one of which is a genetic deficiency of adenosine deaminase (ADA) (5). Therefore, using human T lymphoblastoid and human
histiocytic lymphoma cells treated with deoxycoformycin (dCF) as an in
vitro model of ADA deficiency, we investigated the effect of deoxyadenosine
(dAdo), adenosine (Ado), ara-A or Ado plus L-homocysteine thiolactone on
the 2-5A synthetase activity induced by IFN.

MATERIALS AND METHODS

Cell lines and cell extract preparation

In the present experiments, we used two human T cell lines, (HSB2,
CEM), and naturally occuring ADA deficient cell line, (DHL-9), derived from
the patients with histiocytic lymphoma (6). They were continuously grown
in RPMI medium supplemented with 12% fetal calf serum. In each experiment,
cells were resuspended in fresh medium at a density of 2×10^5/ml 24 hr
before the addition of drugs (200 IU/ml of α-IFN and 5 μM of dCF with dAdo,
Ado, ara-A or Ado plus L-homocysteine thiolactone). After the treatment
of cells with the drugs, the cell extract was prepared by homogenizing the

cells in Buffer A (10 mM Hepes, pH 7.5, 50 mM KCl, 3 mM Mg(OAc)$_2$, 0.3 mM EDTA and 20% glycerol v/v) containing 7 mM 2-mercaptoethanol and 0.5% NP40. The homogenate was centrifuged for 15 min at 17,000 X g and the supernatant was obtained as an enzyme source (7).

Assay for 2-5A synthetase activity

Sokawa's method modified: The assay was performed by the method of Sokawa et al. with minor modifications (8). Briefly, 50 μg of the cell extract was applied to a tube containing poly(I):poly(C) agarose gel. The tube was washed with Buffer A and incubated with [^{3}H]ATP (25 μl, 1 mM, 0.025 mCi/ml) at 33 °C for 2 hr. After elution, the net 2-5A which had formed was separated from ATP in a DEAE cellulose column, and the activity was counted with a scintillation counter.

Baglioni's method modified: The synthesis of 2-5A was performed essentially as described by Baglioni et al. (9). The assay contained 100 μg of the cell extract with 20 μg/ml poly(I):poly(C) or 50 μl poly(I):poly-(C) agarose gel, 2 mM [^{3}H]ATP (20 μl, 2 mM, 0.017 mCi/ml), 4 mM FruP$_2$ and 22 mM Mg(OAc)$_2$ in Buffer A. The separation and determination of 2-5A were carried out as in Sokawa's method modified.

Chemicals

[2,8-^{3}H]ATP was purchased from New England Nuclear. dCF was kindly provided by Yamasa Shoyu Co., and recombinant α-IFN (2.09 X 10^8 IU/mg protein) was the gift of Takeda Chemical Industries, Ltd. Poly(I):poly(C) agarose gel was purchased from P.L. Biochemicals, and other chemicals from Sigma Chemical Co.

RESULTS AND DISCUSSION

Figure 1 shows the effect of varying concentrations of dAdo plus dCF

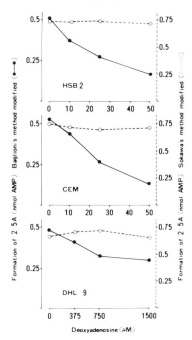

Figure 1.   Effect of dAdo with dCF on induction of 2-5A synthetase by α-IFN

on the induction of 2-5A synthetase by 200 IU/ml α-IFN after 12 hr incubation. The enzyme activity was constant regardless of changes in dAdo concentration by Sokawa's method modified in three different cell lines. However in Baglioni's method modified, the activity decreased as the concentration of dAdo increased. As shown in MATERIALS AND METHODS, Sokawa's assay represents the net 2-5A synthetase activity through semipurification of the enzyme with poly(I):poly(C) agarose gel. On the other hand, Baglioni's assay represents the enzyme activity in the crude cell extract. Therefore, the difference of the activity between two assay systems observed after dAdo treatment strongly suggests the possibility that the crude cell extract contains the substance which inhibits 2-5A synthetase. Moreover, to be noted, the dAdo concentration required for the induction of the inhibitor in each cell line is proportionable to that of ID_{50} as measured at day 3.

In the measurement of 2-5A synthetase, the enzyme activity is expressed as the quantity of synthetized 2-5A. Therefore, the increased activity of 2'-phosphodiesterase which is also induced by IFN and degrades 2-5A into ATP and AMP results in decreased 2-5A synthetase activity as determined by Baglioni's method modified. However, the activity was constant regardless of changes in the dAdo concentration in the treatment of HSB2 with dAdo and dCF (data not shown) (7). Thus, 2'-phosphodiesterase was proven not to participate in the decrease of the activity of 2-5A synthetase in Baglioni's method modified.

Substantial amounts of dATP accumulates intracellularly in dAdo-treated HSB2 cells. Since dATP has been shown to be a competitive inhibitor of 2-5A synthetase, dATP can be another candidate for the inhibitor (10). However, we excluded the possibility from the following reasons (7).

(1) Intracellular levels of dATP after dAdo (50 μM), dCF (5 μM), and α-IFN (200 IU/ml) treatment were approximately 9-10 nmol/10^7 cells as measured by high pressure liquid chromatography on a Whatman Partial SAX-column. Therefore, the estimated final concentration of dATP in Baglioni's assay in dAdo-treated cell extract was 0.02-0.03 mM. When the same concentration of dATP was exogeneously added to the assay, the inhibition of the enzyme activity was negligible (Figure 2).

(2) The inhibitor was not detected only with treatment of dAdo plus dCF, (in other words, without the addition of α-IFN), although the accumulation of intracellular dATP was equivalent.

The effects of agents including ADA substrates other than dAdo on induction of the enzymes were determined next. As you can see from Figure

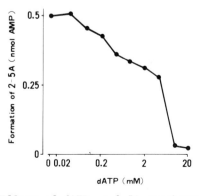

Figure 2.   Effect of dATP on 2-5A synthetase activity

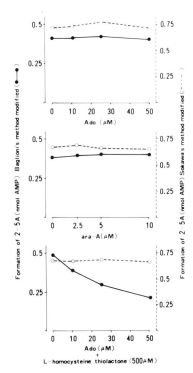

Figure 3.   Effect of Ado, ara-A or Ado plus L-homocysteine thiolactone with dCF on induction of 2-5A synthetase by α-IFN in HSB2

3, Ado or ara-A (up to concentration which had significant effect on cellular growth) failed to induce the inhibitor. Similar inhibitor-like activity was detected after the treatment of Ado plus dCF with 500 μM of L-homocysteine thiolactone or 5'-azacytidine (data not shown). Since we have not measured the actual rate of methylation in DNA, RNA or protein after the drug treatment, our hypothesis is still speculated. However, the following results; (1) dAdo is the well known inhibitor of SAH hydrolase and can inhibit the several methylation reactions through the accumulation of SAH (2) the combination of Ado, dCF and L-homocysteine thiolactone or 5'-azacytidine can induce "similar" inhibitor-like activity and these agents are also proven to inhibit the methylation by different mechanisms (3) α-IFN is necessary for the induction of the inhibitor (4) our preliminary experiments reveal that molecular weight is more than 30,000 and the inhibitory activity is inactivated by heat treatment at 56 °C for 30 min: support the possibility that the nature of the inhibitor would be one of protein newly synthetized with α-IFN, which is modified through the inhibition of methylation.

Finally, this implies that the inhibitor is also induced in patients with ADA-deficient SCID even in vivo and might explain one aspect of the increased susceptibility to viral infections in these patients.

REFERENCES

1. Lengyel, P. (1981) Methods in Enzymology (Pestka, S., ed.) Vol. 78, 135-148, Academic Press, Inc., New York
2. Williams, B.R.G. and Kerr, I.M. (1978) Nature 276, 88-90.
3. Laurent, G.St., Yoshie, O., FloydSmith, G., Samanta, H., Sehgal, P.B.

686

and Lengyel, P. (1983) Cell 33, 95-102.

4. Williams, B.R.G., Read, S.E. and Gelfand, E.M. (1984) Clin. exp. Immunol. 56, 34-38.

5. Giblett, E.R., Anderson, J.E., Cohen, F., Pollara, B., and Meuwissen, H.J. (1972) Lancet2, 1067-1069.

6. Kubota, M., Kamatani, N., Daddona, P.E., and Carson, D.A. (1983) Cancer Res. 43, 2606-2610.

7. T. Heike, K. Katamura, M. Kubota, K. Shinomiya, and H. Mikawa (1985) Biochem. Biophys. Res. Commun. 127, 1019-1025.

8. Shimizu, N. and Sokawa, Y. (1979) J. Biol. Chem. 254, 12034-12037.

9. Minks, M.A., Benvin, S., Maroney, P.A. and Baglioni, C. (1979) J. Biol. Chem. 254, 5058-5064.

10. Minks, M.A., Benvin, S. and Baglioni, C. (1980) J. Biol. Chem. 255, 5031-5035.

2-5A SYNTHETASE ACTIVITIES IN HERPES VIRUS INFECTION AND AFTER INTERFERON TREATMENT

Johannes Mejer*, Just Justesen**, and Peter Ernst***

Department of Medicine* and Department of Chemotherapy***
The Finsen Institute, University of Copenhagen, Copenhagen
and Department of Molecular Biology and Plant Physiology**
University of Aarhus, Aarhus, Denmark

The induction of 2-5A synthetase is one of the biochemical changes in cells caused by all classes of interferon (1). The level of interferon in serum is low and dificult to measure, which may explain why interferon sometimes can not be measured in obvious virus infections (1), and why interferon measurement as a diagnosticum is less usefull. In vitro 2-5A synthetase has been shown to increase after interferon exposure (2). The effect is transient. This focus attention upon the possibility of using 2-5A synthetase as a marker for interferon, and possible also as a diagnostic aid in virus infections. Schatner et al. (3) described the increase of 2-5A synthetase in 85% of investigated patients with acute viral infection.

We have measured 2-5A synthetase in mononuclear cells from patients with herpes zoster and from patients treated with interferon, to evaluate wether interferon and virus infections have the same effect on the level of 2-5A synthetase.

MATERIAL

Eight patients with herpes zoster were investigated. The diagnosis was based on clinical findings of typical skin involvement. Three patients had disseminate disease. One of the patients with disseminate disease died in connection with the zoster despite acyclovir treatment. The 2 other patients with disseminate disease received also acyclovir and recovered. Two females and 6 males were included. Age range from 14 to 63 years. The underlying disease was either Hodgkin or non-Hodgkin lymphoma, 4 patients in each group. Seven patients with either Hodgkin or non-Hodgkin lymphoma but without herpes infection were also investigated, age range 30 to 60 years, 3 females and 4 amles. The interferon treated patients consisted of 7 males with squamous cell broncogenic carcinoma, age range 59 to 67 years. The interferon treatment consisted of 100 x 10^6 units/m² human recombinant interferon alpha, kindly provided by Hoffmann-La Roche, Basel, Switzerland. Venous blood samples were withdrawn before and 1 hour after the interferon treatment. In 4 of the patients samples were also withdrawn 8 and 15 hours after the treatment.

The control group consists of 20 healthy donors age range from 22 to 50 years, 5 females and 15 males.

Table 1.   2-5A synthetase in mononuclear cells from patients
with and without symptoms of herpes zoster infection and in a
group of healthy donors.

Group		2-5A synthetase	
	N	median	range
Lymphoma patients			
with herpes zoster	8	7.4	5.0-10.0
without herpes zoster	7	0.8	0.4-1.6
Controls	20	1.3	0.4-2.5

Mononuclear cells were harvested from peripheral venous blood
using heparin as anticoagulant (100 ie/ml) by isopycnic separation, 5 x 10^5 cells were lysed in 100 ul buffer D (20 mM Tris
HCl, pH 8.0, 5 mM Mg(OAc)$_2$, 25 mM KCl, 1 mM dithiothreitol, 1
mM EDTA, 10% (v/v) glycerol containing 0.5% nonionic detergent
(NP40) for 10 min at room temperature.
The extracts were centrifuged at 10,000 rpm. for 10 min to
remove debris. To 10 ul of supernatant was added the following
to give final concentrations in 20 ul: 5mM ATP including
approximately 0.05uCi alpha (^{32}P)ATP, 15 mM Mg(OAc)$_2$, 100 ug/ml
creatin kinase, 4 mM creatin phosphate in buffer D. Incubation
at 37° C for 2 hours.
5 ul of the reaction mixture was chromatographed on poly-
(ethylene)imine-cellulose thin layer plates with 2 M Tris-HCl
pH 8.65 after wash for 10 min in ethanol and prerun to 3 cm in
ethanol. The spots containing ATP and 2-5A were located by
autoradiography, cut and counted by Cerenkov method.
1 unit of 2-5A synthetase is defined as the amount of enzyme
that can incorporates 1 nmole of AMP from ATP into 2-5A/min.

Statistical tests: Wilcoxon rank sum test, and Wilcoxon test for pair
differences.

RESULTS

   The effect of a virus infection (herpes zoster) on the level of 2-5A
synthetase is given in table 1. All patients were in the vesicular stage
approximately 1 week after onset of symptoms. All the patients exhibited
elevated levels of 2-5A synthetase activities in comparision with 20
healthy donors (p<0.01). In the 7 patients with Hodgkin and non-Hodgkin
lymphoma without symptoms of herpes infection no increase of 2-5A synthetase in mononuclear cells was observed, significantly different from the
level in the group with herpes infection (p<0.01).
   The effect of interferon treatment on the level of 2-5A synthetase in
mononuclear cells is shown in fig. 1 and fig. 2.

The observed increase in 2-5A synthetase after interferon treatment is significant (p<0.01)(Fig. 1) and in the same magnitude as observed in virus infection. Fifteen hours after treatment a gradual drop in the levels occur (Fig. 2), but normal levels have not been attained.

2-5A synthetase

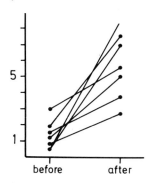

Fig. 1. 2-5A synthetase before and 1 hour after treatment of patients with broncogenic carcinoma with 100 x 10^6 units/m² of recombinant interferon alpha.

2-5A synthetase

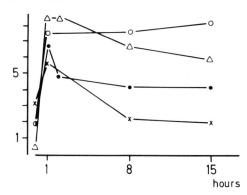

Fig. 2 2-5A synthetase in the first 15 hours after treatment with 100 x 10^6 units/m² of recombinant interferon alpha.

DISCUSSION

The finding of elevated 2-5A synthetase activities in mononuclear cells from patients with herpes zoster infection is in agreement with the findings of others (2). The increase is not likely to be due to altered distribution in subfractions of the mononuclear cells, because previously we observed approximately 100% increase in 2-5A synthetase activity in the monocyte fraction (4) and in herpes zoster the differential counts were unchanged from normal values in the investigated patients. After interferon treatment we found leucopenia with a relative increase in monocyte fraction, approximately 15% (5). But this altered distribution in the subfractions of mononuclear cells can not explain the observed increase in 2-5A synthetase after interferon treatment. Therefore, the presented findings confirm that virus infection and interferon treatment result in an elevation of equal magnitude of the enzyme 2-5A synthetase in mononuclear cells in vivo, Furthermore, this increase occurs within the first hour after interferon presence in the organism, and then a gradual drop takes place.

REFERENCES

1. Sen G. C. Mechanism of interferon action: Progress toward its understanding. Progress in Nucleic Acid Research and Molecular Biology 27: 105 (1982).

2   Schattner A., Merlin G., Levin S. et al. Assay of an interferon-induced enzyme in white blood cells as a diagnostic aid in viral diseases. Lancet i 497 (1981).

3   Lab. M., Thang M. N., Soteriadou K. Regulation of 2-5A synthetase activity in interferon treated chick cells. Biochemical and Biophysical Research Communication. 105: 412 (1982).

4   Justesen J., Hokland P., and Hokland M. The interferon / 2-5A synthetase system in primary preleukemia patients. In Press.

5   Nissen M. H., Plesner T., Larsen J. K. et al. Enhanced expression in vivo of HLA-ABC antigens and B_2-microglobulin on human interferon alpha in patients with lung cancer. Enhanced expression of class I major histocompatibility antigens prior to treatment. Clin. Exp. Immunol. 59: 327 (1985).

CONTRIBUTORS